Human Stem Cell Manual

Human Stem Cell Manual
A Laboratory Guide

Second Edition

Edited By

Jeanne F. Loring

Suzanne E. Peterson

Amsterdam • Boston • Heidelberg • London • New York • Oxford
Paris • San Diego • San Francisco • Singapore • Sydney • Tokyo
Academic Press is an imprint of Elsevier

Academic Press is an imprint of Elsevier
32 Jamestown Road, London NW1 7BY, UK
225 Wyman Street, Waltham, MA 02451, USA
525 B Street, Suite 1800, San Diego, CA 92101-4495, USA

First edition 2007
Second edition 2012

Notice
No responsibility is assumed by the publisher for any injury and/or damage to persons or
property as a matter of products liability, negligence or otherwise, or from any use or operation
of any methods, products, instructions or ideas contained in the material herein.

Because of rapid advances in the medical sciences, in particular, independent verification of
diagnoses and drug dosages should be made.

British Library Cataloguing-in-Publication Data
A catalogue record for this book is available from the British Library

Library of Congress Cataloging-in-Publication Data
A catalog record for this book is available from the Library of Congress

ISBN: 978-0-12-385473-5

Trademarks/Registered Trademarks: Brand names mentioned in this book are protected by their
respective trademarks and are acknowledged.

For information on all Academic Press publications
visit our website at www.elsevierdirect.com

Typeset by MPS Limited, Chennai, India
www.adi-mps.com

Printed and bound by CPI Group (UK) Ltd, Croydon, CR0 4YY

Transferred to Digital Printing, 2013

Contents

FOREWORD by Sherrie Gould.. xvii
FOREWORD by Bob Klein... xix
PREFACE by Jeanne F. Loring xxi
PREFACE by Suzanne E. Peterson... xxiii
LIST OF CONTRIBUTORS.. xxv

Part 1   **Basic Methods**   **1**

CHAPTER 1   Culturing human pluripotent stem cells
             on a feeder layer.. 3
             Editor's commentary 3
             Overview...3
             Procedures 4
             Alternative procedures 9
             Pitfalls and advice 11
             Equipment...12
             Reagents and supplies..13
             Recipes 13
             Reading list 14

CHAPTER 2   Preparation of mouse embryonic fibroblast
             feeder cells... 15
             Editor's commentary 15
             Overview...15
             Procedures 16
             Alternative procedures 23
             Pitfalls and advice 24
             Equipment...24
             Reagents and supplies..25
             Recipes 26
             Reading list 26

CHAPTER 3 Culture of human pluripotent stem cells in feeder-free conditions 29
Editor's commentary 29
Overview 30
Procedures 30
Alternative procedures 34
Pitfalls and advice 36
Equipment 36
Reagents and supplies 37
Recipes 38
Quality control methods 38
Reading list 40

CHAPTER 4 Contamination 41
Editor's commentary 41
Overview 42
Identifying contamination 42
Controlling contamination 47
Reagents and supplies 50
Reading list 51

CHAPTER 5 Cell culture medium 53
Editor's commentary 53
Overview 53
Design of culture medium 53
Commercially available stem cell media 61
Assessing medium efficacy 62
Perspective 64
Reading list 65

CHAPTER 6 Cryopreservation of human embryonic stem cells 71
Editor's commentary 71
Overview 71
Procedures 72
Pitfalls and advice 74
Alternative procedures 74
Equipment 74
Reagents and supplies 74
Recipes 75
Reading list 76

CHAPTER 7 Setting up a research scale laboratory for human pluripotent stem cells 77
Editor's commentary 77
Overview 77

Laboratory 78
Record keeping..86
Equipment maintenance 88
Safety 89
Quality control..89
Cell banking..92
Reading list...102

CHAPTER 8 Biobanks for pluripotent stem cells...........................105
Editor's commentary..105
Overview 105
Banking of human pluripotent stem cells 106
Stem cell providers...114
Future of banking and new iPSC initiatives.......121
Reading list...123

Part 2 **Induced Pluripotent Stem Cell Technology** **127**

CHAPTER 9 Isolation of human dermal fibroblasts from biopsies.. 129
Editor's commentary... 129
Overview 130
Procedure.. 131
Pitfalls and advice... 138
Equipment 139
Reagents and supplies 139
Recipes.. 140
Reading list... 141

CHAPTER 10 Human induced pluripotent stem cell generation: conventional method....................... 143
Editor's commentary... 143
Overview 144
Procedures 145
Alternative procedures... 155
Pitfalls and advice... 156
Equipment 157
Reagents and supplies 157
Recipes.. 159
Quality control methods 162
Acknowledgments 163
Reading list... 163

CHAPTER 11   Integration-free method for the generation
of human induced pluripotent stem cells 165
Editor's commentary 165
Overview 165
Procedures 166
Alternative procedures 169
Pitfalls and advice 170
Equipment 170
Reagents and supplies 171
Recipes 171
Acknowledgments 172
Reading list 173

CHAPTER 12   Methods for evaluating human induced
pluripotent stem cells 175
Editor's commentary 175
Overview 175
Procedures 176
Equipment 183
Reagents and supplies 183
Acknowledgments 184
Reading list 184

Part 3   **Characterization of Human
Pluripotent Stem Cells**   **185**

CHAPTER 13   Classical cytogenetics: karyotyping 187
Editor's commentary 187
Overview 187
Procedures 189
Alternative procedures 196
Pitfalls and advice 198
Equipment 199
Reagents and Supplies 200
Recipes 200
Acknowledgments 201
Reading list 202

CHAPTER 14   SNP Genotyping to detect genomic
alterations in human pluripotent
stem cells 203
Editor's commentary 203
Overview 203
Procedures 205

Alternative procedures..212
Pitfalls and advice..214
Equipment and supplies...215
Quality control methods 216
Reading list..219

CHAPTER 15   Analysis and purification techniques
               for human pluripotent stem cells............................. 223
               Editor's commentary..223
               Overview 223
               Procedures 226
               Alternative procedures..241
               Pitfalls and advice...242
               Equipment 244
               Reagents and Supplies 245
               Reading list..247

CHAPTER 16   Immunocytochemical analysis of
               human stem cells 249
               Editor's commentary..249
               Overview 249
               Procedures 250
               Pitfalls and advice...258
               Digital images of fluorescent cells...........................262
               Equipment 267
               Reagents and supplies 267
               Recipes..268
               Reading list..270

CHAPTER 17   Analysis of genome-wide gene expression
               data from microarrays and sequencing 271
               Editor's commentary..271
               Overview 272
               Experimental design..273
               Critical statistical concepts 275
               Microarray data analysis 278
               RNA sequencing (RNA-seq).....................................282
               Pitfalls and advice...289
               Reading list..290

CHAPTER 18   PluriTest molecular diagnostic assay for
               pluripotency in human stem cells............................. 293
               Editor's commentary..293
               Overview 294
               Procedures 296

Pluritest output 296
Data, experimental design, and formats.............................304
Pitfalls and advice: frequently asked questions 307
Reading list..311

CHAPTER 19 Epigenetics: analysis of histone post-translational modifications by chromatin immunoprecipitation 313
Editor's commentary... 313
Overview 313
Procedures 315
Alternative procedures.. 319
Equipment 320
Reagents and supplies 320
Recipes... 321
Chip antibodies 321
Quality control methods 322
Reading list... 323

CHAPTER 20 Epigenetics: DNA methylation................................. 325
Editor's commentary... 325
Overview 325
Procedures 329
Alternative procedures.. 330
Pitfalls and advice.. 334
Equipment, reagents, and supplies.............................. 334
Quality control methods 335
Reading list... 335

CHAPTER 21 Generation of human pluripotent stem cell-derived teratomas 337
Editor's commentary... 337
Overview 337
Procedures 338
Pitfalls and advice.. 340
Equipment 341
Reagents and supplies 342
Reading list... 342

CHAPTER 22 Characterization of human pluripotent stem cell-derived teratomas..................................... 345
Editor's commentary... 345
Overview 345
Procedures 346

hiPSC-derived teratomas 357
Reading list 359

Part 4   **Differentiation**   **361**

CHAPTER 23   Spontaneous differentiation of human
             pluripotent stem cells via embryoid body
             formation 363
             Editor's commentary 363
             Overview 363
             Procedures 364
             Alternative procedures 368
             Pitfalls and advice 369
             Equipment 369
             Reagents and supplies 369
             Recipes 371
             Reading list 373

CHAPTER 24   Differentiation of human pluripotent
             stem cells into neural progenitors 375
             Editor's commentary 375
             Overview 375
             Procedures 376
             Alternative procedures 380
             Pitfalls and advice 381
             Reagents and supplies 381
             Recipes 383
             Reading list 383

CHAPTER 25   Dopaminergic neuronal differentiation
             of human pluripotent stem cells 385
             Editor's commentary 385
             Overview 385
             Procedures 387
             Alternative procedures 392
             Pitfalls and advice 393
             Equipment 393
             Reagents and supplies 393
             Recipes 395
             Reading list 397

CHAPTER 26   Directed differentiation of human
             pluripotent stem cells to oligodendrocyte
             progenitor cells 399
             Editor's commentary 399

Overview 399
Procedures 400
Alternative procedures.................................. 407
Pitfalls and advice 407
Reagents and supplies 409
Recipes.. 409
Quality control methods 411
Reading list.. 412

CHAPTER 27   Cardiomyocyte differentiation of
              human pluripotent stem cells   413
              Editor's commentary................................... 413
              Overview 413
              Procedures 415
              Alternative procedures................................ 423
              Pitfalls and advice 423
              Equipment 424
              Reagents and supplies 424
              Recipes.. 425
              Quality control methods 430
              Reading list.. 431

CHAPTER 28   Directed differentiation of human
              pluripotent stem cells into fetal-like
              hepatocytes 433
              Editor's commentary................................... 433
              Overview 433
              Procedures 435
              Alternative procedures................................ 438
              Pitfalls and advice 440
              Equipment 440
              Reagents and supplies 440
              Recipes.. 441
              Quality control methods 443
              Reading list.. 446

Part 5   **Genetic Manipulation**   **449**

CHAPTER 29   Transfection of human pluripotent
              stem cells by electroporation 451
              Editor's commentary................................... 451
              Overview 451
              Procedures 452

Alternative procedures...454
Pitfalls and advice..456
Equipment 457
Reagents and supplies 457
Recipes..458
Quality control methods 460
Reading list...461

CHAPTER 30   Lentiviral vector systems for transgene
               delivery 463
               Editor's commentary..463
               Overview 464
               Procedures..464
               Pitfalls and advice..479
               Equipment 480
               Recommended reagents..480
               Recipes..481
               Reading list...483

CHAPTER 31   Gene targeting by homologous recombination
               in human pluripotent stem cells 485
               Editor's commentary..485
               Overview 486
               Procedures..486
               Reagents and supplies 495
               Recipes..496
               Reading list...498

CHAPTER 32   Genetic manipulation of human pluripotent
               stem cells using zinc finger nucleases.......................499
               Editor's commentary..499
               Overview 500
               Procedures..502
               Pitfalls and advice..510
               Reagents and supplies 512
               Recipes..513
               Reading list...513

Part 6   Stem Cell Transplantation   517

CHAPTER 33   Intraspinal transplantation of human
               neural stem cells 519
               Editor's commentary..519
               Overview 519

Procedures 520

Pitfalls and advice 524

Equipment 526

Reagents and supplies 526

Recipes 527

Quality control methods 527

Reading list 528

CHAPTER 34 Transplanting stem cells into the brain 529

Editor's commentary 529

Overview 529

Procedures 530

Pitfalls and caveats 536

Equipment 538

Reagents and supplies 539

Recipes 540

Quality control methods 540

Reading list 541

Part 7 **Derivation of Human Pluripotent Stem Cells** **543**

CHAPTER 35 Development of human blastocysts 545

Editor's commentary 545

Overview 545

Laws and ethics 546

Background 546

Procedures 550

Equipment 554

Reagents and supplies 555

Recipes 555

Reading list 556

CHAPTER 36 Derivation of human embryonic stem cells from blastocysts 557

Editor's commentary 557

Overview 557

Procedures 558

Alternative procedures 569

Pitfalls and advice 570

Equipment 571

Reagents and supplies 571

Recipes 574

Quality control methods 575
Reading list... 576

Part 8   **Stem Cells and Society**   **577**

CHAPTER 37   Stem cell patents: what every researcher
should know 579
Editor's commentary.. 579
Overview 579
What is a patent?... 580
Obtaining a patent on an invention 583
The America Invents Act of 2011 585
Patent licensing policies... 586
Material transfer 587
Patent pools... 588
Stem cell patents... 588
Human embryonic stem cell patents.................................... 589
Induced pluripotent stem cell patents 591
Reading list.. 593

CHAPTER 38   Ethics of human pluripotent stem cell
research and development... 595
Editor's commentary.. 595
Overview 595
Ethical issues for human pluripotent stem cell research ... 596
Informed consent 599
Early clinical trials and stem cell tourism 600
Review committees... 600
Scientific responsibilities.. 602
Reading list.. 602

INDEX.. 605

Foreword

Not since the discovery of levodopa in the 1960s and the use of Deep Brain Stimulation in the 1990s has a potential treatment held so much promise for Parkinson's disease. How ironic it would be if, after millions of dollars invested and countless man-hours devoted to research, the answer were to be found within our own body cells. We must harness the power of healing that exists within all of us – the pluripotent stem cell: something so powerful, so diverse, so genetically perfect, something with the potential to heal not only Parkinson's disease, but innumerable other neurodegenerative diseases as well.

The pain, suffering, and gradual yet inevitable degradation of the body, mind, and soul that Parkinson's unleashes on its victims must soon become an issue of the past. The answer is unfolding and within our reach. Mountains have been conquered; Parkinson's is next.

Sherrie Gould MSN, NP-C
Scripps Clinic, La Jolla, CA,
Creator and Founder of Summit4stemcell

Foreword

During the summer of 2004 Dr Paul Berg (a Nobel Prize winner for Recombinant DNA) responded to my question on his view of whether stem cell research would equal the impact of the hundreds of therapies for chronic illness derived from Recombinant DNA. He said, "Bob, in the history of mankind, we have never before been able to make a human cell or even approach the regeneration of a human tissue or human organ. We could never have rebuilt a human heart ... Twenty years from now you will not recognize how medicine is practiced today." Ironically in 2002 and 2003, the US House of Representatives had voted for the Weldon Bill, subjecting scientists and physicians involved in embryonic stem cell research to a $1 million fine and 10 years in jail. The US Senate rejected this bill, protecting scientists from this intellectual repression, even though President Bush had pledged to sign this bill, if passed.

California represented a potential scientific "safe harbor" for this research, and a critical alternative resource for funding human embryonic stem cell research – the gold standard – for validating pluripotent and progenitor stem cell lines. Recognizing this historic scientific window of opportunity could be slammed shut by federal or state efforts to criminalize the research, before the scarce threads of funding could prove its potential, I wrote Proposition 71: The California Stem Cell Research and Cures Initiative. For patients like my son Jordan, with Type 1 diabetes, and paralyzed friends like Christopher Reeve, for cancer patients, Alzheimer patients like my mother, for patients globally, this was a priceless opportunity for progress, and scientists and physicians could thereby be assured the intellectual freedom and funding – long term – to pursue the remarkable future of stem cell therapies and basic science discoveries.

On November 2, 2004, seven million California voters – 59% – voted to protect science from religiously driven suppression, and the charter for pluripotent and progenitor stem cell research gained constitutional protection. But for patients, it is the dedication, the commitment, and the disciplined excellence of the scientific researchers in California, in Massachusetts, New York, Washington, North Carolina, and Wisconsin – throughout this country and the world – who have rallied to this cause, that hold the key to the future of human suffering and the potential of reducing that historical burden through stem cell research.

The featured researchers and editors of this manual are leaders in the scientific vanguard of this medical revolution. Their discoveries and laboratory manual serve as models of the scientific excellence driving the basic and translational science that daily reveals new possibilities for therapies, enables the commencement of human trials, and brings tangible hope to patients. As a patient advocate, and as the past Chairman of the Governing Board of the California Institute of Regenerative Medicine, the funding agency of Proposition 71, I believe the contributing scientists to this manual have opened a new door for humanity and science, through their discoveries, and their instructions herein, to the path to the future of the Stem Cell Revolution in Medicine.

Bob Klein
Chairman Emeritus,
The California Institute for Regenerative Medicine

Preface

Every book, even a laboratory manual, has a personal story behind it. I wrote the first edition of this book with my friends, Phil Schwartz and Robin Wesselschmidt. Together, in 2004, we developed one of the first NIH-funded human embryonic stem cell courses. At that time, no one had much experience with human pluripotent stem cells, and we were barely ahead of our students. The NIH courses are now history, but with their funding, and support from the California Institute for Regenerative Medicine, we have been able to offer intensive human stem cell courses continuously for the last 8 years. A great deal has changed since our first courses, and yet, as my colleague Suzanne Peterson and I worked on the revision, we realized that some techniques were virtually unchanged. Immunocytochemistry, for example, still requires careful experimental planning, good antibodies, and most importantly, controls. The fundamental principles for culture and characterization of human ES cells have translated well to human induced pluripotent stem cells. And, no matter how much we have ranted about monitoring cultures to avoid contamination, mycoplasma still shuts down human stem cell labs.

The book also has substantial changes. Our methods for culturing cells have become more standardized, and we have much more sophisticated methods for analyzing them. With large-scale high-resolution molecular analysis studies, we have determined that human pluripotent stem cells can vary considerably but have a unique common signature. The early claims for differences between human iPSCs and ESCs have been swamped out by studies that show how similar they are. All of this new knowledge has underscored the message that the culture dish is a hotbed of evolution; as hPSCs proliferate, the fittest cells survive and dominate the culture dish. New methods allow us to detect smaller changes, such as the duplication of just a few genes that convey a growth advantage. We also can now test for epigenetic changes in hPSCs, and find them in abundance. What we cannot yet write is a chapter that teaches a clear cut way to determine what variable characteristics of hPSCs and their derivatives are important for the applications for which we want to use them. Some characteristics are clearly important in some circumstances; for example, epigenetic reactivation of a silent X chromosome in female hPSCs over time in culture will confound studies of recessive X-linked disease. But imprinted genes also change their expression in culture, and we don't know whether or not that will be important for modeling human disease, or affect the safety of cells for clinical cell therapy applications. We also don't know whether duplication of a chromosome or an oncogene in hPSCs really matters for their research or clinical use.

I am thrilled that almost everyone whom I asked to contribute provided a chapter or chapters to the new edition; the book has a stunning cast of contributors. To give the book a sense of coherence, I edited the style of every chapter so that the book has one voice. I want the reader to know how this all got started. Jim Battey and John Thomas of the NIH were prescient about the future of stem cells and helped to establish the first hESC courses. I will never forget that in 2002, John found my home phone number and called me to see if I might be interested in applying for a course grant. I wouldn't have, if Phil and Robin weren't willing to join in. I continue to hope that our courses might be as life-changing for a young scientist as the Cold Spring Harbor course in Developmental Neurobiology was for me over 30 years ago. The students in that course formed enduring friendships and followed each others' careers. I met one of my best friends, Arlene Chiu, at the CSH course, and throughout her career at the NIH, CIRM, and now City of Hope, and mine in several biotechnology companies and now at Scripps, we have remained friends and colleagues. I hope this kind of wonderful relationship blooms in our stem cell courses.

Jeanne F. Loring

Preface

The second edition started with a wonderful idea: "America's Test Kitchen" for stem cells. As an aspiring chef myself, I was very familiar with America's Test Kitchen when Jeanne presented the idea to me. Basically, it is a TV show where a variety of recipes for a particular dish are studied and tested. All the different permutations of each recipe are taste-tested, and through all these iterations a final, perfect recipe is distilled. I hope that we have done the same thing here with stem cells. Each of the protocols presented here was developed by the best in the field and has been tested over and over in the hope of developing the perfect protocol. We truly hope that you will not only enjoy this book, but that it will also make your life easier.

Suzanne E. Peterson

List of Contributors

Lars Ährlund-Richter Karolinska Institute, Department of Woman and Child Health, Stockholm 171 76, Sweden

Gulsah Altun Life Technologies, Inc., Bioinformatics Division, 850 Lincoln Centre Drive, Foster City, CA 94044, USA

Isao Asaka Kyoto University, Center for iPS Cell Research and Applications, Dept of Regulatory Science, 53 Kawahara-cho, Shogoin Yoshida, Sakyo-ku, Kyoto 606-8507, Japan

Marina Bibikova Illumina, Inc., 9885 Towne Centre Drive, San Diego, CA 92121, USA

Mathew Blurton-Jones University of California, Irvine, Institute for Memory Impairments and Neurological Disorders, 1216 Gillespie Neuroscience Research Facility, Irvine, CA 92697, USA

Francesca Boscolo The Scripps Research Institute, Center for Regenerative Medicine, Dept Chemical Physiology, 10550 North Torrey Pines Road, La Jolla, CA 92037, USA

Stefan Braam Department of Anatomy and Embryology, Leiden University Medical Center, Research Building 1, PO Box 9600, 2300 RC Leiden, The Netherlands

Oliver Brustle University of Bonn, Institute of Reconstructive Neurobiology, Sigmund-Freud Straße 25, D-53127 Bonn, Germany

Kevin S. Carbajal University of California, Irvine, Multiple Sclerosis Research Center, Gross Hall Rm 2006, Irvine, CA 92697, USA

Jessica Cedervall Uppsala University, Department of Medical Biochemistry and Microbiology, Box 582, Uppsala 751 23, Sweden

Lu Chen University of California, Irvine, Multiple Sclerosis Research Center, Gross Hall Rm 2006, Irvine, CA 92697, USA

Jonathan Chesnut Life Technologies, Inc., Primary and Stem Cell Systems, 5781 Van Allen Way, Carlsbad, CA 92008, USA

Cleo Choong Nanyang Technological University, School of Materials Science and Engineering, N4.1-02-01, Singapore

Ronald Coleman The Scripps Research Institute, Center for Regenerative Medicine, Dept Chemical Physiology, 10550 North Torrey Pines Road, La Jolla, CA 92037, USA

Hun Chy CSIRO Materials and Science Engineering, Bag 10, Clayton South MDC, Victoria 3169, Australia

Jeremy M. Crook St Vincent's Hospital, The University of Melbourne, Stem Cell Medicine, O'Brien Institute and Department of Surgery, Building 261, 203 Bouverie Street, Parkville, Victoria 3065, Australia

Cheryl Dambrot Department of Anatomy and Embryology, Leiden University Medical Center, Research Building 1, PO Box 9600, 2300 RC Leiden, The Netherlands

Ivan Damjanov University of Kansas Medical Center, Department of Pathology and Laboratory Medicine, 3901 Rainbow Boulevard, Mail Stop 304, Kansas City, KS 66160, USA

Richard Davis Department of Anatomy and Embryology, Leiden University Medical Center, Research Building 1, PO Box 9600, 2300 RC Leiden, The Netherlands

Mary Devereaux University of California, San Diego, Department of Pathology, 9500 Gilman Drive, MC 0612, La Jolla, CA 92093-0612, USA

Mahesh Dodla Stem Cells, Sigma-Aldrich, 2909 Laclede Avenue, St Louis, MO 63103, USA

Biljana Dumevska Genea Ltd, Research Department, Level 3, 321 Kent Street, Sydney, NSW 2000, Australia

Eyitayo S. Fakunle The Scripps Research Institute, Center for Regenerative Medicine, Dept Chemical Physiology, 10550 North Torrey Pines Road, La Jolla, CA 92037, USA

Lisa A. Flanagan University of California, Irvine, School of Medicine, Pathology and Laboratory Medicine, D440 Medical Sciences I, Irvine, CA 92697-4800, USA

Ibon Garitaonandia The Scripps Research Institute, Center for Regenerative Medicine, Dept Chemical Physiology, 10550 North Torrey Pines Road, La Jolla, CA 92037, USA

Karin Gertow Karolinska Institute, Department of Woman and Child Health, Stockholm 171 76, Sweden

Victoria Glenn The Scripps Research Institute, Center for Regenerative Medicine, Dept Chemical Physiology, 10550 North Torrey Pines Road, La Jolla, CA 92037, USA

Johanna E. Goldmann Massachusetts Institute of Technology, Whitehead Institute for Biomedical Research, Nine Cambridge Center, Cambridge, MA 02142, USA

Joel M. Gottesfeld The Scripps Research Institute, Dept Molecular Biology, 10550 North Torrey Pines Road, La Jolla, CA 92037, USA

Jason Gustin Stem Cells, Sigma-Aldrich, 2909 Laclede Avenue, St Louis, MO 63103, USA

Sangyoon Han University of California, San Diego, Department of Medicine, 9500 Gilman Dr., CMM-West, Room 100F, La Jolla, CA 92093-0644, USA

Nick Hannan Department of Surgery, Anne McLaren Laboratory for Regenerative Medicine, University of Cambridge, West Forvie Building, Robinson Way, Cambridge CB2 0SZ, UK

Yvonne Hoang ScienCell Research Laboratories, 6076 Corte Del Cedro, Carlsbad, CA 92011, USA

Natalie Hobson Genea Ltd, Embryology Department, Level 4, 321 Kent Street, Sydney, NSW 2000, Australia

Marlys Houck San Diego Wild Animal Park – CRES, 15600 San Pasqual Valley Road, Escondida, CA 92027-7000, USA

Hans S. Keirstead University of California, Irvine, Reeve Irvine Research Center, 1216 Gillespie Neuroscience Research Facility, Irvine, CA 92657-4265, USA

Philip Koch University of Bonn, Institute of Reconstructive Neurobiology, Sigmund-Freud Straße 25, D-53127 Bonn, Germany

Sherman Ku The Scripps Research Institute, Dept Molecular Biology, 10550 North Torrey Pines Road, La Jolla, CA 92037, USA

Frank M. LaFerla University of California, Irvine, Institute for Memory Impairments and Neurological Disorders, 3212 Biological Sciences III, Irvine, CA92697, USA

Uma Lakshmipathy Life Technologies, Inc., Primary and Stem Cell Systems, 5781 Van Allen Way, Carlsbad, CA 92008, USA

Jack Lambshead CSIRO Materials and Science Engineering, Bag 10, Clayton South MDC, Victoria 3169, Australia

Thomas E. Lane University of California, Irvine, Multiple Sclerosis Research Center, Gross Hall Rm 2006, Irvine, CA 92697, USA

Andrew L. Laslett CSIRO Materials and Science Engineering, Bag 10, Clayton South MDC, Victoria 3169, Australia

Louise C. Laurent University of California, San Diego, Department of Reproductive Medicine, 200 West Arbor Drive, San Diego, CA 92103, USA

Michael Lenz Aachen University, Aachen Institute for Advanced Study in Computational Engineering Science (AICES), Schinkelstrasse 2, 52056, Aachen, Germany

Trevor R. Leonardo The Scripps Research Institute, Center for Regenerative Medicine, Dept Chemical Physiology, 10550 North Torrey Pines Road, La Jolla, CA 92037, USA

Tianjian Li Stem Cells, Sigma-Aldrich, 2909 Laclede Avenue, St Louis, MO 63103, USA

Pauline Lieu Life Technologies, Inc., 5781 Van Allen Way, Carlsbad, CA 92008, USA

Qiuyue Liu The Buck Institute for Research on Aging, 8001 Redwood Boulevard, Novato, CA 94945, USA

Ying Liu University of Texas Health Sciences Center, Houston, Department of Neurosurgery, 1825 Pressler St, Houston, TX 77030, USA

Jeanne F. Loring The Scripps Research Institute, Center for Regenerative Medicine, Dept Chemical Physiology, 10550 North Torrey Pines Road, La Jolla, CA 92037, USA

Mai X. Luong University of Massachusetts Medical School, International Stem Cell Registry, Department of Cell Biology, 222 Maple Avenue, Shrewsbury, MA 01545, USA

Kim Ly ScienCell Research Laboratories, 6076 Corte Del Cedro, Carlsbad, CA 92011, USA

Candace L. Lynch The Scripps Research Institute, Center for Regenerative Medicine, Dept Chemical Physiology, 10550 North Torrey Pines Road, La Jolla, CA 92037, USA

Ian Lyons Stem Cells, Sigma-Aldrich, 2909 Laclede Avenue, St Louis, MO 63103, USA

Steven McArthur Genea Ltd, Embryology Department, Level 4, 321 Kent Street, Sydney, NSW 2000, Australia

Robert E. Morey The Scripps Research Institute, Center for Regenerative Medicine, Dept Chemical Physiology, 10550 North Torrey Pines Road, La Jolla, CA 92037, USA

Franz-Josef Müller University Hospital Schleswig Holstein, Center for Psychiatry, Niemannsweg 147, D-24105, Kiel, Germany

Christine Mummery Department of Anatomy and Embryology, Leiden University Medical Center, Research Building 1, PO Box 9600, 2300 RC Leiden, The Netherlands

Kristopher L. Nazor The Scripps Research Institute, Center for Regenerative Medicine, Dept Chemical Physiology, 10550 North Torrey Pines Road, La Jolla, CA 92037, USA

Joy L. Nerhus University of California, Irvine, Institute for Memory Impairments and Neurological Disorders, Gross Hall Rm 3200, Irvine, CA 92697, USA

Elizabeth Ng Monash University, Level 3, Building 75, Clayton, Melbourne, Victoria 3800, Australia

Kyle S. Nickey The Scripps Research Institute, Center for Regenerative Medicine, Dept Chemical Physiology, 10550 North Torrey Pines Road, La Jolla, CA 92037, USA

Gabriel I. Nistor California Stem Cell, Inc., 18301 Von Karman Avenue, Suite 130, Irvine, CA 92612, USA

Jamison L. Nourse University of California, Irvine, Sue and Bill Gross Stem Cell Research Center, School of Medicine, 3200 Gross Hall, Irvine, CA 92697-1705, USA

Carmel O'Brien CSIRO Materials and Science Engineering, Bag 10, Clayton South MDC, Victoria 3169, Australia

Theo Palmer Stanford University, Department of Neurosurgery, Palo Alto, CA 94305, USA

Mark L. Rohrbaugh National Institutes of Health, Office of Technology Transfer, 6011 Executive Boulevard, Suite 325, Rockville, MD 20852-3804, USA

Oliver Z Pedersen The Buck Institute for Research on Aging, 8001 Redwood Boulevard, Novato, CA 94945, USA

Suzanne E. Peterson The Scripps Research Institute, Center for Regenerative Medicine, Dept Chemical Physiology, 10550 North Torrey Pines Road, La Jolla, CA 92037, USA

Mahendra Rao National Institutes of Health, Center For Regenerative Medicine, 50 South Drive, Bethesda, MD 20892, USA

Julia Schaft Genea Ltd, Research Department, Level 3, 321 Kent Street, Sydney, NSW 2000, Australia

John P. Schell The Scripps Research Institute, Center for Regenerative Medicine, Dept Chemical Physiology, 10550 North Torrey Pines Road, La Jolla, CA 92037, USA

Ulrich Schmidt Genea Ltd, Research Department, Level 3, 321 Kent Street, Sydney, NSW 2000, Australia

Bernhard M. Schuldt Aachen University, Aachen Institute for Advanced Study in Computational Engineering Science (AICES), Schinkelstrasse 2, 52056 Aachen, Germany

Philip H. Schwartz Children's Hospital of Orange County, 455 South Main Street, Orange, CA 92868-3874, USA

Jason Sharp California Stem Cell, Inc., 18301 Von Karman Avenue, Suite 130, Irvine, CA 92612, USA

James Shen ScienCell Research Laboratories, 6076 Corte Del Cedro, Carlsbad, CA 92011, USA

Ronald Simon Scripps Health, Allergy & Immunology, 311 Calley Centre Drive, San Diego, CA 92130, USA

Ileana Slavin The Scripps Research Institute, Center for Regenerative Medicine, Dept Chemical Physiology, 10550 North Torrey Pines Road, La Jolla, CA 92037, USA

Kelly P. Smith University of Massachusetts Medical School, International Stem Cell Registry. Department of Cell Biology, 222 Maple Avenue, Shrewsbury, MA 01545 USA

Elisabetta Soragni The Scripps Research Institute, Dept Molecular Biology, 10550 North Torrey Pines Road, La Jolla, CA 92037, USA

Glyn Stacey National Institute for Biological Standards and Control-HPA, UK Stem Cell Bank, Blanche Lane, South Mimms, Potters Bar, EN6 3QG, UK

Rathi D. Thiagarajan The Scripps Research Institute, Center for Regenerative Medicine, Dept Chemical Physiology, 10550 North Torrey Pines Road, La Jolla, CA 92037, USA

Thomas Touboul   The Scripps Research Institute, Center for Regenerative Medicine, Dept Chemical Physiology, 10550 North Torrey Pines Road, La Jolla, CA 92037, USA

Bhaskar Thyagarajan   Life Technologies, Inc., 5781 Van Allen Way, Carlsbad, CA 92008, USA

Ha T. Tran   The Scripps Research Institute, Center for Regenerative Medicine, Dept Chemical Physiology, 10550 North Torrey Pines Road, La Jolla, CA 92037, USA

Ludovic Vallier   Department of Surgery Anne McLaren Laboratory for Regenerative Medicine, University of Cambridge, West Forvie Building, Robinson Way, Cambridge, CB2 0SZ, UK

Cathelijne Van Den Berg   Department of Anatomy and Embryology, Leiden University Medical Center, Research Building 1, PO Box 9600, 2300 RC Leiden, The Netherlands

Yu-Chieh Wang   The Scripps Research Institute, Center for Regenerative Medicine, Dept Chemical Physiology, 10550 North Torrey Pines Road, La Jolla, CA 92037, USA

Dorien Ward-van Oostwaard   Department of Anatomy and Embryology, Leiden University Medical Center, Research Building 1, PO Box 9600, 2300 RC Leiden, The Netherlands

Jason Weinger   University of California, Irvine, Multiple Sclerosis Research Center, Gross Hall Rm 2006, Irvine, CA 92697, USA

Shinya Yamanaka   Kyoto University, Center for iPS Cell Research and Applications Dept of Reprogramming Science, 53 Kawahara-cho, Shogoin Yoshida, Sakyo-ku, Kyoto 606-8507, Japan

Xianmin Zeng   The Buck Institute for Research on Aging, 8001 Redwood Boulevard, Novato, CA 94945, USA

Qi Zhou   CSIRO Materials and Science Engineering, Bag 10, Clayton South MDC, Victoria 3169, Australia

Boback Ziaeian   Children's Hospital of Orange County, 455 South Main Street, Orange, CA 92868-3874, USA

1

Basic Methods

Culturing Human Pluripotent Stem Cells on a Feeder Layer

Trevor R. Leonardo, John P. Schell, Kyle S. Nickey, and Ha T. Tran

EDITOR'S COMMENTARY

Many technologies for culture and characterization of human pluripotent stem cells (hPSCs) have improved in the 5 years since the first edition of this manual was published. But the underlying principles are the same: culturing human pluripotent stem cells (hPSCs) requires a significant commitment of time and resources. A word of caution for those with experience culturing mouse PSCs: they are not the same! Mouse cells are much more robust and forgiving of variations in technique. They divide quickly and rarely spontaneously differentiate in culture. They can be routinely passaged as single cells, cryopreserved and thawed easily, and can survive even if they are neglected for a day. Human PSCs are quite the opposite, and thrive only with continuous attention.

The basic methods for culturing hPSCs on feeder layers are similar to those described in the earlier edition of the manual. The main differences are in optimizing and perfecting existing techniques that have been proven reliable and effective over the years. This chapter describes the most reliable methods for maintaining healthy cultures of hPSCs; these are the methods that we use in our own lab and that we have taught to the hundreds of students that we have trained in our human stem cell courses over the years. One piece of advice: since the cells need to be fed every day, make friends with your trusted lab mates so that you can care for each others' cultures.

OVERVIEW

Traditionally, hPSCs are cultured on a mitotically inactivated feeder layer such as mouse embryonic fibroblasts (MEFs). Feeder cells provide a suitable substrate for the hPSCs to attach to and help maintain the pluripotency of hPSCs by secreting various proteins into the growth medium. This chapter

3

J.F. Loring & S.E. Peterson (eds): Human Stem Cell Manual, Second edition.
DOI: http://dx.doi.org/10.1016/B978-0-12-385473-5.00001-1

addresses the techniques necessary to culture hPSCs on a feeder layer and will help you gain a better understanding of the basic culture methods of hPSCs. The methods that we will discuss are ones that have been used reliably over time by many researchers in the field. Only time and experience will tell what methods will work best for you and your particular cells. The two fundamental methods that will be covered are manual or "mechanical" passaging and enzymatic dissociation. Generation of feeder cells and their mitotic inactivation are discussed in Chapter 2.

PROCEDURES

Monitoring hPSCs in Culture

In order to successfully maintain hPSCs on a feeder layer over time in culture it is crucial to have a good understanding of their morphological characteristics and to maintain good technique while culturing (Figure 1.1). hPSCs have a unique morphology when grown on a feeder layer. They grow in compact colonies with cleanly defined edges that distinguish the colony from the feeder layer. hPSCs commonly differentiate either on the edges or in the center of the colonies. However, differentiation can occur anywhere in the colony. Cells in the colony should be small, tightly packed together and look uniform (Figure 1.1A). Differentiated areas should be removed daily (if necessary).

Tips for Culturing hPSCs

- Culture medium must be changed every day.
- Observe the hPSCs every day using the 4× and 10× objective lenses on an inverted phase contrast microscope to detect early signs of differentiation or contamination and monitor the morphology and growth rate of the cells.
- Remove differentiated areas of colonies using a p200 pipette tip attached to a sterile 3 mL syringe before changing the medium. It is important to remove differentiated cells from your culture in order to prevent further differentiation of other colonies.
- hPSC medium should be prepared fresh and discarded if not used after approximately 10 days.
- The density and quality of the feeder layer is very important for the maintenance of hPSC cultures. When using inactivated MEFs as a feeder layer, a common density to plate is about 100,000–200,000 cells per well of a 6-well tissue culture plate (see Chapter 2).
- hPSCs grown on a feeder layer should be grown in hPSC medium (see Recipes) which is sometimes referred to as KSR medium or WiCell medium. They should not be grown in media designed for feeder-free culture, such as StemPro or mTeSR.

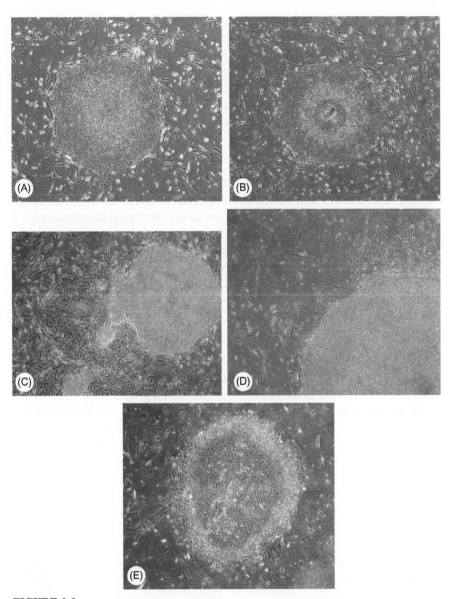

FIGURE 1.1

(A) Example of an undifferentiated hPSC colony. (B) Partially differentiated colony. Differentiated cells in the center appear darker and larger. The outer regions of this colony are salvageable if necessary, but the inner region should be removed. (C) Differentiation on the left side of the edges of the colony. Large, flat differentiated cells migrate out from the colony blurring the line between the colony and the feeder cells. (D) Differentiation on the top of the colony, on the lower right side of the figure. Most of the colony is salvageable though. (E) Differentiation over most of the inside of the colony. The entire colony should be discarded.

■ Differentiation is easier to detect morphologically in hPSCs grown on feeder layers. Because of this we recommend starting off culturing hPSCs on feeder layers until you have learned how to judge them by their morphology.

■ hPSCs may be cultured on mitotically inactivated mouse embryonic fibroblasts (MEFs in Chapter 2), established mouse cell lines like SNL (Chapter 10), or on human fibroblasts such as Hs27 (ATCC) or Nuff (GlobalStem).

Recognizing hPSC Colony Morphology

Being able to identify and discern the difference between colonies with undifferentiated morphology and those that are or are becoming differentiated is a key skill required for culturing hPSCs (Figure 1.1). It is recommended that for routine expansion, the cells be cultured in a medium density so that colony morphology is more easily discerned. While hPSCs can be cultured at a high density (Figure 1.2), this may increase the rate of spontaneous differentiation. Low density cultures may not survive well and may also have problems with spontaneous differentiation.

Passaging hPSCs

For hPSCs derived on feeder layers and routinely grown on feeder layers, it is important to avoid dissociating them into a single cell suspension at any point during culture, to prevent both differentiation and cell death. For this reason, manual or "mechanical" passaging of the cells is the current standard and one of the quintessential hPSC culture techniques. Although this method takes time for those with multiple cultures going at the same time, the method itself is extremely reliable, and is thought to generate the least amount of genetic instability in hPSCs over time in culture. Because it keeps the populations stable, we use this method to generate early stocks

FIGURE 1.2

Colony density. (A) Good density. (B) Too crowded.

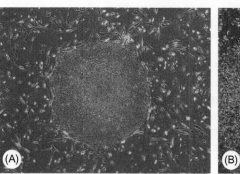
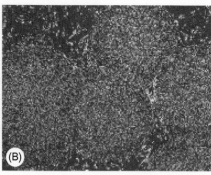

of cryopreserved cells. There are alternative means of passing the cells on a feeder layer, which will be covered in the alternative methods section.

General Passaging Guidelines

- hPSCs grown on feeder layers should be passaged at a dilution of 1:3 every 5 to 7 days (the actual ratio and time in culture may vary depending on the line).
- Prepare the feeder layer at least 24 hours before passaging (see Chapter 2).
- Establishing a routine passaging schedule on the same day of each week will yield high quality cultures over time and prevent differentiation.
- Avoid dissociating the cells into a single cell suspension.
- Observe colonies on a daily basis under a phase contrast microscope to detect any early signs of differentiation.
- Unwanted differentiated colonies should be removed daily if necessary to avoid differentiation of the remaining pluripotent colonies.

Mechanical Dissociation Procedure

■ Note

We highly recommend that mechanical passaging be done in a horizontal laminar flow hood under a dissecting scope with dark field transillumination and an objective range of between 0.8× and 4.5× (Figure 1.3). All other cell culture should take place in a regular biosafety cabinet. Alternatively, use a modified biosafety cabinet with a dissecting microscope fitted inside as described in Chapter 7. ■

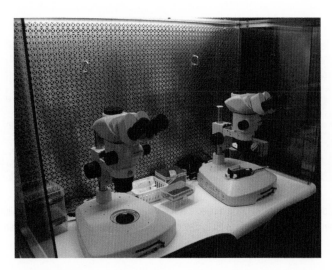

FIGURE 1.3
Set up for manual passaging of hPSC colonies. Dissecting scopes with dark field transillumination are kept in a dissecting hood. Once the colonies are cut, the plate can be returned to a regular biosafety cabinet.

1. Observe cells under 4× and 10× phase contrast microscope (Figure 1.4A).
2. Prepare the appropriate number of feeder layer plates by changing the medium to hPSC medium.
3. In a biosafety cabinet, aspirate hPSC medium from hPSC cultures and replace it with fresh hPSC medium. Then carefully transfer both the new feeder layer dishes and the dish with hPSC colonies to the dissecting scope in the laminar flow hood.
4. Using a cutting implement such as a 25-gauge syringe needle (attached to a syringe to provide a handle), begin cutting the desired undifferentiated colonies into a grid-like pattern. Cut straight parallel lines in one well going in one direction holding the needle at a 45° angle (Figure 1.4B).
5. Cut perpendicular to the first cuts after rotating the plate 90 degrees (Figure 1.4C). The lines made by the needle should very lightly graze the plate so as to not make deep etches into the plate. Each section should be roughly 100–500 cells and uniformity in size is key.
6. Using a less sharp device such as a 200 μL pipette tip, dislodge the individual square cut colonies to lift them from the dish in a "picking" motion, moving down one row of the grid at a time.
7. Once the desired cell clumps are in suspension, transfer them into the new feeder layer plates using a p1000 pipette. It is important to remember to not triturate or pipette too vigorously during this step to avoid breaking the pieces into single cells. Alternatively, all of the medium (containing all the cell clumps) can be transferred to the new well.

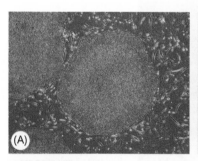 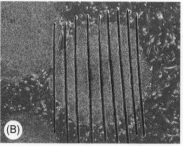 

FIGURE 1.4
(A) Undifferentiated colony ready for passage. (B) Using a needle and syringe, cut the colony by making parallel lines in one direction, then turn the dish 90° and cut again (C). Use a p200 pipette tip to dislodge the pieces of cut colonies from the dish.

ALTERNATIVE PROCEDURES

Enzymatic Dissociation

Enzymatic dissociation is a viable alternative to manual dissociation, especially if the sheer number of cultures makes mechanical methods unreasonable. When using an enzyme, it is crucial to choose one that has been shown to be safe and effective in passing hPSCs. Trypsin, for example, has been shown to have adverse effects on hPSCs over the long term and can lead to genetic drift and aneuploidy. Using an enzyme increases the risk of total cell dissociation, which, as previously discussed, can lead to differentiation and cell death.

Dispase is used in many laboratories and has been effectively used to passage hPSCs for long-term culture. It loosens the colony from the dish around the edges and if used correctly will not allow total cell dissociation. Note that dispase should be made up fresh for each use. This protocol can also be performed using collagenase IV (200 U/mL).

■ Note

Some enzymes on the market are advertised as safe for hPSC culture, but only for feeder-free conditions and may not work as desired if used on cells cultured on a feeder layer. ■

Dispase Dissociation Procedure

1. Remove any unwanted, differentiated colonies so they do not carry over into the next culture.
2. Aspirate culture medium and wash with PBS.
3. Remove PBS and incubate with 0.2 mg/mL dispase diluted in DMEM for 5–10 minutes in the 37°C incubator or until the edges of the colonies begin to curl up off the dish and become phase bright (do not allow the colonies to fully detach) (Figure 1.5).
4. Remove dispase solution with a pipette and gently rinse with fresh culture medium.
5. Remove rinse medium and replace with fresh culture medium.
6. Use a 5 mL pipette to gently dislodge the colonies from the dish and transfer them into a conical tube. If cells begin to detach prior to this step, collect them in a conical tube prior to this step and spin the suspension down in a centrifuge to remove the remaining enzyme.
7. Gently triturate the suspension with a 10 mL pipette and try to achieve relative colony size uniformity.
8. Bring the suspension to a desired final volume and transfer cells into fresh culture dishes.

FIGURE 1.5
Incubate hPSCs in dispase or collagenase until the edges of the colonies become phase bright and begin to come off the dish.

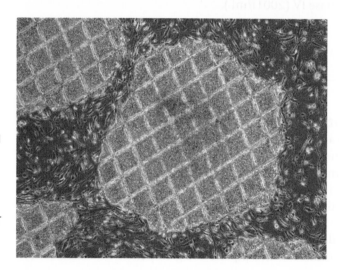

FIGURE 1.6
hPSC colonies cut with the EZ Passage tool. When using a passaging tool, hold it down with even pressure and make sure the roller is flush with the bottom of the plate. Dislodge cut colonies with a pipette tip or by pipetting with a p1000 Pipetman.

Passaging Tools

We usually use 25-gauge syringe needles, but passaging tools, either made by machine shops or purchased commercially (an example is the EZ Passage tool from Life Technologies), are becoming popular. These tools have tiny grooves that cut colonies into evenly sized squares when rolled over the culture (Figure 1.6). Although these tools are often expensive, they can be reused if properly sterilized. Because colonies are cut into standard sized squares, these tools can help with reproducibility.

PITFALLS AND ADVICE

The longer hPSCs are maintained in culture, the more likely they are to succumb to either genetic or developmental drift. In the case of genetic drift, the best defense is to establish a large cryobank of early passage cells (Chapters 7 and 8). We recommend analyzing the cell banks by both karyotyping and SNP genotyping on a routine basis, ideally every 10–15 passages. Karyotyping can determine if a subpopulation of a culture is aneuploid or has gross rearrangements, and SNP genotyping detects subchromosomal abnormalities. For this to be effective, it is important to have a primary analysis done immediately or shortly after you obtain the cell cultures to ensure a proper analytical base line.

Developmental drift refers to changes in the pluripotency of the cells that may occur with time in culture. Specifically, cells may drift towards a more differentiated state. Any change in the morphology of hPSCs that varies from the ideal smooth-edged colony shape is a good indicator that the cells have become at least partially differentiated. It is not uncommon to find some differentiation in many hPSC cultures due to the inherently sensitive nature of the cell type. To prevent this situation from getting worse, identify the unwanted cells and remove them under the dissecting microscope. Changes in pluripotency can be examined by staining for pluripotency markers (Chapter 16), performing teratoma assays (Chapters 21 and 22), testing the cells using the PluriTest molecular diagnostic test (Chapter 18) and characterizing differentiation of embryoid bodies (Chapter 23).

It is important to remember that an hPSC colony that has become highly differentiated may be removed at any stage of the culture by lightly swirling or scratching off the colony with a pipette tip under a dissecting microscope and removing it with a subsequent medium change. Colonies that have a small section of differentiated cells but otherwise look healthy and pluripotent may be rescued by removing only the differentiated section using care not to disturb the remaining cells. In extreme cases, an entire dish may be rescued if a vast majority of the cells begin to differentiate. Under these circumstances, differentiated cells should be removed on a daily basis in order for a successful recovery over the next few passages. For cell lines that have been passaged for long periods of time in the laboratory, it is useful to test the pluripotency of the lines periodically using one or more of the methods listed above.

The decision to remove suspected differentiating cells should be held off for a few days after passaging. Newly cut colonies often take 24–48 hours after attachment to regain the typical hPSC morphology and may be mistaken for differentiated cells.

We strongly recommend that hPSC cultures be maintained in the absence of an antibiotic, so that contamination can be quickly revealed. Proper sterile technique is typically enough to maintain a culture free of contamination. However, if you are just learning to culture the cells, adding an antibiotic to your cultures may help you as your technique improves. Be sure to choose an antibiotic that has been qualified for use with hPSCs.

Keep in mind that antibiotics have no effect on mycoplasma contamination. This is pointed out in several chapters, but cannot be emphasized strongly enough. Mycoplasma can be devastating not only to your cultures, but to an entire lab. If you get mycoplasma in your cultures and spread it to other members of the lab, expect your colleagues to be angry with you. If you do it a second time, your manager should either fire you or retire you from cell culture. The danger is that mycoplasma cannot be seen under a normal cell culture microscope, making it very difficult to detect. We strongly recommend routine mycoplasma testing, usually once a month for all existing cultures in the lab. Any new culture entering the lab from an outside source should be quarantined away from the rest of the lab's cultures until after it has been shown to be mycoplasma-free. Should any cultures become infected, the best option is to destroy the culture immediately. We do not recommend trying to "cure" your cultures of mycoplasma. If you have followed our advice, you will have cryopreserved mycoplasma-free samples of your cell lines, and can restart your experiments easily.

EQUIPMENT

- Tissue culture hood
- Tissue culture incubator, 37°C, 5% CO_2
- Horizontal laminar flow hood
- Inverted Microscope with 4× phase objective
- Dissecting microscope with dark field transillumination and 0.8×–4.5× range
- Pipettes (20 μL, 200 μL, and 1000 μL)
- Pipette aid
- Standard tissue culture equipment
- Centrifuge that can hold 15 mL conical tubes
- Freezers: −20°C, −80°C, liquid nitrogen
- Freezing container, Nalgene® "Mr. Frosty"
- Refrigerator at 4°C
- Water bath, 37°C

REAGENTS AND SUPPLIES

Supplies

- 6-well culture dishes
- 15 mL and 50 mL sterile conical tubes
- 2 mL disposable aspirating pipettes
- Pipette tips for Rainin or similar pipettor with filter
- 25-gauge 5/8″ syringe needles
- 5 mL, 10 mL, 25 mL sterile disposable pipettes

Recommended Reagents

Item	Supplier	Catalog #	Note
DMEM/F-12 + HEPES + L-Glutamine (Dulbecco's Modified Eagle's Medium)	Life Technologies	11330-032	
KnockOut™ Serum Replacement	Life Technologies	108280-028	contains bovine products
GlutaMAX™	Life Technologies	35050-061	
MEM-Non-essential amino acids	Life Technologies	11140-050	
2-Mercaptoethanol	Life Technologies	21985-023	Best if added day of use
PBS (Dulbecco's Phosphate-Buffered Saline without Calcium and Magnesium)	Life Technologies	14190-144	
Recombinant Human bFGF	Life Technologies	13256-029	8 ng–12 ng/mL final
Penicillin/Streptomycin (100x)	Life Technologies	15070-063	Optional
DMEM, High Glucose + GlutaMAX	Life Technologies	10566016	
Fetal Bovine Serum	Life Technologies	10437-028	
Dispase	Life Technologies	17105-041	

RECIPES

hPSC Medium (also known as KSR or WiCell medium)

Ingredient	Amount	Final Concentration
DMEM/F-12 + HEPES + L-Glutamine	200 mL	80%
KnockOut Serum Replacement	50 mL	20%
GlutaMAX 200 mM (100x)	2.5 mL	1x
MEM-Non-essential amino acids 10 mM (100x)	2.5 mL	1x
2-Mercaptoethanol	455 µL	0.1 mM
Recombinant Human bFGF	Use at 12 ng/mL	12 ng/mL

Fibroblast Medium

Item	Stock	Final	Volume
DMEM, High Glucose + GlutaMAX™	100%	90%	450 mL
Fetal Bovine Serum	100%	10%	50 mL
MEM-Non-essential amino acids	10 mM	0.1 mM	5.5 mL

READING LIST

Abraham, S., Sheridan, S.D., Laurent, L.C., Albert, K., Stubban, C., Ulitsky, I., et al., 2010. Propagation of human embryonic and induced pluripotent stem cells in an indirect co-culture system. Biochem. Biophys. Res. Commun. 393 (2), 211–216. (Epub 2010 Feb 1).

Cowan, C.A., Klimanskaya, I., McMahon, J., Atienza, J., Witmyer, J., Zucker, J.P., et al., 2004. Derivation of embryonic stem-cell lines from human blastocysts. N. Engl. J. Med. 350, 1353–1356.

Meng, G., Liu, S., Rancourt, D.E., 2011. Rapid isolation of undifferentiated human pluripotent stem cells from extremely differentiated colonies. Stem. Cells. Dev. 20 (4), 583–591. (Epub 2010 Dec 13).

Peterson, S.E., Westra, J.W., Rehen, S.K., Young, H., Bushman, D.M., et al., 2011. Normal human pluripotent stem cell lines exhibit pervasive mosaic aneuploidy. PLoS ONE 6 (8), e23018. doi: 10.1371/journal.pone.0023018.

Zeng, X., Chen, J., Liu, Y., Luo, Y., Schulz, T.C., Robins, A.J., et al., 2004. BG01V: A variant human embryonic stem cell line which exhibits rapid growth after passaging and reliable dopaminergic differentiation. Restor. Neurol. Neurosci. 22, 421–428.

CHAPTER 2

Preparation of Mouse Embryonic Fibroblast Feeder Cells

Trevor R. Leonardo, John P. Schell, Ha T. Tran, and
Suzanne E. Peterson

EDITOR'S COMMENTARY

Feeder layers are an integral part of deriving and culturing human pluripotent stem cells (hPSCs). Their use seems routine, but their quality is critical, and one of the most common problems that arises for laboratories culturing hPSCs is a failure of their feeder layers to reliably support undifferentiated culture of hPSCs. For most of the methods described in this manual, the feeder cells used are mouse embryo fibroblasts (MEFs). This is a tradition carried over from mouse ES cell culture; since the 1980s, mouse ES cells have been derived and maintained as co-cultures with mitotically inactivated MEFs or an established mouse cell line called STO (SIM immortalized cell line with thioguanine and ouabain-resistance), which was established in the 1970s by Alan Bernstein, and modified by Allan Bradley's lab in 1990 to be neomycin resistant (for use in gene targeting selections) and to produce LIF, a cytokine that helps maintain mouse ES cell pluripotency. SNL76/7 cells (STO-Neo-LIF clones 76/7) are used as feeder layers routinely in some hPSC laboratories (see Chapter 10).

MEFs, like growth factors, have lot-to-lot variability and whether you make the cells in your lab or purchase them from a commercial source, the same rigorous quality controls need to be completed before they are integrated into the lab. The methods described in this chapter are reliable, and, if followed closely, should yield high-quality feeder layers every time. But you must still test them for their effectiveness as described in this chapter.

OVERVIEW

MEFs are primary cells derived from day-13.5 fetuses that do not continue to proliferate indefinitely. Once the cells begin to senesce they lose their capacity to support undifferentiated growth and proliferation of hPSCs, so they are

15

J.F. Loring & S.E. Peterson (eds): Human Stem Cell Manual, Second edition.
DOI: http://dx.doi.org/10.1016/B978-0-12-385473-5.00002-3

used optimally between passage 4 and passage 7. Usually large batches are made, tested, and cryopreserved so that this process needs to be repeated only occasionally.

Each newly prepared batch of MEFs should be tested for robust recovery from cryopreservation, and support of undifferentiated proliferation of hPSC cultures. They also must be tested for mycoplasma and ideally should be subjected to mouse antibody pathogen (MAP) testing, usually by an outside service.

It is important that MEFs be mitotically inactivated before being co-cultured with hPSCs, or the MEFs will become a growing contaminant cell type that is difficult to remove. There are two common ways of inactivating the cells: 1) irradiation and 2) mitomycin C treatment. The cells can be cryopreserved either before or after mitotic inactivation.

PROCEDURES

Isolation of Mouse Embryonic Fibroblasts (MEFs)

You will need to prepare in advance in order to prepare your own batch of mouse embryonic fibroblasts (MEFs). It is important to follow your institutional, local, state, and federal regulations regarding the use of laboratory mice. These guidelines are governed by the Institutional Animal Care and Use Committee (IACUC), which reviews proposed use of vertebrate animals. Obtain IACUC approval, then obtain the desired mouse strain and set up matings as approved by IACUC. Alternatively, timed pregnant mice can be ordered from many mouse providers so that you do not have to maintain a colony or set up the matings yourself.

■ Note

The protocol below is a method that has worked well to produce high quality MEFs from various mouse strains including 129/Sv, 129/B6 F1, B6, Balb/cJ, and CF-1. We typically use CF-1 mice.   ■

Isolation of Mouse Embryos

■ Note

For anatomical pictures and drawings, we recommend a manual such as "Manipulating the Mouse Embryo" (see Reading List).   ■

1. Using an institutionally approved euthanasia method, sacrifice a 13.5 dpc female mouse. At 13.5 dpc, the female should be visibly pregnant.

■ **Note**

Only one mouse should be processed at a time. Embryos from different mice should not be pooled, as they may vary in quality. ■

2. Place the female on her back on a clean bench or inside a dissecting hood. Spray the abdomen with 70% ethanol. Using sterile forceps and scissors make a small lateral incision under the diaphragm.
3. With forceps pull the skin back, revealing the peritoneum. Make an incision in the peritoneum to expose the uterine horns. Pull out and cut off each uterine horn.
4. Using aseptic technique, transfer the uterine horns to a 10 cm dish with 10 mL of ice-cold PBS.
5. Inside a dissecting hood, dissect the embryos from the uterus and carefully remove each from their yolk sac and placenta. Make note of how many embryos were obtained.
6. Place isolated embryos into a fresh 10 cm dish containing 10 mL of PBS.
7. Remove the head and internal organs (dark red tissue in the abdomen) using a pair of small sharp scissors and forceps.
8. Rinse each carcass well, by placing it in a fresh 10 cm dish containing 10 mL of PBS and gently swirling the dish to remove any remaining blood.

Preparation of Cells

■ **Note**

MEFs from each individual impregnated mouse should be processed separately, plated into different flasks, cryopreserved separately, and quality tested separately. ■

1. Place dissected embryos in a new 10 cm dish containing a very small volume (250 μL) of PBS and mince the tissue for 10 minutes using sharp scissors.
2. Add 2 mL of 0.05% trypsin to the tissue and mince for another 5 minutes.
3. Add 5 more mL of trypsin and triturate the solution vigorously using a 10 mL pipette.
4. Place the 10 cm dish into a 37°C incubator for 20–30 minutes.
5. Pipette the tissue up and down with a 10 mL pipette until it has a sludgy consistency. Add 20 mL of fibroblast medium and transfer the cell suspension to a 50 mL conical tube.
6. Spin the tube at 200 g for 3–5 minutes.
7. Aspirate the supernatant and discard.
8. Determine how many T75 flasks to plate based on how many embryos were processed. Use three embryos per T75 flask.

9. Resuspend the cell pellet to the appropriate volume of fibroblast medium (three embryos for each T75 flask). Each T75 flask should contain a total volume of 10 mL.
10. Place flasks in the incubator overnight.
11. Observe the cells the next day using a 4× phase contrast microscope. When they reach 90–95% confluence the cells are ready to be trypsinized and cryobanked.

Cryopreserving MEFs

1. Aspirate the medium from the flask and rinse once with PBS.
2. Add 3 mL of 0.05% trypsin to each flask and place the flasks to the incubator for 3–5 minutes.
3. Agitate the cells by shaking or tapping the flasks gently to ensure they have come off, and add an equal volume of fibroblast medium to the flask. Pipette cells up and down and transfer cells to a 15 mL or 50 mL conical tube.
4. Spin the cells at 200 g for 5 minutes and aspirate the medium leaving only the cells. Add 1.5 mL of fibroblast medium for each T75 flask that was harvested.
5. Add an equal volume of 2× cryopreservation medium and aliquot 1 mL into each labeled cryovial (there should be three cryovials per T75 flask). Freeze at −80°C in a slow-freezing device ("Mr. Frosty") and transfer to liquid nitrogen within 1 week.

■ Note

These are your passage zero (P0) MEF stocks. MEFs from each individual impregnated mouse should be prepared and placed into labeled T75 flasks with the same number. Do not combine MEFs from different mice during the first T75 plating or freezing process. MEFs from each individual impregnated mouse should be quality tested separately.   ■

Initial Thawing of P0 MEFs

1. Warm fibroblast medium in a 37°C water bath.
2. Add 9 mL of fibroblast medium into a 15 mL conical tube.
3. Thaw one vial of P0 MEFs in a 37°C water bath.
4. Once only a small ice pellet remains, remove from the water bath and bring into the biosafety cabinet. Wipe off the cryovial with 70% ethanol.
5. Slowly pipette cells from the cryovial using a 1 mL pipette. Transfer them into the 15 mL conical tube with fibroblast medium.
6. Centrifuge for 5 minutes at 200 g.
7. Aspirate the supernatant leaving the cell pellet.
8. Resuspend the cells in 10 mL fibroblast medium and add to a T75 flask.

Expansion of MEFs

Trypsinize the MEFs and split them 1:3 each time they become 90–95% confluent according to Figure 2.1. MEFs should be split three more times after the initial thaw of the P0 MEFs and then mitotically inactivated and cryopreserved. At the time of inactivation you should have nine confluent T175 flasks, which will yield approximately 150 to 200 million cells.

MEF Inactivation

There are three common ways to inactivate MEFs: 1) gamma irradiation, 2) X-ray irradiation, and 3) mitomycin C treatment. Which method you use will be determined both by the facilities at your institution and by the number of inactivated MEFs required. Typically, if you have access to a gamma irradiator, or even better, an X-ray irradiator, you will want to make use of it because it is cheaper, more reliable and less time-consuming than mitomycin C inactivation. However, if you do not have access to an irradiator or are only inactivating small numbers of MEFs, it may be easier to use mitomycin C.

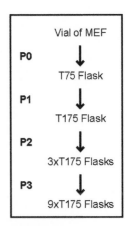

FIGURE 2.1
Expansion scheme for MEFs. One P0 vial should yield nine T175 flasks. At that point, MEFs should be mitotically inactivated and cryopreserved.

■ Note

No matter what method you use, be sure to plate the cells at low density after inactivation to make sure that the cells are truly mitotically inactive. ■

Gamma Irradiation of MEFs

■ Note

At some institutions, authorization to use a gamma irradiator may require a background check and fingerprinting. In the US, approval must come for the Department of Homeland Security. ■

1. Trypsinize cells to be inactivated and add them to a 50 mL conical tube (3 × T175 flasks per conical) and centrifuge for 5 minutes at 200 g.
2. Aspirate the supernatant and resuspend the pellet in 10 mL of fibroblast medium. Triturate gently to get a homogenous mixture.
3. Bring the volume up to 50 mL per conical tube by adding 40 mL of fibroblast medium, mix and then perform a cell count for each conical tube using a hemocytometer or other cell counter.
4. Redistribute cells (spin down if necessary) so each conical tube has 40 mL of medium and less than 50 million cells per tube.
5. Irradiate each conical tube at 3000 rad using a cesium source γ irradiator.

■ Note

Irradiation strength should be "titrated" for each instrument so that the cells are completely inactivated but otherwise viable. ■

6. Keep all 50 mL conical tubes containing MEFs on ice until all have been irradiated.
7. Centrifuge down irradiated MEFs for 5 minutes at 200 g, aspirate off medium leaving only the cell pellets.
8. Freeze down MEFs at an appropriate density using fibroblast medium and 2× cryopreservation medium.

■ Note

It is very common to freeze MEFs in concentrations of 3×10^6 and 10^7 cells per vial. However, this will depend on the feeder requirements for the particular hPSC lines you work with and the number of plates you typically have going at one time. ■

X-ray Irradiation of MEFs

This is our preferred method for inactivating MEFs. We recommend a Faxitron RX-650 (Figure 2.2). The process is essentially the same as gamma inactivation of MEFs, but Step 5 (irradiation) is performed in the X-irradiator. We use 120 kVp for 20 minutes. Only two 50 mL conical tubes are irradiated at one time, on the upper rack. As with gamma irradiation, each conical tube should have no more than 50 million cells in 40 mL of fibroblast medium.

FIGURE 2.2

X-ray irradiator. Place two 50 mL conical tubes in the irradiator at one time and only on the upper shelf in the middle of the shelf.

■ **Note**

As for gamma irradiation, X-irradiation strength should be titrated for each instrument so that the cells are completely inactivated but otherwise viable. ■

Mitomycin C Inactivation

■ **Note**

Mitomycin C is a cytotoxic antitumor agent and must be handled carefully; it works by cross-linking the DNA, which blocks cell division. Follow your institution's rules for safe handling and disposal. One effective method to inactivate the mitomycin C is with an equal volume of household bleach. Inactivation is rapid. ■

1. Dissolve 2 mg of mitomycin C in 200 mL of fibroblast medium and sterile filter to generate a 10 μg/mL solution.
2. Add two thirds the normal volume of mitomycin C-containing medium to the flask, making sure that the cells are adequately covered.
3. Freeze any unused mitomycin C-containing medium and use within 6 months.
4. Incubate cells three hours at 37°C in 5% CO_2.
5. Aspirate the mitomycin C solution and inactivate it with bleach or another procedure recommended by your institution.
6. Wash the inactivated feeder layer three times with the same volume of PBS.
7. Trypsinize the cells to remove them from the flask, count them and cryopreserve them for future use.

■ **Note**

Just as irradiation strength should be tested, time of treatment and concentration of mitomycin C should also be tested to make sure that the cells are completely inactivated but otherwise viable. ■

Substratum Support for Feeder Cells

In order to provide better support for the long culture periods required for hPSC culture, inactivated feeder cells should be plated on gelatinized dishes.

1. Coat culture dishes with 0.1% gelatin solution
2. Incubate 30 minutes to overnight at 37°C
3. Just prior to plating inactivated feeder cells, aspirate gelatin.
4. Plate the inactivated feeders on the gelatin-coated dishes and allow them to attach in the incubator for at least 4 hours before culturing with hPSCs. It is recommended to plate inactivated feeders the day before to ensure that all of the feeder layer cells have attached to the plate.

Quality Controls for MEFs

Once a stock of P0 MEFs is generated it is very important for them to undergo a battery of quality control tests before they are used.

Tests for Mycrobial Contamination

All MEF batches must be tested for microbial contamination, especially mycoplasma (see Chapter 4). If desired, MEF batches can also be tested to be sure that they don't harbor mouse pathogens. Mouse antibody pathogen (MAP) testing is usually performed by a service lab such as RADIL (http://www.radil.missouri.edu/).

Tests for Support of Undifferentiated hPSCs

Each batch must be tested for its ability to support hPSCs. To do this, hPSCs should be grown on the new MEF batch for three passages in parallel with hPSCs grown on a previous batch of MEFs known to work well. After three passages, hPSCs should be analyzed by flow cytometry (Chapter 15) or immunocytochemistry (Chapter 16) for markers of pluripotency such as SSEA4, Tra1-81, or POU5F1/OCT4. New MEF batches should be approved for use only if hPSCs grown on them stain adequately compared to the control MEF batch.

Tests for Mitotic Inactivation

After each new vial of P0 MEFs is expanded and inactivated, tests should be done to confirm that the MEFs are no longer dividing. Although the MEFs will eventually senesce, improperly inactivated MEFs can create a big problem in the culture because the dividing MEFs will alter the normal MEF density and can promote the differentiation of hPSCs. One way to analyze MEF inactivation is by performing a cell-count experiment where cells are counted and multiple dishes are seeded. Every 2 days, one of the dishes should be counted. MEF proliferation can be detected by steadily increasing cell numbers over time. Alternatively, MEFs can be immunostained for Ki67, a marker of dividing cells. In addition, cells can be treated with nucleoside analogs such as BrdU or EdU, and then incorporation can be detected by immunostaining.

Plating hPSCs on Inactivated MEFs

Following inactivation, feeder layers remain adequate for hPSC culture for about 1–2 weeks provided that their medium is changed at least every 3 days. Always observe the feeder layer under the microscope prior to using it for the culture of hPSCs, in order to confirm that the cell layer is still intact and the cells have not begun to deteriorate (see Figure 2.3).

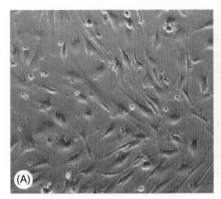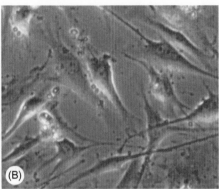

FIGURE 2.3
Low (A) and high (B) magnification photomicrographs showing the morphology of MEF feeder layers. MEFs do not form the whorls of cells that are typical of other fibroblasts used as feeder layers, such as human foreskin fibroblasts.

Since hPSCs are usually passaged every 5–7 days, the feeder layer can start to deteriorate before the hPSCs are ready to passage if "old" feeder layer dishes are used. For best results, plate hPSCs 1–2 days after plating inactivated feeders. If hPSC colonies are taking a long time to get to the point at which they need to be passaged (or when generating iPSCs) new inactivated feeders can be plated on top of the old ones.

Inactivated MEFs are typically plated at 150,000 cells per well of a 6-well tissue culture dish and we recommend this as a starting point for all new lines. However, the optimal number of MEFs can be dependent on the cell line. We have found that some hPSC lines grow best with higher or lower MEF density. If a particular cell line is differentiating a great deal when plated on qualified 150,000 MEFs per well of a 6-well plate, it is important to test different MEF concentrations until an optimum density is reached. If the provider of the hPSCs you are culturing happens to give you a recommended MEF density, follow their instructions at least initially.

ALTERNATIVE PROCEDURES
MEF Cells from Commercial Sources
Many labs simply purchase their MEFs. There are several commercial sources for MEFs from various mouse strains and containing various selectable drug-resistant markers. Depending on the level of use, this can be a convenient alternative to the *de novo* preparation of MEF feeder layers. However, because of lot-to-lot variations in the quality of the cells, be sure to perform the same quality controls as you would for cells made in your own lab.

- ATCC (www.atcc.org)
- Millipore (www.millipore.com)

- Primogenix, Inc. (www.primogenix.com)
- Stem Cell Technologies (www.stemcell.com)
- GlobalStem, Inc (www.globalstem.com)
- ScienCell (http://www.sciencellonline.com)

Human Fibroblasts

Although MEFs are the most common feeder cell type used with hPSCs, human fibroblasts can also be used. Human fibroblasts provide a number of advantages over MEFs in that they can be propagated for many more passages than MEFs before senescing. In addition, transmission of zoonotic viruses is not an issue when using human fibroblasts. Human fibroblast lines can be obtained from many sources (see above). We recommend the human foreskin fibroblast lines Hs27 (ATCC) and NuFF (GlobalStem). These cells are usually used before passage 20 for hPSC culture.

PITFALLS AND ADVICE

General Advice

1. If the provider of an hPSC line recommends a certain MEF plating density, follow their instructions, at least initially.
2. If no instructions are given, a slightly less than confluent layer seems to be optimal. The feeder layer should have healthy cell bodies spread out on the tissue culture plate. Before inactivation, the MEF culture should be doubling every 24–30 hours, requiring passaging at 1:3 every 3 days.
3. MEF feeders should be used between passages 4 and 7. When the MEFs start to slow down in their proliferation or the cultures contain many multinucleate cells or floating debris, dispose of them and thaw a fresh vial.
4. Some laboratories have a strong preference for MEFs derived from particular mouse strains. Others indicate that the strain is unimportant. As a rule, we suggest that the most important characteristic of any cells used for feeder layers is that they be rapidly growing, free of pathogens, and low passage when they are inactivated.
5. As noted often in this chapter, poor MEF quality, which is not a visible characteristic, is the source of many problems in hPSC laboratories. Every source of MEFs must be tested using the rigorous quality control measures described above.

EQUIPMENT

- Class II biosafety cabinet: NuAire or Baker
- Tissue culture incubator: 37°C, 5% CO_2, in humidified air

- Phase contrast microscope, 4×, 10×, 20× objectives
- Low-speed centrifuge (200–3,000 rpm)
- Pipettors: p2, p20, p200, p1000
- Pipette aid
- Refrigerator at 4°C
- Freezers: −20°C, −80°C, and liquid nitrogen

REAGENTS AND SUPPLIES

Feeder Cell Culture Reagents

Item	Supplier	Catalog #	Alternative
DMEM (Dulbecco's Modified Eagle's Medium) High Glucose, L-Glut, no pyruvate	Hyclone	SH30022.02	Life Technologies
FBS (Fetal Bovine Serum)	Hyclone	SH30070.03	BSG (Bovine Growth Serum) Hyclone# SH30541.03
MEM-Non-essential amino acids (100×)	Hyclone	SH30238.01	Life Technologies
Pen/Strep	Hyclone	SV30010	Life Technologies
D-PBS	Hyclone	SH30028.03	Life Technologies
Trypsin/EDTA 0.05%	Hyclone	SH30236.01	Life Technologies
Mitomycin C	Sigma	M 0503	
DMSO	Sigma	D2650	

Tissue Culture Disposables

Item	Supplier	Catalog #	Alternative
10 cm Petri dish	Corning	430167	many
T75 flask	Corning	430641	many
T175 flask	Corning	431079	many
15 mL conical tube	Corning	430053	many
50 mL conical tube	Corning	430291	many
250 mL 0.22 μm, PES filter unit	Corning	431096	many
500 mL 0.22 μm, PES filter unit	Corning	431097	many
2.0 mL cryogenic vials (internal thread)	Corning	2028	many
Permanent cryogenic-storage labels	Diversified Biotech	LCRY-1700	many

RECIPES

Stock Solutions

Component	Amount	Stock Concentration
Mitomycin C	2 mg/200 mL fibroblast medium	10 µg/mL in fibroblast medium

Fibroblast Medium 500 mL

Component	Amount	Final Concentration
DMEM (Formulation contains L-Glutamine)	440 mL	
FBS	50 mL	10%
Non-essential amino acids (100X)	5 mL	0.1 mM
Penicillin/Streptomycin (100X)	5 mL	10 U Penicillin/10 µg Streptomycin

Note: the penicillin/streptomycin should be removed from the medium a few days after isolating MEFs from embryos.

2× Cryopreservation Medium 10 mL

Component	Amount	Final Concentration
Fibroblast medium	8 mL	80%
DMSO	2 mL	20%

0.1% Gelatin Solution 100 mL

Component	Amount	Final Concentration
Gelatin	0.1 g/100 mL water	0.1%

READING LIST

Isolation and Use of Mouse Embryonic Fibroblasts (MEFs) for Culturing hPSCs: Protocols Describing Isolation of MEFs

Nagy, A., Gertsenstein, M., Vintersten, K., Behringer, R. (Eds.), 2002. Manipulating the Mouse Embryo: A Laboratory Manual, third ed. Cold Spring Harbor Laboratory Press.

Robertson, E.J., 1987. Embryo derived stem cell lines. In: Robertson, E.J. (Ed.), Teratocarcinoma and Embryonic Stem Cells: A Practical Approach. IRL Press, Oxford, UK.

Mouse ES Cell Derivation

Evans, M.J., Kauffman, M.H., 1981. Establishment in culture of pluripotent cells from mouse embryos. Nature 292, 154–156.

Martin, G., 1981. Isolation of a pluripotent cell line from early mouse embryos cultured in medium conditioned by teratocarcinoma stem cells. Proc. Natl. Acad. Sci. USA 78, 7634–7638.

STO and SNL76/7 Feeder Cell Lines

STO: Bernstein, A., MacCormick, R., Martin, G.S., 1976. Transformation-defective mutants of avian sarcoma viruses: the genetic relationship between conditional and nonconditional mutants. Virology 70, 206–209.

SNL 76/7: McMahon, A.P., Bradley, A., 1990. The Wnt-1 (int-1) proto-oncogene is required for development of a large region of the mouse brain. Cell 62, 1073–1085.

Web Resources

NIH ES Cell Registry. <http://stemcells.nih.gov/research/registry/>

Culture of Human Pluripotent Stem Cells in Feeder-Free Conditions

Ibon Garitaonandia, Francesca S. Boscolo, and Ha T. Tran

EDITOR'S COMMENTARY

For nearly 20 years before the first human pluripotent stem cell lines were derived, mouse pluripotent stem cells had been routinely cultured on feeder layers of mitotically inactivated mouse embryo fibroblasts or mouse cell lines. The same methods were used for the first hPSC lines, and continue to be used in most routine hPSC culture. As described in Chapter 1, culture of hPSCs on mouse fibroblast feeder layers, with mechanical passaging, is still the best method to prepare cells for banking, since these methods are the least selective for faster growing cells, and thus give the best chance of maintaining genomic and epigenetic integrity of the cells. Feeder-free culture systems have been refined over the past several years, and specialized culture media and specific enzymes have made it straightforward to expand hPSCs using enzymatic passaging in the absence of feeder layers. Although it is far more expensive than the traditional methods, feeder-free culture is important for certain applications. Methods for transduction of hPSCs recommend using feeder-free culture, since the feeder cells compete with the hPSCs for DNA vectors. Feeder cells can interfere with differentiation protocols by producing pluripotency-enhancing factors. Probably the most important application of feeder-free culture methods is for generating cells for cell replacement therapy. These methods make it simpler to scale up production of the cells for clinical use, and mitigate concerns about antigenicity and potential transfer of viruses from mouse cells to human cells. Most laboratories currently use traditional culture methods for expanding their cells, switching to the more expensive feeder-free systems only when it is necessary for differentiation protocols or transductions. A major concern for all hPSC culture methods is the accumulation of genomic and epigenetic errors with expansion of the cells in culture; enzymatic passaging does seem to allow for selection of faster growing cells, and these cells usually have at least small genomic changes. The best way to deal with this concern is to monitor the cells at regular intervals and keep

J.F. Loring & S.E. Peterson (eds): Human Stem Cell Manual, Second edition.
DOI: http://dx.doi.org/10.1016/B978-0-12-385473-5.00003-5

a record of genomic and epigenetic changes. Chapters 14 and 20 describe methods for genomic and epigenetic analysis.

OVERVIEW

There are multiple methods for culturing hPSCs under feeder-free conditions. Here we describe a system using StemPro medium and Geltrex or Matrigel substrata with Accutase passaging. In our hands hPSCs cultured this way remain pluripotent and karyotypically normal for more than 40 passages. The advantage of Accutase passaging is that hPSC cultures are dissociated into single cell suspensions; this allows scale-up and applications such as transfection, FACS sorting, and cloning. Here we explain the appropriate hPSC feeder-free culture technique, recommend cell seeding densities, and show examples of colonies with good and bad morphology. Alternative feeder-free culture methods are discussed, such as alternative media, matrices, and passaging methods. We also provide a protocol for adapting hPSC cultures from feeder layers to feeder-free conditions, which can sometimes be a roadblock for many laboratories. Finally, we describe a method for the successful cryopreservation of hPSCs cultured under feeder-free conditions.

PROCEDURES

In this chapter we describe how to culture hPSCs with StemPro® hESC SFM (Life Technologies) serum-free medium on Geltrex™ (Life Technologies) or Matrigel™ (BD) extracellular matrix substratum. StemPro medium's exact formulation is proprietary, but it contains IGF1, Heregulin1, and Activin A, which act through tyrosine kinase pathways to maintain pluripotency.

Preparation of Geltrex/BD Matrigel Coated Culture Plates

1. Thaw frozen Geltrex/BD Matrigel 2 hours on ice or overnight at 4°C.
2. Dilute 1 mL of Geltrex/BD Matrigel into 29 mL of cold DMEM medium (1:30 final concentration). These matrices polymerize at room temperature, so they should always be kept cold at 4°C. The manufacturer's recommended dilution is 1:30, but hPSCs can be cultured successfully with higher dilutions (1:100–1:300).
3. Add a sufficient amount of diluted Geltrex/BD Matrigel to cover the growth surface area of the dish and coat the dish at 37°C for at least 1 hour.
4. The coated dish can be sealed with Parafilm and stored at 4°C for up to 2 weeks. It is very important to have enough Geltrex/BD Matrigel covering the surface to prevent it from drying.

5. Before use, place the Geltrex/BD Matrigel coated plate at room temperature for at least 15 minutes. Aspirate Geltrex and plate cells in culture medium.

Culture of hPSCs and Accutase Passaging

1. Observe cultures daily. hPSCs cultured on Geltrex/Matrigel have a different morphology than cells cultured on feeder layers. Feeder-free hPSCs do not grow as the characteristic round colonies with defined edges like cells on feeder layers. Instead, they grow as a monolayer until they reach confluency (Figure 3.1).
2. Change the medium daily with StemPro supplemented with 12 ng/mL of bFGF. Lower concentrations of bFGF can be used, but concentrations lower than 8 ng/mL may cause the hPSC culture to differentiate (Figure 3.2).
3. When cells are confluent, passage the cells (Figure 3.3).
4. Aspirate the medium and wash cells with DPBS.
5. Aspirate DPBS and add 2 mL of Accutase per well of a six-well plate.
6. Incubate cells with Accutase at 37°C for 2 to 5 minutes. Monitor the culture and stop incubation when the cells start to round up.
7. Rinse gently to remove the cells from the plate and transfer the cell suspension to a 15 mL conical tube.
8. Add 2 mL of medium to rinse any remaining cells from the dish's surface and transfer to a 15 mL conical tube.
9. Centrifuge the conical tube containing cell suspension at 200 g for 2 to 3 minutes.

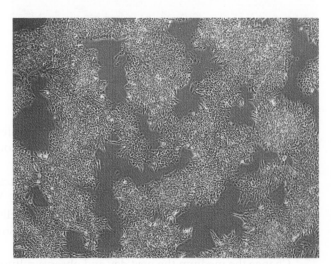

FIGURE 3.1
Example of hPSC culture on Geltrex. Colonies lack the round morphology seen with hPSCs cultured on feeder layers.

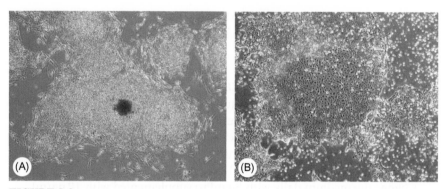

FIGURE 3.2

Examples of differentiated hPSC cultures in feeder-free conditions. Differentiated cells are present at the center of the colony in both A and B. (A) Cells have piled up in the center of the colony making it appear dark. (B) The center of the colony is populated by large, differentiated cells that give the colony a "leopard print" appearance.

FIGURE 3.3

Confluent well of hPSCs on Geltrex. Note that the cells retain a uniform appearance across the well.

10. Aspirate the supernatant and resuspend the cell pellet in StemPro by gently pipetting the cells to break up the clumps. Count the cells and determine cell viability.

11. Plate the desired cell density on Geltrex-coated plates and incubate in a humidified atmosphere of 5% CO_2 in air at 37°C. We recommend passaging the cells at a density of around 20,000 cells/cm^2, but this requires optimization for your particular cell line. Some cell lines might require much higher seeding densities.

12. Change the medium the next day and observe cultures (Figure 3.4).

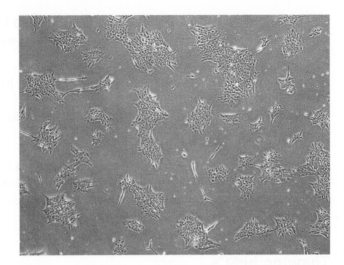

FIGURE 3.4
Morphology of hPSC culture one day after passaging with Accutase. Colonies have a star-shaped morphology.

Adapting hPSCs from Feeder Layers to Feeder-Free Conditions

Adapting hPSC cultures to feeder-free conditions can be a complicated process that requires optimization for each particular cell line. Transferring hPSCs from feeder layers to matrices such as Geltrex or Matrigel requires conditioning the cells with StemPro while still on feeder layers.

1. Three days prior to transferring hPSCs to feeder-free conditions, feed cells with a mixture of 75% standard hESC medium and 25% StemPro, while keeping the final concentration of bFGF the same, in our case 12 ng/mL.
2. Next day, increase the StemPro ratio to 50% when feeding.
3. One day prior to transferring cells to feeder-free conditions, feed the cells with a mixture of 25% standard hESC medium and 75% StemPro.
4. On the day of the transfer, feed cells with 10 µM ROCK Inhibitor (also known as Y27632) in 100% StemPro. ROCK Inhibitor will increase the survival of the cells after the transfer.
5. Passage the cells mechanically with a pipette tip or StemPro EZPassage tool and seed cells on a Geltrex/Matrigel coated plate. Seeding densities should be much higher than usual because a great deal of cell death is expected in the transfer. We recommend a seeding density of at least 100,000 cells/cm^2.
6. Feed cells the next day with 100% StemPro to remove dead cells and debris.
7. Feed cells daily and passage when the culture becomes confluent. At this point cells can be passaged with Accutase but depending on the cell line they might require mechanical passaging for a few more passages. Before passaging, remove differentiated cells from the culture. This can be done mechanically with a pipette tip under the dissecting microscope.

Cryopreservation of hPSCs in Feeder-Free Conditions

To increase the survival of hPSCs after cryopreservation, we recommend using ROCK Inhibitor.

1. One hour before harvesting, refresh the culture with 10 μM ROCK Inhibitor in StemPro.
2. Aspirate the culture medium, wash with DPBS, and harvest cells with Accutase.
3. Centrifuge cell suspension at 200 g for 3 minutes.
4. Resuspend cell pellet with StemPro and count with a hemocytometer.
5. Centrifuge again and resuspend cells at the desired density with 10% DMSO in knockout serum replacement (KSR). We recommend freezing at a density of 10^6 cells/mL.

Thawing hPSCs

1. Immerse a frozen vial of cells in a 37 °C water bath without submerging the vial below the O-ring on the cap.
2. When only a few ice crystals remain, remove the vial from the water bath and spray carefully with 70% ethanol to sterilize.
3. In the hood, slowly transfer the cells from the vial to a 15 mL conical tube containing 10 mL of StemPro.
4. Centrifuge cell suspension at 200 g for 3 minutes to remove DMSO.
5. Plate at the desired density and incubate cells for 24 hours with 10 μM ROCK Inhibitor. We recommend seeding cells at higher density than usual when thawing. If necessary, 100,000 cells/cm^2 can be plated to increase survival.
6. Change medium the next day to remove ROCK Inhibitor. Do not incubate cells with ROCK Inhibitor for more than 24 hours.

ALTERNATIVE PROCEDURES

Alternative Media

mTeSR®1 is also a very widely used medium for culturing hPSCs under feeder-free conditions. mTeSR1 is a complete, serum-free, defined formulation for use with BD Matrigel hESC-qualified Matrix. Unlike StemPro, mTeSR1 already contains a high concentration of bFGF (100 ng/mL), which is much higher than our standard concentration of 12 ng/mL. Other defined media for feeder-free culture include X-VIVO™, NutriStem™, and STEMium™. For clinical applications, xeno-free media such as TeSR2™ or KnockOut™ SR XenoFree CTS™ are available.

Alternative Surfaces

Geltrex and BD Matrigel are matrices derived from animal products. For clinical purposes, xeno-free surfaces such as CELLstart™ CTS and Synthemax™ might be preferable. CELLstart CTS is a defined substrate that contains components of human origin only and has been produced under cGMP (Current Good Manufacturing Practice). Plates are coated with CELLstart the same way as with Geltrex/BD Matrigel. The Corning Synthemax Surface is a synthetic surface coated onto multiple well plates. In Synthemax, vitronectin is covalently linked to a synthetic acrylate polymer surface to mimic biological ligands for cell adhesion. hPSCs grown on CELLstart and Synthemax are reported to remain pluripotent and can be scaled-up.

Alternative Passaging Methods

Other enzymes commonly used in feeder-free culture are dispase, collagenase IV, and Trypsin-like Enzyme (TrypLE). TrypLE is a recombinant fungal serine protease with trypsin-like activity. It has been successfully used for the transition of mechanical to single cell enzymatic passaging. Dispase and collagenase IV, on the other hand, are used to passage cells as clumps.

Collagenase IV Passaging

1. When cells are confluent, aspirate medium and add 1 mL of collagenase IV (1 mg/mL or 200 units/mL) per well of six-well plate.
2. Incubate 5–10 minutes at 37°C and monitor the culture during incubation and stop treatment when the edges of the colony begin to curl up and loosen from the plate.
3. Aspirate collagenase and rinse with 2 mL of PBS taking care not to wash off the loosened colonies.
4. Add 2 mL of medium per well and gently scrape the well with a cell scraper.
5. Collect most of the cells from the well and transfer to a 15 mL conical tube.
6. Gently pipette the cells to break up the clumps but do not make a single cell suspension.
7. Seed the cells at 1:6 dilution onto Geltrex/Matrigel-coated plates.

Non-enzymatic cell dissociation methods are also available, such as Cell Dissociation Buffer. The Cell Dissociation Buffer is an enzyme-free aqueous formulation of salts, chelating agents, and glycerol in Ca^{2+} and Mg^{2+} free phosphate-buffered saline. This buffer is suitable for assays that require intact cell surface proteins such as flow cytometry, ligand binding, and immunocytochemistry. When using this buffer, remove all the protein-containing medium, rinse, and then cover the well with Cell Dissociation Buffer. Monitor the culture under the microscope until the edges of the colonies begin to lift.

Some hPSC lines might not survive with any of the passaging methods described here. The last alternative for passaging the cells under feeder-free conditions is mechanical passage. This can be done with a pipette tip or an EZPassage tool. We recommend the latter because it cuts the cells into uniform pieces. Differentiated cells are first removed, fresh medium is added, and the cells are passaged under the dissecting microscope.

PITFALLS AND ADVICE

Geltrex

Extracellular matrices (ECMs) such as Geltrex and BD Matrigel are undefined mixtures of extracellular matrix proteins isolated from the Engelbreth-Holm-Swarm mouse sarcoma cell line. The major components are laminin, collagen IV, heparin sulfate, proteoglycans, and entactin. Geltrex and BD Matrigel polymerize at room temperature to produce biologically active matrix material resembling the mammalian cellular basement membrane that not only supports cells and cell layers, but also plays an essential role in tissue organization that affects cell adhesion, migration, proliferation, and differentiation.

Unfortunately there is inherent lot-to-lot variability in any cell-derived product even if the manufacturer tries to minimize it through quality control measures. Lot-to-lot variability of Geltrex/BD Matrigel can affect the speed at which the matrix polymerizes, the concentration of the matrix plated, the length of time the plates need to be incubated, how long plates can be stored prior to use, and often, the pluripotency of hPSC cultures.

Growth Factors

Care should be taken when diluting and aliquoting protein growth factors because they can be easily degraded. Coat pipettes and filters with 0.2% BSA in PBS prior to pipetting or filtering growth factors to lower the likelihood of the growth factors sticking to the pipettes and filters. Growth factors should be reconstituted and stored in a solution containing albumin, for example 0.2% BSA, to increase their shelf life.

EQUIPMENT

- Class II biosafety cabinet
- Tissue culture incubator, 37°C, 5% CO_2
- Inverted phase contrast microscope with 4×, 10×, and 20× objectives

- Dissecting Microscope with Dark Field Transillumination and 0.8–4.5× range
- Tabletop centrifuge
- Access to 4°C, −20°C, −80°C, and cryogenic freezers
- Pipette Aid and Micropipettors, p2, p20, p200, and p1000

REAGENTS AND SUPPLIES

Recommended Reagents and Supplies

Item	Supplier	Catalog #	Note
StemPro hESC SFM kit	Life Technologies	A1000701	
Geltrex hESC-qualified Reduced Growth Factor Basement Membrane Matrix	Life Technologies	A10480-02	
hESC-qualified Matrix, 5 ml *LDEV-Free	BD Biosciences	354277	
Dulbecco's Phosphate-Buffered Saline (DPBS) without Calcium and Magnesium	Life Technologies	14190-250	
2-Mercaptoethanol	Life Technologies	21985-023	
StemPro Accutase Cell Dissociation Reagent	Life Technologies	A11105-01	Thaw and warm at room temperature, not 37°C
Recombinant Human bFGF	Life Technologies	13256-029	Use at 12 ng/mL
mTeSR1	Stem Cell Technologies	05850	
TeSR2	Stem Cell Technologies	05860	
KnockOut SR XenoFree CTS Kit	Life Technologies	A1099201	
StemPro EZPassage-Disposable Stem Cell Passaging Tool	Life Technologies	23181-010	
TrypLE	Life Technologies	12563-011	
Collagenase type IV	Life Technologies	17104-019	
Dispase	Life Technologies	17105-041	
Cell Dissociation Buffer	Life Technologies	13150016	
Coastar 6-Well Clear TC-Treated Multiple Well Plates, Sterile	Corning	3506	
Sterile conical tubes, 50 mL	Corning	430291	
Sterile conical tubes, 15 mL	Corning	430053	
Sterile-filtered Dimethyl Sulfoxide	Sigma	D2650	
KnockOut Serum Replacement (KSR)	Life Technologies	10828-028	
500 mL Vacuum Filter/Storage Bottle System, sterile low protein binding 0.22 μm pore, 33.2 cm² PES Membrane	Corning	431097	
CELLstart CTS	Life Technologies	A10142-01	
Stemolecule™ Y-27632	Stemgent	04-0012	
Synthemax Surface	Corning	3876XX1	

RECIPES

Stock Solution Human Basic FGF (FGF2) 12 µg/mL

Component	Amount	Stock Concentration
Human bFGF	10 µg	12 µg/mL in 0.2% BSA in PBS

1. Dissolve 10 µg of human basic FGF in 0.83 mL in 0.2% BSA in PBS.
2. Aliquot in 50 µL samples and store at −20°C or −80°C. When aliquoting, pre-wet pipette tips, tubes, and filters with 0.2% BSA in PBS to prevent loss of growth factor.
3. Store thawed aliquots at 4°C and use them in less than 2 weeks.

Stock Solution Geltrex or Matrigel (1:2)

Component	Amount	Stock Concentration
Geltrex or Matrigel	5 mL	1:2 in DMEM

1. Thaw frozen Geltrex overnight at 4°C.
2. Chill on ice pipettes and 15 mL conical tubes.
3. Resuspend on ice 5 mL of Geltrex with 5 mL of ice cold DMEM. Everything should be cold to avoid polymerization of Geltrex.
4. Aliquot 1 mL of solution in each conical tube and store at −20°C.

Stock Solution Collagenase IV (200 units/mL)

Component	Amount	Stock Concentration
Collagenase IV	20,000 units	200 units/mL

1. Dissolve 20,000 units of Collagenase IV in 100 mL of DMEM which makes a 1 mg/mL solution.
2. Filter solution through a 0.2 µm filter.
3. Aliquot solution and store at −20°C.

QUALITY CONTROL METHODS

Mycoplasma

See Chapter 4 for detailed information about mycoplasma and other contaminating microbes. It is important to monitor periodically for mycoplasma contamination. Mycoplasma contamination can have disastrous effects on hPSCs because it can alter cell metabolism, gene expression, and antigenicity. The experimental results obtained with hPSC cultures contaminated with mycoplasma are unreliable, and these cells are unsafe for therapeutic purposes. If hPSC cultures are not tested for mycoplasma contamination, mycoplasma can remain in culture for a long time because it cannot be detected by visual inspection through a phase contrast microscope. Mycoplasma contamination can be eliminated with reagents such as Plasmocin™ and Plasmocure™ (InvivoGen), but it is recommended to destroy hPSC cultures that are mycoplasma positive because it is highly infectious and cross-contamination is common. Mycoplasma is highly infectious because it can pass through a 0.2 μm filter, is unaffected by standard antibiotics such as penicillin and streptomycin because it lacks a cell wall, and it does not produce turbidity in the media like other bacteria.

Mycoplasma can be detected through simple DAPI or Hoescht staining. In a contaminated culture, the hPSC nuclei are surrounded by multiple fluorescing structures in the cytoplasm. To be absolutely certain there is no mycoplasma, there are other commercial methods such as the MycoAlert™ Mycoplasma Detection Kit (Lonza), MycoFluor™ Mycoplasma Detection Kit (Life Technologies), MycoSensor™ PCR Assay kit (Agilent Technologies), and Mycoplasma Detection Kit (Roche Applied Science). One of the most common sources of contamination is when new cultures are introduced into the laboratory. It is best to quarantine these cultures in a separate incubator until mycoplasma is tested for and proven to be absent. During routine culture, the best way to avoid mycoplasma contamination is through good aseptic technique and periodic testing. Test for mycoplasma when freezing and thawing a cell line and monthly during culture.

Karyotyping

See Chapter 13 for methods of karyotyping. A big concern for stem cell based therapies is that transplanted cells may be tumorigenic (Fox, 2008). With time in culture, hPSCs can accumulate genetic and epigenetic changes that can be associated with cancers, which would make the cell preparations useless for clinical applications (Maitra et al., 2005; Mitalipova et al., 2005). Therefore, when culturing hPSCs it is very important to periodically karyotype the cells for chromosomal abnormalities. Karyotyping should be performed every 10–20 passages and companies such as Cell Line Genetics and WiCell provide these services. Unfortunately, hPSCs can also accumulate subchromosomal genetic changes that cannot be detected by karyotyping. The subchromosomal aberrations can affect regions associated with pluripotency, such as NANOG on chromosome 12 and DNMT3B on chromosome 20 (Hussein et al., 2011; Laurent et al., 2011; Lefort et al., 2008; Mayshar et al., 2010; Narva et al., 2010). To detect subchromosomal genetic changes, the hPSCs can be sequenced or SNP genotyped. DNA sequencing is still expensive, whereas SNP genotyping microarray platforms such as Illumina's HumanOmni5-Quad BeadChip provide a very comprehensive coverage of the genome for a reasonable price (see Chapter 14).

Lot-to-Lot Variability of Reagents

Many of the reagents used in the culture and maintenance of hPSCs are derived from animal sources, so it is important to keep in mind that there is inherent lot-to-lot variability of the reagents. Vendors try to control this as much as possible but even when a product passes inspection, hPSC cultures can still be sensitive to the variability within the inspection-approved range. Record the lot numbers of all the reagents used so in the instance an experiment cannot be replicated or the cells do not survive, the problem can be traced to a bad reagent. Always inform the vendor of the problems with their reagents and request replacements. Vendors need feedback from customers in order to improve their own quality control.

READING LIST

Fox, J.L., 2008. FDA scrutinizes human stem cell therapies. Nat. Biotechnol. 26, 598–599.

Hussein, S.M., Batada, N.N., Vuoristo, S., Ching, R.W., Autio, R., Narva, E., et al., 2011. Copy number variation and selection during reprogramming to pluripotency. Nature 471, 58–62.

Laurent, L.C., Ulitsky, I., Slavin, I., Tran, H., Schork, A., Morey, R., et al., 2011. Dynamic changes in the copy number of pluripotency and cell proliferation genes in human ESCs and iPSCs during reprogramming and time in culture. Cell. Stem. Cell. 8, 106–118.

Lefort, N., Feyeux, M., Bas, C., Feraud, O., Bennaceur-Griscelli, A., Tachdjian, G., et al., 2008. Human embryonic stem cells reveal recurrent genomic instability at 20q11.21. Nat. Biotechnol. 26, 1364–1366.

Li, Y., Powell, S., Brunette, E., Lebkowski, J., Mandalam, R., 2005. Expansion of human embryonic stem cells in defined serum-free medium devoid of animal-derived products. Biotech. Bioeng. 91, 688–698.

Lu, J., Hou, R., Booth, J.C., Yang, S.H., Snyder, M., 2006. Defined culture conditions of human embryonic stem cells. Proc. Natl. Acad. Sci. USA 103, 5688–5693.

Ludwig, T.E., Levenstein, M.E., Jones, J.M., Berggren, T.W., Mitchen, E.R., Frane, J.L., et al., 2006a. Derivation of human embryonic stem cells in defined conditions. Nat. Biotechnol. 24, 185–187.

Ludwig, T.E., Bergendahl, V., Levenstein, M.E., Yu, J., Probasco, M.D., Thomson, J.A., 2006b. Feeder-independent culture of human embryonic stem cells. Nat. Methods. 8, 637–646.

Maitra, A., Arking, D.E., Shivapurkar, N., Ikeda, M., Stastny, V., Kassauei, K., et al., 2005. Genomic alterations in cultured human embryonic stem cells. Nat. Genet. 37, 1099–1103.

Mayshar, Y., Ben-David, U., Lavon, N., Biancotti, J.-C., Yakir, B., Clark, A., et al., 2010. Identification and classification of chromosomal aberrations in human induced pluripotent stem cells. Cell. Stem. Cell. 7, 521–531.

Mitalipova, M.M., Rao, R.R., Hoyer, D.M., Johnson, J.A., Meisner, L.F., Jones, K.L., et al., 2005. Preserving the genetic integrity of human embryonic stem cells. Nat. Biotechnol. 23, 19–20.

Narva, E., Autio, R., Rahkonen, N., Kong, L., Harrison, N., Kitsberg, D., et al., 2010. High-resolution DNA analysis of human embryonic stem cell lines reveals culture-induced copy number changes and loss of heterozygosity. Nat. Biotechnol. 28, 371–377.

Xu, C., Inokuma, M., Denham, J., Golds, K., Kundu, P., Golds, J., et al., 2001. Feeder-free growth of undifferentiated human embryonic stem cells. Nat. Biotechnol. 19, 971–974.

CHAPTER 4

Contamination

Ileana Slavin and John P. Schell

EDITOR'S COMMENTARY

As biomedical research continues to develop, cell culture remains a fundamental part of experimental analysis and biological manufacturing. Biotechnology and pharmaceutical industries rely heavily on sterile, reproducible culturing methods for the production of vaccines, recombinant proteins, antibodies, and more. Research needs an equally high standard for cell culture in order to minimize experimental variability. Cell culture contamination is a constant threat that arises from exposure to microbial contaminants such as bacteria, fungus, and viruses. Contamination can also result from invasive, undesired cell types, and has resulted in frequent misidentification of cell lines. The spread of contamination has the potential to ruin a lab because eradicating it can be very time-consuming and expensive, and any publications or grants that resulted from work on contaminated cells are called into question. In order to avoid contamination, a combination of proper cell culture practice and strict aseptic techniques are required.

Human pluripotent stem cells (hPSCs) are generally cultured without antibiotics, so contamination can be recognized early. The bane of any hPSC lab is mycoplasma. We cannot emphasize strongly enough that ALL cultures in an hPSC lab must be tested for mycoplasma at least once a month. All cells that are brought in from other labs, even if they are commercial sources, need to be quarantined until they are shown to be mycoplasma negative. Mycoplasma is particularly insidious; it is intracellular and is not obvious in cultures until they are badly contaminated. Mycoplasma does not have cell walls, so it can pass through the usual $0.2\,\mu M$ pore filters used to sterilize solutions, and it often spreads through shared hoods and incubators. If you receive a cell sample from another lab and discover that it is contaminated with mycoplasma, inform that lab; they will want to check their other cultures.

J.F. Loring & S.E. Peterson (eds): Human Stem Cell Manual, Second edition.
DOI: http://dx.doi.org/10.1016/B978-0-12-385473-5.00004-7

OVERVIEW

This chapter reviews the signs and sources of cell culture contamination, while identifying strategies for prevention. Contaminants are any undesirable invaders of the culture that have the potential to produce adverse effects. Microbial contaminants have the ability to influence gene expression, proliferation rates, apoptotic response, and morphology, as well as many other subtle differences that can negatively impact experimental continuity. Cross-contamination of cell lines is unfortunately very common but can be easily prevented. It often occurs because of lack of attention; one of the most common ways in which cell cultures are cross-contaminated is by feeding different cell cultures using the same aspiration pipette.

Microbial contaminants can be divided into two groups, based on whether or not they can be visually detected in culture (Ryan, 2008):

- Easily detectable contaminants such as bacteria, molds and yeast.
- Less apparent contaminants such as viruses, protozoa and mycoplasma.

The length of time that a biological contaminant escapes detection will determine the extent of the harmful effects generated. Since contamination will always be a threat, all efforts must be made to reduce its occurrence.

IDENTIFYING CONTAMINATION

Bacterial Contamination

Bacteria include a large variety of unicellular microorganisms. They are typically a few micrometers in diameter, and are found in a range of shapes including spheres, rods, and spirals (Rappe and Giovannoni, 2003). Bacteria are ubiquitous in the environment and are able to rapidly colonize and divide in nutrient rich cell culture media. Their small size and fast growth rates make these microbes the most commonly encountered biological contaminants in cell culture. In the absence of antibiotics, bacteria can be easily detected in a culture within a few days of initial contamination. Bacterial contamination can be detected through direct observation of the cells in culture (Figure 4.1).

Characteristics of bacterial contamination include:

- Infected cultures appear turbid, sometimes with a thin film on the surface.
- At moderate power in a phase contrast microscope (20×) bacteria are easily detected as tiny, moving granules between the cells.
- The pH of the culture medium frequently decreases, thus rendering culture conditions abnormally acidic.
- When an incubator containing cultures infected with bacteria is first opened, there is often a foul scent.

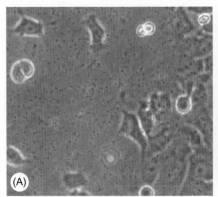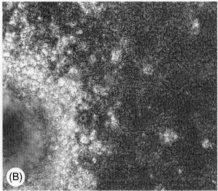

FIGURE 4.1
(A) Early signs of bacterial contamination can be observed under high magnification as tiny actively moving particles. (B) A human embryoid body culture contaminated with bacteria.

It is strongly recommended that cultures contaminated with bacteria are treated with bleach and discarded immediately. While it may be possible in some situations to recover important cells that have become mildly contaminated, it is always a better idea to create stocks of cryopreserved cells and revert back to an uncontaminated stock if possible. If bacterial contamination occurs in a cell line that cannot be replaced, antibiotics can be administered in an attempt to salvage the infected samples (see Chapter 7). Once the cells are free of contamination, they should be QC tested to ensure that the relevant properties of the cell line have been retained.

Fungal Contamination: Yeast and Mold

Yeast and mold are eukaryotic microorganisms of the kingdom Fungi. Their size can vary greatly depending on the species, typically measuring 3–4 μm in diameter, although some yeast can reach over 40 μm. Like bacterial contamination, cultures contaminated with fungi become turbid, especially if allowed to proliferate unchecked. Fungal contamination does not cause a pH change initially, although the pH will increase as the contamination spreads and the culture becomes more basic.

Under the microscope, fungus initially appears as individual ovoid or spherical particles. Fungi may grow as long branching filamentous structures called hyphae. A connected network of these multicellular filaments contains genetically identical nuclei, and is referred to as a colony or mycelium. Under magnification, a connected network of these multicellular filaments (mycelia) usually appears as thin, wisp-like threads (Figure 4.2). Small amounts of contamination can be treated with antimycotics, although cultures exhibiting pronounced fungal contamination should be eliminated immediately by treatment with bleach.

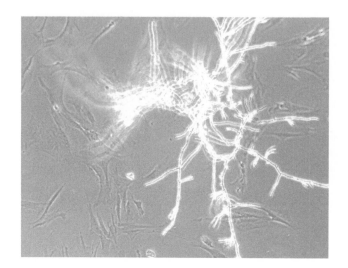

FIGURE 4.2
Branching hyphae, indicative of fungal contamination in the culture.

Viral Contamination

Viruses are microscopic infectious agents that utilize their host cell's machinery in order to self-replicate. Due to their extremely small size (~200 nm), viruses are the most difficult contaminant to detect, as well as the most complicated to remove. Most viruses have very stringent requirements for their original host species' cellular machinery. While this generally limits the spread of viral contaminants to non-host species, it can also present a serious health hazard to the laboratory personnel culturing human cells (Ryan, 2008). Viral infection of cell cultures can be detected by electron microscopy, immunostaining with a panel of antibodies, ELISA assays, or PCR with appropriate viral primers, or discovered through genome sequencing.

Mycoplasma

Mycoplasmas are small (0.2–0.3 μm) intracellular bacteria that attach to the cell membrane, inhibit cell growth, and eventually lead to cell death. They depend on their hosts for many nutrients due to their limited biosynthetic capabilities. Because these parasites do not have cell walls, do not grow in colonies, and do not change the pH of the medium, testing must be done to detect them in culture. In contrast to other bacteria, mycoplasma grows very slowly, ranging from 1 to 3 hours per division. Despite longer replication times, they can still multiply to high concentrations (10^7 to 10^8 organisms/mL) and adversely affect culture conditions. Mycoplasma can alter cell growth characteristics, inhibit cell metabolism, disrupt nucleic acid synthesis, induce chromosomal aberrations, change cell membrane antigenicity, and even alter transfection rates and viral susceptibility (Drexler and Uphoff, 2002; Phelan, 2007).

Mycoplasma contamination of cell cultures is a frequent occurrence that often goes unobserved. Unlike bacterial or fungal contamination, mycoplasma is not visible to the naked eye. Mycoplasma-infected cell lines are themselves the single largest source for perpetuating the spread of the contamination. Mycoplasma is spread through laboratory equipment, media, and reagents that have been contaminated by previous use in processing mycoplasma-infected cells (Drexler and Uphoff, 2002). Their invisibility, the high concentrations of mycoplasma in infected cultures and their ability to survive in a dried state are natural characteristics that perpetuate their continued existence in cell culture laboratories.

Mycoplasma Testing Methods

A number of techniques have been developed in order to detect mycoplasma contamination in cell cultures.

1. *DNA Staining Method:* This method uses Hoechst dye to highlight the A-T rich DNA of mycoplasma, so that mycoplasma appear as bright extranuclear spots in the cytoplasm. Results are quick, but this method is not as sensitive as other methods and can be difficult to interpret due to background bacterial/yeast/fungal contamination, excess debris, reduced or absent live cells, and broken nuclei from dead cells (Phelan, 2007).

2. *Luciferase-Based Screening for Mycoplasma-Specific Enzymes:* The advantage of this method is that it allows the rapid detection of active enzymes that are unique to mycoplasma. With this method, the viable mycoplasma are lysed and their enzymes are allowed to react with a substrate that catalyzes the conversion of ADP to ATP. The ATP is measured and quantified by a bioluminescent reaction between Luciferin and the ATP produced. If the samples show positive or borderline results, the presence of mycoplasma needs to be further validated by PCR.

3. *PCR Methods:* The most reliable detection method uses polymerase chain reaction (PCR) to target mycoplasma's highly conserved genes. This can be done in a single-step PCR or in a two-step (nested) PCR. It is important to note that the PCR may produce false-positive results due to contamination with target DNA or false-negative results caused by inhibition of the Taq enzyme. However, once all PCR-related problems are properly addressed, we find that single-step or double-step PCR is clearly superior to all other mycoplasma detection methods in many respects, as this method combines simplicity and speed with high specificity and extreme sensitivity (Hopert et al., 1993). One drawback, however, is that this method is expensive. Because of the expense, we recommend routine screening using a luciferase-based enzymatic assay and then testing all borderline and positive test results using PCR.

Precautions to Avoid Mycoplasma in the Lab

Mycoplasma contamination is one of the most serious problems in a research laboratory because it is not visible by routine microscopy. Because of this, all cell culture laboratories should have a mycoplasma screening program in operation, especially those collecting tissue for primary culture. There are several precautions that should be taken (Drexler and Uphoff, 2002; Freshney, 2006):

- Treat any new material entering the laboratory from donors or from other laboratories as potentially infected and keep it in quarantine.
- A separate room should be set aside for receiving samples and imported cultures until the contamination status is verified.
- Testing for microbial contamination should be integrated into a cell culture program as part of routine quality control measures. Testing for microbial contamination should be performed after the cells have been cultured in the absence of antibiotics for several weeks.
- The general use of antibiotics is not recommended except in special applications, and then only for short durations. Use of antibiotics may lead to lapses in aseptic technique, selection of drug-resistant organisms, and delayed detection of low-level infection by either mycoplasma or other bacteria.
- Master stocks of mycoplasma-free cell lines should be frozen and stored to provide a continuous supply of cells that are available should working stocks become contaminated.
- Actively growing mycoplasma-infected cell lines should be discarded as quickly as possible in order to prevent lateral spread and should be replaced by fresh stocks known to be mycoplasma-free.
- If multiple cultures appear to have been independently infected, the best procedure is to shut down the lab and thoroughly clean all hoods, incubators, and equipment. Destroy all contaminated cells.

Cell Cross-Contamination

Although most researchers are aware of the possibility of cross-contamination with other cell types and are cautious when handling cells, accidents are inevitable. Cell lines get mislabeled or contaminated with fast-growing cells that can quickly take over the original lines (Chatterjee, 2007). This problem can be solved simply by increasing awareness and introducing regular quality control measures. DNA fingerprinting by STR (short tandem repeat) or SNP (single nucleotide polymorphism) genotyping has become the standard tool for authenticating cell lines. A genetic profile provides scientists with an authentic signature to verify the identity of the lines (Chatterjee, 2007).

Cell cross-contamination in cultures can be caused by a number of possible issues:

- Improperly sterilized solutions and glassware.
- Incorrectly labeled vessels.
- Broken or faulty laminar flow hoods.
- Inadequate aseptic technique.

Strict aseptic technique is the best way to prevent cross-contamination of cell cultures. Here are some important details that must be taken into account (Freshney, 2006):

- Perform all cell culture manipulations in a laminar flow hood or biosafety cabinet.
- Obtain cell cultures from reliable sources.
- Work with only one cell line at a time and completely decontaminate the hood after each use.
- Change gloves when changing from one cell line to another and use a separate bottle of medium and other reagents for each cell line.
- Label all media and reagent bottles with the cell line name and the date the bottle was first used.
- Always label all cell culture vessels and frozen vials of cells clearly with the exact name of the cell line and the date.
- Keep very good records of all cell culture manipulations and dates written in permanent ink in a bound notebook.

CONTROLLING CONTAMINATION

Antibiotics

Antibiotics are often used during collection, transportation, and dissection of biopsy samples because of the intrinsic contamination risk of these operations. However, they should not be routinely used once the primary culture is established. If the culture grows well, then antibiotics can be removed from the bulk of the stocks at the first subculture, retaining one culture in antibiotics as a precaution if necessary (Freshney, 2006). It is important to remember that, unlike bleach and ethanol, antibiotics do not kill microorganisms, they just hide the presence of microbial contamination by retarding or stopping their proliferation. In cell cultures there is no immune system, thus the complete eradication of contamination can be very difficult and sometimes impossible with current techniques.

Aseptic Techniques and General Principles of Handling Cell Cultures

Aseptic technique is a set of *principles* and *practices* used by cell culture workers to reduce the presence of unwanted microorganisms or other cell lines in their cultures. Good aseptic technique is essential for successful long-term cell and tissue culture. Here are some useful recommendations to help improve aseptic technique:

- All supplies and reagents that come into contact with the cultures must be sterile (Phelan, 2007).
- Wash hands before and after handling any cell culture material, even if you are wearing gloves.
- Handle only one cell line at a time. There are intrinsic risks of misidentification or cross-contamination between cell cultures when more than one cell line is in use.
- Quarantine and cautiously handle all incoming cell lines until testing verifies the absence of mycoplasma. One of the most common sources of contamination is the culture given to you by a colleague in another lab. Try to obtain cell lines from repositories that certify that all material is mycoplasma-free prior to distribution.
- Avoid continuous long-term use of antibiotics within cell cultures. The overuse of antibiotics may lead to cytotoxicity and may pose an increased risk of covert mycoplasma contamination within the cell lines.
- When using multiwell plates, if contamination is restricted to one or a few wells, it can be eliminated by simply aspirating the contaminated media, filling the empty well(s) with 10% bleach, aspirating the bleach, then washing the well with 70% ethanol, and aspirating.

Working in Hoods

Laminar flow cabinets (hoods) are physical containment devices that act as primary barriers either to protect the material within the hood from different sources of contamination, or to protect the laboratory worker and laboratory environment from exposure to infectious or other hazardous materials that are present within the hood during routine procedures (Coecke et al., 2005).

Biological safety cabinets provide a clean, safe environment for both the worker and the specimen. Cell culture applications use two types of laminar flow hoods: (a) the horizontal-flow clean bench and (b) the biological safety cabinet. Product selection will depend on the nature of the cell culture work and the biosafety level of the materials being used and processed. Both types of hoods use a high-efficiency particulate air (HEPA) filter and blowers that generate a continuous stream of air.

The horizontal laminar flow clean bench is used to provide a near-sterile environment for the clean handling of non-hazardous material such as sterile media or equipment. Because the air stream pattern directs the flow of air within the hood directly back to the hood operator and the room, horizontal flow hoods are never to be used with infectious agents or toxic chemicals.

The Class II, vertical biosafety cabinet is frequently encountered in cell culture laboratories. The Class IIA biosafety cabinet is suitable for work with low- to moderate-risk biological agents in the absence of toxic or radioactive chemicals.

Here are some important tips when working with hoods:

- Always have the biosafety cabinets certified at the time of installation and if they are ever moved or repaired. It is also recommended to routinely test the quality of the airflow and filter integrity every 6 to 12 months.
- Biosafety cabinets may be equipped with germicidal UV lights for decontaminating work surfaces. However, the efficacy of UV lamps has been challenged. The UV lights must directly strike a microorganism in order to destroy it. Over time, the UV output and germicidal capacity of the tube diminishes. In addition, there are safety concerns related to the exposure to UV light (Phelan, 2007). UV exposure is damaging to the eyes and skin; therefore, the UV light should never be on while the cabinet is in use.
- Biosafety cabinets and hoods should be turned on 15 minutes prior to use each day. Work surfaces should be wiped down with 70% ethanol, before and after each use and between handling different cell lines.
- Wipe down bottles and flasks with 70% ethanol or another suitable disinfectant before placing them in the cabinet.
- Wear a clean lab coat when working in a hood. This coat should be for hood use only and should not be worn anywhere else in the laboratory.
- Limit the number of people in the area around the hood. This reduces levels of airborne contaminants.
- Avoid unnecessary talking while working in the hood. Talking generates microbe-laden aerosols that can then enter into the hood.
- Avoid moving materials in or out of the hood while work is in progress.
- Keep the hood work area clean and uncluttered. Do not use hoods as storage cabinets. Clutter makes it very difficult to clean the work surface properly and can disrupt the laminar flow around the work area.
- Do not use open flames, especially Bunsen burners, in laminar flow hoods.
- Doors in the culture area should be kept closed while the hood is in use. Opening a door can create a back draft that disrupts laminar flow in hoods.

Transporting Cultures
- Minimize transport distances and avoid using common hallways to reduce contact with airborne contaminants. Ideally, incubators should be placed in close proximity to the culture area to restrict the movement of cells within the facility.
- Use flasks with vented caps whenever possible, especially for long-term cultures.
- Transport and incubate unsealed vessels, such as dishes and microplates, in plastic boxes or trays.

Personal Hygiene and Protection
- Wear clean lab coats. For additional protection in the hood use a fresh, closed front lab coat with gloves that overlap the cuffs. Protective eyewear should be used when appropriate. Lab coats used for cell culture should not leave the cell culture area.
- Always wear clean gloves during aseptic procedures.
- Caution should be taken when handling sharp instruments such as needles, scalpels, scissors, and glass pipettes. Sterile disposable plastic supplies should be preferentially used to avoid the risk of broken or splintered glass (Phelan, 2007).

Culture Area
- Ideally, cell and tissue culture should be conducted in a separate, designated room that is exposed to minimal traffic.
- Since microorganisms attach to dust particles, reducing the amount of dust and dirt in the culture area is always a priority. Do not open windows or use window fans that allow in outside air.
- Frequently clean water baths used for warming media or solutions. Alternatively, avoid water baths entirely by using an incubator or a bath filled with metal beads to warm media and solutions.
- Periodically empty and carefully clean incubators.

REAGENTS AND SUPPLIES
Antimicrobials

Component	Catalog #	Supplier
Primocin	ant-pm-1	Invivogen
Penicillin Streptomycin	15070-063	Life Technologies
Antibiotic-Antimycotic	15240096	Life Technologies

Immunofluorescence Kits

- MycoAlert Mycoplasma Detection Kit: LT07-118 (Lonza)
- MycoAlert Mycoplasma Assay Control Set: LT07-518 (Lonza)

PCR Based Kits

- VenorGeM™ Mycoplasma Detection Kit, PCR-based test: MP002 (Sigma-Aldrich)
- JumpStart™ Taq DNA Polymerase: D9307 (Sigma-Aldrich)

READING LIST

Chatterjee, R., 2007. Cell biology. Cases of mistaken identity. Science 315, 928–931.

Coecke, S., Balls, M., Bowe, G., Davis, J., Gstraunthaler, G., Hartung, T., et al., 2005. Guidance on good cell culture practice. A report of the second ECVAM task force on good cell culture practice. Altern. Lab. Anim. 33, 261–287.

Drexler, H.G., Uphoff, C.C., 2002. Mycoplasma contamination of cell cultures: Incidence, sources, effects, detection, elimination, prevention. Cytotechnology 39, 75–90.

Freshney, R.I., 2006. In: Freshney, G.V.-N.a.R.I. (Ed.), Basic Principles of Cell Culture. John Wiley & Sons, Hoboken, NJ.

Hopert, A., Uphoff, C.C., Wirth, M., Hauser, H., Drexler, H.G., 1993. Mycoplasma detection by PCR analysis. In Vitro. Cell. Dev. Biol. Anim. 29A, 819–821.

Phelan, M.C., 2007. Basic techniques in mammalian cell tissue culture. Curr. Protoc. Cell. Biol. (Chapter 1, Unit 11).

Rappe, M.S., Giovannoni, S.J., 2003. The uncultured microbial majority. Annu. Rev. Microbiol. 57, 369–394.

Ryan, J., 2008. Understanding and Managing Cell Culture Contamination. Corning Life Sciences, Technical Literature.

CHAPTER 5

Cell Culture Medium

James Shen, Kim Ly, and Yvonne Hoang

EDITOR'S COMMENTARY

The research and development of new stem cell culture media has been an area of intense interest as the promise of therapeutic applications increases. It would be ideal to have a fully defined medium with control over all aspects of the culture medium. This is a tremendous challenge, however, because such a medium would need to support all of the complex characteristics of hPSCs that are currently maintained in undefined mixtures of growth factors and feeder layer components. If you read the list of ingredients in even the simplest of culture media, you may wonder why all these components are included. Intrigued by what goes into development and testing of a new culture medium, we requested this chapter that describes the elements that have to be kept in balance in order to maintain any mammalian cells.

OVERVIEW

The successful *in vitro* propagation and differentiation of hPSCs depends largely on the culture medium and the surrounding microenvironment. The central aspect to consider in culture medium design is the recapitulation of the *in vivo* physiological conditions, such as providing nutrients, oxygen, proper pH and growth factors.

DESIGN OF CULTURE MEDIUM

An ideal cell culture medium should support the growth and proliferation of cells while allowing them to retain most of their *in vivo* characteristics. Several key factors to consider when designing cell culture media include, but are not limited to: 1) buffers and pH, 2) osmolality, 3) salts, 4) amino acids, 5) vitamins, 6) lipids, 7) trace elements, 8) hormones, 9) growth peptides, 10) oxygen tension, 11) antioxidants, and 12) antibiotics.

53

J.F. Loring & S.E. Peterson (eds): Human Stem Cell Manual, Second edition.
DOI: http://dx.doi.org/10.1016/B978-0-12-385473-5.00005-9

Buffers

Norman E. Good (1917–1992) and his colleagues developed what has been referred to as "good buffers". These are twelve hydrogen ion buffering solutions including Tris, HEPES and MOPS, most of which remain staples in the laboratory. According to their study, a cell culture medium buffer should have the following characteristics:

- Be highly soluble in aqueous solutions
- Be impermeable to cell membranes
- Be easy to prepare
- Be non-toxic
- Not interfere with biochemical reactions
- Have a pKa of 6–8
- Be chemically stable

The most commonly used buffer system in cell culture media is sodium bicarbonate ($NaHCO_3$) and CO_2 gas; however the low pKa of $NaHCO_3$ results in suboptimal buffering throughout the physiological range (6.8–7.3) and medium quickly becomes alkaline when removed from the incubator.

Alternatively, HEPES (4-(2-hydroxyethyl)-1-piperazineethanesulfonic acid) is a zwitterionic organic chemical buffering agent and is effective between pH 6.6 and 8.0 when used at 10–25 mM (Good et al., 1966) and does not require CO_2 to be effective. However, HEPES may be toxic for hPSCs in the absence of serum (Furue et al., 2008). Obviously, care should be taken when selecting the appropriate buffer for each cell type. Other cell culture-tested buffers include: MES, PIPES, MOPS, tricine, glycine, bicine, and beta glycerophosphate. Phenol red is an additive that allows for the visual assessment of pH in which yellow indicates acidic and purple indicates basic conditions. Phenol red has weak estrogenic activity and is often used in low levels in commercially available media formulations specialized for stem cell cultures.

Osmolality

The right osmolality, defined as the concentration of a solution in terms of osmoles of solute per kilogram of solvent, must be empirically determined by the user as an imbalance will adversely affect cell health. In terms of physiological fluids, particularly plasma, osmolality refers to the concentration of dissolved electrolytes such as sodium, chloride, and bicarbonate as well as proteins and sugars. The addition of chemicals, acids, bases, and proteins alters the osmolality of the medium and should be carefully monitored with an osmometer. The normal range of human serum osmolality is 285–295 mOsm/kg and the useful range of culture media used for vertebrate cells is between 260 and 320 mOsm/kg. Osmolality in media is typically between 270 and 330 mOsm/kg for most cell types, but there is a report that the morphology of hESCs can be improved with reduced osmolality (Furue et al., 2008). The combined effects of suboptimal pH and osmolality may decrease proliferation and embryoid body formation in mouse ESCs (Chaudhry et al., 2009). Interestingly, it is widely accepted that the optimal osmolality and salt concentration for germinal and embryonic cells is below that of normal physiological saline (Waymouth, 1970).

Salts

The purpose of cell culture medium is to mimic the physiological salt solution of plasma. Inorganic salt solutions perform the essential function of providing water and salt to the cells in order to provide an environment that maintains the structural and physiological integrity of cells *in vitro*. They maintain membrane potential by providing major ions in the form of sodium, magnesium, potassium, calcium, phosphate, chloride, sulfate, and bicarbonate.

Salts also have buffering effects, which protect cells against metabolite waste-induced fluctuations in pH. In the laboratory, this is achieved using classical balanced salt solutions such as Earle's or Ringer's, which are the salt and glucose basis for several standard media formulations. Earle's balanced salt solution was designed to maintain cells in short-term culture and Ringer's solution was designed to properly balance ions in the medium used for frog heart studies.

Amino Acids

Amino acids are the components of proteins, which are essential energy sources for all cells. Of the 20 amino acids, there are 11 that cannot be synthesized by the human body and are therefore referred to as "essential" amino acids. The seminal studies of Rose and colleagues yielded critical information about the amino acid requirements of humans and rodents (Rose and Cox, 1924) upon which all mammalian cell culture mediums are based. The essential amino acids necessary for culturing mammalian cells and found in classical mediums such as DMEM are: arginine, cysteine, glutamine, histidine, isoleucine, leucine, lysine, methionine, phenylalanine, threonine, tryptophan, tyrosine, and valine. Among these, glutamine is of particular importance as it is required by virtually all mammalian and insect cells.

Glutamine is a precursor for proteins, ribonucleotides, and vitamin synthesis and serves as a major energy source for dividing cells by providing nitrogen when glucose is low. Although it is essential, glutamine is highly unstable in solution and its metabolism results in the formation of ammonia, which is cytotoxic in excess (Ozturk and Palsson, 1990). Glutamate can be substituted for glutamine in cultures of cells that possess sufficient glutamine synthetase activity. Commercially available GlutaMAX™ (Life Technologies) contains dipeptides conjugated with glutamine, which are more heat resistant and stable in culture. DMEM/F-12, commonly used for stem cell cultures, has a high amino acid content and the enriched nutrient mixture found in Ham's F-12. An additional seven non-essential amino acids are found in Ham's F-12: Ala, Asn, Asp, Glu, Gly, Pro, and Ser (Ham, 1965). A formulation that includes both essential and non-essential amino acids is beneficial for cells as it reduces their metabolic burden, allowing the cells to use their energy for other processes such as proliferation.

Vitamins

The term "vitamin" refers to any group of organic substances essential in small quantities for normal metabolism that are found in minute amounts in the diet or are sometimes produced synthetically. Vitamins also have an important role in serving as precursors for many cofactors that are required for essential biochemical reactions. They cannot be synthesized by human cells and are thus obtained from basal media. Serum is a major source of vitamins, but it cannot be used in defined mediums. Early studies by Harry Eagle revealed that the seven vitamins essential for mouse L strain fibroblasts and HeLa cells include choline, folic acid, nicotinamide, panthothenate, pyridoxal, riboflavin, and thiamin (Eagle, 1955). These essential vitamins are present in DMEM/F-12.

Biotin is an essential cofactor for the acetyl SCoA reaction during carbohydrate and lipid metabolism (Wakil et al., 1958), a CO_2 carrier, and is essential for fibroblast culture (Haggerty and Sato, 1969). Biotin is typically provided in basal medium along with the other seven essential vitamins (Bjare, 1992; Butler and Jenkins, 1989). Vitamin B12 is particularly important for cell culture in that it functions as a cofactor for methionine synthase and if deficient, can lead to genome instability and mitochondria-mediated apoptosis. Although poorly soluble in aqueous culture media, Vitamin E or alpha-tocopherol is an important antioxidant that protects cell membranes and other lipids from reactive oxygen species-induced damage.

Lipids

Lipids are essential structural constituents of cellular membranes, energy stores, and serve as transport molecules involved in cell signaling. In terms of cell culture, a practical definition of lipids may be, "water-insoluble molecules that are biosynthetically or functionally related to fatty acids and their derivatives." Commonly used lipids in cell culture include cholesterol and fatty acids, such as linoleate, and phospholipids as well as the precursors choline and inositol. Because mammalian cells cannot synthesize double bonds on fatty acids beyond C9, linoleic and linolenic acids are considered essential. Phospholipid precursors are generally regarded as being essential for cell culture; specifically choline, ethanolamine, and inositol, which are precursors for phosphatidyl-choline, phosphatidyl, and phosphatidyl inositol synthesis, respectively. It is advantageous to use precursors in place of phospholipids themselves as DMSO is required to solubilize them in media.

As stem cell cultures move toward chemically defined conditions, the use of serum and animal-derived carrier proteins for lipid transport is being phased out. Traditionally, serum has been used to supply lipids in cell culture. Therefore, it is desirable to supplement serum-free media with lipid mixtures. Commercially available lipid emulsions contain a mixture of fatty acids and surfactants for solubilization purposes. However, surfactants may prove to be problematic for some adherent cells and may result in toxicity. Therefore, cellular tolerance to such media additives must be empirically determined by the user. Interestingly, the high molecular weight lipid fraction found in KSR is thought to contain the potent factors that help to maintain hESC pluripotency (Garcia-Gonzalo et al., 2008).

Trace Elements

The importance of trace elements was first reported by Shooter and Gey in 1952 who showed that copper, zinc, molybdenum, manganese, cobalt, and selenium derived from serum are required for cell growth. The development of serum-free MCDB 104 (McKeehan et al., 1977) and MCDB 301 (Hamilton and Ham, 1977) mediums included these essential elements. Ham's F-12 element composition includes: iron, manganese, zinc, molybdenum, selenium, vanadium, copper, and vitamin B12. Human mesenchymal stem cell (hMSC) cultures typically require the addition of selenium and iron in media.

As the need to reduce serum dependence and refinement in cell culture media is increasingly evident, the focus of optimization studies has shifted towards trace element mixtures. The composition and levels of essential trace elements can have profound effects on cell performance and biological productivity, which drastically affect scale-up protocols as well as the consistency of experiments. Trace elements are associated with a multitude of proteins in the body. Selenium has long been recognized as an antioxidant involved with oxidation reactions of glutathione. Zinc is involved in antioxidant responses, proliferation, differentiation, and acts as an inhibitor of caspase-3 mediated apoptosis. Copper and zinc are required for the synthesis of super-oxide dismutase, but may also antagonize one another in this process as they compete for the same binding site. Although necessary for cell viability, trace elements may exert synergistic or antagonistic effects on one another. As a result, several formulations containing combinations of elements have been developed.

Hormones

Hormones are physiological constituents of mammalian serum and therefore present another point of batch to batch variation. Supplementation with hormones is a key step toward developing chemically-defined media. Insulin as well as triiodothyronine (T_3), and the glucocorticoids, dexamethasone and hydrocortisone, are common additives in culture media (van der Valk et al., 2010). Insulin is a necessary additive for serum-free cell culture that stimulates proliferation in a variety of somatic cells in culture. *In vitro* its synergistic interactions with other hormones and growth factors, namely FGF, EGF, and PDGF, are potent stimulators of the cell cycle machinery (Straus, 1984). In the absence of serum, insulin contributes to optimal long-term growth. T_3 is a peptide thyroid hormone that stimulates the breakdown of cholesterol, increases rates of protein synthesis, and affects embryonic development. *In vitro*, T_3 regulates cell differentiation and protein expression. Dexamethasone is a synthetic glucocorticoid that enhances EGF binding to its receptor, which in turn, stimulates proliferation (Baker et al., 1978). Hydrocortisone, a naturally occurring glucocorticoid, has a similar effect (Wu et al., 1981).

Growth Factors

Growth factors induce proliferation and differentiation of cells in culture. Maintenance of pluripotency and the suppression of differentiation are key components addressed by the use of growth factors. Growth factors, which may have autocrine, paracrine, or endocrine origins, can be presented to cells in a soluble form from the extracellular fluid, or in an insoluble form from adjacent cells or extracellular matrices. Growth factors operate on cells through specific high-affinity transmembrane receptors that modify key regulatory proteins in the cytoplasm. These in turn affect the decisions controlling proliferation and differentiation, and may lead to changes in gene expression and reactivity to other factors. Growth factors have been repeatedly shown to direct the differentiation of mESCs. For instance, myogenic differentiation is induced by TGF-β1 (Slager et al., 1993). In general, inhibiting bone morphogenetic protein (BMP) signaling in combination with bFGF mediates long-term maintenance of embryonic stem cells in their pluripotent state. bFGF functions to induce proliferation and repress differentiation (Granérus et al., 1996; Xu et al., 2005). bFGF is necessary but insufficient for pluripotency and self-renewal of both hESCs and cells of defined lineages without feeder layers or conditioned medium (Vallier et al., 2005).

A great body of research indicates that hPSCs express many growth factor receptors, including those for FGF, SCF, fetal liver tyrosine kinase-3 ligand (Flt3L), insulin receptor (IR), insulin-like growth factor receptor (IGF1R), and epidermal growth factor receptor (EGFR) family members, including ERBB2 and ERBB3 (Xu et al., 2005, Wang et al., 2007, Ding et al., 2010). bFGF acts as a competence factor for activin/Nodal/TGFβ in the maintenance of pluripotency and self-renewal (Vallier et al., 2005) and stimulates the mitogen-activated protein kinase (MAPK) pathway. bFGF also modulates Wnt signaling in undifferentiated hESCs and iPSCs through activated PI3-K/GSK3β signaling (Kim et al., 2005, Ding et al., 2010). Insulin and insulin-like growth factors activate the PI3 kinase/AKT pathway to maintain the self-renewal of hESCs (Wang et al., 2011). Blocking the IGF-I/II receptor pathway reduces hESC survival and clonogenicity (Bendal et al., 2007). TGFβ potentiates FGF activity (Kirschner et al., 1987) and is necessary for the maintenance of hESC pluripotency (Amit et al., 2003; James et al., 2005). To further compare the effect of these factors on hPSC self-renewal and pluripotency, we compared bFGF, IGF-1 and TGFβ in defined culture conditions and found that bFGF and IGF-I equally promoted hPSC self-renewal and pluripotency. The absence of either bFGF or IGF-I did not decrease the growth rate of hESCs in culture indicating that their role in hPSC self-renewal is exchangeable. Compared to bFGF and IGF-I, TGFβ has less impact on hESC self-renewal in culture.

Heparin

Heparin, a highly sulfated glycosaminoglycan, has the highest negative charge density of any known biological molecule and is widely used as an anticoagulant (Cox and Nelson, 2004). Heparin acts as a cofactor for bFGF, which promotes the growth of hPSCs in defined medium (Furue et al., 2008). The requirement for heparin might be satisfied by the production of heparin sulfate by the hPSCs themselves, or by their derivatives. Heparin-binding EGF-like growth factor also promotes survival and prevents apoptosis of ES cells in culture (Krishnamoorthy et al., 2009). Furthermore, substrates displaying heparin-binding peptides can facilitate the long-term culture of hPSCs (Klim et al., 2010). In chemically-defined medium, heparin, at concentration of 0.2–1 units/mL, improves ES cell morphology.

Oxygen Tension

Cell culture is typically performed at 20% oxygen tension, and in early experiments it was assumed that ambient air was adequate for *in vitro* growth (Shooter and Gey, 1952). However, direct measurements in various tissues revealed a much lower oxygen tension than previously thought (Mitchell and Yochim, 1968). Atmospheric oxygen tension is at supraphysiological levels for some cell types and may therefore lead to oxidative stress and genomic instability. Stem cell niches are known to have decreased oxygen tension (Braun et al., 2001; Cipolleschi et al., 1993; Erecinska and Silver, 2001; Mohyleden et al., 2010). Several investigators have reported that low oxygen tension (1–5% O_2) is beneficial for stem cell cultures. Culture of cells at 11 mm Hg/1.5% O_2 enhanced the generation of induced pluripotent stem cells (iPSCs) (Yoshida et al., 2009) and the rate of hPSC differentiation was dramatically reduced in 3–5% O_2 (Ezashi et al., 2005). This is likely a reflection of the non-vascularized uterine compartment where blastocysts reside.

Antioxidants

Oxidative stress is a consequence of an imbalance between critical processes such respiration and cellular antioxidant defenses. As a result, highly reactive oxygen radicals damage DNA, lipid membranes, proteins, and components in media. The detrimental effects of high oxygen tension *in vitro* may be reduced with the addition of antioxidants to counteract uncontrolled free radicals. Antioxidants commonly used in cell culture include: tocopherol, transferrin, selenium, and reduced glutathione. Tocopherol or vitamin E is a membrane antioxidant with low solubility in water that functions to neutralize lipid peroxides. Transferrin is a chelator that binds iron with such high affinity that virtually no iron is available to generate free radicals. It serves as an extracellular iron transporter and storage molecule. Glutathione is a water-soluble antioxidant tripeptide, which contains a reducing thiol group. It is ubiquitously produced in all cell types and is important for cell proliferation and viability. Selenium is incorporated into glutathione and thioredoxin reductases which reduce glutathione and thioredoxin, respectively. Ascorbate or vitamin C is a water-soluble antioxidant that regenerates reduced tocopherol and is important both *in vivo* and *in vitro*. Most commercially-available basal mediums, including DMEM and Ham's, do not contain antioxidants as they were designed for use in conjunction with serum. Glutathione, tocopherol, transferrin, and selenium are expected to be supplied by serum. Supplementation is recommended for serum-free and chemically defined cultures depending on the cell type.

The benefit of antioxidants in stem cell cultures is controversial because, despite their protective effects, some may affect cellular function and differentiation. Supplementation of muscle-derived stem cells with the glutathione precursor, N-acetylcysteine, improves their tissue regenerative capacity (Drowley et al., 2010). The addition of antioxidants to adipose-derived MSC cultures significantly enhanced their growth rate and longevity (Lin et al., 2005). Vitamin C enhances iPSC generation from both mouse and human somatic cells by inhibiting senescence (Esteban et al., 2010). Conversely, vitamin C induces cardiac differentiation in CGR8 mESCs (Takahashi et al., 2003).

Antibiotics

Protecting irreplaceable cultures from microbial contamination is of utmost importance and the decision to use antibiotics should be based on each investigator's needs and experience. Antibiotic solutions are often used in cell culture with penicillin and streptomycin being the standard, although toxicity can occur at high doses. The use of antibiotics in stem cell cultures should be avoided or minimized when possible as they reduce the growth of hPSC cultures (Cohen et al., 2006) (see Chapter 4).

Albumin

Albumin is the most abundant protein in plasma, comprising 60% of plasma protein. It is a single polypeptide with 585 amino acids. Its molecular weight is approximately 67 kDA. Albumin has a long historical involvement in the design of media for the successful culture of mammalian cells, in both the research and commercial fields. The functions of albumin in cell culture medium include: 1) to promote cell attachment; 2) to act as a carrier; and 3) as an antioxidant. Currently, there are three types of albumin commonly used in cell culture medium: BSA, lipid-rich BSA and HSA. Previously published literature indicates that the influence of albumin on hPSC proliferation can be subject to lot-to-lot variability (Chen et al., 2011). Because BSA and HSA are not identical and yet are used interchangeably in stem cell culture media, we explored the difference between BSA and HSA in terms of their effects on hPSC proliferation and pluripotency (unpublished data). Our preliminary study is in agreement with others which show that albumin is important for self-renewal. Specifically, 1) albumin is required to prevent spontaneous hPSC differentiation; 2) an optimal concentration of albumin is important for hPSC culture, with higher concentrations of albumin causing cytotoxicity; and 3) the species from which the albumin is derived dictates its efficacy, with human albumin being optimal for hPSC culture.

COMMERCIALLY AVAILABLE STEM CELL MEDIA

mTeSR®1 and TeSR™2

The first commercially available growth medium for hESC culture was initially formulated by Tenneille Ludwig and colleagues (2006) and is currently sold by Stem Cell Technologies™ as mTeSR1. This completely defined medium contains no serum but has BSA as a component and can support growth of hPSCs for extended periods of time without feeder layers. TeSR2 is closely related to mTeSR1 but is animal protein-free for more clinically compatible cell culture. When compared to other commercially available stem cell media, mTeSR contains a higher (100 ng/mL) concentration of bFGF, a much higher concentration than that observed physiologically (Ludwig et al., 2006; Rajala et al., 2007). Further product information can be found from the manufacturer's website at http://www.stemcell.com.

StemPro® hESC SFM

StemPro hESC SFM is also a defined, serum-free medium formulated for the feeder-free growth and expansion of hESCs and iPSCs. StemPro hESC SFM has been extensively tested and has been used to scale up production of hPSCs. Similar to mTeSR1, StemPro hESC SFM contains BSA and supraphysiological concentrations of IGF (200 ng/mL). Once assembled, the complete StemPro® hESC SFM (supplement, DMEM/F-12, BSA, and FGF2) is stable for 7 days when stored in the dark at 2–8°C, but 2-mercaptoethanol should be added fresh daily. Further product information can be found from manufacturer website at http://www.lifetechnologies.com.

STEMium™

STEMium is a serum-free medium designed for growth of hPSCs under feeder-free conditions. STEMium is a proprietary serum-free medium containing relatively low amounts of growth factors and proteins, suitable for the stable and long-term propagation of hPSCs. This medium can be used for prolonged passaging of hPSCs with consistent growth rate and normal morphology. Compared to other commercially available stem cell media, STEMium™ contains low concentrations of growth factors (FGF2 and TGFβ1) while maintaining self-renewal and pluripotency. Further product information can be found from the manufacturer's website at http://www.sciencellonline.com.

NutriStem™ XF/FF

This fully defined, xeno-free, low growth factor culture medium is for the maintenance and expansion of hPSCs. It is also suitable for the propagation of hPSCs under feeder-free conditions. NutriStem XF/FF medium offers the ability to culture cells without the need for high levels of bFGF and other stimulatory cytokines and growth factors. Further product information can be found from the manufacturer's website at http://www.stemgent.com.

ASSESSING MEDIUM EFFICACY

hPSCs have the unique ability to differentiate into derivatives of all three germ layers (endoderm, mesoderm, and ectoderm). In addition, they are able to grow indefinitely and form embryoid bodies *in vitro* and teratomas *in vivo*. However, *in vitro* expansion and culture of stem cells can change the characteristics of the cells due to intracellular and extracellular influences. When examining medium efficacy, the medium's ability to support long-term stable hPSC culture should be considered as the most important criterion. It is necessary to demonstrate that when cultured in a particular medium, hPSC characteristics do not change over time. The following criteria should be used to determine the efficacy of a new stem cell medium.

Cell Morphology

Morphological characterization of hPSCs is important in many ways. In culture, the morphology indicates the status of the cells, e.g. undifferentiated or differentiated, and also provides clues about the general health and condition of the cells. When grown on feeder layers, typical hPSC colonies consist of small round cells with a high nucleus to cytoplasm ratio and prominent nucleoli. Cells should be in tightly packed cell clusters with distinct shiny borders (see Chapter 1). When hPSCs are grown without feeder layers, they appear as uniform small, flat cells (see Chapter 3).

Self-Renewal and Pluripotency

hPSCs can renew themselves indefinitely and have the ability to differentiate into all cell types of the body. The easiest way to assess pluripotency and self-renewal capacity is by immunolabeling cells for markers associated with those characteristics. A number of markers are currently used to characterize pluripotency in hPSCs including POU5F1/OCT4, NANOG, SSEA4, TRA-1-60 and TRA-1-80. Analysis can be done via immunocytochemistry or flow cytometry (see Chapters 15 and 16). Pluripotency can also be assessed by gene expression profiling, embryoid body formation and teratoma generation (see Chapters 18, 21–23). hPSCs grown in a new medium should be analyzed periodically for pluripotency and self-renewal capacity.

Gene Expression

Stem cells exhibit unique transcriptional profiles and gene expression analysis can be used to identify pluripotent cells (Bhattacharya et al., 2004, 2009; Mueller et al., 2008, 2011). hPSCs grown in an experimental medium should be analyzed for gene expression changes periodically over long-term culture. See Chapters 15 and 18.

Telomerase Activity

Unlike normal somatic cells, hPSCs can proliferate indefinitely in culture in an undifferentiated state where they do not senescence and yet remain non-transformed. hPSCs exhibit high telomerase activity, and maintain telomere length after prolonged *in vitro* culture. Telomerase activity should be maintained when cells are cultured in a new medium.

Karyotype and Copy Number Variation

The genomic stability of hPSCs has recently received much attention. To determine if stem cells grown in a new medium over a long-term culture can maintain a normal karyotype, cytogenetic analysis should be performed (see Chapter 13). This is typically measured by G-banding, with most hPSCs exhibiting a normal complement of chromosomes. In addition sub-chromosomal abnormalities, copy number variations, should be assessed by Comparative Genome Hybridization (CGH) or Single Nucleotide Polymorphism (SNP) genotype (see Chapter 14). Since stem cells are highly proliferative in nature, the opportunity for spontaneous DNA mutations is high. It is critical to monitor genetic changes in cells before they are applied in a clinical setting particularly because late-passage hPSCs have more genomic alterations commonly found in human cancers relative to early-stage passages.

PERSPECTIVE

hPSCs can proliferate indefinitely and yet maintain the potential to generate derivatives of all three germ layers. These properties make them useful for understanding the basic biology of the human body, for drug discovery and testing, and for regenerative medicine. They can be used for cell replacement therapies to treat many diseases such as Parkinson's disease, spinal cord injury, juvenile diabetes, cancer, and many blood and immune system related genetic disorders. In order to provide cells with defined quality characteristics that are safe for the patient, cGMPs need to be employed throughout the entire process of stem cell isolation, expansion, and differentiation. cGMP, a quality assurance system used by the pharmaceutical industry, is used in the production of stem cells for clinical applications. It ensures that the end products meet preset specifications. cGMP covers both manufacturing and testing of the final product. It requires traceability of raw material and also that production follows validated SOPs. In the US, the Food and Drug Administration (FDA) has issued draft guidelines "Guidance for Reviewers: Instructions and Template for Chemistry, Manufacturing, and Control (CMC) Reviewers of Human Somatic Cell Therapy Investigational New Drug Applications (INDs)" (8/15/2003 and in the 21 CFR Part 1270, and 21 CFR Part 1271 regulations; http://www.fda.gov/OHRMS/DOCKETS/98fr/03d0349gdl.pdf) to regulate cellular therapies to ensure that they are safe and effective, and that persons enrolled in clinical trials using cellular products are protected from undue risk (Unger et al., 2008).

In the development of a clinical-grade hPSC line, it will be important to show that all steps in the derivation, passaging, and culturing of hPSC are completely free of animal products and human proteins. Animal products that may introduce unknown animal pathogens into patients during cell transplantation therapy should be avoided in stem cell medium. Ideally, such hPSC lines would be derived from a mechanically isolated inner cell mass in a culture system which uses synthesized extracellular matrices and chemically defined medium that is animal and human protein-free, with non-enzymatic passaging. During the past several years, there have been substantial improvements in the methods used to grow cultures of hPSCs. These improvements include the use of defined medium without any animal and human derived proteins. For example, OsrHSA, a transgenic rice seed-derived human serum albumin, has been introduced to replace human serum albumin. Our current study shows that OsrHSA, at the same concentration as HSA, maintained hPSCs self-renewal and pluripotency in culture. But, further study is needed to confirm of the long-term stability of hPSCs in the presence of OsrHSA.

READING LIST

Amit, M., Margulets, V., Segev, H., Shariki, K., Laevsky, I., Coleman, R., et al., 2003. Human feeder layers for human embryonic stem cells. Biol. Reprod. 68 (6), 2150–2156.

Bain, G., Kitchens, D., Yao, M., Huettner, J.E., Gottlieb, D.I., 1995. Embryonic stem cells express neuronal properties in vitro. Dev. Biol. 168 (2), 342–357.

Baker, J.B., Barsh, G.S., Carney, D.H., Cunningham, D.D., 1978. Dexamethasone modulates binding and action of epidermal growth factor in serum-free cell culture. Proc. Natl. Acad. Sci. USA 75 (4), 1882–1886.

Bendal, S.C., Stewart, M.H., Menendez, P., George, D., Vijayaragavan, K., Werbowetski-Ogilvie, T., et al., 2007. IGF and FGF cooperatively establish the regulatory stem cell niche of pluripotent human cells in vitro. Nature 448 (7157), 1015–1021.

Bhattacharya, B., Miura, T., Brandenberger, R., Mejido, J., Yongquan, L., Yang, A.X., et al., 2004. Gene expression in human embryonic stem cell lines: Unique molecular signature. Blood 103 (8), 2956–2964.

Bhattacharya, B., Puri, S., Puri, R.K., 2009. A review of gene expression profiling of human embryonic stem cell lines and their differentiated progeny. Curr. Stem Cell Ther. 4 (2), 98–106.

Bjare, U., 1992. Serum-free cell culture. Pharmacol. Ther. 53 (3), 355–374.

Boeuf, H., Hauss, C., Graeve, F.D., Baran, N., Kedinger, C., 1997. Leukemia inhibitory factor-dependent transcriptional activation in embryonic stem cells. J. Cell Biol. 138 (6), 1207–1217.

Braam, S.R., Zeinstra, L., Litjens, S., Ward-van Oostwaard, D., van den Brink, S., van Laake, L., et al., 2008. Recombinant vitronectin is a functionally defined substrate that supports human embryonic stem cell self-renewal via alphavbeta5 integrin. Stem Cells 26 (9), 2257–2265.

Burridge, P.W., Thompson, S., Millrod, M.A., Weinberg, S., Yuan, X., Peters, A., et al., 2011. A universal system for highly efficient cardiac differentiation of human induced pluripotent stem cells that eliminates interline variability. PLoS ONE 6 (4), e18293.

Burridge, P.W., Keller, G., Gold, J.D., Wu, J.C., 2012. Production of de Novo cardiomyocytes: human pluripotent stem cell differentiation and direct reprogramming. Cell Stem Cell 10 (1), 16–28.

Butler, M., Jenkins, H., 1989. Nutritional aspects of growth of animal-cells in culture. J. Biotechnol. 12 (2), 97–110.

Chambers, S.M., Fasano, C.A., Papapetrou, E.P., Tomishima, M., Sadelain, M., Studer, L., 2009. Highly efficient neural conversion of human ES and iPS cells by dual inhibition of SMAD signaling. Nat. Biotechnol. 27, 275–280.

Chaudhry, M.A., Bowen, B.D., Piret, J.M., 2009. Culture pH and osmolality influence proliferation and embryoïd body yields of murine embryonic stem cells. Biochemical. Bioeng. J. 45, 126–135.

Chen, G., Gulbranson, D.R., Hou, Z., Bolin, J.M., Ruotii, V., Probasco, M.D., et al., 2011. Chemically defined conditions for human iPSC derivation and culture. Nat. Methods 8 (5), 424–429.

Cheng, L., Hammond, H., Ye, Z., Zhan, X., Dravid, G., 2003. Human adult marrow cells support prolonged expansion of human embryonic stem cells in culture. Stem Cells 21 (2), 131–142.

Cipolleschi, M.G., Dello Sbarba, P., Olivotto, M., 1993. The role of hypoxia in the maintenance of hematopoietic stem cells. Blood 82 (7), 2031–2037.

Cohen, S., Samadikuchaksaraei, A., Polak, J.M., Bishop, A.E., 2006. Antibiotics reduce the growth rate and differentiation of embryonic stem cell cultures. Tissue Eng. 12 (7), 2025–2030.

Conti, L., Cattaneo, E., 2010. Neural stem cell systems: Physiological players or in vitro entitites? Nat. Rev. Neurosci. 11 (3), 176–187.

Cox, M., Nelson, D., 2004. Lehninger, Principles of Biochemistry. W.H. Freeman, New York, NY. p. 1100.

Ding, V.M.Y., Ling, L., Subaashini, N., Yap, M.G.S., Cool, S.M., Choo, A.B.H., 2010. FGF-2 modulates Wnt Signaling in undifferentiated hESC and iPS cells through activated PI3-K/GSK3β signaling. J. Cell Physiol. 225 (2), 417–428.

Dings, J., Meixensberger, J., Jäger, A., Roosen, K., 1998. Clinical experience with 118 brain tissue oxygen partial pressure catheter probes. Neurosurgery 43 (5), 1082–1095.

D'Ippolito, G., Diabira, S., Howard, G.A., Roos, B.A., Schiller, P.C., 2006. Low oxygen tension inhibits osteogenic differentiation and enhances stemness of human MIAMI cells. Bone 39 (3), 513–522.

Draper, J.S., Smith, K., Gokhale, P., Moore, H.D., Maltby, E., Johnson, J., et al., 2004. Recurrent gain of chromosomes 17q and 12 in cultured human embryonic stem cells. Nat. Biotechnol. 22, 53–54.

Drowley, L., Okada, M., Beckman, S., Vella, J., Keller, B., Tobita, K., et al., 2010. Cellular antioxidant levels influence muscle stem cell therapy. Mol. Ther. 18 (10), 1865–1873.

Eagle, H., 1955. The minimum vitamin requirements of the L and HeLa cells in tissue culture, the production of specific vitamin deficiencies, and their cure. J. Exp. Med. 102 (5), 595–600.

Erecińska, M., Silver, I.A., 2001. Tissue oxygen tension and brain sensitivity to hypoxia. Respir. Physiol. 128 (3), 263–276.

Esteban, M.A., Wang, T., Qin, B., Yang, J., Qin, D., Cai, J., et al., 2010. Vitamin C enhances the generation of mouse and human induced pluripotent stem cells. Cell Stem Cell 6 (1), 71–79.

Evans, M.J., Kaufman, M.H., 1981. Establishment in culture of pluripotential cells from mouse embryos. Nature 292 (5819), 154–156.

Ezashi, T., Das, P., Roberts, R.M., 2005. Low O_2 tensions and the prevention of differentiation of hES cells. Proc. Natl. Acad. Sci. USA 102 (13), 4783–4788.

Fehrer, C., Brunauer, R., Laschober, G., Unterluggauer, H., Reitinger, S., Kloss, F., et al., 2007. Reduced oxygen tension attenuates differentiation capacity of human mesenchymal stem cells and prolongs their lifespan. Aging Cell 6, 745–757.

Fu, X., Toh, W.S., Liu, H., Lu, K., Li, M., Hande, M.P., et al., 2010. Autologous feeder cells from embryoïd body outgrowth support the long-term growth of human embryonic stem cells more effectively than those from direct differentiation. Tissue Eng. C Methods 16 (4), 719–733.

Furue, M.K., Na, J., Jackson, J.P., Okamoto, T., Jones, M., Baker, D., et al., 2008. Heparin promotes the growth of human embryonic stem cells in a defined serum-free medium. Proc. Natl. Acad. Sci. USA 105 (36), 13409–13414.

Garcia-Gonzalo, F.R., Izpisúa Belmonte, J.C., 2008. Albumin-associated lipids regulate human embryonic stem cell self-renewal. PLoS One 3 (1), e1384.

Good, N.E., Winget, G.D., Winter, W., Connolly, T.N., Izawa, S., Singh, R.M., 1966. Hydrogen ion buffers for biological research. Biochemistry 5 (2), 467–477.

Granerus, M., Engström, W., 1996. Growth factors and apoptosis. Cell Prolif. 29 (6), 309–314.

Grant, W.C., Root, W.S., 1947. The relation of O2 in bone marrow blood to post-hemorrhagic erythropoiesis. Am. J. Physiol. 150 (4), 618–627.

Grayson, W.L., Zhao, F., Bunnell, B., Ma, T., 2007. Hypoxia enhances proliferation and tissue formation of human mesenchymal stem cells. Biochem. Biophys. Res. Commun. 358 (3), 948–953.

Greber, B., Coulon, P., Zhang, M., Moritz, S., Frank, S., Müller-Molina, A.J., et al., 2011. FGF signaling inhibits neural induction in human embryonic stem cells. EMBO J. 30 (24), 4874–4884.

Haggerty, D.F., Sato, G.H., 1969. The requirement for biotin in mouse fibroblast L-cells cultured on serumless medium. Biochem. Biophys. Res. Commun. 34 (6), 812–815.

Ham, R.G., 1965. Clonal growth of mammalian cells in a chemically defined, synthetic medium. Proc. Natl. Acad. Sci. USA 53, 288–293.

Hamilton, W.G., Ham, R.G., 1977. Clonal growth of chinese hamster cell lines in protein-free media. In Vitro 13 (9), 537–547.

Harb, N., Archer, T.K., Sato., N., 2008. The Rho-Rock-Myosin signaling axis determines cell-cell integrity of self-renewing pluripotent stem cells. PLoS ONE 3 (8), e3001.

Hoffman, L.M., Carpenter, M.K., 2005. Characterization and culture of human embryonic stem cells. Nat. Biotechnol. 23 (6), 699–708.

Hovatta, O., Mikkola, M., Gertow, K., Strömberg, A.M., Inzunza, J., Hreinsson, J., et al., 2003. A culture system using human foreskin fibroblasts as feeder cells allows production of human embryonic stem cells. Hum. Reprod. 18 (7), 1404–1409.

Itsykson, P., Ilouz, N., Turetsky, T., Goldstein, R.S., Pera, M.F., Fishbein, I., et al., 2005. Derivation of neural precursors from human embryonic stem cells in the presence of noggin. Mol. Cell Neurosci. 30 (1), 24–36.

James, D., Levine, A.J., Besser, D., Hemmati-Brivanlou, A., 2005. TGFβ/activin/nodal signaling is necessary for the maintenance of pluripotency in human embryonic stem cells. Development 132 (6), 1273–1282.

Kim, H.S., Oh, S.K., Park, Y.B., Ahn, H.J., Sung, K.C., Kang, M.J., et al., 2005. Methods for derivation of human embryonic stem cells. Stem Cells 9, 1228–1233.

Kimelman, D., Kirschner, M., 1987. Synergistic induction of mesoderm by FGF and TGF-beta and the identification of an mRNA coding for FGF in the early Xenopus embryo. Cell 51 (5), 869–877.

Klim, J.R., Li, L., Wrighton, P.J., Piekarczyk, M.S., Kiessling, L.L., 2010. A defined glycosaminoglycan-binding substratum for human pluripotent stem cells. Nat. Methods 7 (12), 989–994.

Krishnamoorthy, M., Heimburg-Molinaro, J., Bargo, A.M., Nash, R.J., Nash, R.J., 2009. Heparin binding epidermal growth factor-like growth factor and PD169316 prevent apoptosis in mouse embryonic stem cells. J. Biochem. 145 (2), 177–184.

Laflamme, M.A., Chen, K.Y., Naumova, A.V., Muskheli, V., Fugate, J.A., Dupras, S.K., et al., 2007. Cardiomyocytes derived from human embryonic stem cells in pro-survival factors enhance function of infarcted rat hearts. Nat. Biotechnol. 25 (9), 1015–1024.

Lee, J.B., Song, J.M., Lee, J.E., Park, J.H., Kim, S.J., Kang, S.M., et al., 2004. Available human feeder cells for the maintenance of human embryonic stem cells. Reproduction 128 (6), 727–735.

Lin, T.M., Tsai, J.L., Lin, S.D., Lai, C.S., Chang, C.C., 2005. Accelerated growth and prolonged lifespan of adipose tissue-derived human mesenchymal stem cells in a medium using reduced calcium and antioxidants. Stem Cells Dev. 14 (1), 92–102.

Ludwig, T.E., Levenstein, M.E., Jones, J.M., Berggren, W.T., Mitchen, E.R., Frane, J.L., et al., 2006. Derivation of human embryonic stem cells in defined conditions. Nat. Biotechnol. 24 (2), 185–187.

Maitra, A., Arking, D.E., Shivapurkar, N., Ikeda, M., Stastny, V., Kassauei, K., et al., 2005. Genomic alterations in cultured human embryonic stem cells. Nat. Genet. 37 (10), 1099–1103.

Martin, G.R., 1981. Isolation of a pluripotent cell line from early mouse embryos cultured in medium conditioned by teratocarcinoma stem cells. Proc. Natl. Acad. Sci. USA 78 (12), 7634–7638.

Martin, M.J., Muotri, A., Gage, F., Varki, A., 2005. Human embryonic stem cells express an immunogenic nonhuman sialic acid. Nat. Med. 11 (2), 228–232.

McKeehan, W.L., McKeehan, K.A., Hammond, S.L., Ham, R.G., 1977. Improved medium for clonal growth of human diploid fibroblasts at low concentrations of serum protein. In Vitro 13 (7), 399–416.

Melkoumian, Z., Weber, J.L., Weber, D.M., Fadeev, A.G., Zhou, Y., Dolley-Sonneville, P., et al., 2010. Synthetic peptide-acrylate surfaces for long-term self-renewal and cardiomyocyte differentiation of human embryonic stem cells. Nat. Biotechnol. 28 (6), 606–610.

Mohyeldin, A., Garzón-Muvdi, T., Quiñones-Hinojosa, A., 2010. Oxygen in stem cell biology: A critical component of the stem cell niche. Cell Stem Cell 7 (2), 150–161.

Nagaoka, M., Si-Tayeb, K., Akaike, T., Duncan, S.A., 2010. Culture of human pluripotent stem cells using completely defined conditions on a recombinant E-cadherin substratum. BMC Dev. Biol. 10, 60.

Nandivada, H., Villa-Diaz, L.G., O'Shea, K.S., Smith, G.D., Krebsbach, P.H., Lahann, J., 2011. Fabrication of synthetic polymer coatings and their use in feeder-free culture of human embryonic stem cells. Nat. Protoc. 6 (7), 1037–1043.

Ozturk, S.S., Palsson, B.O., 1990. Chemical decomposition of glutamine in cell culture media: Effect of media type, pH, and serum concentration. Biotechnol. Prog. 6 (2), 121–128.

Pera, M.F., Andrade, J., Houssami, S., Reubinoff, B., Trounson, A., Stanley, E.G., et al., 2004. Regulation of human embryonic stem cell differentiation by BMP-2 and its antagonist noggin. J. Cell Sci. 117 (Pt 7), 1269–1280.

Perrier, A.L., Tabar, V., Barberi, T., Rubio, M.E., Bruses, J., Topf, N., et al., 2004. Derivation of midbrain dopamine neurons from human embryonic stem cells. Proc. Natl. Acad. Sci. USA 101 (34), 12543–12548.

Price, P.J., Goldsborough, M.D., Tilkins, M.L., 1998. Embryonic stem cell serum replacement. International Patent Application WO 98/30679.

Rajala, K., Hakala, H., Panula, S., Aivio, S., Pihlajamäki, H., Suuronen, R., et al., 2009. Testing of nine different xeno-free culture media for human embryonic stem cell cultures. Hum. Reprod. 22 (5), 1231–1238.

Rajala, K., Pekkanen-Mattila, M., Aalto-Setälä, K., 2011. Cardiac differentiation of pluripotent stem cells. Stem Cells International. Volume 2011 (Article ID 383709).

Redon, R., Ishikawa, S., Fitch, K.R., Feuk, L., Perry, G.H., Andrews, T.D., et al., 2006. Global variation in copy number in the human genome. Nature 444 (7118), 444–454.

Richards, M., Fong, C.Y., Chan, W.K., Wong, P.C., Bongso, A., 2002. Human feeders support prolonged undifferentiated growth of human inner cell masses and embryonic stem cells. Nat. Biotechnol. 20 (9), 933–936.

Rodin, S., Domogatskaya, A., Ström, S., Hansson, E.M., Chien, K.R., Inzunza, J., et al., 2010. Long-term self-renewal of human pluripotent stem cells on human recombinant lamini-511. Nat. Biotechnol. 28 (6), 611–615.

Roobrouck, V.D., Vanuytsel, K., Verfaillie, C.M., 2011. Concise review: Culture mediated changes in fate and/or potency of stem cells. Stem Cells 29, 583–589.

Rose, C.W., Cox, G.J., 1924. The relation of arginine and histidine to growth. J. Biol. Chem. 61, 747–773.

Saxena, D., Qian, S., 2008. Laminin/entactin complex: A feeder-free surface for culture of human embryonic stem cells. BD Biosciences Technical Bulletin #464.

Shin, S., Dalton, S., Stice, S.L., 2005 Junn. Human motor neuron differentiation from human embryonic stem cells. Stem Cells Dev. 14 (3), 266–269.

Shooter, Ra, Gey, Go, 1952. Studies of the mineral requirements of mammalian cells. Br. J. Exp. Pathol. 33 (1), 98–103.

Siemssen, S.O., Sjøntoft, E., Kofoed, H., Olesen, H.P., 1985. On the blood supply of muscle-pedicled bone: An experimental study. Br. J. Plast. Surg. 38 (4), 506–511.

Slager, H.G., van Inzen, W., Freund, E., van den Eijnden-van Raaij, A.J.M., 1993. Transforming growth factor-b in the early mouse embryo: Implications for the regulation of muscle formation and implantation. Dev. Gen. 14, 212–224.

Smith, J.R., Vallier, L., Lupo, G., Alexander, M., Harris, W.A., Pedersen, R.A., 2008. Inhibition of Activin/Nodal signaling promotes specification of human embryonic stem cells into neuroectoderm. Dev Biol. 313, 107–117.

Straus, D.S., 1984. Growth-stimulatory actions of insulin in vitro and in vivo. Endocr. Rev. Spring 5 (2), 356–369.

Takahashi, T., Lord, B., Schulze, P.C., Fryer, R.M., Sarang, S.S., Gullans, S.R., et al., 2003. Ascorbic acid enhances differentiation of embryonic stem cells into cardiac myocytes. Circulation 107 (14), 1912–1916.

Takei, S., Ichikawa, H., Johkura, K., Mogi, A., No, H., Yoshie, S., et al., 2009. Bone morphogenetic protein-4 promotes induction of cardiomyocytes from human embryonic stem cells in serum-based embryoid body development. Am. J. Physiol. Heart Circ. Physiol. 296, H1793–H1803.

Thomson, J.A., Itskovitz-Eldor, J., Shapiro, S.S., Waknitz, M.A., Swiergiel, J.J., Marshall, V.S., et al., 1998. Embryonic stem cell lines derived from human blastocysts. Science 282, 1145–1147.

Tran, T.H., Wang, X., Browne, C., Zhang, Y.l., Schinke, M., Izumo, S., et al., 2009. Wnt3a-induced mesoderm formation and cardiomyogenesis in human embryonic stem cells. Stem Cells 27, 1869–1878.

Unger, C., Skottman, H., Blomberg, P., Diber, M.S., Hovatta, O., 2008. Good Manufacturing practice and clinical human embryonic stem cell lines. Human Mol. Genet. 17 (R1), R48–R53.

Vallier, L., Alexander, M., Pedersen, R.A., 2005. Activin/Nodal and FGF pathways cooperate to maintain pluripotency of human embryonic stem cells. J. Cell Sci. 118 (Pt 19), 4495–4509.

van der Valk, J., Brunner, D., De Smet, K., Fex Svenningsen, A., Honegger, P., Knudsen, L.E., et al., 2010. Optimization of chemically defined cell culture media–replacing fetal bovine serum in mammalian in vitro methods. Toxicol In Vitro 24 (4), 1053–1063.

Villa-Diaz, L.G., Nandivada, H., Ding, J., Nogueira-de-Souza, N.C., Krebsbach, P.H., O'Shea, K.S., et al., 2010. Synthetic polymer coatings for long-term growth of human embryonic stem cells. Nat. Biotechnol. 28 (6), 581–583.

Wakil, S.J., Titchener, E.B., Gibson, D.M., 1958. Evidence for the participation of biotin in the enzymic synthesis of fatty acids. Biochim. Biophys. Acta. 29 (1), 225–226.

Wang, L., Schulz, T.C., Sherrer, E.S., Dauphin, D.S., Shin, S., Nelson, A.M., et al., 2007. Self-renewal of human embryonic stem cells requires insulin-like growth factor-1 receptor and ERBB2 receptor signaling. Blood 110, 4111–4119.

Waymouth, C., 1970. Osmolality of mammalian blood and of media for culture of mammalian cells. In Vitro 6 (2), 109–127.

Xu, C., Inokuma, M.S., Denham, J., Golds, K., Kundu, P., Gold, J.D., et al., 2001. Feeder-free growth of undifferentiated human embryonic stem cells. Nat. Biotechnol. 19, 971–974.

Xu, C., Jiang, J., Sottile, V., McWhir, J., Lebkowski, J., Carpenter, M.K., 2004. Immortalized fibroblast-like cells derived from human embryonic stem cells support undifferentiated cell growth. Stem Cells 22, 972–980.

Xu, C., Rosler, E., Jiang, J., Lebkowski, J.S., Gold, J.D., O'Sullivan, C., et al., 2005. Basic fibroblast growth factor supports undifferentiated human embryonic stem cell growth without conditioned medium. Stem Cells 23, 315–323.

Yang, L., Soonpaa, M.H., Adler, E.D., Roepke, T.K., Kattman, S.J., Kennedy, M., et al., 2007. Human cardiovascular progenitor cells develop from a KDR+ embryonic-stem-cell-derived population. Nature 453, 524–528.

Yeo, G.W., Coufal, N., Aigner, S., Winner, B., Scolnick, J.A., Marchetto, M.C., et al., 2008. Multiple layers of molecular controls modulate self-renewal and neuronal lineage specification of embryonic stem cells. Hum. Mol. Genet. 17 (R1), R67–R75.

Yoshida, Y., Takahashi, K., Okita, K., Ichisaka, T., Yamanaka, S., 2009. Hypoxia enhances the generation of induced pluripotent stem cells. Cell Stem Cell 5 (3), 237–241.

CHAPTER 6

Cryopreservation of Human Embryonic Stem Cells

Trevor R. Leonardo

EDITOR'S COMMENTARY

Cryopreservation allows researchers to bank human pluripotent stem cells (hPSCs) at specific passages to preserve their genetic integrity so that they can be thawed and used at a later time. hPSCs that are continuously cultured for long periods acquire genomic aberrations and can potentially form heterogeneous populations that do not reflect the characteristics of the original cell line. It is always a good idea to cryopreserve cells to have a backup if the cells become contaminated. Because of staged hPSC differentiation protocols, it is sometimes possible to literally freeze an experiment in progress and continue it later. The greatest concern with cryopreservation is that the cells don't recover when thawed, or recover at such a low rate that there is selection for subpopulations after thawing the cells. This chapter gives detailed methods that will ensure the maximum survival of cells that are cryopreserved. Following these procedures may allow you to occasionally take a vacation from the lab without worrying about your cells!

OVERVIEW

Human pluripotent stem cells (hPSCs) are prone to apoptosis and differentiation when subjected to disassociation and detachment from culture plates. It is therefore important to have a method to preserve the pluripotent state of these cells while minimizing cell death. The two methods that are conventionally used to freeze down hPSCs are a slow freezing method and an open pulled straw vitrification method. The open pulled straw vitrification method has been shown to be highly efficient for the cryopreservation of hESCs (Reubinoff et al., 2001). However, vitrification is limited to only small numbers of cells, requires direct exposure to liquid nitrogen, and is a very time-consuming process. Advances in slow freezing methods have increased the ability to freeze bulk quantities of hPSCs efficiently and in a

J.F. Loring & S.E. Peterson (eds): Human Stem Cell Manual, Second edition.
DOI: http://dx.doi.org/10.1016/B978-0-12-385473-5.00006-0

relatively short period of time. Slow freezing can be significantly improved with the treatment of the cells with a small molecule, Y27632, also known as ROCK inhibitor, before and after freezing (Li et al., 2008; Martin-Ibanez et al., 2008). A 1-hour treatment has been shown to inhibit apoptosis of dissociated hPSCs grown on feeders or in feeder-free conditions to increase the survival rate during freezing without affecting their pluripotency (Li et al., 2009; Mollamohammadi et al., 2009; Watanabe et al., 2007). ROCK inhibitor blocks the activity of Rho Kinase and it is thought to increase hPSC survival by altering the adhesive properties of the cells (Krawetz et al., 2009). This chapter focuses on a method that works well in our laboratory, using ROCK Inhibitor both during slow freezing and during thawing of hPSCs.

PROCEDURES

Slow Freezing Method

■ Note

This method uses ROCK inhibitor to improve recovery from freezing.   ■

1. Expand a homogeneous population of hPSCs to approximately 90% confluence.
2. Mechanically remove all differentiated areas of colonies using a sterile pipette tip and wash with PBS to remove the unwanted cells.
3. Change medium to hESC medium supplemented with $10\,\mu M$ ROCK inhibitor and incubate for at least 1 hour.
4. During this incubation period, precool Nalgene freezing containers, prepare freezing medium (see Recipes), and label 1.8 mL cryovials with cell line name, passage number, type of passaging, date, and your initials.
5. Under the dissecting scope, mechanically scrape up colonies using a 1 mL pipette tip. Alternatively, if you are growing your cells feeder free, enzymatically passage them with Accutase (see Chapter 3).
6. For each well of a 6-well plate, collect the cells in 2–3 mL of medium in a 15 mL conical tube.
7. Centrifuge for 5 minutes at $200\,g$.
8. Gently resuspend pellet in $500\,\mu L$ of KnockOut™ Serum Replacement (Life Technologies) using a 1 mL pipette to triturate the cells.
9. Add $500\,\mu L$ of resuspended cells to each cryovial using a 1 mL pipette.
10. Add $500\,\mu L$ of premade $2\times$ freezing medium (see Recipes) drop-wise into each cryovial using a 5 mL pipette and mix gently by tapping the tube.

11. Quickly transfer the vials to the precooled Nalgene freezing container and place in a freezer at −80°C overnight.
12. Transfer the cryovials to liquid nitrogen the next day for long-term storage.

Thawing

1. Add 10 mL of hESC medium to a 15 mL conical tube (used to dilute DMSO in the freezing medium).
2. Remove cryovial from liquid nitrogen and rapidly transfer to 37°C water bath.
3. Once only a small ice pellet remains, remove the cryovial from the water bath, spray thoroughly with 70% ethanol and bring into the biosafety cabinet.
4. Carefully add the cells from the cryovial to the 15 mL conical tube using a 1 mL pipette.
5. Rinse the cryovial with 1 mL of hESC medium and add to the 15 mL conical tube to ensure all of the cells are collected from the cryovial.
6. Centrifuge for 5 minutes at 200 g.
7. Aspirate the medium and gently resuspend pellet into 2 mL hESC medium supplemented with 10 μM ROCK inhibitor using a 5 mL pipette.
8. Add the 2 mL of cell suspension to one well of a 6-well plate (with appropriate substrate) and place in 37°C incubator.
9. On the following day, change medium to hESC medium without ROCK inhibitor (see Figure 6.1)

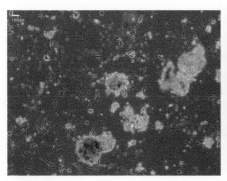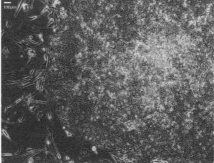

FIGURE 6.1

Images of a hPSC line thawed after cryopreservation using the slow freezing method. The image to the left shows attached clumps of cells 1 day post-thawing and the image to the right is the same culture approximately 8 days post-thawing.

PITFALLS AND ADVICE

- It is important to have a confluent well of undifferentiated hPSCs that are growing well to ensure high quality colonies upon thawing.
- Removing the differentiated colonies prior to freezing will help prevent differentiation of colonies upon thawing.
- Centrifugation of the cells at the suggested speeds, in the specified tube sizes and amount of medium will prevent damage to the hPSCs.
- Once the freezing medium is added to the cryovials, it is crucial that the cells are not kept at room temperature for more than 5–10 minutes as the freezing medium contains DMSO, which is toxic to the cells.
- Freezing down more cells than necessary is a better policy than barely freezing down enough cells.
- Sometimes thawed hPSC colonies may take a couple of passages before they show good morphology. Be patient!

ALTERNATIVE PROCEDURES

An alternative to the slow freezing method is the open pulled straw vitrification method. Please see the reading list below and also Chapter 35 for a detailed protocol on this method.

EQUIPMENT

- Tissue culture hood
- Tissue culture incubator, 37°C, 5% CO_2
- Dissecting microscope
- Pipettes (20 μL, 200 μL, and 1000 μL)
- Pipette aid
- Standard tissue culture equipment
- Centrifuge that can hold 15 mL conical tubes
- Freezers: −80°C, liquid nitrogen
- Freezing container, Nalgene® "Mr. Frosty"

REAGENTS AND SUPPLIES

- 2 mL cryogenic vials
- 6-well culture dishes
- 15 mL sterile conical tubes
- Sterile 9″ pasture pipettes
- Pipette tips for Rainin or similar pipettor with filter
- 5 mL, 10 mL, 25 mL sterile disposable pipettes

Recommended Reagents

Item	Supplier	Catalog #	Note
DMEM/F-12 + HEPES + L-Glutamine (Dulbecco's Modified Eagle's Medium)	Life Technologies	11330-032	
KnockOut Serum Replacement	Life Technologies	108280-028	Contains bovine products
GlutaMAX™	Life Technologies	35050-061	
MEM-Non-essential amino acids 10 mM (100×)	Life Technologies	11140-050	
2-Mercaptoethanol	Life Technologies	21985-023	
PBS (Dulbecco's Phosphate-Buffered Saline without Calcium and Magnesium)	Life Technologies	14190-144	
Recombinant human bFGF	Life Technologies	13256-029	Use at 20 ng/mL
Dimethyl Sulfoxide (DMSO)	Sigma-Aldrich	D2650	
ROCK Inhibitor (Y-27632)	Millipore	SCM075	Use at 10 μM
StemPro® Accutase™ Cell Dissociation Reagent	Life Technologies	A11105-01	

RECIPES

2× Freezing Medium

Ingredient	Amount	Final Concentration
KnockOut™ Serum Replacement	40 mL	80%
DMSO	10 mL	20%

hESC Medium

Ingredient	Amount	Final Concentration
DMEM/F-12 + HEPES + L-Glutamine	200 mL	80%
KnockOut Serum Replacement	50 mL	20%
GlutaMAX 200 mM (100×)	2.5 mL	1x
MEM-Non-essential amino acids 10 mM (100×)	2.5 mL	1x
2-Mercaptoethanol	455 μL	0.1 mM
Recombinant Human bFGF	Use at 12 ng/mL	12 ng/mL

READING LIST

Krawetz, R.J., Li, X., Rancourt, D.E., 2009. Human embryonic stem cells: Caught between a ROCK inhibitor and a hard place. BioEssays 31 (3), 336–343. doi: 10.1002/bies.200800157.

Li, X., Meng, G., Krawetz, R., Liu, S., Rancourt, D.E., 2008. The ROCK inhibitor Y-27632 enhances the survival rate of human embryonic stem cells following cryopreservation. Stem Cells Dev. 17 (6), 1079–1085.

Li, X., Krawetz, R., Liu, S., Meng, G., Rancourt, D.E., 2009. ROCK inhibitor improves survival of cryopreserved serum/feeder-free single human embryonic stem cells. Hum. Reprod. 24 (3), 580–589. doi: 10.1093/humrep/den404 (Research Support, Non-U.S. Gov't).

Martin-Ibanez, R., Unger, C., Stromberg, A., Baker, D., Canals, J.M., Hovatta, O., 2008. Novel cryopreservation method for dissociated human embryonic stem cells in the presence of a ROCK inhibitor. Hum. Reprod. 23 (12), 2744–2754.

Mollamohammadi, S., Taei, A., Pakzad, M., Totonchi, M., Seifinejad, A., Masoudi, N., et al., 2009. A simple and efficient cryopreservation method for feeder-free dissociated human induced pluripotent stem cells and human embryonic stem cells. Hum. Reprod. 24 (10), 2468–2476.

Reubinoff, B.E., Pera, M.F., Vajta, G., Trounson, A.O., 2001. Effective cryopreservation of human embryonic stem cells by the open pulled straw vitrification method. Hum. Reprod. 16 (10), 2187–2194.

Watanabe, K., Ueno, M., Kamiya, D., Nishiyama, A., Matsumura, M., Wataya, T., et al., 2007. A ROCK inhibitor permits survival of dissociated human embryonic stem cells. Nat. Biotech. 25 (6), 681–686.

CHAPTER 7

Setting up a Research Scale Laboratory for Human Pluripotent Stem Cells

Cleo Choong, Mahendra Rao, and Ian Lyons

EDITOR'S COMMENTARY

We have distilled our collective experience of setting up over ten stem cell laboratories on three continents, with the intention of lessening the pain of others who are setting up such laboratories for the first time. We are sure of two things: first, the effort extended in planning and setting up systems and processes for a laboratory will be repaid many fold in future efficiencies, and, second, the effort to set systems in place from the very beginning is much less than retro-fitting them. So if you are fortunate to be setting up a laboratory from scratch, our advice is to try to get it right, right from the beginning.

The editor has set up three stem cell laboratories herself and has long wished for a comprehensive source of information like this chapter. There are so many mistakes to be made in such a complex operation, and the stakes are so high because of the great potential of hPSCs. Following the advice in this chapter can ensure that a lab will work well (and be free of mycoplasma!).

OVERVIEW

The recommendations outlined in this chapter are for a small-to-medium scale academic or biotech research laboratory; those setting up manufacturing or therapeutic laboratories will have a different set of regulatory and logistic constraints. Most of the design features for a human pluripotent stem cell (hPSC), human embryonic stem cell (hESC) and/or induced pluripotent stem cell (iPSC) laboratory are the same for any other cell culture laboratory. The physical laboratory space needs to be set up in a way that efficiently reflects the expected workflows, taking into account the number of workers who will be sharing the space and the equipment and the expected work throughput. Like any well-run cell culture facility, materials and supplies need to be monitored to ensure that quality and amounts match the demands, and

J.F. Loring & S.E. Peterson (eds): Human Stem Cell Manual, Second edition.
DOI: http://dx.doi.org/10.1016/B978-0-12-385473-5.00007-2

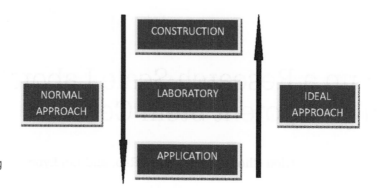

FIGURE 7.1
Ideal approach in designing hPSC culture laboratory.

waste is disposed of appropriately. Similarly, the ancillary activities for molecular analysis of cell phenotype are not unique to hPSCs. However, given the need for a regular supply of cells with defined characteristics and uncompromised differentiation potential, a careful system of cell banking is even more critical for hPSCs than for other cell types.

The critical issue in setting up the laboratory is planning for long-term success (Figure 7.1). This will rest on the functional design and utilization of the space and the appropriate equipment. Most important is the implementation of quality control systems that allow a continuous supply of validated cells and reagents. This book provides most of the protocols and methods for these systems, and other resources that are also available; however, there is no substitute for hands-on training, and wherever possible, we recommend that key personnel receive specific training in hPSC manipulation, and that the laboratory develops a network of communication and collaboration with other hPSC groups.

LABORATORY

General Considerations

- hPSC culture laboratories are similar to other cell culture research facilities. Cell culture laboratories should be located in an environment that is tidy and not crowded. Layout should allow for tissue culture, standard biochemistry and molecular biology, and cell imaging experiments. Incoming material and waste disposal logistics need to be accommodated.
- When setting up a laboratory, we typically define two distinct work areas: a cell culture area with rooms that can be dedicated for different types of work (Figure 7.2) and a molecular biology area with laboratory benches and appropriate equipment organized for common work flows (Figure 7.3).

FIGURE 7.2
Compartmentalized cell culture facility showing adjacent but separate rooms that are used for culturing different types of stem cells (adult versus embryonic, human versus animal). Arrow shows second biological safety cabinet in adjacent room.

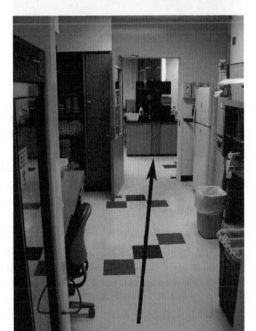

FIGURE 7.3
Arrow shows the molecular biology area that is adjacent to but outside of the cell culture facility, allowing easy access but maintaining the integrity of the cell culture area.

If possible, construction should consist of a modular framework that allows the laboratory to be reconfigured easily, if needed.

■ A centralized CO_2 gas system with all CO_2 gas cylinders located outside the laboratories saves valuable laboratory space and simplifies the monitoring of gas levels, while the exchange of cylinders does not compromise cleanliness of the culture space. Drop lines carry CO_2 gas to

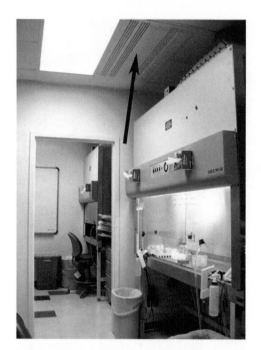

FIGURE 7.4

The air handling system of the cell culture rooms are HEPA-filtered (arrow) and maintain positive pressure with respect to adjacent non-cell culture areas.

equipment in each tissue culture room and in the fluorescence activated cell sorting (FACS) and microscopy rooms, if required for live cell imaging.

■ High quality water for the laboratory is supplied through a laboratory-specific Milli-Q or similar system and regular testing procedures for the quality of the water are needed, as water quality is critical to all cell culture.

■ In order to facilitate cleaning and maintenance, vinyl flooring, gloss paint (low in volatile organic compounds), and bench tops of impermeable materials should be used as appropriate. Ideally, the rooms should be lit with fluorescent light and ambient temperature should be constant (typically 18–22°C), while circulating air should be passed through 0.22 μm HEPA filters (Figure 7.4). Cell culture laboratories should be under positive pressure, while molecular biology areas or those used for viral work should be under negative pressure.

■ Standard safety requirements for laboratories, conforming to safety standards in the local jurisdiction, need to be implemented, but should include eye-washes, safety showers, first aid kits, fire extinguishers, and chemical spill kits at appropriate sites along the corridors. Sinks are typically located at the end of each workbench and, ideally, the taps will be foot activated (Figure 7.5).

FIGURE 7.5
All sinks in the cell culture areas are foot-pedal controlled (arrow), allowing hand-washing without contamination of or by fixtures.

- Common work areas for procedures such as media preparation will be conveniently located in the middle of the laboratory. Computers, internet connection points, and telephones are provided at the end of designated workbenches for communal use.
- When the layout of the laboratory has been determined and equipment locations decided, electrical wiring needs can be planned. Depending on local supply voltage and supply type, special circuits will probably be needed to supply higher voltage or three-phase power to some of the large equipment. In addition, certain equipment, such as ice machines and autoclaves, require easy access to floor drains and water inlet points.
- We recommend that all storage freezers and incubators should be connected to an emergency back-up supply and that a centralized alarm system, linked to key equipment such as freezers and incubators, should be considered.

Cell Culture Area
- Depending on available space, the cell culture areas should be functionally separated into dedicated rooms (or areas) for hPSCs, for non-human cells (e.g. mouse feeder cells), and for viral work. Ideally, each would have automatic doors such as elbow-activated glass sliding

doors for easy access, and cell culture rooms should be under positive air pressure, while the room for viral work should be under negative air pressure.

- We have found it useful to equip each tissue culture room with independent workstations that consist of a biological safety cabinet (BSC), a CO_2 incubator, a vacuum aspiration system, a complete set of personal equipment such as pipetting aids, micropipettes (2–1000 μL), and a drawer unit containing a selection of serological pipettes, filters, syringes, needles, culture vessels (flasks, plates, dishes), and cryopreservation supplies. From these workstations, there is easy access to shared equipment such as sinks, water baths, vortex mixers, centrifuges, and inverted microscopes for bright field/phase contrast visualization of cells in culture. Microscopes should be physically isolated from centrifuges and other vibrating equipment.

Set up for Manual Manipulation of hPSC Cultures

At some point, hPSCs always require some manual manipulation under low power microscopy, whether it is observing and collecting colonies lifting from dishes by enzymatic treatment, clearing cultures of differentiated areas, or manually passaging cells by cutting colonies into pieces. Differentiation protocols also often require selecting cell groups with particular morphology from cultures by dissection or by removal of unwanted cells. See Chapters 1, 10–12, 23–28 for more information about these methods. There are two types of set up for manual manipulations of cultures in a sterile environment.

Set up for Class II Biosafety Cabinets

- One option is to use a BSC equipped with a heated base and a stereomicroscope to facilitate colony picking or other manipulations, and also germicidal ultraviolet lamps (under timer control), outlets for CO_2 and vacuum, 2 L collection flasks, and a power supply for equipment such as pipetting aids.
- Fitting a zoom dissecting microscope to a class II BSC is not trivial. hPSC colony picking requires a relatively large working distance and a heated stage is useful for the extended time the cultures are manipulated. A low microscope base with sloping sides allows easier transfer of the cell culture dish from the worktop to the stereomicroscope with minimal movement, and reduces chances of spillage or contamination.
- We have found Gelman BH Class II Series (type A/B3) BSCs coupled with the Leica MZ6 to be a good combination. The microscope can be fitted with extended eyepiece tubes, and the BSC is modified to allow the extendable eyepieces to protrude from the sash toward the user (Figure 7.6). A pneumatic window sash is available and swings forward and back,

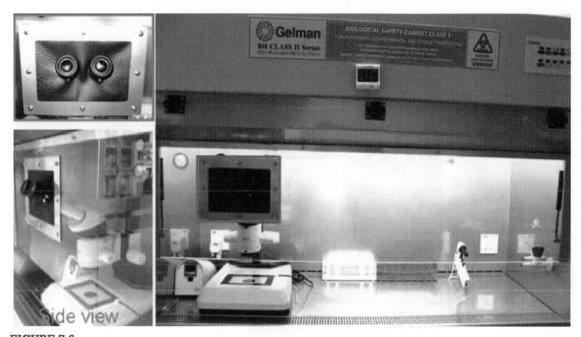

FIGURE 7.6
Biological safety cabinet with fitted dissecting microscope.

allowing access to the BSC interior for equipment transfer and cleaning without having to move the stereomicroscope.

Set up for Laminar Flow Workbench

■ A less expensive alternative is to simply use a dissecting microscope in a laminar flow hood for colony dissection and an unmodified BSC for all other cell culture (Figure 7.7). This provides adequate elbow room for manually manipulating cultures while maintaining sterility. After manipulation, the plate is returned to a regular BSC for transfers and medium changes. We find that a Forma Laminar Airflow Workstation works well with an Olympus SZX stereomicroscope with a heated base and dark field illumination. To keep the laminar flow hood sterile, leave the air vents on at all times and wipe the cabinet down with 70% ethanol after each use.

Incubators

■ The CO_2 incubators should be conveniently situated beside the BSCs instead of behind the user, minimizing movement of culture dishes from BSC to incubator and reducing the chance of accidental spills.

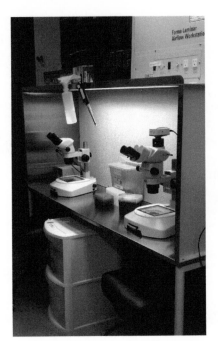

FIGURE 7.7
A laminar flow hood with a dissecting microscope can be used for manually cutting colonies. Afterwards, cells are returned to a regular BSC for further manipulation.

Electric outlets in some of the incubators allow temporary installation of equipment inside the incubator (for example an orbital shaker for embryoid body formation).

■ There are many CO_2 incubator options available. Important criteria include: readily cleaned internal surfaces without cracks or seams, monitored CO_2 level and temperature with tight tolerances, and alarms. We recommend the Binder CB 150 with a data recording system which allows CO_2 levels and changes in temperature levels to be recorded for up to 14 days. Installation of tri-gas control incubators should be considered if the effects of manipulation of oxygen tension are likely to be investigated at some stage. The extra cost is incremental, since these incubators can also be used conventionally at ambient oxygen.

Molecular Biology Area

The organization of the molecular biology area should be determined by both the space available and the number of workers employed. For greatest efficiency in high density laboratory areas, the space can be organized into functional areas for specific workflows. In this system, all the equipment and reagents are readily at hand for specific processes and workers can come to a dedicated and shared area and find all that is needed for their task. Examples of this would be an area for immunostaining, with all equipment and antibodies easily at hand. Another example is a separate bench (preferably a laminar flow cabinet) for RNA preparation and for polymerase chain reaction (PCR) set-up which is physically remote from the area for PCR product manipulation and analysis. The latter has the advantage of separating the PCR set-up from analysis, reducing the chances of reagent contamination with PCR products.

If space is less limited, a separate bench or area can be assigned for each worker. Each person will have access to a basic set of tools including micropipettes (2–1000 μL), a PCR machine, centrifuge, and vortex mixer, and an under-the-counter refrigerator and freezer.

In addition to the functional areas described above, we have found it convenient to have "communal areas" with more generally used equipment such as analytical balances, centrifuges, spectrophotometers, gel imaging equipment, pH meters, plate readers, ultracentrifuges, and power supplies for electrophoresis.

Additional Considerations

In addition to laboratory areas dedicated to cell culture and to molecular cell analysis, there are a number of other activities that might require dedicated areas. These may be part of the hPSC culture facility itself or may be part of core institute facilities, but if access is needed, it should be taken into account during the planning phase.

Dedicated rooms for microscopy, image analysis and flow cytometry equipment will reduce dust and traffic flow. Depending on the equipment used, purpose-built benches with vibration isolation are a good investment.

Sufficient storage space needs to be available to the laboratory. In particular, disposable plastic cell culture supplies are bulky and having a separate storage room helps avoid bringing potentially dusty cartons into the laboratory. Storage for flammable liquids must be available in fireproof cabinets. Fridges, freezers, and ultra-low freezer units should be located carefully. Apart from under-bench fridges and freezers, it is good to have these units housed separately wherever possible, as they are a major source of ambient noise in laboratories.

Long-term cell storage has traditionally been done in liquid nitrogen freezers. If these are to be used, they should be housed separately from the laboratory in a well-ventilated space with low oxygen alarms to reduce danger of asphyxiation. This could be in the same area used to store consumables. Ultra-low (−152°C) freezers are an alternative to liquid nitrogen storage; they are reported to have become increasingly reliable and do not need a constant supply of liquid nitrogen, the regular replenishment of which brings dirt into the laboratory and is also expensive. All cryopreserved cell lines should be stored in two different locations, so that the catastrophic loss of one freezer does not mean loss of all cell stocks. In the past, earthquakes, tornados, hurricanes, flooding, and severe storms have made laboratories inaccessible, so be prepared for the worst.

Hazardous biological waste needs to be disposed of according to local regulations. In most regions this will require access to an autoclave, and this has to be taken into account when the laboratory is being planned.

RECORD KEEPING

General Procedures

Good record keeping is crucial for proper operation of a stem cell laboratory; paperwork ranging from quotations through purchase orders, delivery orders, invoices, certificates of analysis (CofA), material safety data sheets (MSDS), material transfer agreements (MTAs), research agreements, and licenses should be filed and organized in an easy to understand system. This is particularly important in the case of hPSCs, for which regulations can be stringent and vary by country. It is critical that the laboratory has the proper permissions to use hPSC lines; these may be different for academic and commercial labs. Careful documentation from the start will save considerable time and stress in the future.

Reagent and Consumable Tracking

A reliable supply of validated reagents is critical to hPSC research quality. In general, reagents from major suppliers can be used with confidence; however, some are known to be subject to lot-to-lot variation. There are a number of reasons for this, including the unfortunate fact that some reagents are quality tested against cell types less fastidious than hPSCs. An example of this is KnockOut™ Serum Replacement (KSR; Life Technologies catalog no. 10828-028), a product that is reliable for the cells it was designed for, mouse embryonic stem cells (mESCs), but for which lot-to-lot variation has been described for hPSCs. Antibodies are particularly troublesome, because vendors sometimes change lots without notice to the consumers. To maintain a continuity of supply of reliable reagents, it is important to have a system that:

- Ensures that the laboratory does not exhaust any reagent.
- Allows timely testing of incoming reagents to ensure comparability with those being replaced.

For hPSC culture, some of the critical reagents that should be tested well in advance of being brought into general use include fetal bovine serum (FBS), KSR, and mouse embryonic fibroblasts (MEFs). Other reagents are less variable and are usually used untested. If a reagent fails your quality control, contact the vendor. It is important to vendors that they maintain a reputation for consistently high quality reagents, and you will be doing them a favor by reporting problems. Always request replacements for reagents that fail in your hands.

Stock and Reagent Inventory

The levels of stocks of reagents that need to be held to ensure continuity of supply will depend upon the rate of usage by the laboratory, on the delay between ordering and delivery, the probability of an exceptional delay, and the risk aversion of the laboratory director. For a guideline, we recommend that stocks of all reagents be maintained at a minimum of three times the delivery lead-time. So for reagents that typically can be delivered 1 week after ordering, we recommend that a minimum of 3 weeks' normal usage should be maintained. If these reagents are being tested to validate lots, then the holding time should be increased to account for the time it takes to test. Maximum holding times will depend on the storage capacity of the facility and the shelf life of the reagents.

For all reagents, it is important to keep track of the manufacturers' stock keeping units (SKUs) and lot numbers. These numbers should be maintained as part of the laboratory record keeping and inventory records. Each worker should record these details in their laboratory notebook for every experiment or process undertaken. It is relatively rare to have a lot fail, especially if rigorous lot matching and testing is undertaken systematically; however, rare, sporadic lot failure will occur, and recovery from a cell culture "crash" or unexplained failure will be more efficient if all reagents can be traced. The savings in time and money will justify the upfront time and effort to establish such a system. All members of the lab should take responsibility for helping to maintain the tracking system, from postdoctoral staff to lab technicians or interns.

Lab Index System

In addition to labeling the outside of drawers and cabinets with their contents, it is convenient to number each of the laboratory drawers and cabinets and to keep a database of their contents. This is particularly useful in a laboratory that is likely to have frequent visitors or changes of staff. This also facilitates stock-taking and auditing. The laboratory manager should ensure that these lists are updated regularly so that finding irregularly used items is quick and easy.

Freezer Inventory Systems

Keeping track of the stored cell stocks is critical for the smooth operation of the facility on a day-to-day basis, and also allows for long-term planning. Whether cell stocks are stored in an ultra-low temperature mechanical freezer or in liquid- or vapor-phase nitrogen, a system of cell labeling and record keeping is essential to allow retrieval of cells, and to keep track of the rate of cell stock usage. The cell freezer will typically come with a system of racks to store the cryovials and allow them to be quickly located. This must be backed up by an inventory system that keeps track of each cryovial. This can be done electronically, using a purpose-designed system or more generic database software, or it can be done manually using a dedicated laboratory notebook or ring binder.

In either case, the data need to be regularly backed up. Each cryovial must be unambiguously labeled either using printable cryoresistant printer labels, or with a cryoresistant marker pen. Details should include cell type and passage number, date, technician, and laboratory notebook reference. Like any system, this will fail if there is not compliance. Every vial added or removed should be noted immediately. Proper adherence by the users is essential for the smooth functioning of the system, so creating an environment where users have ownership over their work is important.

Examples of cryostorage record systems can be found at http://www.cryotrack.com, and http://www.nalgenelabware.com/techdata/technical/manual.asp.

EQUIPMENT MAINTENANCE

Equipment maintenance is essential for the smooth operation of the laboratory and for the generation of reproducible data. We recommend providing all equipment and accessories with maintenance schedules and a local and alternate contact person for sales and service. All major equipment must be tested and commissioned before use, and maintained on a regular basis by the vendor or a reputable service organization. It is important to keep track of dates and records of these procedures in order to prolong the lifespan of each piece of equipment and ensure its proper use. Usage of common large equipment such as centrifuges and spectrophotometers should be tracked using a logbook. For regular maintenance of equipment, there should be clear ownership and visibility.

- *Incubators* should be checked on a weekly basis using a digital thermometer and a Fyrite device. The readings obtained can be recorded on the door of the incubator so that all users are aware of any fluctuations.
- *Water baths* are a common source of contamination. To minimize this, the water should be changed at least weekly. Siting the water baths adjacent to the sinks simplifies the changing of water and cleaning. If possible, eliminate water baths entirely and dry heat media using heat-conducting beads. Keep in mind that even dry heaters must be cleaned regularly to avoid contamination.

- *Microscopes* are often used while not optimally aligned. They should be adjusted regularly, and all scientists and staff using them should be trained in their correct use. Phase contrast should be optimized, and the light field should be even across the field. Photodocumentation is important, and every photo should be good enough for publication.
- *BSCs* should have their interior wiped down after every use with 70% ethanol. The germicidal ultraviolet lamps should be switched on at the end of each day. For this reason, it is important that the equipment kept inside the BSC is able to withstand exposure to UV light; remove all vulnerable equipment and supplies before operating the UV light. Replace the UV light regularly, because its power fades with time.

SAFETY

Staff should be trained in the safe handling of all equipment, reagents, and biological materials and in the use of first aid and safety equipment. Laboratory gowns and safety glasses should be supplied and be worn at all times. Personal protective equipment (PPE) should be available for all pertinent laboratory work. In particular, full-face shields should be worn when removing and thawing ampules from cryostorage.

Biological Hazards

Human and other primate cells should be handled using Biosafety Level 2 practices and containment appropriate for the institute and local authorities. All work should be performed in a BSC with appropriate PPE (gloves, glasses, laboratory coats), and all waste material should be decontaminated by autoclaving or disinfection before discarding. For a discussion of recommendations and requirements refer to "Biosafety in Microbiological and Biomedical Laboratories" at http://www.cdc.gov/biosafety/publications/bmbl5/index.htm.

QUALITY CONTROL

The aim of the laboratory set-up is to have systems in place that lead to reliable and reproducible experimental outcomes (Figure 7.8). This is dependent on four critical factors:

1. *Reliable techniques* that yield a minimum of variability between processes. Use of methods developed by others and shown to be reliable is a good way to establish a laboratory, and to have the results of that laboratory comparable to the work of others.
2. *Validated reagents*. As this is critical, we described this in stock keeping and discuss later a validation system for laboratory reagents.

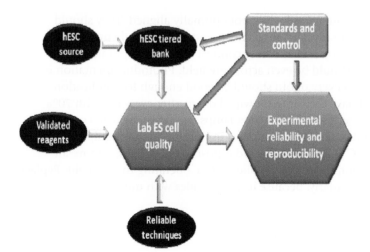

FIGURE 7.8
hPSC quality systems: shown for hESCs but applicable to iPSCs.

3. *The quality of the hPSCs.* A suitable method for generating a structured stock of hPSCs that have been validated molecularly and functionally, and which has sufficient size to cover the demands of the laboratory is also described (see also Chapter 8 for banking practices).
4. *A system of testing* and standards that confirms comparability of hPSCs being used. They need to be free of contamination, genetically stable, to express defined markers, and capable of differentiation into representatives of the three germ lineages.

Reagent Testing

Because lot-to-lot variation has been described for FBS, KSR, and MEFs, we recommend that a regimen be introduced to test lots before they are brought into general use. Other reagents are less variable and are usually used untested.

Lot Testing FBS for MEFs

Almost all FBS lots are able to support MEFs, but they still need to be tested in advance, because the cost may be significant. We confirm FBS lot suitability on the basis of cell expansion over two passages, but do not monitor biological activity. A positive control in this assay is either an FBS sample that has been shown previously to support MEFs, or if unavailable, a commercially available FBS validated for support of mESCs (e.g. Life Technologies catalog no. 10439-016). mESC-qualified FBS can be used routinely for MEF propagation, but is significantly more expensive than unvalidated FBS.

Protocol

1. Make medium with each of the test FBS samples, and with the control.
2. Seed MEFs at usual density in triplicate 10 cm dishes in medium.
3. Harvest cells and count after 5 days (but before the control reaches confluence).
4. Count cells and replate a second passage in each medium.
5. Again harvest cells and count after 5 days (but before the control reaches confluence).
6. Suitable batches of FBS will have cell yields in excess of 80% of that of the control lot.

Testing FBS and KSR for hPSCs

When testing lots of serum or KSR for use with hPSCs, we passage cells four times before confirming morphology and marker expression patterns, since many medium components do not have an obvious effect until at least three passages. Ideally, one would use a positive control serum or KSR batch, but these are not commercially available. Typically most laboratories use a batch that has been previously used successfully. A new laboratory that is being set up will not have that option, but may be able to obtain samples from an established laboratory.

Protocol (see Chapter 1 for Medium Recipe)

1. Make medium with the test serum or KSR batches and with a positive control, if available.
2. Passage cells in the usual manner for three passages in the test media.
3. Passage into appropriate culture systems for immunocytochemistry for SSEA-4, POU5F1/OCT4, and alkaline phosphatase staining, and for expansion to extract RNA for microarray or PCR analysis.
4. Analysis: serum or KSR batches are acceptable if they support cells as well as the control lot that has been used successfully and/or have these minimal performance characteristics:
   - Immunocytochemistry: 85% of cells express SSEA-4 and POU5F1/OCT4.
   - RT-PCR: conventional and quantitative PCR is helpful to assess expression of genes that are markers of undifferentiated cells and to confirm the absence of markers of differentiation for each of the three germ lineages.
   - Morphology and alkaline phosphatase: more than 95% of colonies express alkaline phosphatase and are morphologically undifferentiated.

Lot Testing MEFs

During preparation, MEFs should be tested for microbiological contamination. Testing new batches of MEFs before they are brought into general use is a wise precaution. The testing procedures and readouts are essentially the same as those described above for serum and KSR. Cells are maintained on test and control MEFs for four passages and then marker expression is analyzed as in "Testing FBS and KSR for hPSCs" above.

CELL BANKING

Rationale

A well-structured cell bank is crucial for successful cell culture in general, but is critical for hPSCs, with their inherent karyotypic instability and tendency to differentiate. A tiered or master cell banking system brings validated and reproducible cells to each experiment and procedure (see Chapter 8 as well). Cells will be:

- At a defined range of passage number for all experiments
- Of normal karyotype
- Free of microbiological contamination
- Capable of differentiation to desired lineages
- Of undifferentiated phenotype with expression of appropriate molecular markers.

Establishment of a Tiered Banking System

A tiered banking system typically consists of three levels of cryopreserved cells: the earliest "mother stock," the critical "master cell bank," and the "working cell bank" (Figure 7.9). The number of vials of each depends on the cell type and projected usage, and on the extent of the testing required for that cell type. For hPSCs, the testing recommended is extensive and relatively expensive, and so we suggest that large banks of cells be frozen.

At each level of banking, the cells need to be tested for:

- Karyotype
- Genotype
- Differentiation capacity
- Marker expression (both the presence of markers of undifferentiated cells and absence of markers of differentiation)
- Microbial contamination.

Other aspects of cell status can also be tested since the cells have been shown to drift with extended culture (e.g. imprinting status); however, that drift has not been shown to be associated with systematic loss of pluripotency and so may be seen as less critical.

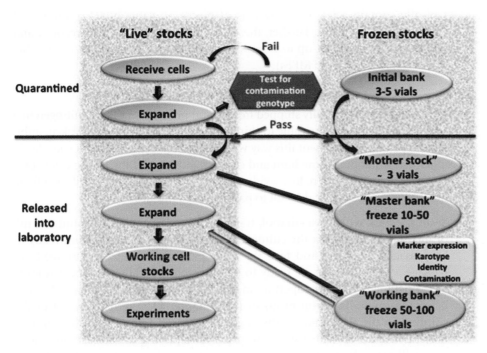

FIGURE 7.9
hPSC workflow and tiered cell bank.

Incoming Cells

When any cells first arrive, it is prudent to treat them as if they were micro-biologically contaminated. Preferably, they should be handled in a separate quarantine facility, although in usual practice this is not always possible and they are handled carefully in the same laboratory, with care to avoid aerosols that are capable of distributing contaminant microorganisms. Biosafety cabi-nets are diligently swabbed after handling new cells, and if a separate incuba-tor is not available, the culture vessels are placed in an uncovered, or vented, plastic box to contain any inadvertent spills.

Thawing and Plating Cells

If the cells arrive on dry ice, transfer to liquid nitrogen storage immediately. Some distributors recommend immediate thawing of samples shipped on dry ice rather than transfer to liquid nitrogen. Cells can then be recovered after all reagents have been assembled and validated as appropriate.

Initial cell recovery and maintenance should be done using the instructions of the cell providers, since troubleshooting by the providers is difficult if they are not familiar with your methods. In general, cells should be recov-ered as gently as possible, and patience must be exercised. Cryopreserved vials typically contain a few to a few tens of pieces of viable hPSC colonies and these can easily be lost. Further, these colony pieces are slow to recover, and often do not appear for up to 2 weeks (or even after a passage is required due to the exhaustion of the MEFs).

Recovery of Cells Shipped "Live"

Although cells are typically shipped frozen, and preferably in liquid nitrogen in a "dry shipper", it is also possible to ship them as living cultures in a flask at ambient temperatures. Cells sent this way will usually arrive in a T25 flask completely filled with equilibrated medium and closed tightly to give the flask greater compressive strength and to reduce swirling that might dislodge cells. Cells in a flask, however, are much more difficult to passage mechanically.

Swab the flask with 70% ethanol, transfer to the BSC, remove all but 6–8 mL of medium, and inspect the cultures microscopically. If they are not ready to be passaged, incubate and feed the cells, per usual, until they are ready to passage. When the cells are ready to be passaged, carefully use a heated knife or scalpel blade to cut an aperture in the top of the flask and passage the cells using the microdissection protocol (see Chapter 1). Alternatively, all of the colonies in the flask may be dislodged with a cell scraper and the colonies may be broken up with trituration prior to re-plating. This technique, however, will generally lead to the plating of colonies that are larger than desired and may lead to greater than normal levels of differentiation.

Initial Expansion: "Mother Stock"

The aim of this phase is to establish the earliest bank of frozen cells as soon as possible after receipt, and to confirm that they are not contaminated with microorganisms.

Expansion and Cryopreservation

About three vials should be frozen within two or three passages of the cells' arrival. For this phase, the cells should be maintained using the protocols provided by the supplier of the cells. Cells will be maintained in culture and further expanded to generate the next level of stocks, while simultaneously some cells will be tested for contamination and genotyped to confirm the identity of the cells.

Cells can be cryopreserved using variations on commonly used methods; however, for the first freeze, the methods recommended by the supplier of the cells needs to be followed. Again, this will allow more efficient troubleshooting, if necessary. One way to confirm cryopreservation viability of another cryopreservation method is to thaw one of the vials from the first freezing to confirm recovery before the freezing of additional cells for the master cell bank.

Testing
Microbial Contamination

Cultures must be monitored for mycoplasma, bacteria, and fungi on a regular basis. The most frequent contaminants of hPSC cultures are mycoplasma, which are small prokaryotes without cell walls. Since most antibiotics do not kill mycoplasma, it often persists at low levels without developing the turbidity that indicates bacterial contamination, and may induce a variety of deleterious effects, ranging from changing cell metabolism to chromosomal changes. If the cells are contaminated at this early stage, abandon the expansion and get fresh cells. See Chapter 4 for detailed information about contamination.

Genotype

Cells should be genotyped for positive identification because, although rare, confusion can occur. There are a number of methods available, including single nucleotide polymorphism (SNP), human leukocyte antigen (HLA), and short tandem repeat (STR). These can be carried out in house or by a contract organization such as a paternity testing lab. See Chapter 14.

Secondary and Tertiary Expansion: "Master Cell Bank" and "Working Cell Bank"

The master and working cell banks are the two critical levels of the tiered cell bank and utmost care should be taken during the expansion of the cells for them. Diligently change medium as recommended, passage using best practice manual dissection, and dissect away any regions of differentiation. Limits are set as to the number of times that cells are serially passaged during normal laboratory activities, and cells must, therefore, be recovered from the working cell bank. As the working cell bank is depleted, it is replaced by expansion of cells from the master cell bank. At each banking step, the cells are extensively validated as described below.

The number of dishes of cells required for this freeze depends on the size of the cell bank planned, but as a rule of thumb, one to two vials per 60 mm dish or three to six vials per six-well plate are needed. So, for a bank of 20 vials, plan to have 10–20 60 mm dishes at confluent density.

Cells can be cryopreserved by a variety of methods (see Chapter 6). We recommend thawing a vial of cells previously frozen by your laboratory to ensure that appropriate recovery rates can be achieved, and expanding these cells to undertake the battery of quality control (QC) tests, including microbial contamination, genotype (genetic identity and karyotype), and phenotype (molecular marker expression and differentiation capacity). These tests are described below in the section on QC standards.

QC Standards: Routine Monitoring of Cell Status

Although hPSC lines are highly similar to each other in their expression of cell surface antigens and markers characteristic of the hPSC state and are relatively stable over time in continuous culture, like all other cell types hPSCs can undergo changes in culture. Most dramatic are changes in the chromosomal composition of the cells, but other forms of genetic and epigenetic drift have also been detected in hPSCs after extended culture. These changes include accumulation of DNA damage, oxidative damage, erosion of telomeres, acquisition of mitochondrial mutations, loss of imprinting, and reactivation of inactive X chromosomes in female hPSCs.

The potential for these changes makes it important to monitor cells regularly to ensure maintenance of baseline hPSC characteristics. There are many aspects of cell status that could be meaningfully tested; however, to carry out all of these tests while freezing regularly through extended passage would become prohibitively expensive and time consuming. The choice of tests should be most complete at the time of laying down each layer of the tiered frozen cell bank, and less extensive at the time of regular tests through the period of expansion of the cells in culture. We recommend that these regular tests be carried out at least every ten passages and ideally at every fifth.

Basic Set of Tests for Monitoring Passaged Cells
Karyotype

Generally SNP analysis or G-banding karyotyping is used to monitor genomic stability of hPSCs. Although fluorescence *in situ* hybridization (FISH), spectral karyotyping (SKY), and comparative genome hybridization (CGH) are also possibilities, these require longer time periods, larger cell numbers or cost much more, so may or may not be appropriate for routine monitoring. They nevertheless remain important second stage tests for verification or more precise definition of an abnormality detected by SNP or G-banding. See chapters on karyotyping and SNP genotyping (Chapters 13 and 14).

Marker Expression: RT-PCR, Immunocytochemical Analysis, and Telomerase Activity

A combination of PCR and quantitative PCR (qPCR) is an effective approach to assess cell populations. PCR is used to confirm expression of markers of undifferentiated cells and to demonstrate the absence of markers of differentiation and the lack of contaminating feeder populations. If these are detected, qPCR can be used to assess the degree of contamination by feeders and the extent of differentiation.

While PCR or qPCR give an impression of the average cell population, immunocytochemistry allows assessment of the status of individual cells. POU5F1/ OCT4 or SOX2 and cell surface epitopes such as SSEA-4 (and binding of lectins such as UEA-1) are useful for gaining information on the status of the cells.

While flow cytometry may be a more quantitative way to assess the proportion of cells carrying different cell surface epitopes, obtaining accurate numbers is difficult as undifferentiated hPSCs attach to each other by tight junctions and so obtaining a representative single cell suspension is problematic. Moreover, the total amount of cell sample required is much higher than that required for other tests with equivalent or greater sensitivity.

Telomerase activity is a defining characteristic of hPSCs and levels should be assessed regularly.

Additional Tests for Frozen Cell Banks
To the tests described above for routine assessment of cells in culture, the following are added to confirm the identity and characteristics of the cells frozen at each level of the tiered cell bank.

Identity
At each major freezing step or before a critical series of experiments, the identity of the cells should be confirmed by STR or HLA typing. SNP genotyping also provides a definitive test of identity. This is particularly important when more than one line is being maintained in the laboratory. Mix-ups are inevitable and are most often attributable to human error. However, such mistakes can be readily determined if a genomic fingerprint of the cells is on file for comparison. STR testing can be readily done using off-the-shelf kits, as can HLA typing. SNP genotyping is available as a service, and both STR and HLA testing are offered as services by forensic and pathology labs. The cost is easily justified by the alternative greater cost of the consequences of working with misidentified cells.

Demonstration of Differentiation Potential
In addition to the cells being relatively uncontaminated by differentiated cells, it is critical that they retain the ability to differentiate to representative cells of the three germ lineages (endoderm, ectoderm, and mesoderm) and to trophoblast lineages. This can be ascertained most readily by *in vitro* differentiation to embryoid bodies (see Chapter 23), followed by molecular analysis (RT-PCR or immunocytochemical). Alternatively, differentiation capacity can be demonstrated by the *in vivo* induction of teratomas in immunocompromised mice followed by histological analysis (see Chapters 21 and 22).

Other Tests

Whenever a new line is derived, it should be characterized even more stringently. In addition to the tests described above, mitochondrial sequences, mRNA expression (Chapter 17), global methylation profile (Chapter 20) and methylation-specific PCR to assess key developmental and imprinted genes should be considered. Our data suggest that although these aspects of cell biology might be important when cell lines are developed for clinical use, they do not appear to affect most cell culture differentiation protocols, and so these may not need to be included in the routine tests. One caveat: *in vitro* hPSC models of X-linked or imprinted disease can be easily misinterpreted because of epigenetic changes over time in culture. In these cases, it is critical to check the methylation status and allele-specific expression of the relevant genes *at the time* of analysis for disease-associated phenotype.

Standards

The ideal standard to compare gene expression patterns of hPSC cultures may be a perfectly defined culture of hPSCs with no differentiation at all, but this is difficult to generate in sufficient quantities to allow the use of the same standard sample across time. Although there is no perfect cellular control for the validation of hPSC cultures, there are cells that may be used for this purpose because they approximate the expression patterns of hPSCs, but are more stable in culture. These include embryonal carcinoma cells (N-Tera 2) and aneuploid hESCs (BG01v).

Microbial Contamination of Cultures

Chapter 4 provides detailed descriptions of contaminations.

Mycoplasma Testing

Mycoplasma is a common contaminant in mammalian cell culture, and can be very insidious. Cultures can be contaminated with mycoplasma from the reagents or by inadequately trained technicians, and once contaminated, low levels may persist but be undetected (see Chapter 4).

Fungal and Bacterial Sterility Testing

This is usually done by classical culture techniques that use fluid thioglycollate medium (Sigma catalog no. 90404) for detection of aerobic and anaerobic bacteria, and soy bean/casein broth (Sigma catalog no. S1674) for detection of aerobes, facultative anaerobes, and fungi. Samples are inoculated into the broths in duplicate, with one being cultured at 22°C and the other at 32°C for 14 days, after which they are examined for turbidity.

Recovery of Normal Populations

While it is possible to recover cells that have become contaminated with microorganisms, the treatments are often harsh, and we strongly recommend going back to an uncontaminated stock if possible. If there has been a significant amount of work invested in the generation of a clone or sub-line, for example, then the following methods might be attempted; however, once the cells are free of contamination, they should be QC tested to ensure that the relevant properties have been retained.

When using multiwell plates, contamination is generally restricted to only one or a few wells, unless the contamination results from a severe compromise of the medium used to feed all the wells. One option, therefore, is simply to aspirate the contaminated well(s), fill the now-empty well(s) with full-strength bleach, aspirate the bleach, wash the well(s) with 70% ethanol, and aspirate the well(s) to dryness. The remaining uncontaminated well(s) are then likely to remain so. This consideration, by the way, highlights one main advantage of using multiple wells of multiwell plates rather than equivalent surface area in a single dish – the entire culture may not be lost in the former situation while it is much more likely to be lost in the latter.

It is also important to remember that, unlike bleach and ethanol, antibiotics do not kill microorganisms, they merely retard or stop their proliferation. In the intact animal this allows the immune system time to do its job of actually eradicating the invading microorganisms. In cell culture there is no immune system, so complete eradication of contamination is very difficult and sometimes, at best, all one can hope for is to save the culture with continuous antibiotic supplementation, keeping the contamination at a very low level – it may not ever be possible to go back to an antibiotic-free medium. While this may be acceptable for research purposes, it is clearly unacceptable for clinical purposes, so in this latter case complete clearing of the contamination must be proven by an extended period of continuous antibiotic-free culture.

Mycoplasma Eradication from Labs

The best defense against mycoplasma is mandatory testing of all of the cultures in the lab on a monthly basis. See Chapter 4 for a description of mycoplasma testing. Mycoplasma infection is a serious threat in hPSC laboratories; if a mycoplasma infection is discovered, the best course is to destroy the cultures, then retest all the other cultures in the lab. We cannot recommend methods that are reported to eradicate mycoplasma in cultured cells because they have not been successfully shown to be free of adverse effects on hPSCs. There should be frozen mycoplasma-free stocks for all important cultures.

Recovery from Bacterial Contamination

Antibiotics that target bacteria are routinely used in many cell culture situations, and some laboratories use them routinely in hPSC culture. Bacterial contamination is usually first recognized by cloudiness in the medium. Contaminated cultures should be immediately isolated from other cultures, remembering that bacteria cannot jump from culture to culture, but rather are spread by spills and aerosols. It is recommended that the cells be handled in separate incubators, and kept in a disposable outer container such as a plastic box. Media and reagents for these cells should be entirely separate from those used with other cells; whenever the contaminated cells are handled, the BSC should be swabbed thoroughly with 70% ethanol before other cells are handled there. We recommend a regimen in which these cells are manipulated last thing at night and, after swabbing the work surfaces with 70% ethanol, the UV lights are left on overnight in the BSC.

1. Working as carefully as possible to prevent aerosols, wash away as much of the bacteria as possible.
2. Aspirate the medium, and replace with (Ca/Mg-free) D-PBS. Aspirate and replace the Ca/Mg-free D-PBS six times. This must be done gently as the Ca/Mg- free buffer loosens the attachment of the hPSC colonies to the substrate.
3. Aspirate and replace with hPSC medium containing penicillin (100 U/mL) and streptomycin (100 µg/mL). Aspirate and replace with hPSC medium containing penicillin (100 U/mL) and streptomycin (100 µg/mL) a second time.
4. If possible, passage the colonies by manual dissection to a fresh dish in order to reduce the burden of bacteria further.
5. Put cultures into the incubator and change the medium twice daily for 2 days and then daily until ready to passage.
6. Keep the cells in antibiotic-containing medium for at least two passages before attempting to wean them from the antibiotics, then monitor carefully by microscope, and, if bacteria reappear, repeat this entire process. It may be prudent to use sterile-filtered MEF-conditioned medium for these latter feeding steps.

Eradication of Yeast and other Fungi
Option 1

There are a number of antimycotics available, including amphotericin B ("Fungizone") and mycostatin ("Nystatin") (Life Technologies) but these drugs should only be used as a last resort. We are not aware of any groups having successfully recovered fully functional hPSC cultures after yeast or fungus contamination and most laboratories simply destroy yeast- or fungus-contaminated cultures.

Contamination is usually recognized microscopically as chains of yeast or filaments of fungus in the culture. Contaminated cultures should be immediately isolated from other cultures, remembering that pre-sporulating fungi and yeast are spread by spills and aerosols. It is recommended that the cells be handled in separate incubators, and kept in a disposable outer container such as a plastic box. Media and reagents for these cells should be entirely separate from those used with other cells, and whenever they are handled, the BSC should be swabbed thoroughly with 70% ethanol before other cells are handled there. We, again, recommend a regimen in which these cells are manipulated only at the end of the day and after swabbing the work surfaces with 70% ethanol, the UV lights are left on in the BSC overnight.

1. In the case of fungal contamination, first carefully aspirate the visible fungal colonies with a Pasteur pipette.
2. Then, working as carefully as possible to prevent aerosols, wash away as much of the fungus as possible. Aspirate the medium, and replace with (Ca/Mg-free) D-PBS. Aspirate and replace the Ca/Mg-free D-PBS six times. This must be done gently as the Ca/Mg-free buffer loosens the attachment of the hPSC colonies to the substrate.
3. Aspirate and replace with hPSC medium containing the commercial antimycotic. Aspirate and replace with hPSC medium containing the antimycotic a second time. If possible, passage the colonies by manual dissection to a fresh dish in order to further reduce the burden of fungi.
4. It is possible to use two to three times the usual concentrations of antimycotics during this process but antibiotics must not be used simultaneously with the higher antimycotic concentrations as the toxicity of the antimycotics to the cultured cells is greatly increased in their presence.
5. Put cultures into the incubator and change the medium twice daily for 2 days and then daily until passage.
6. Keep the cells in antimycotic-containing medium for at least two passages before attempting to wean them from the antimycotics, then monitor carefully by microscope, and, if yeast or fungi reappear, repeat this entire process. It may be prudent to use sterile-filtered MEF-conditioned medium for these latter feeding steps.

Option 2

It is also possible, but very difficult, to recover cultures contaminated with fungus without the use of antimycotics. Visible fungal colonies are removed and the cultures are washed repeatedly as described above, first with PBS and then with medium to remove the bulk of the contamination. The colonies are then passaged by microdissection under a high-powered dissecting microscope, taking care to remove pieces of colony without bringing along fungal cells. Those skilled with a pipette can even "clean off" fungal cells from pieces of colony. Individual pieces of colony are transferred to small culture wells. As mentioned above, splitting the colonies up into multiple wells allows for the possibility that separate uncontaminated and contaminated cultures may be generated.

Cultures are monitored daily for contamination. Using this method it is possible to recover about a quarter of the transferred colony pieces. A proportion will carry the fungi, however, and can be discarded, or treated with the antimycotic agent as described above.

Recovery of Karyotypically Normal Cells

hPSC populations are prone to accumulation of karyotypic abnormalities over time. This does not happen spontaneously for all the cells of a given culture, but, presumably, by the appearance of cells with abnormal karyotype which have a growth advantage. The appearance of karyotypic abnormalities has been hypothesized to be associated with the method of passaging; passage by microdissection appears to be less prone to abnormalities than enzymatic passage. The appearance of karyotypic abnormalities is not sudden, but rather there is accumulation of abnormal cells and, by the time the abnormality is recognized, abnormal cells will represent a significant proportion of the culture. A carefully constructed tiered cell bank, as described above, can reduce the impact of aneuploid cells by allowing the researcher, with relatively little cost, to go back to a validated cell population with normal karyotype and with known differentiation capacity.

Sometimes, however, the amount of effort that has been invested in the generation and characterization of a population of cells makes it cost effective to clone out normal cells from the population.

This process is not very efficient, however, and with normal culture conditions a plating efficiency of less than 1% is expected. Cloning is generally done by FACS sorting directly into 96-well trays and it may be made more efficient using medium supplemented with neurotrophins and/or ROCK inhibitor. In this way, it is quite feasible to clone and expand 10–50 clones and to identify those with normal karyotype.

READING LIST

General
European Collection of Cell Cultures, 2001. The Fundamentals of Cell Culture – A Laboratory Handbook. Sigma-Aldrich, St. Louis, MO.

Overview of General Cell Culture Methods
Freshney, R.I., 1999. Freshney's Culture of Animal Cells: A Multimedia Guide. John Wiley & Sons, New York, NY.

General Cell Culture Methods and Laboratory Set-up

Centers for Disease Control and Prevention and National Institutes of Health, 1999. Biosafety in Microbiological and Biomedical Laboratories (BMBL), fourth ed. <http://www.cdc.gov/od/ohs/biosfty/bmbl4/bmbl4toc.htm>.

Specific References

Amit, M., Carpenter, M.K., Inokuma, M.S., Chiu, C.P., Harris, C.P., Waknitz, M.A., et al., 2000. Clonally derived human embryonic stem cell lines maintain pluripotency and proliferative potential for prolonged periods of culture. Dev. Biol. 227, 271–278.

Bhattacharya, B., Miura, T., Brandenberger, R., Mejido, J., Luo, Y., Yang, A.X., et al., 2004. Gene expression in human embryonic stem cell lines: unique molecular signature. Blood 103, 2956–2964.

Brandenberger, R., Khrebtukova, I., Thies, R.S., Miura, T., Jingli, C., Puri, R., et al., 2004. MPSS profiling of human embryonic stem cells. BMC Dev. Biol. 4, 10.

Buzzard, J.J., Gough, N.M., Crook, J.M., Colman, A., 2004. Karyotype of human ES cells during extended culture. Nat. Biotechnol. 22, 381–382.

Cai, J., Chen, J., Liu, Y., Miura, T., Luo, Y., Loring, J.F., et al., 2006. Assessing self-renewal and differentiation in human embryonic stem cell lines. Stem Cells 24, 516–530.

Carpenter, M.K., Rosler, E.S., Fisk, G.J., Brandenberger, R., Ares, X., Miura, T., et al., 2004. Properties of four human embryonic stem cell lines maintained in a feeder-free culture system. Dev. Dyn. 229, 243–258.

Cervantes, R.B., Stringer, J.R., Shao, C., Tischfield, J.A., Stambrook, P.J., 2002. Embryonic stem cells and somatic cells differ in mutation frequency and type. Proc. Natl. Acad. Sci. USA 99, 3586–3590.

Department of Health and Human Services, 2001. Stem cells: scientific progress and future research directions. <http://stemcells.nih.gov/info/scireport>.

Draper, J.S., Smith, K., Gokhale, P., Moore, H.D., Maltby, E., Johnson, J., et al., 2004. Recurrent gain of chromosomes 17q and 12 in cultured human embryonic stem cells. Nat. Biotechnol. 22, 53–54.

Ludwig, T.E., Levenstein, M.E., Jones, J.M., Berggren, W.T., Mitchen, E.R., Frane, J.L., et al., 2006. Derivation of human embryonic stem cells in defined conditions. Nat. Biotechnol. 24, 185–187.

Maitra, A., Arking, D.E., Shivapurkar, N., Ikeda, M., Stastny, V., Kassauei, K., et al., 2005. Genomic alterations in cultured human embryonic stem cells. Nat. Genet. 37, 1099–1103.

Martins-Taylor, K., Xu, R.-H., 2012. Concise review: genomic stability of human induced pluripotent stem cells. Stem Cells 30, 22–27.

Miura, T., Luo, Y., Khrebtukova, I., Brandenberger, R., Zhou, D., Thies, R.S., et al., 2004. Monitoring early differentiation events in human embryonic stem cells by massively parallel signature sequencing and expressed sequence tag scan. Stem Cells Dev. 13, 694–715.

Noaksson, K., Zoric, N., Zeng, X., Rao, M.S., Hyllner, J., Semb, H., et al., 2005. Monitoring differentiation of human embryonic stem cells using real-time PCR. Stem Cells 23, 1460–1467.

Pyle, A.D., Lock, L.F., Donovan, P.J., 2006. Neurotrophins mediate human embryonic stem cell survival. Nat. Biotechnol. 24, 344–350.

Rosler, E.S., Fisk, G.J., Ares, X., Irving, J., Miura, T., Rao, M.S., et al., 2004. Long-term culture of human embryonic stem cells in feeder-free conditions. Dev. Dyn. 229, 259–274.

Rugg-Gunn, P.J., Ferguson-Smith, A.C., Pedersen, R.A., 2005. Epigenetic status of human embryonic stem cells. Nat. Genet. 37, 585–587.

Sidhu, K.S., Tuch, B.E., 2006. Derivation of three clones from human embryonic stem cell lines by FACS sorting and their characterization. Stem Cells Dev. 15, 61–69.

CHAPTER 8

Biobanks for Pluripotent Stem Cells

Mai X. Luong, Kelly P. Smith, Jeremy M. Crook, and Glyn Stacey

EDITOR'S COMMENTARY

Since the first derivation of human embryonic stem cell (hESC) lines in 1998 and human induced pluripotent stem cells (iPSCs) in 2007, thousands of pluripotent cell lines have been established around the world. Many research programs have been initiated to investigate the properties of these cells and their application for cell therapy, drug discovery, toxicology, and the study of developmental biology. To facilitate the effective use of these cells in research and clinical applications, and to prevent contamination and deterioration, stem cell banking facilities have been established which rigorously apply quality control tests to cell banks for identity, sterility, viability, and other important characteristics such as pluripotency and genetic stability. This chapter provides an overview of the fundamental principles of stem cell banking. A detailed section on current practices by stem cell providers, including methods for establishing key characteristics of pluripotent cells, and a discussion about the future of banking are also included. Other chapters in this book provide detailed methods for generation and analysis of the cells.

OVERVIEW

To date, over 700 hESC lines have been used in over 1450 published studies (International Stem Cell Registry (ISCR)), signifying remarkable progress in a rapidly growing field. However, the findings in a number of these reports have been controversial, with some difficult to reproduce. This situation may have resulted from many factors, including varying competencies for cell line handling, differing culture conditions and usage of cells that have undergone deleterious changes (Buzzard et al., 2004; Mitalipova et al., 2005).

J.F. Loring & S.E. Peterson (eds): Human Stem Cell Manual, Second edition.
DOI: http://dx.doi.org/10.1016/B978-0-12-385473-5.00008-4

With over 400 hESC lines publicized as available for distribution from a variety of sources (including the cell line originator, stem cell cores, banks and commercial providers; ISCR; hESCreg registry) and an increasing demand for pluripotent stem cells for research and clinical applications, the need for national and international cooperation has never been greater. As methods are rapidly evolving, it will not be easy in the short term to achieve consensus on a universal standard for hPSC cell culture. However, implementing "best practice" is important to ensure that a cell line is maintained for research in accordance with recognized practices. In 2009, the International Stem Cell Banking Initiative (ISCBI) published a consensus guidance document for banking and supply of hESCs, which is broadly applicable to iPSCs (ISCBI, 2009). As a first formal attempt to highlight the need for coordination among distribution centers, and draw wider attention to the importance of quality assurance for basic and translational research, the guidance will be an indispensable instrument for harmonizing hESC and iPSC banking and use by the wider stem cell community. The document has already been translated into Chinese and Japanese (translations prepared for publication, 2012).

This chapter provides an overview of the fundamental principles of hPSC banking. A detailed section on current practices by stem cell providers, including methods for establishing key characteristics of pluripotent cells, and a discussion about the future of banking are also included.

BANKING OF HUMAN PLURIPOTENT STEM CELLS

The goal of banking or long-term storage of these cells is to preserve their characteristics, prevent contamination and deterioration, and facilitate their effective use in research and clinical applications. Relevant guidance on good cell culture practice (Coecke et al., 2005) and cell banking (ISBER, 2007; ISCBI, 2009; OECD, 2007) has been developed. Briefly, the fundamental principles of banking include the creation of a quality controlled Master Cell Bank (MCB) and Distribution/Working Cell Banks (D/WCBs), supported by traceability of procured cell lines and thorough documentation of the entire banking process (Figure 8.1). Ideally banks should minimally operate in "the spirit" of good laboratory practice (GLP) or if their purpose requires it, certified GLP or good manufacturing practice (GMP) that complies with national and/or international quality systems standards (e.g. OECD, 1997; Eudralex, 2011; FDA, 2010).

Master Cell Banks and Distribution/Working Cell Banks

Upon arrival at a banking facility, a newly acquired cell line is quarantined until it is shown to be clear of contamination (such as mycoplasma). Clean cells are expanded to create the Master Cell Bank (MCB), which is used to replenish Distribution/Working Cell Banks (D/WCBs) only, and not for

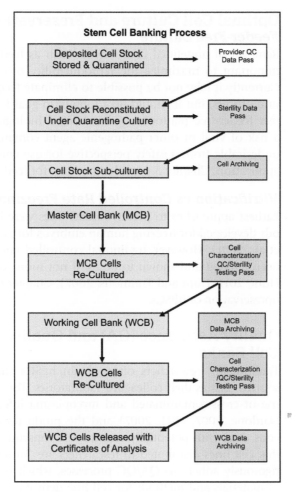

Stem Cell Banking Process

Deposited Cell Stock Stored & Quarantined → Provider QC Data Pass

Cell Stock Reconstituted Under Quarantine Culture → Sterility Data Pass

Cell Stock Sub-cultured → Cell Archiving

Master Cell Bank (MCB)

MCB Cells Re-Cultured → Cell Characterization/ QC/Sterility Testing Pass

Working Cell Bank (WCB) → MCB Data Archiving

WCB Cells Re-Cultured → Cell Characterization /QC/Sterility Testing Pass

WCB Cells Released with Certificates of Analysis → WCB Data Archiving

FIGURE 8.1
Outline of the process for banking of hESC and hiPS cell lines, from the initial deposit of cells to their distribution with a Certificate of Analysis.

working cell cultures. D/WCBs are best maintained at low passage to avoid phenotypic and genetic variation. In order to minimize the risk of complete loss of a cell line due to, for example, contamination, equipment or facility failure, it is critical that frozen stocks are also stored at a second site (Stacey and Sheeley, 1994).

A D/WCB can be prepared from live cultures or a single frozen vial from the MCB. Cells are grown for several passages, allowing for minimal necessary characterization and scale-up prior to aliquoting into vials and freezing. Cells from the D/WCB are normally used for distribution to researchers. Thus, creation of a two-tiered frozen cell bank enables a supply of reliable quality-controlled cells at the same passage over a long period of time, which are the principal requirements for biomedical research and product development (Hay, 1988; Stacey, 2004).

Optimal Cell Culture and Preservation/Freezing
Feeder-Free Culture

Removal of undefined components such as feeder cells from cell culture is important to maximize the reproducibility of cells supplied to customers. Currently it may not be possible to eliminate feeder cells from the entire culture process, but it should be a long-term goal for Stem Cell Banks to establish feeder-free platforms. As undefined biologicals, feeder cells may carry a risk of viral or other pathogenic agent contamination. They may also be undesirable from a safety perspective for use with cells intended for clinical application. Chapter 3 provides methods for feeder-free culture of hPSCs.

Vitrification vs Controlled Rate Freezing

Earliest approaches to preserve hESC lines were based on vitrification methods developed for freezing human embryos for *in vitro* fertilization (Reubinoff et al., 2001). However, traditional controlled rate freezing has more recently been applied and shown to be as, if not more, effective (Crook et al., 2007; Hunt, 2011; Hunt and Timmons, 2007). Chapter 6 provides a method for cryopreservation of hPSCs.

Quality Assurance (QA) and Quality Control (QC) of Cell Banks

Traditionally, researchers often obtain hESCs and hiPSCs from unqualified sources such as their colleagues' laboratories. This can result in the widespread use of cross-contaminated and mycoplasma infected cells (Chatterjee, 2007; Nardone, 2007; NIH, 2007) and the production and publication of erroneous data, with potentially catastrophic consequences for the field. To ensure that cell lines distributed to researchers are "fit for purpose", cell banks must rigorously adhere to QA/QC processes, which include regulating the quality of materials and methods for cell line derivation and/or culture and banking, and applying quality control tests to cell banks for identity, sterility, viability, and other important characteristics such as pluripotency and genetic stability. A vital part of QA/QC is to confirm that a MCB and D/WCB are representative of the originally deposited material (i.e. the banking process has not altered the cell line). Thus, by applying systematic monitoring and evaluation of the various aspects of the banking process and cells received for banking, the bank can state, with certain qualifications, that distributed cells have the correct identity, are devoid of contamination, and are "fit for purpose".

Cell Characterization

The Consensus Guidance for Banking and Supply of hESC Lines for Research (ISCBI, 2009) recommends meticulous assessment of key characteristics such as the phenotype, genotype, genetic stability, pluripotency and sterility, as summarized below. Although a variety of methods can be used to evaluate each characteristic, each method has advantages and limitations (Table 8.1). Detailed methods for characterization are provided in Chapters 4, 13–17, 21 and 22.

Table 8.1 Methods for Cell Characterization

Tests	Description	Advantages	Limitations
Mycoplasma Detection			
Mycoplasma contamination is a serious cell culture issue. These assays provide assurance that banked cells are free of contamination.			
Bioluminescent reporter	Detects ATP production by mycoplasma enzymes using bioluminescent assay.	Rapid, kit-based assay which can be done on-site, doesn't require significant culture.	May not detect nascent contamination, requires luminometer.
DNA fluorochrome	Fluorescent dye binds to DNA. Mycoplasma detected by microscopy.	Rapid, can be done on site or by sending samples to vendor.	May not detect nascent contamination. Requires experience to determine mycoplasma contamination by location and morphology of DNA signals.
Direct culture	Culture of aliquots of sample media in a variety of liquid media and on agar plates. Detection by fluorochrome assay.	Most sensitive for detection of nascent contamination.	Usually requires sample to be sent to vendor for testing. Very lengthy test, samples may be cultured for up to 28 days. Some species, e.g. *M hyorhinis*, difficult to grow.
Cell Line Identity			
Confirmation that cell line is not cross-contaminated or misidentified.			
DNA fingerprint	Short tandem repeat (STR) analysis.	Standard set of forensic markers in widespread use.	Samples usually sent to vendor.
SNP genotype	Whole genome analysis for common single nucleotide polymorphisms.	Large number of probes (millions). Data also informative for genomic stability.	Core lab or service usually required.
Genomic Stability			
Assessment of changes in genome during culture			
Karyotyping	G-banding of cells in metaphase.	Provides well recognized analysis of chromosome count and morphology.	Normally requires independent lab to perform and cytogeneticist to analyze. Can miss small changes such as deletions (2–10 Mb resolution)
SNP genotype	Whole genome analysis for common single nucleotide polymorphisms.	Large number of probes (millions). Data also informative for genomic stability.	Core lab or service usually required.

(Continued)

Table 8.1 Methods for Cell Characterization *(Continued)*

Tests	Description	Advantages	Limitations
Comparative genomic hybridization	Microarray based analysis of chromosomes.	Very high resolution (4–20 Kb). Large no. of cells analyzed – more representative of culture population.	Test can be more expensive and requires bioinformatic analysis.

Pluripotency Marker Expression

Determine that cells are undifferentiated by assaying protein biomarkers or mRNA profiles.

qRT-PCR	Quantitative real time PCR to assess mRNA levels of specific marker genes.	Accurate, reasonably rapid testing. qRTPCR is a fairly standard lab technique. Some kits are available.	Requires deciding which markers to use. Markers may be expressed in differentiating cell populations.
Microarray analysis	mRNA expression is determined by hybridization to microarray chips.	Not biomarker-based. Assesses expression of whole genome. Databases are available online.	Core lab or service usually required.
Flow cytometry	Specific markers are detected using fluorescent antibodies or lectins. Positive cells are counted by flow cytometer.	Allows accurate quantification of cell numbers to assess differentiation levels in culture.	Usually requires a core laboratory.
Immunocytochemical analysis	Marker proteins are detected using specific fluorescent antibodies (or by alkaline phosphatase assay). Cultures are fixed and examined with fluorescent microscope.	Relatively easy. Equipment is reasonably common. Allows determination of intact colonies to see where differentiation occurs.	Not readily quantifiable. Staining can be variable.

Differentiation Capacity

Determine ability of cell lines to develop into specific cell lineages and cell types.

Embryoid body formation	Pluripotent cells form clumps in suspension that begin to spontaneously differentiate into multiple lineages.	Assays ability of cells to form cell types representative of the three germ layers.	EBs do not necessarily demonstrate tissue or structure formation.
Directed differentiation	Differentiation of cells along cell lineages and/or into cell types by modifying culture conditions and using specific growth factors.	Can determine a cell line's capacity to form specific cell types.	Protocols can be time consuming and difficult. There are few definitive biomarkers (e.g. gene expression profiles) for specific cell types. Lack of consensus means that multiple methods may be needed to thoroughly determine the cell line's capacity to develop into a particular cell type.
Teratoma formation	Injection of cells into specific location in immunocompromized mice to form tumors.	Complex structures specific to cell lineages of each germ layer can be visualized.	Teratomas are time consuming and subject to variability. Requires pathology expertise to evaluate.

Gross Morphology

Undifferentiated hESCs and hiPSCs have a distinct morphology and grow in compact colonies with clear borders. When grown on feeder layers, cells at the center of the colony tend to exhibit a high nucleus/cytoplasm ratio with prominent nucleoli, are rounded and lie tightly packed suggesting close cell membrane contact. Cells at the periphery may exhibit epithelial cell morphology, with a lower nucleus/cytoplasm ratio. When the cells are cultured without a feeder layer, the cells and thus the colonies may appear to be flatter. If cells are maintained under suboptimal conditions (inappropriate culture medium or substrate, feeding cells too infrequently), some of the colonies in a dish lose their distinct borders due to the presence of differentiated cells at the colony periphery. These differentiated cells continue to divide and must be eliminated during passage or by discarding the cultures. For each cell line, it is important to note the morphology and determine the doubling time (generally 20–40 hours) or split ratio (Crook et al., 2007), and the best practice is to feed cultures every day.

Biomarkers for Pluripotent Cells

hESCs and iPSCs express a particular set of cell surface antigens. According to a standardized international study of a large number of hESC lines (Adewumi et al., 2007), a typical surface marker profile for these cells analyzed by flow cytometry (FC) is SSEA-1 (negative or very low), SSEA-3 "positive," SSEA-4 "positive," TRA-1-60 "positive", and TRA-1-81 "positive." In the same study, microfluidic arrays for qRTPCR showed that transcripts for *NANOG*, *POU5F1/OCT4*, *DNMT3B* (DNA (cytosine-5-)-methyltransferase 3 beta), *TDGF1* (teratocarcinoma-derived growth factor 1), *GABRB3* (gamma-aminobutyric acid (GABA) A receptor, beta 3), and *GDF3* (growth differentiation factor 3) were highly expressed in undifferentiated hPSCs. A recent study showed that in addition to cell surface antigens, a lectin, UEA1, also shows specificity for undifferentiated pluripotent cells (Wang et al., 2011).

Identity and Genomic Stability

It has been estimated that 30% of cell lines reported in published work have been misidentified, cross-contaminated with another cell line, or contaminated with mycoplasma (Capes-Davis et al., 2010; MacLeod et al., 1999). Use of these lines in research decreases the quality and reproducibility of published findings (Dunham, 2008; Nardone, 2007). To ensure that the correct cell line is banked and distributed, and that no cross-contamination has occurred during the banking process, identity testing must be performed on early stage pre- and post-banked material (early passage cell DNA). DNA profiling by Short Tandem Repeat (STR) analysis is most commonly used to confirm cell line identity (Barallon et al., 2010), and SNP genotyping (Chapter 14) is also used for identification of cells as well as for monitoring their genomic stability.

It is strongly recommended that the genomic stability of cell lines be monitored at regular intervals. Chapters 13 and 14 describe methods for karyotyping and single nucleotide polymorphism (SNP) genotyping. Karyotyping, with Giemsa-banding (G-banding) gives an overall view of the integrity of the chromosomes. SNP genotyping provides a higher resolution (currently up to 5 million SNPs per genome) and can detect small duplications and deletions that are undetectable by karyotyping. However, SNP genotyping cannot identify inversions or translocations, because it detects difference in DNA sequence, not chromosomes. Other complementary techniques include "spectral karyotyping" (SKY), which is particularly good for locating translocations, and comparative genome hybridization (CGH) microarrays, which gives a snapshot of the integrity of the whole genome.

Pluripotency

A defining characteristic of hESCs and iPSCs is their potential to differentiate into all cells of the body (Hanna et al., 2010), and hence the bank should provide evidence that each stem cell line is pluripotent, i.e. has the potential to at least produce cells representative of the three germ layers after banking. Pluripotency should be confirmed at least once after banking a cell line. As pluripotent cells have a tendency to spontaneously differentiate in culture, the bank should ensure that pluripotency is not lost by *in vitro* expansion and cryopreservation. The historical "gold standard" for assessing pluripotency has been the formation of teratomas in immunodeficient mice that show tissues from all three germ layers (Smith et al., 2009). Chapter 22 details the analysis of hPSC-derived teratomas. There are newer *in vitro* methods for demonstrating pluripotency (for example, gene expression profiles, Chapter 18) and analysis of *in vitro* differentiation (Chapter 23). But most cell banks continue to use teratoma assays.

Sterility

Microbial contamination can change cell characteristics and function without causing obvious changes in cell growth or appearance. The use of contaminated cells could result in the generation of misleading data and expose the unsuspecting researcher to infectious agents. Even with very good sterile technique, it is possible to cross-contaminate cultures with mycoplasma when they are manipulated in the same hood or grown in the same incubator. It is vital that banks perform comprehensive microbiological testing based on the infectious risks to which the cells may have been exposed, to ensure that all MCBs and W/DCBs are free of microbial contamination, as far as can be reasonably determined.

Mycoplasma species pose the greatest risk of contamination in mammalian cell cultures, with several studies estimating a 15–30% incidence in cell lines that are shared between labs (Bolske, 1988; Rottem and Naot, 1998; Timenetsky et al., 2006). Mycoplasma contamination can significantly alter cell phenotypes by affecting metabolism and growth. Although less prevalent, bacterial and fungal contaminations may occur from time to time and viral contaminations may be established without noticeable cytopathic effect, which poses a hazard to transplant recipients. As a minimum standard, it is recommended that early archive or master stocks of cells be tested for serious and primarily blood-borne viruses such as HepB, HepC, HIV, HTLV I/II, EBV, and hCMV (Cobo et al., 2006). In addition, all banks should be tested for bacterial and fungal contamination.

Microbiological testing should be performed using methods that are qualified for testing cell lines. Inoculation of microbiological media is a recommended method to detect the growth of bacteria and fungus. Methods for the detection of mycoplasma contamination include direct culture in broth or agar and indirect tests using indicator culture or DNA dye (Young et al., 2010).

Traceability and Documentation

Comprehensive documentation of the banking process is essential for demonstrating appropriate legal, ethical and perhaps regulatory standards, to address queries about cell line performance from recipients, and to troubleshoot internal failures in QA/QC. Banks should obtain documentation of the informed consent from the original donor of cells used to derive the cell line. Information obtained from the depositor of the cell line should also include: the name of the line, documentation of IRB (Institutional Review Board) and ESCRO (Embryonic Stem Cell Research Oversight) approvals (if relevant) or their national equivalents outside of the USA, cell identity, contamination status, viral testing, passage level, mycoplasma testing, and molecular marker tests. Data from QA, including characterization assays, should be meticulously recorded, and materials and procedures used to prepare each bank should be traceable. Standard operating procedures (SOPs) for all key processes should be established and documented.

It is important that the correct cell line name is obtained from the depositor, as confusion has arisen from multiple names being applied to individual hESC and hiPSC lines, as well as use of the same name for different cell lines. Furthermore, cell line names are not used consistently and may often have a different placement of a hyphen or inconsistent inclusion of a zero (e.g. "ABC2," "ABC 2," "ABC-2" or "XYZ3," "XYZ03," "XYZ003"). To address these issues, a standard nomenclature (Figure 8.2) for human iPSC and hESC lines has been proposed (Luong et al., 2011).

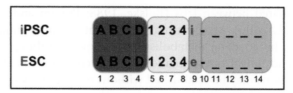

FIGURE 8.2

Recommendation for a standardized nomenclature. Blue box: contains a group of letters to represent source reference, such as laboratory (for iPSC lines) or institution (for hESC lines). Yellow box: may contain a sequential number to identify specific cell lines. Green box: contains an "i" or "e" to represent "iPSC" or "ESC," respectively. Orange box: begins with a dash and is followed by up to 12 letters and/or numbers. This box can be used to note specific characteristics such as disease, reporter genes, patient number, and clone number. The green and orange boxes are optional. While the exact number of characters for any element can vary, we recommend that names have no spaces and be limited to 14 characters.

Taken together, assurance of cell line traceability combined with a standardized nomenclature will ensure that the correct cells are banked and supplied to an end-point user. This will circumvent a common, but under-recognized, problem of supplying incorrect cells by informal and uncontrolled distribution practices (MacLeod et al., 1999).

STEM CELL PROVIDERS

A bank of stem cells is a uniform population of cells that has been preserved and maintained in a secure, controlled, and monitored storage environment. These cells have been handled in a manner that maintains their integrity and ensures they are readily available for distribution. hESCs and hiPSCs are currently distributed by three types of stem cell providers, including the cell line originator (e.g. individual investigators, biotechnology companies), stem cell cores/research institutes and formal banks. As there are hundreds of stem cell originators across the globe, their "banking" practices vary according to local norms, research goals and funding availability. For the purpose of this chapter, we have focused on pluripotent stem cell cores and banks, which are facilities solely dedicated to banking and distribution (Table 8.2).

Existing Cores and Banks

Core facilities generally distribute cell lines that originate from their own institution, and their services often include the supply of live cultures for internal research purposes. In contrast, cell banks generally distribute cryopreserved cell lines that originate from various institutions. Although cores and banks may have different missions and operate in various regulatory environments, they each test lines with a variety of assays to ensure cell

Table 8.2 Stem Cell Banks and Cores

Distributor	Location	Website
Massachusetts Human Stem Cell Bank	University of Massachusetts Medical School/Massachusetts, USA	www.umassmed.edu/mhscb/index.aspx
National Research Center of Human Stem Cells HESC Bank	Hunan, China	www.hescbank.cn
Spanish Stem Cell Bank	Grenada, Spain	www.juntadeandalucia.es/bancoandaluzdecelulasmadre/
Taiwan Stem Cell Bank	Hsinchu, Taiwan	www.tscb.bcrc.firdi.org.tw/index.do
UK Stem Cell Bank	Hertfordshire, UK	www.ukstemcellbank.org.uk/
WISC Bank	WiCell Research Institute/ Wisconsin, USA	www.wicell.org/
Core Facilities		
Children's Hospital Boston hESC Core Facility	Boston Children's Hospital/ Massachusetts, USA	http://stemcell.childrenshospital.org/about-us/stem-cell-program-labs/the-hesc-core-facility/
Harvard HUES Facility	Harvard University/ Massachusetts, USA	www.mcb.harvard.edu/melton/hues/
Harvard Stem Cell Institute iPSC Core Facility	Harvard University/ Massachusetts, USA	www.hsci.harvard.edu/ipscore/
Stanford Institute for Stem Cell Biology and Regenerative Medicine	Stanford University, California, USA	http://stemcell.stanford.edu/
UCONN Stem Cell Core	University of Connecticut/ Connecticut, USA	http://stemcellcore.uchc.edu/
Commercial Banks		
Cellartis	Gothenburg, Sweden	www.cellartis.com/
GlobalStem	Maryland, USA	https://globalstem.com/
Reproductive Genetics Institute (Stemride International)	Illinois, USA	www.reproductivegenetics.com/

Banking facilities that do not distribute cells internationally were not included in this table.

identity, genomic stability, pluripotency, and sterility (Table 8.3). Over 470 hESC and iPSC lines are currently available for research from stem cell cores and banks, and many additional cell lines are anticipated (Table 8.4).

Choosing a Cell Provider

In order to achieve the most desirable research outcome in the least amount of time, care must be taken when choosing a cell provider. Factors that should be considered include: suitability of cell lines offered, effectiveness of technical support and training, replacement policy, qualified testing of cell banks, intellectual property and ownership. Chapter 37 explains the patenting of hESCs and methods.

Table 8.3 Current Quality Control Testing by Banks and Cores

Markers (Assay)	Pluripotency	Genotype	Genomic stability	Sterility	Feeder Conditions	Post-thaw Recovery
Massachusetts Human Stem Cell Bank						
FACS: SSEA-4, SSEA-1 (MCB and lots) qRT-PCR: 6 pluripotency markers and positive 3 differentiation markers compared to EB control (MCB and lots)	qRT-PCR of EB: markers for 3 germ layers compared to undifferentiated controls (MCB)	STR (MCB and lots)	G-banding (MCB and lots)	Mycoplasma: DNA fluorochrome staining (MCB and lots) Mycoplasma: Culture broth and agar plates (lots)	MEF	≥15 colonies with ≤30% differentiation
WISC Bank (National Stem Cell Bank (NSCB) prior to 2010)						
FACS: Oct 3/4, SSEA-4, Tra 1-60, Tra 1-81, SSEA-1 (NSCB MCB only)	Teratoma (NSCB MCB only)	STR (NSCB MCB & Lots, & 2010 post-NSCB) HLA, ABO testing (NSCB MCB only)	Karyotype by G-banding. Count 20 cells, analyze minimum of 4. (NSCB MCB Lots & 2010 post-NSCB) Array CGH (NSCB MCB only)	Mycoplasma: Culture broth and agar plates. (MCB and Lots) Viral Testing (NSCB MCB only): HIV, HTLV, HBV, HCV, CMV, EBV, HHV, HP B19, bovine and porcine pathogens, MAP, adventitious virus, TSEM, Co-cultivation (2010 post NSCB; one lot from each cell line): John Cunningham virus, BK Virus, Herpesvirus, HP B19, EBV, Hepatitis, HBV, Human T-lymphotropic virus, Human cytomegalovirus, HIV, Adeno-associated virus, Human Foamy Virus, LCMV, Hantavirus	MEF, feeder-independent, feeder-free	2010 post-NSCB: ≥15 Undifferentiated Colonies prior to split; ≤30% Differentiation in culture NSCB Lots: >30 undifferentiated colonies, no inactivated MEFs; <10% differentiation
UCONN Stem Cell Core Facility						
ICC (MCB only)	Teratoma (MCB only)	STR (MCB only)	G-banding (MCB and lots)	Mycoplasma: DNA fluorochrome staining (MCB and lots) Viral Testing: (GlobalStem) (MCB only)	MEF	

	Pluripotency markers	Differentiation	Identity	Karyotype	Microbial/Viral	Feeder cells	Cryopreservation/recovery
National Research Center of Human Stem Cells Human Embryonic Stem Cell Bank (China)							
	ICC, RT-PCR, telomerase activity (MCB only) FACS: SSEA-4 (Lots)	EB formation Teratoma	STR (MCB and Lots) HLA, ABO testing (MCB)	G-banding (MCB and lots)	Mycoplasma (MCB and Lots)	MEF	
Spanish Stem Cell Bank							
	FACS, IF and RT-PCR	IF of EB: markers for 3 germ layers Teratoma	STR		Mycoplasma: PCR based assay	Human foreskin, feeder-free, MSC	
Sydney IVF (Genea)							
	HCA analysis for nanog, Tra1-60, Tra1-81 and SSEA4	Teratoma analysis (positive for all three germlayers) Alkaline phosphatase staining. In vitro differentiation (not applicable to all cell lines)	DNA fingerprinting (STR) Disease status test by PCR, CGH or linkage analysis	Karyotype analysis (G banding)	Mycoplamsa free. Embryo donors generally tested for Hep B&C and HIV1&2	Human fetal fibroblasts (from ATCC). Some are derived feeder free.	Successful (recovery depends on freeze method but is generally around 80–90%) Freeze methods include single cell freezing or OPS vitrification
UK Stem Cell Bank							
	FACS and immunocyto-chemistry for canonical markers SSEA-3, SSEA-4, TRA-1-81, TRA-1-60, etc.	Currently qualifying directed differentiation & EB based methods	STR and isoenzyme analysis for non human lines MCB & DCB	G-Banding on MCB and DCBs	HTLV, HIV, CMV, HBV, HCV, EBV & mycoplasma by PCR. Mycoplasma and sterility testing following European Pharmacopeia. MCB & DCB	Standard on MEFs. Some banks also on human feeders or feeder-free	Traditional preservation methods employed give good viability typically >60%
GlobalStem							
	FACS: SSEA-4		STR	G-banding	Mycoplasma: PCR based assay. Viral Testing: HIV, Human T-Cell Lymphotrophic Virus I/II, HSV, EBV, CMV, HBV and Hepatitis C		2×10^6 viable cells

Table 8.4 Available hESC and iPSC Lines

Available Lines	Research Grade				Clinical Grade
(total # banked lines)	hESC		iPSC		hESC
Banks					
Massachusetts Human Stem Cell Bank					
15 (51)	CHB-2	CHB-11	hFib2-iPS5	29d	
	CHB-4	HUES-6	SBDS-iPS2	18b	
	CHB-6	HUES-7	18c	27e	
	CHB-8	HUES-8			
	CHB-9				
National Research Center of Human Stem Cells Human Embryonic Stem Cell Bank (China)					
190[a]	chHES-3	chHES-7–103			
	chHES-4				
	chHES-5	chHES-105–195			
Spanish Stem Cell Bank					
50	AND-1	VAL-10B	hiPSC clone 1	[GA]FiPS4F	
	AND-2	VAL-11B	hiPSC clone 4	HKiPS4F	
	AND-3	ES[2] to ES[10]	iPS-SCU-CD34+ #1	XF-iPSF44-3F-1	
	AND-4	ES[11]-EM	iPS-SCU-CD34+ #2	XF-iPSF44-3F-2	
	VAL-3	pES[12]	CBiPS6-2F-4	cFA404-KiPS4F-1	
	VAL-4	HVR-1	CBiPS30-4F-3	cFA404-KiPS4F-3	
	VAL-5	HVR-2	CBiPS30-4F-5	KiPS3F-7	
	VAL-6M	HVR-3	CBiPS32-2F-2	KiPS4F-1	
	VAL-7	RiMi1	CBiPS32-3F-10	KiPS4F-8	
	VAL-8	VAL-9-GFP	CBiPS32-3F-12	MSUH-001	
	VAL-9		FiPS-3F-1		
Sydney IVF (Genea)					
72 (plus 8 GMP lines- ESI/Biotime)	SIVF001-SIVF072				8 through ESI/Biotime only
Taiwan Stem Cell Bank					
3 (8)	TW01	TW05			
	TW03	NTU1			

(Continued)

Table 8.4 Available hESC and iPSC Lines *(Continued)*

Available Lines (total # banked lines)	Research Grade				Clinical Grade
	hESC		iPSC		hESC
UK Stem Cell Bank					
17 (40)	EDI-4 KCL003 KCL002 MAN-1 MEL-1 NCL-1 to NCL-5 NOTT-1	NOTT-2 SHEF-1 SHEF-2 SHEF-3 SHEF-6 SHEF-7			
WISC Bank					
39	SA01 SA02 ES01 to ES06 BG01 BG02 BG03 TE03 TE04 TE06 UC01 UC06 H1	H7 H9 H13 H14 H1 OCT4-EGFP H9 Cre-LoxP H9 hNanog-pGZ H9 hOct4-pGZ H9 inGFPhES H9 Syn-GFP h9-hTnnTZ-pGZ-D2	iPS(Foreskin) iPS(IMR90) clone 1 iPS(IMR90) clone 4 iPS - DF19-9 (clone 7T) iPS - DF19-9-7T-MCB-01	iPS - DF19-9 (clone 11T.H) iPS-DF19-9-11T.H-MCB-01 iPS - DF4-3 (clone 7T.A) iPS-DF4-3-7T.A-MCB-01 iPS - DF6-9 (clone 9T.B)	WA09
Core Facilities					
Children's Hospital Boston hESC Core Facility					
6 (11)	CHB-2 CHB-4 CHB-6 CHB-8	CHB-9 CHB-10			
Harvard HUES Facility					
59	HUES 1-24 HUES 26-29	HUES 50 HUES 51			

(Continued)

Table 8.4 Available hESC and iPSC Lines *(Continued)*

Available Lines	Research Grade			Clinical Grade
(total # banked lines)	hESC		iPSC	hESC
	HUES 31-44 HUES 46 HUES 47	HUES 52 HUES 54-61 HUES 67-70		
Harvard Stem Cell Institute iPSC Core Facility				
29			hFib2-iPS4 PD-iPS11 hFib2-iPS5 SBDS-iPS2 ADA-iPS2 SBDS-iPS3 ADA-iPS3 GD-iPS3 BMD-iPS1 11b BMD-iPS4 15b BMD2-iPS1 17a BMD2-iPS5 18a DS1-iPS4 18b HD1-iPS1 18c HD1-iPS4 20b JDM-iPS1 27b JDM-iPS10 27e PD-iPS1 29d 29e	
Stanford hESC Center				
2	LSJ-1	LSJ-2		
UCONN Stem Cell Core				
15	CT-1 CT-2 CT-3 CT-4	HUES 1 HUES 3 HUES 9 HES3-GFP	HDFa-YK26 IMR90-YZI HDFa-YK27 ES4skin clone 1 IMR90-TZI ESIMR90 clone 1 hFIB2-IPS4 F4	
Commercial Banks				
GlobalStem				
2	BG01v	2102Ep		
Reproductive Genetics Institute (Stemride International)				
166[a] (244)	RG 1-300			

Note: *Not all cell lines publicized as "available" for distribution have been characterized as described in Table 8.1.*

Suitability of Cell Lines

The cell lines should meet minimal relevant eligibility criteria for reputable research. The criteria vary by nation and occasionally by regions within a nation. However, most funding agencies require evidence of the donor's free and voluntary consent. Different legal stipulations are associated with each cell line and may include restrictions of research activities. For example, the NIH requires that use of the HUES-1 to HUES-17 cell lines derived at Harvard University in funded research be limited "to study the embryonic development of endoderm." Furthermore, inherent cell line characteristics such as the gender, ethnicity and detected genetic mutations may affect the suitability of using certain lines for addressing specific questions. Depending on the differentiation protocol, certain cell lines may show a preference for distinct differentiation pathways (Chang et al., 2008; Sivarajah et al., 2010; Tavakoli et al., 2009).

Technical Support and Replacement Policy

Prior to requesting cells from a provider, it is important to determine that technical support is effective and prompt, and that culture instructions are easily accessible. If cells are purchased, ensure that there is a clear replacement policy for cultures that fail to thrive. To avoid frustration it is also helpful to verify that there is a complaints procedure in place.

Certificate of Analysis

Most dedicated stem cell facilities provide recipients with a statement or certificate listing and qualifying the testing performed on cells supplied including the test material used. This documentation should be informative and obtained prior to requesting cells because it will influence the choice of stem cell provider. It is important to verify the use of appropriate controls, which are good indicators of a rigorous QC process. For feeder cell-dependent cultures, all analytical work should minimize feeder cell contamination from cell samples to avoid interference in data on stem cell lines. As cores and banks are dedicated to banking and distribution, cells obtained from these providers have generally undergone a more comprehensive and qualified testing protocol.

FUTURE OF BANKING AND NEW iPSC INITIATIVES

Banking and distribution of hPSCs is an expensive endeavor requiring a significant investment initially and thereafter. However, the benefits of dedicated banking operations are many, and include the collection and storage of qualified and traceable seed stock cells that have the correct identity and appropriate cell characteristics necessary for research and clinical applications. The absence of such resources can lead to waste of research funding on the order of millions of dollars and euros. Banking initiatives have rightly been backed in many instances by the support of public funding via national governments. There are estimates that over 7000 vials of properly banked hESC and iPSC cells have been distributed to researchers over the past decade.

As inroads are made to advance pluripotent cells toward therapeutic and pharmaceutical applications, the availability of large banks of clinical grade cells for use in pre-clinical research and clinical trials becomes essential. For example, companies based in the UK, the USA and elsewhere are beginning trials of hESC-based therapies for the treatment of macular degeneration and spinal trauma. In 2010 Geron, a California-based company, initiated the first clinical trial for the treatment of spinal cord injury using WA01 (H1) hESCs. Importantly, GMP-grade clinically-compliant hESC banks are available (Crook et al., 2007), and have been selected by the California Institute of Regenerative Medicine (CIRM) and University of California System (UC-system) institutions to streamline their efforts to translate hPSCs to human therapies. This is complemented by the establishment of formal purpose-built biobanks in the US, the UK, and Singapore to expand cells under GLP and/or GMP (Sivarajah et al., 2010).

A significant limitation of cell therapy using allogeneic transplants is the need for chronic immunosuppressive medications. One strategy to reduce the likelihood of graft rejection is the creation of a bank of HLA-typed hESCs and iPSCs from which a best match could be selected. Banks of HLA-matched hESCs have been in the planning phase for a number of years. Taylor et al. estimated that approximately 150 consecutive blood group compatible donors could provide the maximum practical benefit for HLA matching in the UK (Taylor et al., 2005), while Nakajima et al. estimated that from a bank of hESC lines derived from 170 randomly selected donated embryos, 80% of patients in Japan were expected to find at least one cell line with only a single mismatch at one HLA locus (Nakajima et al., 2007). These feasibility studies have practical, political, and ethical implications for the establishment of clinical grade hESC banks.

hiPSCs offer another source of clinically relevant hPSCs. Over 40 hiPSC lines are offered by current banking facilities and many more will become available as the National Institutes of Health, CIRM, and European and Asian funding agencies provide resources for hiPSC generation and distribution. It is theoretically possible to generate hiPSCs from a patient and use them for cell therapy for the same patient; but this is currently impractical (Nishikawa et al., 2008) and no public funding has been appropriated for patient-specific banks.

The availability of well-characterized hPSCs from qualified sources is vital, requiring formal national and/or international biobanks. However, the greatest barrier to exploiting hPSCs for improving human health remains the technical investment in culturing and characterizing hPSC derivatives, and it will take innovative approaches to deliver all of their benefits.

READING LIST

Adewumi, O., Aflatoonian, B., Ahrlund-Richter, L., Amit, M., Andrews, P.W., Beighton, G., et al., 2007. Characterization of human embryonic stem cell lines by the international stem cell initiative. Nat. Biotechnol. 25, 803–816.

Barallon, R., Bauer, S.R., Butler, J., Capes-Davis, A., Dirks, W.G., Elmore, E., et al., 2010. Recommendation of short tandem repeat profiling for authenticating human cell lines, stem cells, and tissues. In Vitro Cell Dev. Biol. Anim. 46, 727–732.

Bolske, G., 1988. Survey of Mycoplasma infections in cell cultures and a comparison of detection methods. Zentralbl. Bakteriol. Mikrobiol. Hyg A. 269, 331–340.

Buzzard, J.J., Gough, N.M., Crook, J.M., Colman, A., 2004. Karyotype of human ES cells during extended culture. Nat. Biotechnol. 22, 381–382 (author reply 382).

Capes-Davis, A., Theodosopoulos, G., Atkin, I., Drexler, H.G., Kohara, A., MacLeod, R.A., et al., 2010. Check your cultures! A list of cross-contaminated or misidentified cell lines. Int. J. Cancer 127, 1–8.

Chang, K.H., Nelson, A.M., Fields, P.A., Hesson, J.L., Ulyanova, T., Cao, H., et al., 2008. Diverse hematopoietic potentials of five human embryonic stem cell lines. Exp. Cell Res. 314, 2930–2940.

Chatterjee, R., 2007. Cell biology. When 60 lines don't add up. Science 315, 929.

Cobo, F., Talavera, P., Concha, A., 2006. Diagnostic approaches for viruses and prions in stem cell banks. Virology 347, 1–10.

Coecke, S., Balls, M., Bowe, G., Davis, J., Gstraunthaler, G., Hartung, T., et al., 2005. Guidance on good cell culture practice. A report of the second ECVAM task force on good cell culture practice. Altern. Lab. Anim. 33, 261–287.

Crook, J.M., Peura, T.T., Kravets, L., Bosman, A.G., Buzzard, J.J., Horne, R., et al., 2007. The generation of six clinical-grade human embryonic stem cell lines. Cell Stem Cell 1, 490–494.

Dunham, J.H., Guthmiller, P., 2008. Doing good science: Authenticating cell line identity. Cell Notes 22, 15–17.

EU Directive, 2004/23/EC: on setting standards of quality and safety for the donation, procurement, testing, processing, preservation, storage and distribution of human tissues and cells. <http://www.eur-lex.europa.eu/LexUriServ/LexUriServ.do?uri=OJ:L:2004:102:0048:0058:EN:PDF>.

Eudralex, 2011. EU Guidelines to Good Manufacturing Practice Medicinal Products for Human and Veterinary Use. <http://ec.europa.eu/health/documents/eudralex/vol-4/index_en.htm>.

FDA, 2010. 21CFR58. Good Laboratory Practice for Nonclinical Laboratory Studies. <www.accessdata.fda.gov/scripts/cdrh/cfdocs/cfcfr/cfrsearch.cfm?cfrpart=58&showfr=1>.

Hanna, J.H., Saha, K., Jaenisch, R. Pluripotency and cellular reprogramming: Facts, hypotheses, unresolved issues. Cell 143, 508–525.

Hay, R.J., 1988. The seed stock concept and quality control for cell lines. Anal. Biochem. 171, 225–237.

Hunt, C.J., 2011. Cryopreservation of human stem cells for clinical application: A review. Transfus. Med. Hemother. 38, 107–123.

Hunt, C.J., Timmons, P.M., 2007. Cryopreservation of human embryonic stem cell lines. Methods Mol. Biol. 368, 261–270.

ISBER, 2008. Best practices for repositories: Collection, storage, retrieval and distribution of biological materials for research. Cell Preserv. Technol. 6 (1), 3–58.

ISCBI, 2009. Consensus guidance for banking and supply of human embryonic stem cell lines for research purposes. Stem Cell Rev. 5, 301–314.

Luong, M.X., Auerbach, J., Crook, J.M., Daheron, L., Hei, D., Lomax, G., et al., 2011. A call for standardized naming and reporting of human ESC and iPSC lines. Cell Stem Cell 8, 357–359.

MacLeod, R.A., Dirks, W.G., Matsuo, Y., Kaufmann, M., Milch, H., Drexler, H.G., 1999. Widespread intraspecies cross-contamination of human tumor cell lines arising at source. Int. J. Cancer 83, 555–563.

Mitalipova, M.M., Rao, R.R., Hoyer, D.M., Johnson, J.A., Meisner, L.F., Jones, K.L., et al., 2005. Preserving the genetic integrity of human embryonic stem cells. Nat. Biotechnol. 23, 19–20.

Nakajima, F., Tokunaga, K., Nakatsuji, N., 2007. Human leukocyte antigen matching estimations in a hypothetical bank of human embryonic stem cell lines in the Japanese population for use in cell transplantation therapy. Stem Cells 25, 983–985.

Nardone, R.M., 2007. Eradication of cross-contaminated cell lines: A call for action. Cell Biol. Toxicol. 23, 367–372.

National Institutes of Health, 2007. Notice regarding authentication of cultured cell lines. <http://www.grants.nih.gov/grants/guide/notice-files/NOT-OD-08-017.html>.

Nishikawa, S., Goldstein, R.A., Nierras, C.R., 2008. The promise of human induced pluripotent stem cells for research and therapy. Nat. Rev. Mol. Cell Biol. 9, 725–729.

OECD, 1997. OECD Principles on Good Laboratory Practice. <http://www.oecd.org/officialdocuments/displaydocumentpdf?cote=env/mc/chem(98)17&doclanguage=en>.

OECD, 2007. OECD Best Practice Guidelines for Biological Resource Centres. < http://www.oecd.org/dataoecd/7/13/38777417.pdf>.

Reubinoff, B.E., Pera, M.F., Vajta, G., Trounson, A.O., 2001. Effective cryopreservation of human embryonic stem cells by the open pulled straw vitrification method. Hum. Reprod. 16, 2187–2194.

Rottem, S., Naot, Y., 1998. Subversion and exploitation of host cells by mycoplasmas. Trends Microbiol. 6, 436–440.

Sivarajah, S., Selva Raj, G., Mathews, A.J., Sahib, N.B., Hwang, W.S., Crook, J.M., 2010. The generation of GLP-grade human embryonic stem cell banks from four clinical-grade cell lines for pre-clinical research. In Vitro Cell Dev. Biol. Anim. 46 (3–4), 210–216.

Smith, K.P., Luong, M.X., Stein, G.S., 2009. Pluripotency: Toward a gold standard for human ES and iPS cells. J. Cell Physiol. 220, 21–29.

Stacey, G.N., 2004. Cell line banks in biotechnology and regulatory affairs. In: Fuller, B, Benson, E.E., Lane, N. (Eds.), Life in the Frozen State. Taylor-Francis, London, UK.

Stacey, G.N., Sheeley, H.J., 1994. Have bio-safety issues in cell culture been overlooked? J. Chem. Technol. Biotechnol. 61, 95–96, CRC Press, Boca Raton, Florida.

Tavakoli, T., Xu, X., Derby, E., Serebryakova, Y., Reid, Y., Rao, M.S., et al., 2009. Self-renewal and differentiation capabilities are variable between human embryonic stem cell lines I3, I6 and BG01V. BMC Cell Biol. 10, 44.

Taylor, C.J., Bolton, E.M., Pocock, S., Sharples, L.D., Pedersen, R.A., Bradley, J.A., 2005. Banking on human embryonic stem cells: Estimating the number of donor cell lines needed for HLA matching. Lancet 366, 2019–2025.

Timenetsky, J., Santos, L.M., Buzinhani, M., Mettifogo, E., 2006. Detection of multiple mycoplasma infection in cell cultures by PCR. Braz. J. Med. Biol. Res. 39, 907–914.

Young, L., Sung, J., Stacey, G., Masters, J.R., 2010. Detection of mycoplasma in cell cultures. Nat. Protoc. 5, 929–934.

Wang, Y.-C., Nakagawa, M., Garitaonandia, I., Slavin, I., Altun, G., Lacharite, R.M., et al., 2011. Specific lectin biomarkers for isolation of human pluripotent stem cells identified through array-based glycomic analysis. Cell Res. 21, 1–13.

PART 2

Induced Pluripotent Stem Cell Technology

CHAPTER 9

Isolation of Human Dermal Fibroblasts from Biopsies

Victoria L. Glenn, John P. Schell, Eyitayo S. Fakunle, Ronald Simon, and Suzanne E. Peterson

EDITOR'S COMMENTARY

Since human embryonic stem cell (hESC) lines were first reported in 1998, dozens of laboratories have been studying their remarkable properties of self-renewal and differentiation potential. In 2007, remarkably similar cells, called induced pluripotent stem cells (iPSCs), were generated by transient expression of reprogramming factors in human dermal fibroblasts. hESCs and iPSCs both self-renew and can differentiate into widely diverse cell types, and there is now considerable debate about which pluripotent cells are preferable for particular research and clinical studies. While hESCs have been in laboratories for much longer and researchers have more experience in their culture and characterization, there are several reasons why iPSCs may find a wider range of uses. Since hESCs are not generated from living individuals, there is no clinical information associated with them; for example, as models for studying human genetic disease, hESCs are limited to those derived from embryos that have been screened using prenatal genetic diagnostics. Given that most diseases cannot be identified prenatally, this limits the utility of these cells for disease modeling. Although there are now hundreds of hESC lines available to researchers, they are still dominated by particular ethnicities, which will likely limit their use for drug screening and toxicology applications as well as immunological matching for cell therapy.

Over the past several years, we have been building collections of iPSCs for three purposes: collections for studies of human genetic disease, collections of ethnically diverse cells for screens of drug toxicity, and collections for specific cell transplantation therapies. We have chosen dermal fibroblasts derived from skin biopsies as the most practical source of cells for reprogramming. This chapter describes methods we developed for producing and banking fibroblasts from donor skin biopsies.

J.F. Loring & S.E. Peterson (eds): Human Stem Cell Manual, Second edition.
DOI: http://dx.doi.org/10.1016/B978-0-12-385473-5.00009-6

OVERVIEW

Although new methods for generating iPSCs have been a hot topic for the past several years, less attention has been focused on the cells being reprogrammed. Fibroblasts have been the most common source of reprogrammable cells, probably because there are several resources that have already banked fibroblasts from both diseased and normal donors. However, there is a growing need for primary cells that are not available from a company or a somatic cell bank. When considering banking their own collections of somatic cells for reprogramming, researchers need to consider which somatic cell type would be the best for their particular application. Over the years, a number of different somatic cell types have been successfully reprogrammed into iPSCs with varying efficiency, including fibroblasts, keratinocytes from hair and skin, various cells from the blood, and urothelial cells from urine (Aasen, 2009; Stadtfeld, 2010; Zhou, 2011). Each cell type has different advantages and disadvantages. For example, isolating keratinocytes from the outer root sheath of hair does not require a physician's assistance, is painless for the donor, and the cells reprogram with high efficiency. However, primary keratinocytes do not expand well nor do they recover well after cryopreservation. Blood is fairly easy to acquire but the reprogramming efficiency is low and since normal blood cells do not proliferate well *in vitro*, there is limited potential for banking cells for further reprogramming. In our lab, we have generated a number of iPSC lines using different types of primary cells that we isolated ourselves. We have weighed the pros and cons of each cell type and have found that human dermal fibroblasts (HDF) are the most robust cell type for reprogramming purposes with the fewest drawbacks.

In this chapter we present reliable techniques for isolating HDFs from a skin punch biopsy. Note that before any experimentation begins proper approval must be obtained from an IRB (institutional review board; see Chapter 38). This will involve the generation of an appropriate informed consent document that must be signed by each donor. Biopsies must be performed by a licensed medical practitioner in a clinical setting.

Isolation of HDFs begins by performing a 3 mm punch biopsy with local anesthesia, usually on the inner forearm or shoulder of the donor. Before isolation of the cells, it is important to keep the biopsy specimen submerged in ice-cold fibroblast medium containing an antibiotic and antimycotic. This is critical because the skin naturally harbors a variety of native bacteria and microorganisms on its surface that can lead to contamination. The tissue sample is then cut into pieces and enzymatically digested. Cells and minced tissue are plated on fibronectin-coated tissue culture plates and allowed to settle and grow. Once the fibroblasts have become confluent, they are trypsinized and passaged to larger flasks for expansion, cryobanking, and experimentation.

FIGURE 9.1
A 3 mm punch biopsy.

PROCEDURE

It is important to use sterile technique throughout this protocol. For best results, process the biopsy immediately after it is taken. Alternatively, biopsies can last for a few days if they are kept in fibroblast medium on ice. If biopsies are shipped from a remote location make sure they are kept on ice continuously.

1. In a laminar flow hood, remove the 3 mm punch biopsy from the conical tube in which it was transported to the lab using sterile forceps.
2. Transfer the biopsy to a 60 mm dish containing 5 mL of 1× PBS with Penicillin/Streptomycin and Primocin to wash (Figure 9.1).
3. Transfer the biopsy to a new 60 mm dish containing only a small volume of PBS with Penicillin/Streptomycin and Primocin. Using forceps and very sharp scissors cut the tissue into approximately 10–15 equally sized small pieces (Figure 9.2).

■ Note

It is very important to use sharp scissors to cut the tissue. Do not shred or tear the tissue. Also, 10–15 pieces are appropriate for a 3 mm biopsy. If you are using larger biopsies, scale up the number of pieces to cut. ■

4. Make up collagenase B solution immediately before use (See Recipes).
5. Gather the pieces together in a clump and use forceps to transfer them to a 1.5 mL microfuge tube. Make sure that the pieces are transferred to the bottom of the tube where they will come into contact with the collagenase B solution added in the next step.

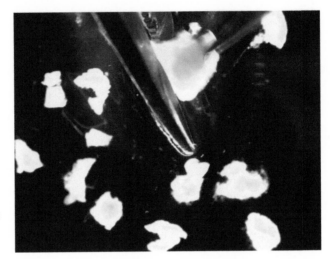

FIGURE 9.2
Hold the biopsy with forceps (right) and cut it into 10–15 pieces with scissors (left).

■ **Note**

This step can be frustrating because sometimes the tissue sticks to the forceps or to the top of the microfuge tube. It is easiest to transfer the pieces in a clump rather than individually. Be patient. ■

6. Add 1 mL of sterile filtered collagenase B to the microfuge tube. Make sure the pieces of tissue are completely covered.
7. Place the tube in a 37°C incubator.
8. About once each hour, gently agitate the tissue pieces in the tube, being careful not to let any pieces stick to the side of the tube above the collagenase B solution.
9. After 3–6 hours, spin the tube at 1150 g for 5 minutes.
10. Carefully pipette off the supernatant, making sure not to disturb the pellet.
11. Resuspend the pellet in 1 mL of fibroblast medium, pipetting up and down several times to ensure that the pellet is broken up as much as possible.

■ **Note**

Fibroblast medium is fine for use in this protocol. However, specialty media designed for fibroblasts will generate higher cell yields. We recommend the fibroblast medium from ScienCell or Cell N Tec. ■

12. Transfer the cell suspension to one fibronectin-coated well of a 12-well plate (see Recipes).

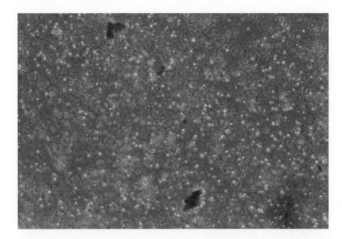

FIGURE 9.3
Skin biopsy plated after
5 hours of incubation in
collagenase B.

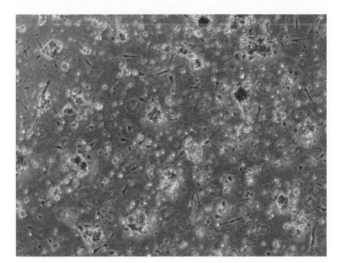

FIGURE 9.4
Skin biopsy 3 days after
plating on fibronectin.
Note the dark attached
fibroblast cells.

■ Note

Under a phase contrast microscope, many isolated cells should be apparent (Figure 9.3). There may also be a few larger pieces of tissue. ■

13. Incubate the cells at 37°C, 5% CO_2.
14. Leave the plate undisturbed for 2–3 days, then examine it under the microscope, looking for small patches of attached cells (Figure 9.4).
15. Add 1 mL of medium to the well without aspirating the original medium.
16. Two days later, aspirate all the medium and replace it with 1 mL of fibroblast medium.

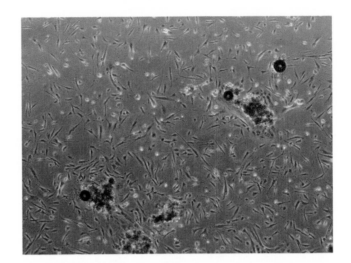

FIGURE 9.5
Skin biopsy dissociated in collagenase B and grown on fibronectin for 1 week.

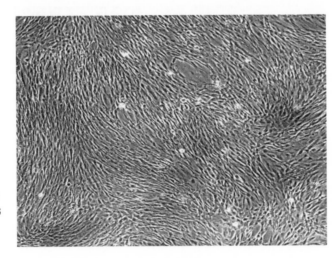

FIGURE 9.6
Confluent well of fibroblasts from a skin biopsy. The cells are ready to be passaged into a T25 flask.

17. Change the medium every other day. You should have nice outgrowth by 1 week (Figure 9.5).

18. When the well is confluent (Figure 9.6), trypsinize the cells and transfer them to a larger flask.

■ Note

Most biopsies will be ready to passage in 1–2 weeks. We typically passage one 12-well plate well into a T25 flask and then expand from there for banking and reprogramming. ■

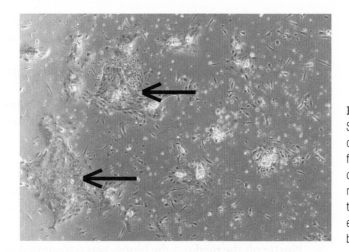

FIGURE 9.7
Skin biopsy culture with contaminating non-fibroblast cell types. These cell types can easily be removed during the first trypsinization and will eventually be outcompeted by the fibroblasts.

■ Note

Occasionally, contaminating non-fibroblast cells are seen in the original biopsy well (Figure 9.7). These cells are more resistant to trypsinization so monitor the cells during the trypsinization process and stop it when only the fibroblasts are detached. Even if some contaminating cells persist past the first trypsinization, they will be out-competed by the fibroblasts quickly.   ■

Passaging HDFs

1. Aspirate the medium from the 12-well plate well.
2. Rinse with 1 mL PBS.
3. Add 0.5 mL 0.05% Trypsin-EDTA.
4. Place the plate under a microscope and monitor the cells every minute or so.
5. When the fibroblasts have come off but the contaminating cell types still remain attached, add 0.5 mL of fibroblast medium to the well.
6. Pipette the cells up and down and transfer to a 15 mL conical tube.
7. Spin the tube at $200\,g$ for 5 minutes.
8. Carefully remove the supernatant and resuspend the cells in 5 mL of fibroblast medium.
9. Add the cell suspension to a T25 flask (does not need to be coated with fibronectin) and return it to the incubator. Check cell growth in 2 days and do not disturb for at least 24 hours after passaging to ensure cell attachment to the flask.

■ Note

Keep antibiotics and antimycotics in the medium only for the first week or two of culture.   ■

10. Replace the medium every 2 days until the flask becomes confluent, then passage again. Split the cells at a 1:3 ratio (e.g. T25 flask into T75 flask, T75 flask into T175 flask).
11. Cryopreserve at least two vials of cells before they are cultured antibiotic-free.
12. Make a larger cryobank after the cells have been cultured without antibiotics for a week and have been mycoplasma tested.

ALTERNATIVE PROCEDURES

Although the first fibroblast isolation protocol described in this chapter is robust, this alternative protocol may also be used. Cells usually take longer to reach confluency using this protocol.

1. Pipette off the medium the biopsy was transported in, being careful not to aspirate the biopsy.
2. Rinse the biopsy 2 times in 1× PBS with Penicillin/Streptomycin and Primocin by manual pipetting.
3. Add 2–5 mL sterile filtered dispase at 2 units/mL (in PBS with Primocin and Penicillin/Streptomycin) and incubate at 4°C for 18 hours.
4. Rinse the biopsy twice in 1× PBS with Penicillin/Streptomycin and Primocin by manual pipetting.
5. Prepare one 12-well plate well per biopsy:
   - With a sterile needle, etch the bottom of one well of a 12-well plate in a cross-hatched pattern.
   - Dilute 12 μL of bovine fibronectin (1 mg/mL stock) in 800 μL of PBS and add to the etched well and incubate at room temperature for 45 minutes. Wash once with PBS before use.
6. In a laminar flow hood, transfer the biopsy to a 60 mm culture dish containing 1–2 mL 1× PBS (with Penicillin/Streptomycin and Primocin) to keep it from drying out.
7. Using sterile instruments, separate the dermis and epidermis (Figure 9.8).
   - Use larger forceps with a flat broad end to hold the dermal layer (Figure 9.8A).
   - Hold epidermal layer using thin forceps with pointed tips (Figure 9.8A).
   - Pull to separate. It should come apart very easily (Figure 9.8B).
8. Transfer dermal tissue to a 35 mm dish and add 2 mL of AccuMax™.

■ Note

The epidermal layer may be discarded at this point or it can be used to isolate keratinocytes. ■

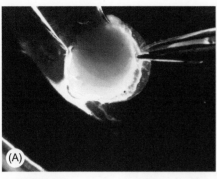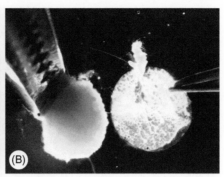

FIGURE 9.8

Separation of the dermal and epidermal layers. (A) Hold the dermal layer with large forceps that have a broad tip and the epidermal layer with pointed tipped forceps. (B) Pull the two layers apart. They should separate easily.

9. Using two pairs of sterile dissecting scissors (one in each hand), manually dissociate tissue repeatedly for 3 minutes.
10. Transfer the 35 mm dish containing tissue in AccuMax to a 37 °C incubator for 20 minutes.
11. Remove the dish from the incubator and transfer the contents of the dish to a 15 mL conical tube.
12. Neutralize AccuMax by adding fibroblast medium, bringing the total volume to 15 mL.
13. Spin down by centrifugation for 5 minutes at 200 g.
14. Carefully aspirate the supernatant and resuspend in 2 mL of fibroblast medium.
15. Transfer the cell suspension to a prepared well of a 12-well plate and place in a CO_2 incubator at 37 °C.
16. Optional: Spin the plate at 2100 g for 5 minutes at room temperature in order to help the cells adhere to the substratum.

■ Note

Once plated, do not remove from the incubator for 2 days (do not even touch the dish) to allow the larger undigested tissue to settle and allow cells to migrate out. ■

17. Add 1 mL of fibroblast medium to the well on day 3; handle the plate gently to avoid dislodging the cells.

■ Note

At this point, some of the pieces of tissue should have attached to the etched areas in the wells (Figure 9.9). ■

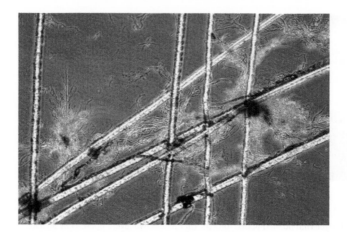

FIGURE 9.9
Skin biopsy 2 days after harvest. Pieces of tissue from the biopsy tend to attach to the plate in the etched areas.

18. On day 5, medium may be collected and biopsy fragments that did not attach may be seeded onto a new well.
    - Prepare a new well of the 12-well plate by etching and coating with fibronectin.
    - Pipette up the medium from the original well and put it in a 15 mL conical tube (there will be pieces of the biopsy which did not attach to the plate in this medium).
    - Add 2 mL of fibroblast medium to the original well.
    - Bring up the volume in the 15 mL conical tube to 10 mL with fibroblast medium. Spin at 200 g for 5 minutes, aspirate and resuspend in 2 mL of fibroblast medium.
    - Add 2 mL of cell suspension to the new (etched and coated) well (on the same plate if possible). Treat the new well the same as the original biopsies – do not disturb for 2 days, and change medium on the same schedule.

19. Change the medium on the cells every other day until the well is confluent.

■ Note

This protocol is much more variable in regard to when the cells will start to grow out of the tissue. Outgrowths can sometimes be seen as early as 1 week but other times it may take several weeks. ■

PITFALLS AND ADVICE

Manual Dissociation

When manually dissociating the biopsy into smaller pieces, make clean cuts. Minimizing tearing and shredding of the tissue will greatly improve primary cell outgrowth from the dissociated tissue. To generate the cleanest cuts, make sure that the scissors you use are sharp enough to cleanly slice rather than shred the tissue.

Enzyme Exposure

When using enzymes to dissociate tissue or to detach cells for passaging it is important not to leave the cells in active enzymes for longer than necessary. Enzymes used for cell culture work by breaking down connecting proteins and matrix components, such as collagen, but can begin to act on the cellular membrane proteins if given enough time. Serum used in culture medium can be used to decrease enzyme activity to prevent cell death.

Contamination

Since many microbes can live on human skin, it is important to test your fibroblasts for contaminants before cryobanking them. Remove the Primocin and Penicillin/Streptomycin from the medium 1–2 weeks after plating the cells and watch for bacteria and fungi (see Chapter 4). In addition, it is important to test your fibroblasts for mycoplasma (see Chapter 4).

EQUIPMENT

- Class II biosafety cabinet
- Tissue culture incubator, 37°C, 5% CO_2
- Inverted phase contrast microscope with 4×, 10×, and 20× objectives
- Dissecting microscope with Dark Field Transillumination and 0.8–4.5× range
- Tabletop centrifuge
- Scale capable of measuring in mg
- Access to 4°C, −20°C, −80°C, and cryogenic freezers
- Pipette aid and micropipetters, p-2, p-20, p-200, and p-1000 µL

REAGENTS AND SUPPLIES

Recommended Reagents and Supplies

Item	Supplier	Catalog #	Note
Dulbecco's Phosphate-Buffered Saline (DPBS) without Calcium and Magnesium	Life Technologies	14190-250	
Dispase	Life Technologies	17105-041	
Costar 12-Well Clear TC-Treated Multiple Well Plates, Sterile	Corning	3512	

Item	Supplier	Catalog #	Note
Sterile conical tubes, 50 mL	Corning	430291	
Sterile conical tubes, 15 mL	Corning	430053	
Bovine Fibronectin	Calbiochem	341631	
Collagenase B	Roche	11088807001	
DMEM + GlutaMAX™	Life Technologies	10566-016	
MEM NEAA	Life Technologies	11140-050	
Fetal Bovine Serum	Gemini	900-108	10% total volume
Penicillin/Streptomycin	Life Technologies	15140	10 µL per 1 mL of medium or PBS
Primocin	InvivoGen	Ant-pm-1	2 µL per 1 mL of medium or PBS
500 mL Vacuum Filter/ Storage Bottle System, sterile low protein binding 0.22 µm pore, 33.2 cm^2 PES membrane	Corning	431097	
0.22 µm syringe filter	Millipore	SLGP033RS	
10 mL syringe	BD	309604	
60 mm dish	Sarstedt	83.1801	
Micro dissecting scissors	World Precision Instruments	501263-G	
Thumb dressing forceps	Roboz	RS-8122	
1.5 mL microfuge tube	Eppendorf	022363204	
0.05% Trypsin-EDTA	Life Technologies	25300-054	
T25 flask	Corning	430168	
T75 flask	Corning	430641	
T175 flask	Corning	431080	
Sterile needle and syringe	BD	309569	
AccuMax	Stemgent	01-0007	
Fibroblast Medium	Sciencell	2301	
Fibroblast Medium	Cell nTec	CnT-05	

RECIPES

Preparation of Fibroblast Medium

1. To make 500 mL:
    Sterile filter
    - 445 mL DMEM with GlutaMAX
    - 50 mL FBS (Fetal Bovine Serum)
    - 5 mL MEM NEAA (Non-Essential Amino Acids).
2. Once filtered, aliquot medium for use in 50 mL conical tubes.
3. Add Primocin and Penicillin/Streptomycin to the aliquot.

■ Note

Maintain primary fibroblast cultures in Primocin and Penicillin/Streptomycin for only the first couple of weeks to prevent contamination. ■

Preparation of Bovine Fibronectin Coated Wells

1. Coat one well of a 12-well plate (1 well per biopsy) with bovine fibronectin.
2. For each well, dilute 12 μL of bovine fibronectin (1 mg/mL stock) in 800 μL of PBS and add to the well.
3. Allow the plate to incubate at room temperature for 45 minutes.
4. Wash well once with PBS just prior to plating the cells

■ Note

Aliquot fibronectin to avoid repeated freeze thaw cycles. ■

Preparation of Collagenase B

1. Make up collagenase B immediately before use.
2. Collagenase B is used at 2 units/mL; however, units of activity can vary between lots so be sure to check the lot specifications prior to use.
3. Bring collagenase B to room temperature before opening the container.
4. Once at room temperature, measure the appropriate amount needed (make 1 mL per biopsy).
5. Add PBS and collagenase B to a conical tube.
6. Sterile filter mixture into a sterile 15 mL conical tube with a 0.22 μm filter and a 10 mL syringe.

READING LIST

Aasen, T., Belmonte, J.C., 2010. Isolation and cultivation of human keratinocytes from skin or plucked hair for the generation of induced pluripotent stem cells. Nat. Protoc. 5 (2), 371–382.

Stadtfeld, M., Hochedlinger, K., 2010. Induced pluripotency: History, mechanisms, and applications. Genes. Dev. 24 (20), 2239–2263.

Zhou, T., Benda, C., Duzinger, S., Huang, Y., Li, X., Li, Y., et al., 2011. Generation of induced pluripotent stem cells from urine. J. Am. Soc. Nephrol. 22 (7), 1221–1228.

Human Induced Pluripotent Stem Cell Generation: Conventional Method

Isao Asaka and Shinya Yamanaka

EDITOR'S COMMENTARY

The first edition of this book was published in 2007; had it come out a year later, it would have included the most remarkable advance in the human pluripotent stem cell (hPSC) field since the first report of human embryonic stem cell (hESC) line derivation in 1998. Before the method for reprogramming human somatic cells to pluripotency was reported in 2007, embryo-derived cells were our only choices. Even as the number of hESC lines grew, they were still almost all of Caucasian ethnicity, so the possibility of using hPSC-derived cells for screening genomics-associated adverse drug effects seemed impossibly remote. The prospects for using hESCs to model genetic diseases were limited to those rare occasions when mutation-carrying embryos were identified by preimplantation genetic diagnosis and were donated for hESC derivation. For potential cell-replacement therapy applications, many hundreds of hESC lines would be required in order to collect the range of haplotypes required for matching a significant proportion of the population. Human induced pluripotent stem cells (hiPSCs) have made all of these hPSC applications possible. Ethnically diverse collections of hiPSCs are being developed for integration into drug development pathways; large collections of iPSC lines carrying known mutations are being used to understand the underlying mechanism of diseases; haplotype-matched iPSC banks for cell therapy are growing.

There is a growing weight of evidence that hiPSCs and hESCs are indistinguishable for all practical purposes. Both hESC and hiPSC lines currently vary in how easily they differentiate into the particular cell types of interest; both pluripotent cell types acquire genetic and epigenetic changes as they are expanded in culture. It seems almost certain that improvements in derivation, expansion, and differentiation of pluripotent cells will make these variations less important. This chapter, written by the researchers who first made iPSCs, provides the most basic and reliable methods for reprogramming human

143

J.F. Loring & S.E. Peterson (eds): Human Stem Cell Manual, Second edition.
DOI: http://dx.doi.org/10.1016/B978-0-12-385473-5.00010-2

cells, and should allow many more researchers to integrate iPSCs into their particular area of interest.

OVERVIEW

In 2006, it was first reported that mouse embryonic and adult fibroblasts could obtain properties similar to those of embryonic stem (ES) cells after the introduction of genes encoding four transcription factors (Oct3/4 (Pou5F1), Sox2, Klf4, and c-Myc (Myc))(Takahashi and Yamanaka, 2006). These ES-like cells were named "induced pluripotent stem cells (iPSCs)". Mouse iPSCs were similar to ES cells in morphology, proliferation, expression of ES cell marker genes, and ability to form teratomas in immune compromised mice. In 2007, iPSCs were also generated from human fibroblasts using the same transcription factor combination. iPSCs have since proven valuable not only as material for developmental biology research, but also for medical research.

Many iPSC lines have been established from patients with various diseases; for example, amyotrophic lateral sclerosis (ALS), Parkinson's disease and Fanconi anemia. The pathology of spinal muscular atrophy (SMA) has also been recapitulated in motor neurons derived from disease-specific iPSCs. The ability of these disease-specific iPSCs to differentiate into the various tissues affected by different diseases has provided a new research tool for investigating the mechanisms of disease pathogenesis in a human system. Disease-specific iPSCs are also useful for screening for drug toxicity caused by genetic mutations, such as long QT syndrome.

Reprogramming methods using retroviral vectors cannot be used directly for the production of tissues or organs for transplantation due to the potential for genomic modification as a result of transgene integration into host chromosomes. However, such disease-specific iPSC lines are tremendously valuable for research on disease pathogenesis and drug discovery.

The method of generating human iPSCs described below is highly reproducible using basic molecular and cell biology techniques, without otherwise requiring any specialized equipment. We recommend retroviral transduction of the four reprogramming factors as a conventional and robust method for the generation of human iPSC lines that can be used for various avenues of research.

In this chapter we provide a conventional method for establishing human iPSC lines using retroviral vectors, as well as an alternative method for retroviral packaging and transduction. We recommend for the safety of personnel that the human dermal fibroblast cells be first transfected with the mouse ecotropic retrovirus receptor into human dermal fibroblasts. This step is necessary to enhance viral transduction efficiency and increase the level of safety for the investigators. Subsequently, the transcription factors (Oct3/4 (Pou5f1), Sox2, Klf4, and c-Myc (Myc)) may be introduced into fibroblasts expressing the ecotropic receptor. For transduction of the transcription factors, a combination of the pMXs retroviral vectors, a proven vector for the generation of mouse iPSCs, and PLAT-E packaging cells, which produce high titers of retrovirus, is used. We also make use of SNL cells as a feeder layer to maintain human iPSCs. The SNL (STO-Neo-LIF) cell line was derived from mouse embryos and expresses neomycin resistance as well as the leukemia inhibitory factor (LIF) gene. This cell line has two significant advantages: it is not subject to the same batch to batch variability as mouse embryo fibroblasts (MEFs) and it has a more extended period of proliferation than primary MEFs.

We provide an alternative viral packaging method, using PLAT-GP system, which allows direct transfection of human cells, but requires a higher level of safety measures for laboratory personnel.

The process of establishing human iPSCs thus consists of the following main steps:

- Introducing the mouse ecotropic receptor for retroviruses (Slc7a1) into human cells by lentiviral vectors
- Introducing four reprograming factors (Oct3/4, Sox2, Klf4, c-Myc) into human cells expressing the mouse ecotropic receptor using retroviruses
- Reseeding the cells onto a feeder layer
- Picking up the ES cell-like colonies
- Expanding and preserving cultured cells from each colony.

PROCEDURES

Introduction of the Mouse Ecotropic Receptor for Retroviruses (Slc7a1) into Human Cells by Lentiviral Vector

Preparation of 293FT Cells

1. Remove a vial of frozen cells from liquid nitrogen and thaw quickly in a 37°C water bath.
2. Just before the cells are completely thawed, decontaminate the outside of the vial with 70% ethanol, and transfer the cells to a sterile 15 mL tube containing PBS. Briefly centrifuge the cells at 150–200 g and resuspend them in 2 mL of 293FT medium without antibiotics.
3. Transfer the cells to a 100 mm dish containing 10 mL of 293FT medium without antibiotics.
4. Incubate the dish overnight at 37°C to allow the cells to attach to the bottom of the dish.

5. The next day, aspirate the medium and replace it with fresh, 293FT medium with antibiotics.
6. Incubate the cells and check them daily until they are 80–90% confluent.
7. When the 293FT cell culture reaches 80–90% confluence, aspirate the medium and wash the cells with DPBS($-$) (without Ca++/Mg ++).
8. Add 1 mL of 0.25% trypsin, 1 mM EDTA solution and incubate the 100 mm dishes for 2 minutes at room temperature.
9. Add 10 mL of medium and dissociate the cells by pipetting up and down (about 10 times).
10. Collect the cell suspension into a 15 mL conical tube, and count the cells.
11. Adjust the concentration to 4×10^5 cells/mL using 293FT medium without antibiotics. Seed the cells at 4×10^6 cells in 10 mL/100 mm dish, and incubate overnight at 37°C in a 5% CO_2 incubator.

Lentiviral Vector Production

1. Dilute 9 μg of Virapower™ packaging mix (pLP1, pLP2 and pLP/VSVG mixture) and 3 μg of pLenti6/UC encoding the mouse Slc7a1 gene in 1.5 mL of OPTI-MEM® I, and mix gently by finger tapping.
2. In a separate tube, dilute 36 μL of Lipofectamine™ 2000 in 1.5 mL of Opti-MEM I. Mix gently by finger tapping and incubate for 5 minutes at room temperature.
3. After incubation, combine the diluted DNA with the diluted Lipofectamine 2000. Mix gently by finger tapping, and incubate for 20 minutes at room temperature.
4. During incubation, remove the medium from 293FT dishes, and add 9 mL of fresh medium to each dish.
5. Next, add 3 mL of the DNA-Lipofectamine 2000 complexes to each dish. Mix gently by rocking the dish back and forth. Incubate the dish overnight at 37°C in a 5% CO_2 incubator.
6. 24 hours after transfection, aspirate the medium containing the transfection cocktail, and add 10 mL of fresh FP medium. Incubate the dish overnight at 37°C in a 5% CO_2 incubator.

■ Note

When working with lentivirus, always a wear a suitable lab coat and disposable gloves, and work in a biological safety cabinet.   ■

Preparation of Human Fibroblasts

1. Aspirate the medium and wash the cells with DPBS($-$).
2. Add 0.5 mL of 0.25% trypsin, 1 mM EDTA solution to each well, and incubate the dish for a few minutes at room temperature.

3. Add 4.5 mL of FP medium and dissociate the cells by pipetting up and down.
4. Collect the cell suspension into a conical tube. Count the cell number. Adjust the concentration to 8×10^4 cells/mL by adding an appropriate volume of FP medium.
5. Seed the cells at 8×10^5 cells in 10 mL of cell suspension/100 mm dish. Incubate overnight at 37°C in a 5% CO_2 incubator.

Lentiviral Vector Infection
1. 48 hours after transfection, collect the supernatant from the 293FT culture with a 10–50 mL disposable syringe, and then filter it with a 0.45 μm cellulose acetate filter.

■ **Note**
The lentivirus-containing medium can be stored at −80°C. To avoid reducing the viral titer, do not repeat freeze/thaw cycles. ■

2. Replace medium with 10 mL/dish of the virus-containing supernatant, supplemented with 4 μg/mL polybrene. Incubate the dishes for at least 5 hours (usually dishes can be incubated for 24 hours) at 37°C in a 5% CO_2 incubator.

■ **Note**
If the cells die due to viral toxicity, run a double dilution of viruses, or use a shorter viral infection time (about 5 hours). ■

3. 5–24 hours after transduction, aspirate the virus-containing medium and add 10 mL of fresh FP medium.

■ **Note**
This is key for the next step to ensure lentiviral infection and thus expression of the mouse Slc7a1 gene in fibroblasts. ■

■ **Note**
EGFP or DsRed expressing virus should be included as an infection control. ■

■ **Note**
pLenti6/UbC/mSlc7a1 contains the Blasticidin S resistance gene. Culturing cells in medium containing Blasticidin S will confirm and enrich for lentiviral infection. ■

Introduction of the Four Reprogramming Factors (Oct3/4, Sox2, Klf4, c-Myc) by Retroviral Vectors into Human Cells Expressing the Mouse Ecotropic Receptor
Preparation of PLAT-E Cells

1. Prepare 9 mL of FP medium in a 15 mL conical tube.
2. Take out PLAT-E frozen stocks from the liquid nitrogen tank and put the vials in a 37°C water bath until most (but not all) cells are thawed.
3. Wipe the vial with ethanol, open the cap, and transfer the cell suspension to the tube prepared in Step 1.
4. Centrifuge at 180 g for 5 minutes, then discard the supernatant.
5. Resuspend the cells with 10 mL of FP medium, and transfer to 0.1% gelatin-coated 100 mm dishes. Incubate the cells in a 37°C 5% CO_2 incubator.
6. The following day, replace the medium with new media supplemented with 1 µg/mL of puromycin and 10 µg/mL of Blasticidin S. Continue to incubate the cells in a 37°C, 5% CO_2 incubator until they are 80–90% confluent.
7. After the cells reach 80–90% confluence, subculture them at a 1:4–1:6 split ratio. The cultures should reach confluence 2–3 days after seeding.
8. The optimal seeding density for PLAT-E cells in different sized dishes or plates is as follows:

Dish	Cell Number
100 mm diameter	3.6×10^6 cells
60 mm diameter	1.5×10^6 cells
35 mm diameter (6–well plate)	6.0×10^5 cells

Note: *FP medium without puromycin or Blasticidin S should be used.*

Retrovirus Production and Transfection into PLAT-E Cells

1. Transfect one plasmid each into separate 100 mm dishes. Transfection of more than two plasmids per dish results in a reduction of the efficiency of virus production. If using the Yamanaka factors (Oct3/4, Sox2, Klf4 and c-Myc) and EGFP, prepare five dishes of PLAT-E cells.
2. Transfer 0.3 mL of OPTI-MEM I into each 1.5 mL tube.
3. Transfer 27 µL of Fugene® 6 transfection reagent into the tube prepared in Step 2, mix gently by finger tapping, and incubate the tubes for 5 minutes at room temperature.
4. Add 9 µg of the appropriate plasmid DNA (1 µg/mL) to each tube, and add one plasmid per tube. Mix by finger tapping, and incubate tubes for 15 minutes at room temperature.
5. Add the DNA/Fugene 6 mixture to one of five separate cultures of PLAT-E cells. Incubate the dishes overnight at 37°C in a 5% CO_2 incubator.
6. The following day, replace the medium containing DNA and Fugene 6 with fresh FP medium without puromycin or Blasticidin S, and place the dishes back in the incubator.

■ Note

Transfect one dish with a suitable control. We use the pMXs retroviral vector expressing EGFP or DsRed to monitor the transfection efficiency. We routinely obtain efficiencies above 60%. High efficiency is essential for human iPSC induction. ■

Preparation of Ectopic Receptor-Expressing Human Fibroblasts

1. When human fibroblasts expressing the mouse Slc7a1 gene from p.147, Step 3 in the section "Lentiviral vector infection" reach 80–90% confluence, aspirate the medium and wash the cells once with 10 mL of DPBS($-$).
2. Discard the DPBS($-$), add 0.5 mL/dish of 0.25% trypsin, 1 mM EDTA solution, and incubate the dish for 2–3 minutes at room temperature.
3. Add 4.5 mL of FP medium per dish and generate a single cell suspension by pipetting. Transfer cells to a 50 mL conical tube.
4. Count the cells, with an expected yield of ~2×10^6 cells/dish. Adjust the concentration to 8×10^4 cells/mL with FP medium. Transfer 10 mL of the cell suspension (8×10^5 cells) to 100 mm dishes. Incubate the dishes overnight at 37°C in a 5% CO_2 incubator.

Retrovirus Infection

1. Collect the medium from each PLAT-E dish by using a 10 mL sterile disposable syringe, filter through a 0.45 μm cellulose acetate filter, and transfer into a 15 mL conical tube.
2. Add 5 μL of 8 mg/mL polybrene solution into the filtered virus-containing medium, and mix gently by pipetting up and down. The final concentration of polybrene is 4 μg/mL.
3. Make a mixture of equal parts of the medium containing Oct3/4-, Sox2-, Klf4-, and c-Myc-retroviruses.

■ Note

Retroviruses should be used fresh. Do not freeze retrovirus-containing medium. The titer of retrovirus is critical for successful iPSC generation. Freeze/thaw cycles decrease the titer of the retrovirus. ■

4. Aspirate the medium from fibroblast dishes, and add 10 mL per dish of the polybrene/virus-containing medium. Incubate the cells from 4 hours to overnight (24 hours) at 37°C in a 5% CO_2 incubator.
5. Following the culture period, aspirate the medium from the transduced fibroblasts, and add 10 mL per dish of fresh FP medium. Change the medium every second day until reseeding (it is usually 5 or 6 days after retrovirus infection).

Reseeding on the Feeder Layer
Preparation of SNL Feeder Cells

1. Prepare 9 mL of SNL feeder cell medium in a 15 mL conical tube.
2. Take a vial of frozen SNL cells out of the liquid nitrogen tank, and incubate the vial in a 37°C water bath until almost all cells are thawed.
3. Wipe the vial with alcohol, open the cap, and immediately transfer the cell suspension into the tube prepared in Step 1.
4. Centrifuge the tube at 160 g for 5 min, and discard the supernatant.
5. After dissociating the pellet of SNL cells by finger tapping, resuspend them with 1–2 mL of fresh SNL feeder cell medium. Take out a small aliquot of the suspension for trypan blue staining, and count the number of total and living cells using a counting chamber or cell counter.
6. Dilute the cells with SNL feeder cell medium so that the number of living cells is adjusted to $0.4–1 \times 10^4$ cells/mL. Plate 10 mL of the cells on a 0.1% gelatin-coated 100 mm dish, and incubate the cells at 37°C in an incubator supplemented with 5% CO_2.
7. Passage cells every 3–4 days at a 1:8–1:16 dilution (the maximum dilution is 1:16). When the cells reach 80–90% confluence, discard the medium and wash the cells once with 5 mL of DPBS(−).
8. Aspirate the DPBS(−), add 0.5 mL of 0.25% trypsin, 1 mM EDTA solution evenly over the dish and incubate for approximately 1 minute at room temperature.
9. After cells are detached from the dish, add 4.5 mL of SNL feeder cell medium to stop the trypsinization, resuspend the cells as a single cell suspension, and transfer them into a new conical tube.
10. Remove a small aliquot (10 μL per count) of the cell suspension and add an equal volume of trypan blue. Count the number of total and living cells.
11. Dilute the cells with SNL feeder cell medium to adjust the cell number to $0.4–1 \times 10^5$ cells/mL. Plate 10 mL of the cells for each 0.1% gelatin-coated 100 mm dish, and incubate the dishes at 37°C with 5% CO_2.
12. When the cells reach 80–90% confluence, add 310 μL of 0.4 mg/mL mitomycin C solution per 100 mm dish directly into the SNL feeder cell medium, and incubate for 2.25 hours at 37°C in a 5% CO_2 incubator.
13. Aspirate the medium containing mitomycin C, and wash the cells twice with 5 mL of DPBS(−).
14. After aspirating the last DPBS(−) wash, add 0.5 mL of 0.25% trypsin, 1 mM EDTA per 100 mm dish, and incubate for approximately 1 minute at room temperature.
15. After the cells are dissociated, add 4.5 mL of SNL feeder cell medium to stop the trypsinization, resuspend the cells to make a single cell suspension, and transfer them into a conical tube.

16. Remove a small aliquot ($10 \mu L$ per count) of the cell suspension and add an equal volume of trypan blue. Count the number of total and living cells.

17. Passage or cryopreserve the cells:

For iPSC Culture

Dilute cells with SNL feeder cell medium to adjust the concentration to $1.25 - 1.5 \times 10^5$ cells/mL and plate them on 0.1% gelatin-coated dishes, and incubate at 37°C in a 5% CO_2 incubator. SNL feeder cells can be used 24 hours after plating.

Dish Size	Cell Number
100 mm diameter	1.5×10^6 cells
60 mm diameter	5.0×10^5 cells
35 mm diameter (6–well plate)	2.5×10^5 cells
24–well plate	6.3×10^4 cells

Note: *use feeder cells within 3 days.*

For Cryopreservation

After centrifugation at $160 g$ for 5 minutes, aspirate the supernatant. Dilute cells with SNL feeder cell medium to adjust the concentration to $1-2 \times 10^7$ cells/mL, and add the same volume of $2\times$ SNL cell freezing medium to make 1 mL frozen stocks. After closing the cap, transfer the vials into a "Mr. Frosty" container (Nalgene), and store in a $-80°C$ freezer. Once the cells are frozen completely, transfer them into a liquid nitrogen tank for long-term storage.

Reseed the Transduced Human Fibroblasts onto Mitomycin C-Treated SNL Feeder Cells

1. Aspirate the culture medium and wash with 10 mL/dish DPBS($-$).
2. Discard the DPBS($-$), add 0.5 mL/dish of 0.25% trypsin, 1 mM EDTA solution, and incubate the dish for 2–3 minutes at 37°C
3. Add 4.5 mL/dish of FP medium, resuspend the cells as a single cell suspension, and transfer to a 50 mL conical tube.
4. Count cell numbers, and adjust the concentration to 5×10^3 or 5×10^4 cells/mL. Transfer 10 mL of cell suspension (5×10^4 or 5×10^5 cells) to 100 mm dishes with mitomycin C-treated SNL cells. Incubate the dishes overnight at 37°C in a 5% CO_2 incubator.

■ Note

When seeding retrovirally-transduced fibroblasts on feeder cells, the cell density is critical for human iPSC generation. Overgrowth of fibroblasts inhibits the generation of iPSC colonies. On the other hand, if the cell number is too low, no colonies emerge. The optimal cell densities are different for each fibroblast culture. Two or three different densities should be tested among $0.5-5 \times 10^5$ cells/dish.   ■

5. The next day and every second day thereafter, change the medium to 10 mL of fresh human ES cell medium until the colonies become large enough to be picked. Colonies should become visible 2 to 3 weeks after retroviral infection. They can be picked at around day 30 when they become large enough (visible to the naked eye).

■ Note

Even if the same fibroblasts are used for iPSC generation, the timing of iPSC colony emergence may be different between experiments. The culture should be checked for colonies at every medium change.   ■

Picking ES Cell-Like Colonies

1. Aliquot 200–300 μL of human ES cell medium/well in 24-well plates.
2. Pick colonies from the dish under a stereomicroscope using a pipetteman, and transfer them into the 24–well plate prepared in Step 1. Pipette up and down 2–3 times to gently break the colonies into small clumps (20–30 cells).

■ Note

Do not break up the cell clumps into single cells.   ■

3. Transfer the cell clumps into 6–well plates with mitomycin C-treated SNL feeder cells, and incubate in a 37°C, 5% CO_2 incubator until they form colonies of 1–2 mm in diameter. This usually takes 5–7 days.

Expansion and Preservation of Colonies
Passaging Cells

Volumes listed in the following protocol are per 60 mm dish.

1. Bring the dissociation solution (CTK solution) (0.5 mL is required per 60 mm dish) to room temperature.

■ Note

The CTK solution may become less effective if it remains for more than 2 hours at room temperature.   ■

2. Bring the PBS (2 mL is required per 60 mm dish) and human ES cell medium to room temperature.
3. Aspirate the culture medium.
4. Wash cells with 2 mL of DPBS(−) and aspirate.
5. Add 0.5 mL of the CTK solution to the dishes, and incubate them at room temperature for 1–2 minutes.

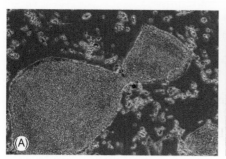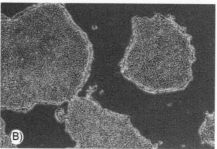

FIGURE 10.1
Cultured iPSCs after the feeder layer has been detached and removed. (A) Feeders are detached from the iPSC culture by CTK treatment. (B) iPSC culture after feeder removal by DPBS(−) wash.

■ Note

SNL feeder cells attach to the surface more weakly than human iPSCs. However, extensive CTK incubations or aggressive washing in subsequent steps will also detach human iPSC colonies. Constant microscopic observation is required since the incubation time may vary depending on the feeder cell and colony density (Figure 10.1). ■

6. After the feeder cells are detached, wash the cells with 2 mL of DPBS(−) as gently as possible. Any remaining SNL feeders should detach.
7. Aspirate the PBS and wash the cells again with 2 mL of DPBS(−).
8. Aspirate the DPBS(−) so that almost all of the feeder cells are removed, then add about 3 mL of human ES cell medium to each dish.
9. Detach the iPSCs using a sterile cell scraper and dissociate the colonies into small clumps with an average size of 50–100 μm by pipetting up and down.

■ Note

Do not break colonies into smaller clumps or single cells. ■

10. Transfer the cell suspension into 15 mL conical tubes and then add 9 mL of fresh human ES cell medium. Resuspend gently.
11. Aspirate the medium from three 60 mm dishes seeded previously with mitomycin C-treated SNL feeder cells (p.151, section Preparation of SNL feeder cells Step 17 – For iPSC culture), and add 4 mL of the suspension into a dish. Distribute the clumps evenly.

■ Note

Be careful not to allow the human iPSC colonies to collect in the center of the dishes. ■

12. Incubate the cell suspension at 37 °C in a 5% CO_2 incubator. Most clumps will attach within the first few hours, and the culture should be ready to passage again within 3–5 days.

■ Note
Be careful when handling the incubator doors to prevent vibration and poor colony distribution. ■

Cryopreservation
1. Incubate the CTK solution, PBS, and human ES cell medium at room temperature.
2. Thaw the iPSC freezing medium (DAP213 solution) and keep at 4°C until use.
3. Aspirate the culture medium from human iPSCs cultured in a 100 mm dish.
4. Wash cells with 4 mL of DPBS(−) and aspirate the DPBS(−).
5. Add 1 mL of CTK solution to each dish, and incubate at 37°C for 2–5 minutes. Observe each minute.
6. After the feeder cells are detached, wash the cells with 4 mL of DPBS(−).
7. Aspirate the DPBS(−), and wash cells again with 4 mL of DPBS(−).
8. Aspirate the DPBS(−) so that all of the feeder cells are removed, and add 6 mL of human ES cell medium.
9. Detach the human iPSC colonies using a cell scraper, then transfer the cell suspension to a 15 mL conical tube, and then divide the cell suspension into two 15 mL conical tubes (each 4 mL).

■ Note
Do not break up the colonies. ■

10. Centrifuge the suspension at 160 g for 5 minutes and aspirate the supernatant.
11. Resuspend the pellet in 0.2 mL of DAP213 solution, and transfer the cells into 2 mL cryovials.

■ Note
Do not break up the colonies. ■

12. Close the cap and quickly put the vials into liquid nitrogen.

■ Note
Because the DAP213 preservation solution is toxic to human iPSCs at room temperature, Steps 10–11 should be performed within 15 seconds to ensure good cell viability. ■

Thawing and Plating of Human iPSCs
1. Aliquot 10 mL of prewarmed (37°C) human ES cell medium in a 15 mL conical tube.
2. Take the vials of frozen human iPSCs out of the liquid nitrogen tank. Keep the vials in liquid nitrogen during the transfer to maintain the temperature as low as possible.

3. In a tissue culture hood, wipe the outside of the vials with alcohol and open the caps.
4. Add 0.8 mL of the medium prepared in Step 1 into the vials, and quickly thaw the frozen cells by pipetting up and down gently with a 1000 µL pipette.
5. Transfer the thawed cells into the tube prepared in Step 1, rinse the vial briefly, and close the tube cap.

■ Note

Do not break up the cell clumps.   ■

6. Centrifuge the tubes at 160 g for 5 minutes, and aspirate the supernatant.
7. Add 4 mL of human ES cell medium into each tube.
8. Seed the cell suspension on mitomycin C-treated SNL feeder cells in 60 mm dishes, and incubate at $37\,^{\circ}C$ with 5% CO_2.

■ Note

Do not move dishes within 48 hours after seeding, because the human iPSCs have difficulty attaching just after thawing.   ■

9. Change the medium 2 days after plating, and continue to incubate the cells until they reach 80–90% confluence.

ALTERNATIVE PROCEDURES

Direct Transfection of Human Cells with PLAT-GP Packaging System

The PLAT-GP packaging system can be used to introduce the Yamanaka factors instead of the PLAT-E system. If PLAT-GP is used for human iPSC generation, the steps used to introduce the Slc7a1 receptor into the cells can be omitted.

■ Note

Because this pantropic system is able to transfect human cells directly, laboratory staff must take care when working with the PLAT-GP cells and tissue culture media.   ■

1. Seed 3.6×10^6 cells in a 100 mm culture dish without antibiotics, including puromycin and Blasticidin S, 1 day before transfection.
2. After 24 hours, introduce one plasmid per 100 mm dish. Transfection of more than two pMXs plasmids into a single dish causes a reduction in the efficiency of iPSC generation. If using the Yamanaka factors (Oct3/4 (Pou5f1), Sox2, Klf4, and c-Myc (Myc)) and EGFP, prepare five dishes of PLAT-GP cells.

3. Transfer 0.3 mL of Opti-MEM I into each 1.5 mL tube.
4. Deliver 27 µL of Fugene 6 transfection reagent into the tubes prepared in Step 3, mix gently by finger tapping, and incubate the tubes for 5 minutes at room temperature.
5. Add 6 µg each of pMXs plasmid DNA (1 µg/mL) and 3 µg pCMV-VSV-G to each tube. Mix by finger tapping, and incubate the tubes for 15 minutes at room temperature.
6. Add the DNA/Fugene 6 mixture to one of five separate cultures of PLAT-GP cells. Incubate the dishes overnight at 37°C in a 5% CO_2 incubator.
7. On the following day, replace the medium containing DNA and Fugene 6 with fresh FP-medium without puromycin or Blasticidin S, and place the dishes in the incubator.

■ Note

Transfect the cells with a suitable control. We use a pMXs retroviral vector expressing EGFP or DsRed to monitor the transfection efficiency. We routinely obtain efficiencies above 60%. High efficiency is essential for human iPSC induction.   ■

8. Harvest the retroviral supernatant 24 hours after the medium change.

■ Note

When working with the PLAT-GP cells and their medium, always a wear a suitable lab coat and disposable gloves, and work in biological safety cabinet to prevent potential infection.   ■

PITFALLS AND ADVICE

- Because transduction with viral vectors may affect fibroblast viability, do not use fibroblasts with a high passage number.
- 3–5 days after infection, the transgene expression of the Slc7a1 lentivirus should reach the maximum level.
- Introduce one plasmid into one PLAT-E dish. Transfection of more than two plasmids into a dish causes a reduction in the efficiency of iPSC generation.
- Retroviruses should be used fresh! Do NOT freeze them, or you will not obtain iPSCs.
- The quality of the feeder cells is crucial. Aspects such as their density and freshness are important for maintaining undifferentiated colonies. We have always used SNL cells before passage 20.
- The incubation time for enzymatic dissociation may depend on the cell density. It should be checked visually every minute.

- iPSC clumps should be treated gently during subculture and freezing.
- Freezing of iPSC clumps should be performed within 15 seconds to ensure good cell viability.
- To ensure the viability of the iPSCs, frozen iPSCs should be thawed as quickly as possible. Do not break up cell clumps into single cells.

EQUIPMENT

- Tissue culture hood, class II A/B3
- Tissue Culture Incubator: 37°C, 5% CO_2, in humidified air
- Inverted phase/contrast microscope with 4×, 10×, and 20× objectives
- Stereomicroscope
- Cell counter or hemocytometer
- Centrifuge: Low–speed centrifuge 300–1000 rpm
- Water bath, 37°C
- Pipette aid: automatic pipettor for use in measuring and dispensing media
- Micropipette set, such as the Eppendorf p-2, p-20, p-200, and p-1000
- Aspirator: for use in the biosafety cabinet hood, with collection flask
- Refrigerator, 4°C
- Freezers: −20°C, −80°C, and −140°C
- Liquid nitrogen tank, for long term storage of cell stocks. Recommended storage in gas phase.
- Liquid nitrogen dewar: for transferring and holding vials during cryopreservation and thawing

REAGENTS AND SUPPLIES

Cells and Genetic Materials

- 293FT cells for producing the lentivirus (Life Technologies; see the manufacturer's protocol for the culture conditions)
- Human fibroblast cells are available from the following sources:
  - Cell Applications Inc. (http://www.cellapplications.com/)
  - Lonza (http://www.lonza.com/group/en.html)
  - American Type Culture Collection (ATCC, http://www.atcc.org/)
  - European Collection of Cell Cultures (ECACC; http://www.ecacc.org.uk/)
  - Riken Bioresource Center (http://www.brc.riken.jp/)
  - Coriell Institute for Medical Research (http://ccr.coriell.org/)

■ Note

Primary fibroblasts, such as those obtained from skin biopsies of consenting patients, can also be used for iPSC generation (see Chapter 9). ■

- PLAT-E packaging cells
- PLAT-GP packaging cells
- pLenti6/UbC containing the mouse Slc7a1 gene (Addgene; http://www.addgene.org/Shinya Yamanaka)
- pMXs retroviral vectors encoding OCT3/4, SOX2, KLF4, and/or c-myc (Addgene; http://www.addgene.org/Shinya Yamanaka):
  - pMXs-hOCT3/4
  - pMXs-hSOX2
  - pMXs-hKLF4
  - pMXs-hc-MYC
- pMXs retroviral vector encoding the green fluorescence protein (GFP) to monitor the transfection efficiency and serve as a negative control for iPSC induction (Cell Biolabs, Inc.)

Supplies

- 5 mL, 10 mL, and 25 mL sterile disposable pipettes
- 60 mm and 100 mm tissue culture dishes
- 24-, 12-, and 6-well culture plates
- 15 mL and 50 mL sterile conical tubes
- Sterile 9″ pasture pipettes
- 0.22 μm pore size filters (Millex GP, SLGP033RS, Millipore)
- 0.45 μm pore size cellulose acetate filters (FP30/0.45 CA-S, Schleicher & Schuell)
- 10 mL disposable syringes (SS-10ESZ, Terumo, Japan)

Reagents

Item	Supplier	Catalog #	Note
Primate ES medium	ReproCELL	RCHEMD001	From Japan
DMEM (High Glucose, with L-Glutamine)	Nacalai tesque	08459-35	From Japan
DMEM/F12 1:1 (Glutamax™, no HEPES)	Life Technologies	10565-018	
Knockout™ Serum Replacement (contains bovine products)	Life Technologies	10828-028	
Fetal Bovine Serum (FBS)	Japan Bio Serum	Lot 5145, etc.	From Japan
L-Glutamine 200 mM	Life Technologies	25030-081	
MEM-Non-essential amino acids (100× = 10 mM)	Life Technologies	11140-050	

Item	Supplier	Catalog #	Note
2-Mercaptoethanol (55 mM in D-PBS)	Life Technologies	21985-023	
Sodium Pyruvate	Sigma	S8636	
D-PBS (Dulbecco's Phosphate-Buffered Saline without Calcium and Magnesium)	Nacalai tesque	14249-95	From Japan
Recombinant Human bFGF	WAKO	064-04541	From Japan
Bovine Serum Albumin (BSA)	MP Biomedical	810-661	
2.5% Trypsin	Life Technologies	15090-046	
0.25% Trypsin, 1 mM EDTA solution	Life Technologies	25200-056	
Collagenase, Type IV	Life Technologies	17104-019	
Dimethyl Sulfoxide (DMSO)	Sigma	D2650	
Acetamide	WAKO	015-00115	From Japan
Propylene glycol	WAKO	164-04996	From Japan
Penicillin/Streptomycin (100x)	Life Technologies	15140-122	
G418 Sulfite, 50 mg/mL solution	Life Technologies	10131-035	
Puromycin	Sigma	P7255	
Blasticidin S Hydrochloride	Funakoshi	KK-400	From Japan
Mitomycin C	Kyowa Hakko Kirin	MITOMYCIN Injection 2mg	From Japan
Virapower Lentiviral Expression System	Life Technologies	K4990-00	
Fugene 6 Transfection Agent	Promega	E2691	
Lipofectamine 2000	Life Technologies	11668-019	
Opti-MEM I	Life Technologies	31985-062	
Trypan Blue	Life Technologies	15250061	
Hexadimethrine Bromide (Polybrene)	Nacalai tesque	17736-44	From Japan

RECIPES

Stock Solution of 10 μg/mL Human Basic FGF (bFGF) (1 mL)

Component	Amount	Stock Concentration
Human bFGF	50 μg	10 μg/mL PBS with 0.2% BSA

1. Dissolve 50 μg of human basic FGF in 5 mL PBS containing 0.1% BSA.
2. Aliquot into 200 μL samples.
3. Store frozen aliquots at $-20\,°C$ or $-80\,°C$.
4. Store thawed aliquots at $4\,°C$ for up to 2 weeks.

■ Note

For all growth factors manipulations, pre-wet all pipette tips, tubes, and filters with PBS + 0.1% BSA to reduce the loss of growth factor.   ■

Human ES Cell Medium (500 mL)

Component	Amount	Final Concentration
DMEM/F12 1:1 (GlutaMAX, no HEPES)	394 mL	
Knockout Serum Replacement (KSR)	100 mL	20%
100× MEM-Non-essential Amino Acids	5 mL	0.1 mM
2-Mercaptoethanol (55 mM)	1 mL	0.1 mM
Human bFGF (10 µg/mL)	200 µL	20 ng/mL

1. Prepare all media in the tissue culture hood using aseptic techniques.
2. To prepare 500 mL of human ES cell medium, mix 394 mL of DMEM/F12, 100 mL of KSR, 5 mL of non-essential amino acids and 1 mL of 2-Mercaptoethanol.
3. Add 200 µL of 10 µg/mL bFGF to the 500 mL human ES cell medium.
4. Store at 4°C for up to a month.

■ Note

This medium can be replaced with Primate ES cell medium (ReproCELL: RCHEMD001) containing 20 ng/mL bFGF.   ■

FP Medium (DMEM + 10% FBS; for Human Fibroblasts and PLAT-E cells) (500 mL)

Component	Amount	Final Concentration
DMEM (high glucose, with L-glutamine)	447.5 mL	
FBS	50 mL	10%
Penicillin/Streptomycin (100×)	2.5 mL	50 U/mL/ 50 µg/mL

1. To prepare 500 mL of FP medium, mix 447.5 mL of DMEM, 50 mL of FBS and 2.5 mL of 100× penicillin/streptomycin (containing 10,000 U penicillin and 10,000 µg/mL streptomycin). Store at 4°C for up to a month.
2. For PLAT-E cells, add 1 µL of 10 mg/mL puromycin stock and 10 µL of 10 mg/mL of Blasticidin S stock to 10 mL of 10% FBS medium.

293FT Medium (500 mL)

Component	Amount	Final Concentration
DMEM (high glucose, with L-glutamine)	437.5 mL	
FBS	50 mL	10%

Component	Amount	Final Concentration
100× MEM-non essential amino acids	5 mL	0.1 mM
Sodium pyruvate	5 mL	1 mM
Penicillin/Streptomycin (100×)	2.5 mL	50 U/mL/50 µg/mL

1. To prepare 500 mL of the 293FT medium, mix 437.5 mL of DMEM, 50 mL of FBS, 5 mL of non-essential amino acids, 5 mL of sodium pyruvate and 2.5 mL of penicillin/streptomycin.
2. Store at 4°C for up to 1 month.
3. Add 0.1 mL of 50 mg/mL G418 into 10 mL of the 293FT medium.

SNL Medium (500 mL)

Component	Amount	Final Concentration
DMEM (high glucose, with L-glutamine)	462.5 mL	
FBS	35 mL	7%
Penicillin/Streptomycin (100×)	2.5 mL	50 U/mL/50 µg/mL

1. To prepare 500 mL of the SNL medium, mix 462.5 mL of DMEM, 35 mL of FBS and 2.5 mL of penicillin/streptomycin.
2. Store at 4°C for up to 1 month.

400 µg/mL Mitomycin C (5 mL)

Component	Amount	Stock Concentration
Mitomycin C	2 mg	400 µg/mL in SNL medium

1. Dissolve 2 mg of Mitomycin C in 5 mL of SNL medium.

CTK Solution (50 mL)

Component	Amount	Final Concentration
2.5% Trypsin	5 mL	0.25%
1 mg/mL Collagenase, Type IV	5 mL	0.1 mg/mL
Knockout Serum Replacement (KSR)	10 mL	0.1 mM
0.1 M CaCl$_2$	0.5 mL	1 mM
Distilled water	29.5 mL	

1. To prepare 50 mL of the CTK solution, mix 29.5 mL of distilled water, 5 mL of 2.5% trypsin, 5 mL of 1 mg/mL collagenase, 10 mL of KSR and 0.5 mL of 0.1 M CaCl$_2$.
2. Aliquot in 1–5 mL portions, and store at −20°C for up to 1 year. Do not repeat freeze/thaw cycles.

0205

DAP213 (50 mL)

Component	Amount	Final Concentration
DMSO	7.1 mL	2 M
10 M acetamide	5 mL	1 M
Propylene glycol	11 mL	3 M
Human ES cell medium	26.9 mL	

1. To prepare 50 mL of the DAP213 solution, mix 26.9 mL of human ES cell medium, 7.1 mL of DMSO, 5 mL of 10M acetamide and 11 mL of propylene glycol.
2. Sterilize the mixture with a 0.22 μm Nylon membrane filter.
3. Aliquot into 1–2 mL tubes and store at −20°C for up to 1 year.

QUALITY CONTROL METHODS

Titer Analysis of the Lentiviral Vector

The introduction of the Slc7a1 gene into human fibroblasts is a critical step for retrovirus vector infection. The titer of the lentiviral vector solution can be estimated with the p24 antigen ELISA kit (Zeptometrix: RETRO-TEK HIV-1 p24 Antigen ELISA). We have confirmed that a lentiviral solution with a titer of at least 1×10^7 TU/mL is able to effectively introduce the Slc7a1 gene into human fibroblasts using the conditions described herein.

Transfection Efficiency of Retroviral Infection

A high transfection efficiency is essential for iPSC induction. The efficiency of transfection can be monitored with a GFP-encoding pMXs vector. Our laboratory has confirmed a transduction efficiency higher than 60% using the conditions outlined in this protocol.

Monitoring for Mycoplasma Contamination

Mycoplasma infections drastically affect cell metabolism, gene expression and antigenicity, and can be devastating to a human iPSC laboratory. Infections are difficult to eradicate once they take hold, and some tissue culturists recommend that contaminated cells be destroyed as soon as mycoplasma is detected. Mycoplasma is highly infectious, and cross-contamination commonly occurs when new cells are introduced into laboratories. Testing for mycoplasma can be done by enzymatic, PCR or fluorescent staining methods. Examples of the kits and reagents that can be used for these various methods are listed below.

Enzymatic Assays
- Lonza Corp. (www.Lonza.com) MycoAlert® Mycoplasma Detection Assay
- Life Technologies (www.lifetechnologies.com) MycoTect™ Kit

PCR
- Stratagene (www.stratagene.com), MycoSensor™ PCR Assay Kit

Fluorescence
- Sigma (www.sigmaaldrich.com) Mycoplasma Stain Kit
- Life Technologies (www.lifetechnologies.com) MycoFluor™ Mycoplasma Detection Kit

ACKNOWLEDGMENTS

We would like to thank the members of the Yamanaka and Asaka laboratory and CiRA for their encouragement and administrative support, and especially thank Dr. Kazutoshi Takahashi for scientific comments, and Dr. Knut Woltjen for his critical reading of the manuscript.

READING LIST

Human iPSC Generation and Culture Maintenance
Takahashi, K, Yamanaka, S., 2006. Induction of pluripotent stem cells from mouse embryonic and adult fibroblast cultures by defined factors. Cell 126, 663–676.

Takahashi, K., Tanabe, K., Ohnuki, M., Narita, M., Ichisaka, T., Tomoda., K., et al., 2007. Induction of pluripotent stem cells from adult human fibroblasts by defined factors. Cell 131, 861–872.

Takahashi, K., Okita, K., Nakagawa, M., Yamanaka, S., 2007. Induction of pluripotent stem cells from fibroblast cultures. Nat. Protoc. 2, 3081–3089.

Ohnuki, M., Takahashi, K., Yamanaka, S., 2009. Generation and characterization of human induced pluripotent stem cells. Curr. Protoc. Stem Cell Biol. (Chapter 4, Unit 4A 2).

Cryopreservation of Human iPSCs
Fujioka, T., Yasuchika, K., Nakamura, Y., Nakatsuji, N., Suemori, H., 2004. A simple and efficient cryopreservation method for primate embryonic stem cells. Int. J. Dev. Biol. 48, 1149–1154.

CHAPTER 11

Integration-Free Method for the Generation of Human Induced Pluripotent Stem Cells

Isao Asaka and Shinya Yamanaka

EDITOR'S COMMENTARY

Soon after the retroviral methods for making human iPSCs became established in multiple laboratories, researchers began to search for methods that would not require integration of exogenous sequences into the genome. In most, but not all, cases, the integrated sequences are epigenetically inactivated and remain inactive through further expansion and differentiation of the iPSCs. However, there have been instances in which mouse iPSCs reactivated the genes and became tumorigenic in transplantation models. Among the methods that have been explored are direct delivery of reprogramming proteins, use of modified mRNA transcripts, DNA plasmids, microRNA constructs, and mature microRNAs. None of these methods has proved to be as effective or efficient as virus-based delivery methods. There has been more success with non-integrating virus (i.e. Sendai) and episomal vectors. This chapter describes a method developed by the authors that achieves high reprogramming efficiency using a combination of non-integrating episomal vectors.

OVERVIEW

Various types of somatic cells derived from human induced pluripotent stem cells (iPSCs) can be used in regenerative medicine to repair tissues damaged through disease or injury. In Chapter 10, we described a method using retrovirus-mediated transduction of the four reprogramming factors as a conventional and robust method for the derivation of research-grade human iPSC lines. However, genomic integration of retrovirally delivered transgenes raises safety concerns for cells that are produced for cell replacement therapy and better methods for integration-free reprogramming are needed.

Most non-integrating methods have shown low reprogramming efficiencies, making them impractical for use in the clinical setting. Recently, we

165

J.F. Loring & S.E. Peterson (eds): Human Stem Cell Manual, Second edition.
DOI: http://dx.doi.org/10.1016/B978-0-12-385473-5.00011-4

developed a method using episomal plasmid vectors that shows efficiency nearing that of retroviral systems. This method is enhanced through the use of TP53 (p53) suppression and non-transforming MYCL (L-Myc), in addition to OCT4 (POU5F1), SOX2, KLF4, and LIN28. Human iPSCs have been established from multiple donors using these new episomal vectors, including two putative human leukocyte antigen (HLA)-homozygous donors. Since the episome is maintained extra-chromosomally and not replicated during cell division, in nearly all cases, the iPSCs generated are integration-free. Human iPSC generation using these enhanced episomal vectors encoding OCT4 (POU5F1), SOX2, KLF4, MYCL(L-Myc), LIN28 and TP53 (p53) shRNA comprises a robust method for generating integration-free human iPSC lines.

For this method, cultured somatic cells are dissociated into single cells and the episomal vectors are introduced into the somatic cells by electroporation. Following a few days of recovery in culture, the transformed somatic cells are reseeded onto feeder layers. The resulting embryonic stem (ES) cell-like colonies are picked and expanded, similar to the process described in Chapter 10.

The protocol for establishing integration-free human iPSCs involves the following steps:

- Preparation of somatic cell material
- Introduction of reprogramming factors (OCT4, SOX2, KLF4, MYCL, LIN28 and TP53 (p53) shRNA) into human cells by episomal vectors using a microporator
- Plating electroporated cells onto a feeder layer
- Picking ES cell-like colonies and expanding each clone.

We will also outline an alternative episome transfection method using a cuvette-based system.

PROCEDURES

Preparation of Somatic Cell Material

1. Let reagents stand at room temperature.
2. Add 2 mL of FP medium without antibiotics into each well of a 6 well plate, and equilibrate the plate in a 37°C, 5% CO_2 incubator.
3. Aspirate the medium from the human fibroblast culture.
4. Wash cells with 10 mL of DPBS(−).
5. Aspirate the DPBS(−), add 0.5 mL of 0.25% trypsin, 1 mM EDTA solution per well, and incubate the dish for 2–3 minutes at room temperature.

6. Add 4.5 mL of the medium to each well, and make single cell suspensions by pipetting.
7. Transfer the cell suspension to a 15 mL conical tube.

Introduction of Reprogramming Factors into Human Cells by Episomal Vectors Using a Microporator

1. Count the cell numbers and transfer cell suspensions of $1.5-3 \times 10^5$ cells into new 15 mL conical tubes.
2. Adjust the volume to 10 mL with PBS, and centrifuge at $160\,g$ for 5 minutes.
3. While the sample is being centrifuged:
   a. Fill the Neon™ Tube with 3 mL of electrolytic buffer and insert the Neon Tube into the Neon Pipette Station.
   b. Transfer 3 μL of plasmid DNA mixture (See Figure 11.1: pCXLE-hOCT3/4-shp53-F, pCXLE-hSK and pCXLE-hUL; 1 μg each) to a new 1.5 mL tube.

■ Note

The transfection should be monitored with a suitable control; we use pCXLE EGFP to monitor the transfection efficiency (Figure 11.2). ■

4. After centrifugation, aspirate the supernatant completely using an aspirator and pipetman.
5. Insert a Neon Tip into the Neon Pipette and apply Solution E to the Neon Tip.

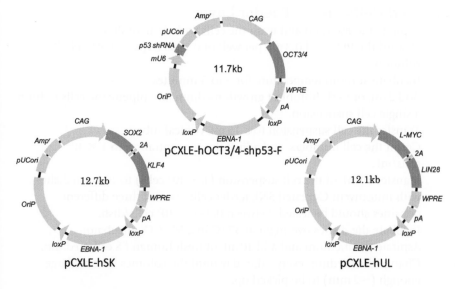

FIGURE 11.1
Diagrams of OCT4 (POU5F1), SOX2, KLF4, MYCL, LIN28 and TP53/p53 shRNA) episomal vectors.

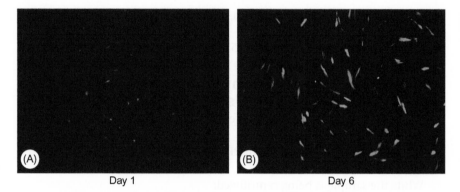

FIGURE 11.2

Expression of EGFP fluorescence by the pCXLE-EGFP episomal vector.

(A) Day 1   (B) Day 6

6.   Suspend the cells with 110 μL of Solution R and transfer them to a 1.5 mL tube containing 3 μL of the plasmid mixture.
7.   Draw up 100 μL of the sample into the Neon™ Tip.
8.   Insert the Neon Pipette with the sample vertically into the Neon Tube placed in the Neon Pipette Station.
9.   Select the appropriate electroporation protocol (Voltage: 1650, Width: 10 ms, Pulse Number: 3) and press Start on the touchscreen.
10. Transfer the sample into the prepared culture plate containing prewarmed medium.
11. After 24 hours, aspirate the medium and add 2 mL of fresh medium per well. Change the medium every 2 days until cells are reseeded on SNL feeder cells.

Reseed Cells on the Feeder Layer
1.   Aspirate the medium and wash the cells with 2 mL of PBS.
2.   Discard the PBS, add 0.3 mL per well of 0.25% trypsin, 1 mM EDTA solution.
3.   Incubate at room temperature for 3 to 5 minutes.
4.   Add 2 mL of fresh fibroblast growth medium and pipette the cells to form a single cell suspension.
5.   Transfer the cell suspension to a 15 mL conical tube.
6.   Count the cell numbers, and adjust the concentration to 1×10^4 cells/mL.
7.   Transfer 10 mL of the cell suspension (1×10^5 cells) to 100 mm dishes with mitomycin C-treated SNL feeder cells. Two or three different densities should be tested, between $0.2-1 \times 10^5$ cells/dish.
8.   Incubate the dishes overnight at 37°C in a 5% CO_2 incubator.
9.   Aspirate the medium and add 10 mL of fresh human ES cell medium. Change the medium every other day until the colonies become large enough (~2 mm) to be picked up.

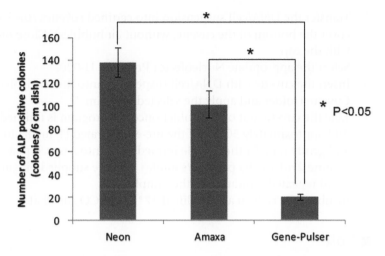

FIGURE 11.3
Comparison of iPS cell colony formation among three electroporation systems with the number of alkaline phosphatase (ALP) colonies per dish as the read-out.

Pick up the ES Cell-Like Colonies and Expand Each Colony

The process following colony picking can be performed according to the procedure described in Chapter 10.

ALTERNATIVE PROCEDURES

The steps described above for introducing the episomal vectors using Neon™ electroporation can be replaced with the following procedure using the Nucleofector® system (Kit#: VPD-1001). However, we recommend that standard optimization procedures be followed before attempting iPSC generation with other electroporation systems (Figure 11.3).

1. Prepare 6 well plates by filling an appropriate number of wells with 2 mL of FP medium and equilibrate plates in a humidified 37°C, 5% CO_2 incubator.
2. Cells harvested by centrifugation (1.5–3×10^5 cells) are resuspended carefully with 100 μL room temperature Nucleofector Solution.

■ Note

Avoid storing the cell suspension for longer than 15 minutes in Nucleofector Solution as this reduces the cell viability and transfection efficiency. ■

3. Combine 100 μL of cell suspension with 3 μg of the plasmid DNA mixture (pCXLE-hOCT3/4-shp53-F, pCXLE-hSK and pCXLE-hUL; 1 μg each).

4. Transfer the DNA/cell suspension into certified cuvettes (the sample must cover the bottom of the cuvette, without air bubbles). Close the cuvette with the cap.
5. Select the appropriate Nucleofector Program, U-020.
6. Insert the cuvette with DNA/cell suspension into the Nucleofector Cuvette Holder and apply the selected program.
7. Take the cuvette out of the holder once the program is finished.
8. Add approximately 500 μL of the pre-equilibrated FP media to the cuvette, and gently transfer the sample immediately into a 6-well plate (final volume 2 mL media per well/sample). Use the supplied pipettes and avoid repeated aspiration of the sample.
9. Incubate the cells in a humidified $37\,°C$, 5% CO_2 incubator.

■ Note

The reseeding process after electroporation can be performed according to the procedure described above. ■

PITFALLS AND ADVICE

- Avoid storing the cell suspension for more than 15–30 minutes at room temperature, as this reduces the cell viability and transfection efficiency.
- When preparing for electroporation, avoid air bubbles during pipetting, as air bubbles cause arcing during electroporation, leading to lowered or failed transfection.
- The reprogramming efficiency varies depending on the properties of the fibroblasts. Because dissociation using trypsin is stressful for the iPSCs in the process of reprogramming, the timing of reseeding should be optimized.

EQUIPMENT

- Microporator: Neon system (Life Technologies)
- Tissue culture hood, Class II A/B3
- Tissue culture incubator, $37\,°C$, 5% CO_2, in humidified air
- Inverted phase/contrast microscope with 4×, 10×, and 20× objectives
- Low-speed centrifuge 300–1000 rpm
- Water bath, $37\,°C$
- Pipette aid: automatic pipettor for use in measuring and dispensing media
- Aspirator in the hood, with flask
- Refrigerator at $4\,°C$
- Freezers: $-20\,°C$, $-80\,°C$, and $-140\,°C$

REAGENTS AND SUPPLIES

Supplies

- 5 mL, 10 mL, 25 mL sterile disposable pipettes
- 60 mm, 100 mm tissue culture dishes
- 24-, 12-, 6-well culture plates
- 15 and 50 mL sterile conical tubes
- Sterile 9″ pasture pipettes
- Pipettors, such as the Eppendorf p-2, p-20, p-200, p-1000 and tips

Reagents

Item	Supplier	Catalog #	Note
Primate ES Medium	ReproCELL	RCHEMD001	From Japan
DMEM (High Glucose, with L-glutamine)	Nacalai tesque	08459-35	From Japan
DMEM/F12 1:1 (GlutaMAX, no HEPES)	Life Technologies	10565-018	
Knockout™ Serum Replacement (contains bovine products)	Life Technologies	10828-028	
Fetal Bovine Serum (FBS)	Japan Bio serum	Lot 5145, etc.	From Japan
L-Glutamine 200 mM	Life Technologies	25030-081	
MEM-Non-essential Amino Acids (100× = 10 mM)	Life Technologies	11140-050	
2-Mercaptoethanol (55 mM in D-PBS)	Life Technologies	21985-023	
Sodium Pyruvate	Sigma	S8636	
D-PBS (Dulbecco's Phosphate-Buffered Saline without Calcium and Magnesium)	Nacalai tesque	14249-95	From Japan
Recombinant human bFGF	WAKO	064-04541	From Japan
0.25% Trypsin, 1 mM EDTA solution	Life Technologies	25200-056	
Penicillin/Streptomycin (100×)	Life Technologies	15140-122	
Neon Transfection System Kit (100 μL)	Life Technologies	MPK10096	

RECIPES

Stock Solution of 10 μg/mL Human Basic FGF (bFGF) (1 mL)

Component	Amount	Stock Concentration
Human bFGF	50 μg	10 μg/mL PBS with 0.2% BSA

1. Dissolve 50 μg of human basic FGF in 5 mL PBS containing 0.1% BSA.
2. Aliquot into 200 μL samples.
3. Store frozen aliquots at $-20°C$ or $-80°C$.
4. Store thawed aliquots at 4°C for up to 2 weeks.

■ **Note**

For all growth factors, pre-wet all pipette tips, tubes, and filters with PBS +0.1% BSA to lessen the loss of the growth factor. ■

Human ES Cell Medium (500 mL)

Component	Amount	Final Concentration
DMEM/F12 1:1 (GlutaMAX, no HEPES)	394 mL	
Knockout Serum Replacement (KSR)	100 mL	20%
100× MEM-Non-essential Amino Acids	5 mL	0.1 mM
2-Mercaptoethanol (55 mM)	1 mL	0.1 mM
Human bFGF (10 µg/mL)	200 µL	20 ng/mL

1. Prepare all media in the tissue culture hood using aseptic techniques. To prepare 500 mL of the human ESC medium, mix 394 mL of DMEM/F12, 100 mL of KSR, 5 mL of non-essential amino acids and 1 mL of 2-Mercaptoethanol.
2. Add 200 µL of 10 µg/mL bFGF to 500 mL human ESC medium.
3. Store at 4°C for up to a month.

■ **Note**

This medium can be replaced with Primate ES cell medium (ReproCELL: RCHEMD001) containing 20 ng/mL bFGF. ■

FP Medium (DMEM + 10% FBS; for Human Fibroblasts and PLAT-E Cells) (500 mL)

Component	Amount	Final Concentration
DMEM (high glucose, with L-glutamine)	447.5 mL	
FBS	50 mL	10%
Penicillin/Streptomycin (100×)	2.5 mL	50 U/mL/50 µg/mL

1. To prepare 500 mL of FP medium, mix 447.5 mL of DMEM, 50 mL FBS and 2.5 mL of 100× penicillin/streptomycin (containing 10,000 U penicillin and 10,000 µg/mL streptomycin).
2. Store at 4°C for up to a month.

ACKNOWLEDGMENTS

We would like to thank the members of the Yamanaka and Asaka laboratory and CiRA for encouragement and administrative support, and especially thank Dr. Keisuke Okita for the scientific comments, and Dr. Knut Woltjen for his critical reading of the manuscript.

READING LIST

Hong, H., Takahashi, K., Ichisaka, T., Aoi, T., Kanagawa, O., Nakagawa, M., et al., 2009. Suppression of induced pluripotent stem cell generation by the p53–p21 pathway. Nature 460 (7259), 1132–1135.

Nakagawa, M., Takizawa, N., Narita, M., Ichisaka, T., Yamanaka, S., 2010. Promotion of direct reprogramming by transformation-deficient. Myc. Proc. Natl. Acad. Sci. USA 107, 14152–14157.

Okita, K., Nakagawa, M., Hyenjong, H., Ichisaka, T., Yamanaka, S., 2008. Generation of mouse induced pluripotent stem cells without viral vectors. Science 322, 949–953.

Okita, K., Hong, H., Takahashi, K., Yamanaka, S., 2010. Generation of mouse-induced pluripotent stem cells with plasmid vectors. Nat. Protoc. 5 (3), 418–428.

Okita, K., Matsumura, Y., Sato, Y., Okada, A., Morizane, A., Okamoto, S., et al., 2011. A more efficient method to generate integration-free human iPS cells. Nat. Methods.

Yu, J., Hu, K., Smuga-Otto, K., Tian, S., Stewart, R., Slukvin, I.I., et al., 2009. Human induced pluripotent stem cells free of vector and transgene sequences. Science 324, 797–801.

READING LIST

CHAPTER 12

Methods for Evaluating Human Induced Pluripotent Stem Cells

Isao Asaka and Shinya Yamanaka

EDITOR'S COMMENTARY

As is evident from the detailed methods provided in this manual, human pluripotent stem cells (hPSCs) require special methods for their culture and characterization, and more attention to detail than has been necessary for the immortalized human cell lines that have been the main cellular tools for decades. Improved reliability of methods for making induced pluripotent stem cells (iPSCs) means that many more laboratories with no previous experience in human embryonic stem cell (hESC) culture or characterization would like to generate iPSCs to use in their own areas of research interest and expertise. Many of the methods for their culture and characterization are identical to those used for hESCs. But iPSCs also require special attention because they are made by genetic manipulation of cells. It is necessary not only to determine that they resemble hESCs in multiple characteristics such as pluripotency, but also to determine for integrating reprogramming methods what exogenous genes have integrated into the genome, and to establish that they have been silenced. For non-integrating methods, it is critical to show that the genome is free of exogenous sequences. This chapter provides specific methods that are necessary to validate iPSCs and show that they are stable pluripotent cell lines.

OVERVIEW

Human iPSC lines have been established by laboratories worldwide. Methods for evaluating iPSC pluripotency are similar to those used for hESC evaluation, and are described in Chapters 13–22. However, the derivation of human iPSC lines by reprogramming factors requires specialized characterization, especially when the cells are produced through retroviral transduction. The characteristics of iPSC lines vary based on transgene silencing, transgene integration sites and copy number, even within a

175

J.F. Loring & S.E. Peterson (eds): Human Stem Cell Manual, Second edition.
DOI: http://dx.doi.org/10.1016/B978-0-12-385473-5.00012-6

single reprogramming experiment. For example, if the exogenous reprogramming sequences are not completely silenced, the differentiation capacity of an iPSC clone is often reduced. As described in Chapter 11, non-integrating methods, including enhanced episome systems, require screening to ensure that the reprogramming vectors have not been randomly integrated. Thus, to screen valuable human iPSC lines, some specialized evaluation methods are required.

In this chapter we provide the following evaluation methods, which we developed specifically to screen new human iPSC lines:

- Evaluating reprogramming efficiency using embryonic stem (ES) cell-like colony formation
- Quantitative PCR for evaluating transgene silencing (for use with the viral vector method described in Chapter 10)
- Methodology for detecting transgene integration into chromosomes (for the episomal vector method described in Chapter 11).

PROCEDURES

Methodology for Evaluating Reprogramming Efficiency Using ES Cell-Like Colony Formation

ES cell-like colony formation may be estimated by alkaline phosphatase (ALP) staining and counting ALP positive colonies. Our conventional procedure is described below.

1. Add 2–5 mL of paraformaldehyde solution to a culture dish (100 mm) and incubate cells at room temperature for 10 minutes to fix.
2. Wash cells with distilled water.
3. Mix 100 μL of Sodium Nitrite Solution and 100 μL of FRV-Alkaline Solution provided in the Alkaline Phosphatase (AP) Leukocyte kit by inverting tubes, and incubate for 2 minutes.
4. Add 4.5 mL of distilled water and 100 μL of Naphthol AS-BI Alkaline Solution as a substrate solution.
5. After washing, add the solution prepared in Step 4 and incubate at room temperature for 15 minutes.
6. Aspirate the substrate solution and wash cells with distilled water. After aspirating the water, add fresh distilled water and incubate for 2 minutes. After shaking gently, dry the cells.
7. ALP positive colonies will be stained red/violet (Figure 12.1A), while non-reprogrammed or differentiated colonies often display weak or mosaic staining. Take a picture of the ALP stained dishes using a high-resolution digital documentation system.

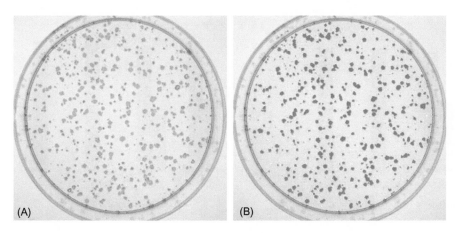

FIGURE 12.1
Analysis of iPSC generation efficiency using Alkaline Phosphatase (ALP) staining. (A) ALP stained colonies. (B) Counted colonies using IMA program (Area: 40-10000 pixel, Threshold: 105) of Metamorph®.

8. The ALP positive colonies in the digital image may be estimated using an image analysis software program (for example, the Integrated Morphometry Analysis (IMA) program of Metamorph® or the Analyze Particles command of ImageJ) (Figure 12.1B).

■ Note
ALP staining is indicative, but not a definitive marker of pluripotency. Human fibroblasts normally have undetectable ALP activity, but other cell types have ALP activity that will be detectable by this method. It is advised that you check the cells that you plan to reprogram to be sure that they are negative. ■

Methodology for Evaluating Transgene Silencing
■ Note
This protocol is for use with the viral vector method of iPSC generation outlined in Chapter 10. ■

The silencing of reprogramming transgenes (see Figure 12.2) is evaluated by comparing the level of mRNA transcribed specifically from transgene constructs to the total gene expression (transgenes + endogenous genes) using a quantitative PCR method.

Preparation of Cell Lysates
1. Wash the cells cultured in a 6-well plate once with 2 mL of DPBS(−).
2. Aspirate the DPBS(−) completely, add 1 mL of Trizol reagent, and incubate for 5 minutes at room temperature.
3. Collect the 1 mL of cell lysate in a 1.5 mL microcentrifuge tube, and homogenize the sample by pipetting.

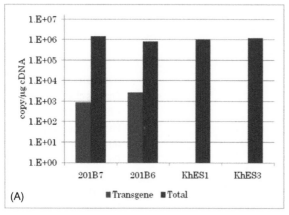

(A)

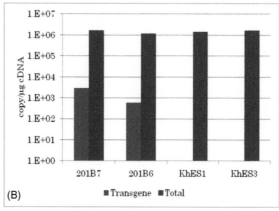

(B)

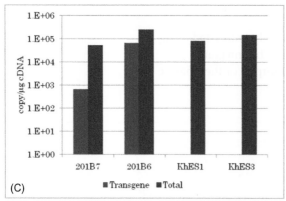

(C)

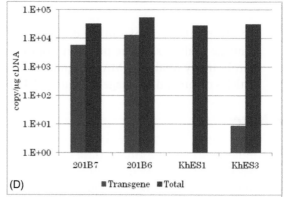

(D)

FIGURE 12.2

Real time PCR evaluation of transgene silencing. (A) POU5F1/OCT3/4, (B) SOX2, (C) KLF4, (D) MYC/cMYC. Transgenes are all silenced in the KhES1 clone, and the KhES3 clone has a small amount of expression from the MYC transgene. The other clones tested have active transgene expression.

You can pause the experiment after completing this step by storing cell lysates at $-80°C$ until further analysis.

Purify the RNA

1. Thaw lysates if required. Add 200 μL of chloroform to the lysate and mix vigorously by shaking.
2. Centrifuge for 5 minutes at 15,000 g at room temperature.
3. Transfer the aqueous phase (500 μL) to a new 1.5 mL microcentrifuge tube, add 400 μL of isopropanol, and mix well by inversion for 20 minutes.
4. Centrifuge the tube for 5 minutes at 15,000 g at room temperature to pellet the RNA.

5. Remove the supernatant, add 500 μL of 70% ethanol, and centrifuge for 5 minutes at 15,000 g at room temperature.
6. Remove the ethanol completely and air dry the pellet at room temperature for 2–3 minutes.
7. Resuspend the pellet in 26 μL of RNase-free water.

■ Note

You can pause the experiment after completing this step. Purified RNA samples should be stored at −80°C.   ■

Remove Genomic DNA Contamination by DNase Treatment

1. Add 3 μL of 10× DNase I buffer and 1 μL of DNase I (from the Turbo DNA-free kit) to the RNA sample, mix gently by finger tapping, and incubate for 30 minutes at 37°C.
2. Add 3 μL of DNase Inactivation Reagent (from the Turbo DNA-free kit), and mix well.
3. Incubate for 3 minutes at room temperature with occasional mixing by finger tapping.
4. Centrifuge for 3 minutes at 15,000 g at room temperature. Transfer the supernatant carefully to a new 1.5 mL microcentrifuge tube.

Determine the RNA Concentration

1. Use 1 μL of DNase-treated sample to determine the RNA concentration of samples by measuring the A260/A280 with an optical spectrometer (e.g. NanoDrop™), and adjust the concentration of each sample to an appropriate concentration. Samples should contain >100 ng/μL RNA for RT-PCR.

■ Note

You can pause the experiment at this step. Purified RNA samples should be stored at −80°C.   ■

Perform Reverse Transcription

1. Prepare 20 μL of reaction mixture by mixing the reagents (from the ReverTra Ace® kit) for each sample in the amounts listed below:
   - 4 μL 5× reverse transcription buffer
   - 2 μL 10 mM dNTPs
   - 1 μL ribonuclease inhibitor
   - 1 μL ReverTra Ace (reverse transcriptase)
   - 1 μL 10 μM oligo dT20 primer
   - 1 μg DNase-treated total RNA
   - Nuclease-free water up to 20 μL

■ **Note**

You should prepare reactions containing no reverse transcriptase as negative controls for each sample to ensure the complete removal of genomic DNA contamination.   ■

2. Incubate the mixture in a thermal cycler using the following conditions:
   ▪ 60 minutes at 42°C
   ▪ 5 minutes at 95°C
   ▪ Indefinitely at 4°C.

■ **Note**

You can pause the experiment at this step. The cDNA samples should be stored at −20°C or below. Note that it is preferable to use fresh cDNA in subsequent analyses.   ■

Amplify the Products by qPCR

1. Prepare 20 μL of PCR mixture by mixing the reagents listed below in a 0.2 mL PCR reaction tube:
   ▪ 10 μL SYBR® Premix Ex Taq™ II
   ▪ 0.4 μL ROX reference dye
   ▪ 0.4 μL Primer 1 (10 μM, see table below)
   ▪ 0.4 μL Primer 2 (10 μM, see table below)
   ▪ 1 μL cDNA template
   ▪ Nuclease-free water up to 20 μL.

■ **Note**

The pMXs-hOCT3/4, hSOX2, hKLF4, and hc-MYC plasmids are mixed as standard templates instead of cDNA templates. Dilution of standard plasmids should be started with 5×10^7 copies/μL, and each subsequent sample diluted 10-fold.   ■

The suggested primers are as follows:

For Transgene	Forward (Primer 1)		Reverse (Primer 2)
hOCT4-S1061	GCT CTC CCA TGC ATT CAA ACT GA	TgAS3	CTT ACG CGA AAT ACG GGC AGA CA
hSOX2-S875	TTC ACA TGT CCC AGC ACT ACC AGA	pMXs-TgAS	GAC ATG GCC TGC CCG GTT ATT ATT
hKlf4-S1380	CCA CCT CGC CTT ACA CAT GAA GA	pMXs-TgAS	GAC ATG GCC TGC CCG GTT ATT ATT
hcMycS1203	ATA CAT CCT GTC CGT CCA GCA GA	pMXs-TgAS	GAC ATG GCC TGC CCG GTT ATT ATT

For Total	Forward (Primer 1)		Reverse (Primer 2)
hOct3/4-S944	CCC CAG GGC CCC ATT TTG GTA CC	hOct3/4-as	ACC TCA GTT TGA ATG CAT GGG AGA GC
hSOX2-S875	TTC ACA TGT CCC AGC ACT ACC AGA	HsSox2-AS	TCA CAT GTG TGA GAG GGG CAG TGT GC
hKlf4-S1094	CAT GCC AGA GGA GCC CAA GCC AAA GAG GGG	hKLF4-AS1225	CGC AGG TGT GCC TTG AGA TGG GAA CTC TTT
hcMYC-S1203	ATA CAT CCT GTC CGT CCA AGC AGA	hcMyc-AS1297	ACG CAC AAG AGT TCC GTA GCT G

2. Incubate the mixture in a real-time PCR system using the following cycling parameters:

	For Total KLF4	For Other Primers	
Predenaturation	–	95°C, 30 seconds	
Denaturation	95°C, 10 seconds	95°C, 5 seconds	
Annealing	60°C, 10 seconds	60°C, 30 seconds	50 cycles
Extension	72°C, 40 seconds		

Methodology for Detecting Transgene Integration into Chromosomes

■ Note

This protocol is for use with the episomal vector method outlined in Chapter 11.   ■

Although episomal vectors are maintained extrachromosomally and lost over time, we recommend screening all new iPSC lines to ensure the absence of episomes and exclude the possibility of rare random integration events.

DNA in total lysates from established iPSC cultures are subjected to quantitative PCR using a PCR primer pair specific for the Epstein-Barr virus EBNA-1 sequence to calculate the episomal vector copy number. For reference, a second primer pair amplifies the endogenous *FBXO15* (F-box protein 15) locus to estimate the genome copy number, and thus infer the cell number in the sample.

This assay is able to detect as little as one copy of episomal DNA in 4000 cells.

1. Aspirate the culture medium from iPSCs cultured to confluency in 6-well plates.

2. Wash the cells with 2 mL of DPBS(−), then aspirate off the DPBS(−).
3. Add 0.5 mL of CTK solution to the dish and incubate at room temperature for 1–2 minutes.
4. After the feeder cells are detached, wash cells with 2 mL of DPBS(−) as gently as possible.
5. Aspirate DPBS(−) and wash cells again with 2 mL of DPBS(−).
6. Aspirate the DPBS(−) so that almost all of the feeder cells are removed, add about 3 mL of human ES cell medium. Then, detach the iPSCs using a cell scraper.
7. The cells are then placed into tubes and centrifuged, and the cell pellets are lysed with 200 μL of lysis solution containing the following components:
   - 24 μL 10× ExTaq buffer
   - 4 μL 10 mg/mL Proteinase K
   - 212 μL Nuclease-free water.
8. The lysates are incubated at 55°C for 3 hours, and proteinase K is inactivated by incubation at 95°C for 3 minutes
9. The lysates are then used for a quantitative PCR analysis containing the following reagents (per reaction):
   - μL SYBR Premix Ex Taq II
   - 0.4 μL ROX reference dye
   - 0.2 μL Primer F (20 μM)
   - 0.2 μL Primer R (20 μM)
   - 1 μL genomic DNA template
   - Nuclease-free water up to 20 μL.

■ Note

The pCXLE-Fbx15-cont2 plasmids are mixed as standard templates instead of cDNA templates. Dilution of standard plasmid should be started with 1.1 pg/μL, and each sample diluted 10-fold.   ■

The primer sequences are:

hFbx15-2F	GCC AGG AGG TCT TCG CTG TA
hFbx15-2R	ATT GCA CGG CTA GGG TCA AA
EBNA-183F	ATC AGG GCC AAG ACA TAG AGA TG
EBNA-243R	GCC AAT GCA ACT TGG ACG TT

■ Note

These primers do not detect the mouse Fbx gene sequence, and thus are insensitive to SNL feeder DNA contamination.   ■

10. Incubate the mixture in a real-time PCR system using the following cycling parameters:

	Temperature	Time	
Initial denaturation	94°C	2 minutes	
Denaturation	94°C	10 seconds	
Annealing	60°C	10 seconds	50 cycles
Extension	72°C	30 seconds	
Melting	95°C	15 seconds	
	60°C	1 minute	

- The pCXLE-hFbx15-cont2 plasmid is used to generate a standard curve to determine the correlation between the copy number and threshold cycle (C_t) values for FBXO15 or EBNA-1.
- The copy numbers of FBXO15 and EBNA-1 in each iPSC sample are then estimated from the observed C_t values.
- The cell number in each reaction is estimated by dividing the estimated copy number of FBXO15 by two, since each cell has two FBXO15 alleles.
- One reaction includes up to 1.2×10^4 cells. The total copy number of EBNA-1 is measured in $\sim 5 \times 10^4$ cells by repeating six or seven reactions.

EQUIPMENT

- Real-time PCR system: StepOne Plus™
- Optical spectrometer: Nanodrop
- Inverted phase/contrast microscope with 4×, 10×, and 20× objectives
- Low-speed centrifuge (300–1000 rpm)
- High-speed microfuge (up to 13,000 rpm)
- Water bath, 37°C
- Pipette aid: automatic pipettor for use in measuring and dispensing media
- Pipettors, such as Eppendorf p-2, p-20, p-200, and p-1000
- Aspirator in the hood, with flask
- Refrigerator at 4°C
- Freezers: −20°C, −80°C, and −140°C
- Image analysis software: Metamorph® or ImageJ

REAGENTS AND SUPPLIES

Supplies
- 5 mL, 10 mL, and 25 mL sterile disposable pipettes
- 6-well culture plates
- 15 mL and 50 mL sterile conical tubes
- Sterile 9″ pasture pipettes

- 1.5 mL microfuge sample tubes
- 0.2 mL PCR tubes/8-well strips/96-well plates

Reagents

Item	Supplier	Catalog #	Note
Alkaline Phosphatase (AP) Leukocyte Kit	Sigma	86-R	
4% paraformaldehyde Solution	Wako	163-20145	From Japan
Trizol Reagent	Life Technologies	15596-026	
Chloroform	Sigma	C7559	Also available from Nacalai Tesque in Japan
Isopropanol	Sigma	I9516	Also available from Nacalai Tesque in Japan
Turbo DNA-free Kit	Life Technologies	AM1907	
ReverTra Ace-α- kit	Toyobo	FSK-101	From Japan
SYBR Premix Ex Taq II	Takara	RR081A	From Japan
Proteinase K	Takara	9033	From Japan
ExTaq kit	Takara	RR001A	From Japan

Please refer to Chapters 10 and 11 concerning other reagents.

ACKNOWLEDGMENTS

We would like to thank the members of the Yamanaka and Asaka laboratory and CiRA for encouragement and administrative support, and are especially grateful to Dr. Keisuke Okita, Dr. Megumu Saito and Takayuki Tanaka for their scientific comments, and to Dr. Knut Woltjen for his critical reading of the manuscript.

READING LIST

Ohnuki, M., Takahashi, K., Yamanaka, S., 2009. Generation and characterization of human induced pluripotent stem cells. Curr. Protoc. Stem Cell Biol. (Chapter 4, Unit 4A. 2).

Okita, K., Hong, H., Takahashi, K., Yamanaka, S., 2010. Generation of mouse-induced pluripotent stem cells with plasmid vectors. Nat. Protoc. 5 (3), 418–428.

Okita, K., Matsumura, Y., Sato, Y., Okada, A., Morizane, A., Okamoto, S., et al., 2011. A more efficient method to generate integration-free human iPS cells. Nat. Methods.

Yu, J., Hu, K., Smuga-Otto, K., Tian, S., Stewart, R., Igor I. Slukvin, I.I., et al., 2009. Human induced pluripotent stem cells free of vector and transgene sequences. Science 324, 797–801.

PART

3

Characterization of Human Pluripotent Stem Cells

CHAPTER 13

Classical Cytogenetics: Karyotyping

Marlys Houck

EDITOR'S COMMENTARY

Human pluripotent stem cells (hPSCs) are arguably the most stable normal diploid cells that can be maintained in long-term culture. However, aneuploidies and other chromosomal abnormalities do occur with time in culture. Aneuploidy in mouse ESC lines invariably results in their failure to contribute to the germline in chimeric animals. But since this ultimate test of normalcy cannot be applied to human PSCs, the cultures must be routinely evaluated for chromosomal abnormalities.

Current methods allow hPSCs to be maintained for long periods of time without developing gross genomic abnormalities. However, if abnormalities give cells a growth advantage, the abnormal cells can rapidly take over a culture. While we do not know when accumulated genomic changes tip the scales to make a hPSC culture no longer useful for experimental or therapeutic applications, we do know that the higher the percentage of abnormal cells in our cultures, the more cells are drifting towards an abnormal phenotype, and the less reproducible and dependable are the results. This chapter is updated from the chapter published in our 2007 edition, and we thank Robin Wesselschmidt for her contributions to writing the earlier version.

OVERVIEW

There are many methods for assessing chromosomal stability, including the classical cytogenetic approaches described here, SNP genotyping described in Chapter 14, and the SKY and FISH methods described in the previous edition of this manual. These methods differ in resolution and the types of abnormalities they can detect. The normal resolution obtainable by classic cytogenetic methods is estimated to be in the megabase range, while single nucleotide polymorphism (SNP) can give 2 kb resolution (Chapter 14). However, more

J.F. Loring & S.E. Peterson (eds): Human Stem Cell Manual, Second edition.
DOI: http://dx.doi.org/10.1016/B978-0-12-385473-5.00013-8

resolution is not necessarily better; for example, SNP genotyping cannot be used to detect balanced translocations or inversions.

Cytogenetics is currently the most accessible method for detecting chromosomal abnormalities, but there are shortcomings to this technique. First, cytogenetic analysis can only be applied to metaphase-stage cells, so a rapidly dividing population is required. And second, while most hospitals have laboratories that perform karyotyping as a service, such laboratories routinely examine metaphases from 20 cells; at most 5 of the cells are fully analyzed and the remaining cells are only counted. This gives only a hint of the composition of the cell population and many hPSC researchers prefer analysis of 100 metaphases. For this reason, and to lower costs, some research laboratories are learning to perform their own karyotyping, at least at the gross level of counting chromosomes to detect aneuploidies.

It is relatively simple to count the chromosomes to determine the modal chromosome number and, with training, one may be able to identify chromosomes by their individual size and banding pattern. It is unlikely that an untrained eye will be able to identify the translocations or deletions that do not change the chromosome count, but may drastically modify the genome. So while we suggest that a research laboratory should routinely count chromosomes, we recommend that the detailed G-banded karyotype of the culture be obtained from a trained cytogeneticist every 10–15 passages.

The basic conventional cytogenetic method involves chromosome harvest, slide preparation, banding of the chromosomes, analysis of banding patterns, karyotyping and interpretation of the results. In this chapter we will describe:

- How to prepare a culture to maximize the number of metaphase chromosomes
- How to prepare slides containing chromosome spreads
- Three methods used to stain chromosomes
- Interpreting the cytogenetic report.

Chromosome harvesting consists of arresting the cell cycle at metaphase, hypotonic treatment of the cells, and their fixation. After fixation, the cell pellets containing chromosomes are dropped onto glass slides, air-dried, and aged before banding. Banding is a staining method that produces a continuous series of light and dark bands and is used for visualizing the unique pattern of each chromosome pair.

A band is defined as that part of a chromosome that is clearly distinguishable from its adjacent segments by appearing darker or lighter. Slide preparation profoundly affects the quality of banding and it is one of the most challenging steps in chromosome preparation and analysis. Different banding patterns, such as G-, Q-, R-, C-, T-, or NOR-banding, can be generated for analysis.

The G-banding method (using Giemsa stain) is the most commonly used staining method, and it can generate up to 1000 bands per haploid human genome. Each band has a specific number assigned to indicate its location on the human chromosome. The nomenclature of band assignment and chromosome aberrations is summarized by the International System for human Cytogenetic Nomenclature (ISCN, 2009).

PROCEDURES

Metaphase Harvest of hPSCs for Karyotyping

A culture with actively dividing human pluripotent stem cells (hPSCs) provides the best conditions for obtaining high quality metaphase spreads. During metaphase of mitosis the chromosomes are condensed to the optimal length for visualizing and banding the individual chromosomes. Because hPSCs are usually actively dividing, it is relatively easy to obtain a high mitotic index (many quality metaphase chromosomes) from a culture. We suggest harvesting the cells for karyotyping roughly 3 days before they would normally be passaged. This strategy should yield a high number of dividing cells and therefore a sufficient number of metaphase spreads in order to make an accurate analysis of the culture.

This procedure describes harvesting cells from a 35 mm dish or one well of a 6-well plate.

1. Add colcemid at a final concentration of 0.1 µg/mL to the culture.
2. Return the culture to the incubator for 1–3 hours.

■ Note

The timing is important; longer time in colcemid increases the mitotic index, but also decreases chromosome length. Short chromosomes provide less information during analysis. If several small cultures are available each can be treated for different amounts of time in colcemid and then pool the harvest before making slides.   ■

3. Aspirate the medium and add 1 mL of PBS. Aspirate PBS.
4. Add 0.3 mL trypsin-EDTA until cells detach, recover with 0.7 mL complete medium, and transfer into a microfuge tube.

■ Note

Obtaining good spreads requires single cells. Trypsinize the cell colonies to single cells.   ■

5. Spin in a microfuge at 1000–3000 rpm for 3–5 minutes at room temperature.

■ **Note**

Use the lowest speed that will pellet the cells without leaving a "trail" of cells on the side of the tube.   ■

6. Aspirate the medium carefully, leaving about 50–100 μL. Resuspend the cell pellet by tapping the tube.
7. Add 1.5 mL of 0.075 M KCl (hypotonic solution) and incubate at 37 °C for 15–30 minutes.

■ **Note**

The timing of the hypotonic solution is important; if the incubation time is too long the cells may burst and the chromosomes will spill out of the membrane. If the incubation is not long enough, the chromosomes may be too tightly packed to analyze.   ■

8. Gently invert the tube several times in order to resuspend the cells and then add 3 drops of freshly made fixative (3:1 methanol:glacial acetic acid) one drop at a time, tapping the tube in between drops to agitate the pellet. Fixative is very hygroscopic so make small batches fresh each day and keep the container tightly capped.
9. Mix by inverting the tube several times. Spin in a microfuge at 1000–3000 rpm for 3–5 minutes at room temperature.
10. Aspirate the hypotonic/fix solution carefully, leaving about 50–100 μL. Resuspend the cell pellet well by tapping the tube.
11. Add 1 mL of fixative; mix well.
12. Spin in a microfuge at 1000–3000 rpm for 3–5 minutes at room temperature.
13. Aspirate the fixative carefully, leaving about 50–100 μL. Resuspend the cell pellet well.
14. Repeat Steps 11–13. Add 1 mL fixative, mix well, and then spin in a microfuge at 1000–3000 rpm for 3–5 minutes at room temperature. This will result in a total of three washes in the fixative after the hypotonic step.
15. Add an appropriate amount of fixative to the pellet to obtain the desired concentration of cells for making slides.

■ **Note**

Fixed cells may be stored at 4 °C for up to a year before slides are made.   ■

Slide Preparation: Making Chromosome Spreads

There are many ways to make slides. The protocol below has been used successfully to make high quality chromosome spreads and will provide a good starting point from which one can develop an individual method.

Set up for Slide Preparation

1. Prepare a Coplin jar with slides soaking in 100% methanol. Use slides with a white area at one end for easier labeling, but avoid those that have been sand blasted to create the labeling area because the slides will be lightly pitted over the entire surface. Colored art pencils are recommended instead of lead pencils to eliminate lead particles drifting onto the spreads during the slide making process.

2. Make fresh fixative (3:1 methanol:glacial acetic acid).

3. Prepare a slide-making area with 2–3 sheets of paper towels, a folded Kimwipe™ for slide cleaning, and a pencil for marking the slides. Place the Coplin jar with slides soaked in 100% methanol and a beaker of deionized or distilled water on one side and fixative with a Pasteur pipette on the other side (if you are right-handed, the slides and water should be on the left and the fixative should be on the right). The chromosome harvest can be placed either in the middle or on the same side as the fixative, with a Pasteur pipette alongside.

4. Remove a slide from the Coplin jar of methanol and polish the surface of the slide to be used with a folded Kimwipe. It is important to keep track of the polished or labeled side of the slide, so the chromosome harvest is dropped on the correct side of the slide.

■ Note

The quality of the slides is important. There should be few imperfections when viewed under a phase microscope after polishing. Debris embedded in the slide will interfere with chromosome analysis. Also, the water should sheet evenly across the slide and not form pools or droplets. For best results use slides specific for cytogenetics such as Fisherbrand® Superfrost® CyGen™ microscope slides. ■

Slide Preparation

If you are right-handed, hold the slide with your left hand so that you can hold a pipette in your right hand. The descriptions below are for a right-handed person.

■ Note

The following steps will be performed at nearly the same time, so you will need to coordinate left and right hands. We recommend that you practice these steps several times before using cell samples that may be limited. ■

1. *Left hand:* dip the slide back into the methanol jar briefly, remove and swirl the slide in a beaker of deionized or distilled water. The rinse should be just long enough for a uniform film of water to coat the polished/label side of the slide. You will be able to observe this easily as you lift the slide out of the water.

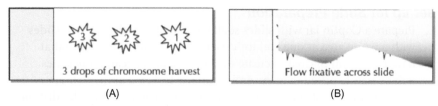

FIGURE 13.1
The two steps for making slides of chromosome spreads. (A) Apply chromosome harvest in three drops, starting at the far end of the slide. (B) Immediately flow the fixative across the top of the slide.

2. *Right hand:* gently but thoroughly mix the prepared cells with a glass Pasteur pipette.
3. *Right hand:* draw the cell mixture into the Pasteur pipette and allow the suspension to sit in the pipette, so that it is ready for dropping onto the slides. At the same time, proceed with the next step with the other hand.
4. *Left hand:* hold the label end of the slide between fingers and thumb and keep the polished/label side up as you lift the slide out of the beaker of water. Place the long edge of the slide in contact with the paper towel and tilt the top edge of the slide forward quickly to drain off the excess water, then tilt the top edge backwards until there are approximately 30 degrees between the back side of the slide and the paper towels.
5. *Right hand:* hold the Pasteur pipette horizontally about 2–7 cm above the slide. Drop three drops of chromosome harvest, evenly spaced, along the slide, starting from the free end of the slide and moving toward the end you are holding. The drops should land on the slide slightly above the midline of the length of the slide – about one-third of the slide width from the top of the slide (Figure 13.1A).
6. *Right hand:* fill a Pasteur pipette with fresh fixative and flow it across the top of the slide immediately after dropping the chromosome harvest (Figure 13.1B).
7. *Left hand:* tilt the slide forward and tap gently on the paper towels to drain off the fixative.
8. *Right hand:* wipe off the back of the slide with a Kimwipe and dab the long edges on a paper towel quickly, and mark the slide with the harvest identification and date by using a pencil on the label end.
9. Evaluate the density, contrast, spreading and drying time of the first slide on the microscope using a 40× phase contrast objective and adjust conditions following the evaluation instructions below.

Evaluating Slides and Adjusting Conditions
■ Density: the number of drops per slide can be adjusted according to the density of the chromosome harvest after the test slide is evaluated. If it

requires more than 4 drops of chromosome suspension, spin down the suspension in the microfuge and reduce the amount of fixative accordingly. The density will also affect spreading. If the chromosomes are too tightly packed together try diluting the pellet to decrease the density.

■ Drying: the drying process will affect how the chromosomes spread and also the banding quality. In general, it is sufficient to dry the slide without much manipulation when the humidity is about 40–45% and the ambient temperature is between 22–26°C. If necessary, the local humidity can be manipulated by using a damp paper towel or hot plate as a drying surface as needed. Warming or cooling the water in the beaker will also alter the drying time.

■ Contrast: the contrast of the chromosomes is determined by the drying time. The quality of banding is affected by the slide drying conditions. Under phase contrast the chromosomes should appear dark gray. In general, a glowing purple (refractile) appearance usually indicates that the slide dried too quickly, while flat, pale gray spreads can indicate that the slides dried too slowly. Adjust temperature and humidity to achieve the optimal drying time.

■ Spreading: the chromosomes from each nucleus should be spread out enough so that each element is distinct, yet not spread so much that some chromosomes are drifting away from the main group. If necessary, 5:1 methanol:glacial acetic acid fixative can be substituted for 3:1 fixative at Step 6 in the slide preparation procedure to tighten the spreading. If the opposite condition (spreads are too tightly clumped to identify individual chromosomes) is seen on the first test slides, the culture may have to be reharvested with a longer hypotonic incubation time.

Giemsa-Banding of Chromosomes

Chromosomes can be stained with dyes that result in a specific banding pattern on each chromosome. This banding pattern is used to determine the identity and integrity of individual chromosomes and the karyotype of a cell. Giemsa ("G") is the dye mixture that is most commonly used to stain chromosomes. G-banding allows the specific identification of individual chromosomes as well as segments of each individual chromosome.

The "bands" are differently stained light and dark regions (and sub-regions) that are recognizable in chromosomes, and are given numerical designations based on their distance from the centromere, from proximal to distal on the chromosome arms. The short and long arms of chromosomes are designated as p and q, respectively.

A cytogeneticist can identify deletions, translocations, inversions, and duplications of chromosomes by analyzing G-banded chromosome spreads. G-bands are provided in reference materials that describe genes (for example,

NIH's NCBI Entrez Gene: www.ncbi.nlm.nih.gov/entrez/query.fcgi?DB_gene, and the European Bioinformatics Institute: http://www.ebi.ac.uk/), and the Online Mendelian Inheritance in Man database (http://www.ncbi.nlm.nih. gov/entrez/query.fcgi?DB_omim) and provide disease information linked to banding data. Standard nomenclature of chromosome number, band assignment and chromosome aberrations is provided by the International System for human Cytogenetic Nomenclature (ISCN, 2009).

Chromosome Staining: Giemsa-Banding with Trypsin Pretreatment (GTG Method)

1. Prior to staining, bake the slides at 90°C for 30 minutes. For best results the slides are usually "aged" by holding them at room temperature 1–3 days prior to baking.
2. Set up four Coplin jars and label them:
   - **Coplin jar 1:** trypsin (prepare stock solution of 0.1% trypsin in 1× PBS)
     - 25 mL trypsin stock solution
     - 25 mL 1× PBS
     - 25 mL sterile H_2O
   - **Coplin jar 2:** 1× PBS rinse
   - **Coplin jar 3:** 1× PBS rinse
   - **Coplin jar 4:** Giemsa stain
     - 2 mL Gurr's Giemsa stain
     - 4 mL Harleco-Wright Giemsa stain
     - 74 mL Gurr's buffer
3. Dip the dried slides into Coplin jar 1, trypsin solution; timing generally varies from 5 to 60 seconds depending on the cell type and length of the chromosomes.

■ Note

Too long in the trypsin and the chromosomes become "ragged" (over trypsinized), too short a time and the bands will not be visible (under trypsinized). Observe one slide on a bright field microscope at 100× and adjust time accordingly for next slide.   ■

4. Rinse quickly in Coplin jar 2 then jar 3 containing 1× PBS.
5. Stain in Giemsa solution (Coplin jar 4) for 3–7 minutes.
6. Rinse the slide under running dH_2O water for about 10 seconds and then gently dry the slide by blowing air from the lab house air system connected to tubing and a pipette or with lens paper placed on top of the slide while resting it on a flat dry surface and allowing it to air dry.
7. Cover with mounting medium and a coverslip (this step is optional).
8. Observe chromosomes using a bright field 100× oil immersion lens.

Interpreting Results

Professional cytogeneticists examine at least 20 metaphases. Generally, at most 5 metaphases are analyzed under the microscope by identifying each chromosome by banding pattern and the remaining 15 are counted to verify the diploid number. Two of the analyzed metaphases are karyogrammed and printed. However, if an abnormal chromosome is observed, the cytogeneticist will search the slide for this and other abnormalities and may end up evaluating more than 20 spreads to determine whether this particular abnormality represents clonal expansion of an abnormal cell altering the majority of cells within the culture or a random culture artifact in a single cell that does not represent a significant change in the culture.

The International System for human Cytogenetic Nomenclature (ISCN, 2009) establishes the rules for identifying and naming whole chromosomes and banded sections numerically and is used by cytogeneticists to arrange the karyotype. ISCN also provides the convention for naming any abnormalities in a karyotype. Updates of this information and a schematic diagram of the banding patterns are published periodically.

Karyotype

The karyotype, by convention, provides the following information: modal number, sex chromosomes, and abnormality abbreviation if an abnormality is present.

- Modal number: total count of number of chromosomes in each cell of a given cell line
- Sex chromosomes: complement of X and Y chromosomes
- Band number: numerical description of the location on a chromosome arm, in order from the centromere out to the end of the chromosome. The short arm is designated "p" and the long arm is "q". These numbers are a standard determined by the ISCN, revised in 2009.

Example	Interpretation
47, XY + 12	Male with an extra copy of chromosome 12
46, XX, del (5p15.2)	Female with a deletion of band 15.2 in the p-arm of chromosome 5 (Cri du chat syndrome)
47, XXY	Male with an extra X chromosome (Klinefelter syndrome)

A karyogram is created by taking a photograph of a G-banded metaphase. Then the individual chromosomes are cut out of the photograph (originally, the "cuts" were made by scissors, but now software is usually used) and arranged in a standardized method based on size, centromere location and

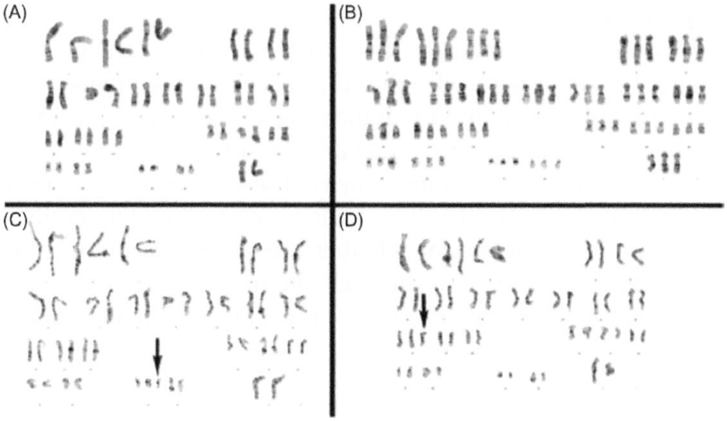

FIGURE 13.2

Human karyograms:
(A) Normal female:
46, XX. (B) Trisomy of
all chromosomes: 69,
XXX, +1–23. (C) Trisomy
21: 47, XX, +21. (D)
Trisomy 13: 47, XX, +13.

specific banding pattern. By convention, the short (p) arm is at the top of the chromosome image (Figure 13.2).

Resolution

One of the variables in classical karyotyping by G-banding is the "resolution." The resolution of the karyotype is a measure of the number of bands that are visible and therefore the smallest segment of the genome that can be detected using this method. Resolution increases with the overall length of the chromosome spread, thus it is desirable to obtain long chromosomes to yield more detailed information during karyotype analysis. The most common method to determine the resolution of banded chromosomes is to count the number of bands visible on chromosome 10 and then read off the resolution from the chart in Table 13.1.

ALTERNATIVE PROCEDURES

Other Species (Non-Human)
Harvest

Harvesting chromosomes from non-human species can generally be accomplished following the same procedures detailed above with minor alterations. It is important to determine the expected diploid number and chromosome morphology for the species before beginning. Species with high diploid numbers (mammals range from 2n = 6 to 106) may require a longer time in hypotonic solution to achieve sufficient chromosome spreading. Harvesting species with a high percentage of acrocentric (single-armed) chromosomes, such as mice may require reducing the strength and amount of time in colcemid to avoid producing overly short chromosomes with low resolution

Table 13.1 The Resolution of a Karyotype Is Determined by Counting the Bands Found on Chromosome 10

Average Number of Bands in Chromosome 10	Estimate of the Total Number of Bands in One Haploid Set (=Resolution)
12	375
13–14	400
15–16	425
17–18	450
19–21	475
22–23	500
24–25	525
26–28	550
29	575
30	600
31	625
32	650
33	675
34	700
35	725
36	750
37	775
38	800
39	825
40–41	850

G-bands. Bi-armed chromosomes with the centromere at (metacentric) or near (submetacentric) the middle are usually not affected by colcemid as profoundly as acrocentric chromosomes.

Karyotyping

Information about the chromosome number, morphology and standard karyotype for a wide variety of species can be found individually in journal publications or in compilations such as the *Atlas of Mammalian Chromosomes*. We recommend examining the chromosomes initially by conventional Giemsa staining without banding. Numeric and morphologic abnormalities are easier to detect in non-banded preparations because the chromosomes are easier to count and the centromere is more visible than in banded chromosomes. After obtaining knowledge of the basic chromosome number and morphology, banding will allow positive identification of individual chromosomes and detection of rearrangements such as deletions and translocations.

Staining

These are two simple methods used to stain DNA, count chromosomes and observe centromere location, but they will not allow positive identification of individual chromosomes.

DAPI Staining

1. Add a drop of mounting medium containing DAPI to the slide.
2. Seal the coverslip.
3. Count chromosomes under UV light using 100× oil immersion lens.

Conventional Giemsa Stain (Non-banded)

1. Follow Steps 5–8 of procedure for GTG banding of chromosomes.

PITFALLS AND ADVICE

Cultures

The cultures should be subconfluent and actively dividing for best results. One effective method of obtaining enough metaphase chromosomes is to harvest hPSCs for karyotyping 3 days before they would be passaged.

During the metaphase stage of mitosis, the chromosomes reach their highest level of condensation and become identifiable under the microscope. The chromosomes are less condensed at early metaphase and become more condensed as the cell progresses towards the end of metaphase. Because the goal of harvesting the cells is to obtain as many quality metaphase chromosomes as possible in order to make an accurate analysis of the culture, colcemid is added to the cultures as it arrests the cells in metaphase. Longer treatment with colcemid will increase the mitotic index and improve chromosome spreading, but prolonged treatment will lead to a higher fraction of cells with condensed, short chromosomes, and the resolution of G-bands will be low. The optimum length of time the cells are incubated with colcemid can be determined empirically in order to obtain both a good mitotic index and good chromosome length.

Chromosome Spreads

Unlike mouse chromosomes, human chromosomes generally have distinct arms visible on both sides of the centromere (bi-armed). However, chromosomes often overlap in the metaphase spread, so it can be difficult for untrained researchers to identify individual chromosomes and obtain an accurate modal number; this is especially difficult for the smaller human chromosomes of 21, 22, and Y.

Hypotonic solution is used to swell the cells and allow the chromosomes to separate. If the cells are left too long in the hypotonic solution the cells will burst and it becomes impossible to determine which chromosomes belong together, but insufficient time in the solution yields spreads with chromosomes too tightly clumped to count. Determining the optimal hypotonic incubation time is an important step in the harvesting process.

One of the most critical variables in making good slides is the amount of time it takes a slide to dry immediately after the cell suspension is dropped onto it. Best results are generally obtained when drying in an atmosphere of 40–45% humidity and 22–26°C. This can be achieved by monitoring the humidity and temperature in the working area using a portable hygrometer/thermometer and adjusting humidity by adding or removing wet paper towels in the immediate slide-making area along with adjusting the temperature of the water used in slide preparation. Even after one becomes proficient in making slides, variables such as temperature and humidity may be difficult to control and can vary seasonally. In general, a "test" slide is made to determine whether the density of the cell suspension is adequate and if the condition of slide drying is appropriate for the specific harvest and for the given day. If the amount of pellet is limited, use only a single drop of cell suspension on the test slide. Slide drying chambers with adjustable humidity and temperature can be purchased to help standardize the drying procedure.

EQUIPMENT

- Tissue culture incubator, 37°C (for most mammalian species), 5% CO_2 in humidified air
- Tissue culture hood, class II
- Microcentrifuge
- Microscope with 100× oil immersion objective and 40× phase contrast objective
- Camera for photographing spreads (mounted on microscope)
- Coplin jars with lids
- Forceps
- Water bath, 37°C
- Oven heated to 90°C
- Portable hygrometer/thermometer for monitoring the humidity and temperature in the working area
- Slide drying chamber (optional)
- Software for karyotype preparation (optional)

REAGENTS AND SUPPLIES

Item	Supplier	Catalog no.	Alternative
Acetic Acid, Glacial	Gallade Chemical, Inc.	2504-14	Many
Colcemid, KaryoMax® (Gibco)	Life Technologies	15212-012	
Coplin jars	VWR	25460-000	Many
Coverslips no. 1	VWR	48393-081	Many
Gurr's Buffer Tablets	Life Technologies	10582-013	
Gurr's Giemsa Stain	CTL Scientific Supply Corp	35086HE	Many
Harleco Wright-Giemsa stain, EM Science	VWR	15204-232	
Methyl Alcohol, Anhydrous (99.8% min.) (Mallinckrodt Chemicals) (methanol)	Gallade Chemical, Inc.	301606	Many
Microcentrifuge tubes, 1.5 mL	VWR	20170-355	Many
Mounting Medium	VWR	48212-290	Many
Mounting Medium with DAPI	Vector Laboratory	H-1200	
Pasture pipettes 5″ glass	VWR	14672-608	Many
Pipettes: 5 mL, 10 mL	Corning	4487, 4488	Many
Potassium Chloride (KCl)	Fisher Scientific	P217-500	Many
Slides, microscope Fisherbrand Superfrost CyGen	Fisher Scientific	22-034-730	
Trypsin-EDTA (TE) for cell culture, Gibco	Life Technologies	25300-062	TrypLE Express™ Cat.#12604
Trypsin 250 (for banding)	Becton Dickinson	215240	

RECIPES

Stock Solutions

Component	Amount	Stock Concentration
Colcemid	10 mL	10 µg/ml
Potassium chloride	100 mL	0.075 M
Fixative	40 mL	3:1 Methanol: acetic acid
Gurr's buffer solution pH 6.8	1000 mL	1 tablet/1000 mL dH$_2$O

Fixative: Methanol:Acetic Acid 3:1 (40 mL)

Make fresh at time of use.

Component	Amount	Final Concentration
Methanol	30 mL	75% by volume
Acetic acid	10 mL	25% by volume

Fixative: Methanol:Acetic Acid 5:1 (30 mL)

Make fresh at time of use.

Component	Amount	Final Concentration
Methanol	25 mL	83.3% by volume
Acetic acid	5 mL	16.7% by volume

Hypotonic Solution (0.075 M KCl, 100 mL)

Alternatively, this solution can be purchased pre-made and ready to use.

Component	Amount	Final Concentration
KCl	0.56 g	0.075 M
MilliQ water	100 mL	Pure

Gurr's Buffer Stock Solution (1000 mL)

Component	Amount	Final Concentration
Gurr's buffer tablet	1 tablet	
MilliQ water	1000 mL	Pure

Gurr's Buffer/Giemsa Stain Solution (80 mL)

Component	Amount	Final Concentration
Gurr's Giemsa stain	2 mL	0.025% by volume
Harleco-Wright Giemsa stain	4 mL	0.05% by volume
Gurr's buffer	74 mL	92.5% by volume

ACKNOWLEDGMENTS

The authors and editors thank Dr. Chih-Lin Hsieh, University of Southern California and Dr. Touran Zadeh, The Genetics Center, Orange CA, for helpful discussions and comments.

READING LIST

Historical Perspective and Basic Techniques

Anon, 2004. Counting chromosomes in ES cells. In: Manipulating the Mouse Embryo, third ed. Cold Spring Harbor Laboratory Press, Cold Spring Harbor, NY.

Arsham, M., Barch, M., Lawce, H. (Eds.), The Association of Genetic Technologists (AGT) Cytogenetics Laboratory Manual, fourth ed. Springer, New York, NY (In press).

Hogan, B., Beddington, R., Costantini, F., Lacy, E., 1994. Karyotyping mouse cells. In: Manipulating the Mouse Embryo, second ed. Cold Spring Harbor Laboratory Press, Cold Spring Harbor, NY.

Hsieh, C.L., 1998. Basic cytogenetic techniques: culturing, slide making, and G-banding. In: Cell Biology: A Laboratory Handbook, second ed. Vol. 3. Academic Press, London.

Hsu, T.C., Benirschke, K. (Eds.), 1967–1977. An Atlas of Mammalian Chromosomes. Springer-Verlag, New York, NY.

An International System for human Cytogenetic Nomenclature (ISCN), 2009. Recommendations of the International Standing Committee on Human Cytogenetic Nomenclature. Karger AG, Basel.

O'Brien, S.J., Menninger, J.C., Nash, W.G. (Eds.), 2006. Atlas of Mammalian Chromosomes. Wiley-Liss, Hoboken, NJ.

Popescu, P., Hayes, H., Dutrillaux, B. (Eds.), 2000. Techniques in Animal Cytogenetics. Springer-Verlag, Berlin.

Rowley Janet, D., 2001. Chromosomal translocations: Dangerous liaisons revisited. Nat. Rev. Cancer 1, 245.

Reports on hPSC Karyotypic Stability

Buzzard, J.J., Gough, N.M., Crook, J., Colman, A., 2004. Karyotype of human ES cells during extended culture. Nat. Biotechnol. 22, 381–382.

Draper, J.S., Smith, K., Gokhale, P., Moore, H.D., Maltby, E., Johnson, J., et al., 2004. Recurrent gain of chromosomes 17q and 12 in cultured human embryonic stem cells. Nat. Biotechnol. 22, 53–54.

Pera, M., 2004. Unnatural selection of cultured human ES cells. Nat. Biotechnol. 22, 42–43.

Peterson, S.E., Westra, J.W., Rehen, S.K., Young, H., Bushman, D.M., Paczkowski, C.M., et al., 2011. Normal human pluripotent stem cell lines exhibit pervasive mosaic aneuploidy. PLoS One 6, e23018.

CHAPTER 14

SNP Genotyping to Detect Genomic Alterations in Human Pluripotent Stem Cells

Francesca S. Boscolo, Candace L. Lynch, Robert E. Morey,
and Louise C. Laurent

EDITOR'S COMMENTARY

There is a great deal of interest in using hPSC-derived differentiated cells for cell replacement therapies, and for these applications, safety is the highest priority. The major concern for stem cell-based therapies is the possibility that they may become tumorigenic (Fox, 2008). Because cancer is associated with genetic aberrations, it is logical to assume that cells destined for clinical use should be free from cancer-associated genomic alterations. There are many methods for examining abnormalities in the human genome, ranging in resolution from the chromosomal level to the single-base level. Classic G-banding karyotyping (Chapter 13) can routinely identify megabase-sized aberrations and can detect some common breakpoints at 15 kb resolution. Comparative genomic hybridization (CGH) can typically resolve 50–100 kb, and high density SNP (single nucleotide polymorphism) genotyping can detect copy number variations as small as 2 kb. Each method has advantages and disadvantages. SNP genotyping cannot resolve translocations of chromosomes or detect rare variants in populations of cells, but it is the highest resolution technique that is currently practical. Full genome sequencing is the only method that has higher resolution, but is much more expensive and data analysis-intensive for routine use. Exon sequencing is easier than full genome sequencing, but it doesn't pick up aberrations in non-coding regulatory regions. It is important to keep in mind that although we frequently detect genomic aberrations in hPSCs, there is no consensus about the impact of these variations, whether they should be reason for concern or will prove to be harmless.

OVERVIEW

Using traditional cytogenetic methods, karyotyping and fluorescence *in situ* hybridization (FISH), human embryonic stem cell (hESC) lines have been

J.F. Loring & S.E. Peterson (eds): Human Stem Cell Manual, Second edition.
DOI: http://dx.doi.org/10.1016/B978-0-12-385473-5.00014-X

shown to accumulate aneuploidies including changes (trisomies of chromosomes 12 and 17) that are frequently found in malignant germ cell tumors (Atkin and Baker, 1982; Rodriguez et al., 1993; Skotheim et al., 2002). The advantage of these cytogenetic methods is that they allow the inspection of individual cells, and therefore enable the identification of a subset of genetically abnormal cells in a mosaic population. However, the resolution of karyotyping is low, limited by the number of visible bands per chromosome (up to 400–800, depending on the chromosome). The average chromosome is 130 Mb, and G-banding can reliably detect abnormalities in the Mb range. FISH provides much greater resolution, with the ability to identify regions as small as a few kb. However, for FISH, each interrogated region requires a specific probe, and therefore FISH is only useful when a specific genetic aberration is suspected.

Recently, several groups have reported that small genetic aberrations, including subchromosomal duplications, deletions, copy neutral loss-of-heterozygosity (LOH), and point mutations, are frequently found in hPSCs. These studies have used a variety of molecular techniques, including gene expression profiling, array CGH, SNP genotyping arrays, and exome sequencing.

The accuracy and resolution of gene expression profiling is limited by the fact that genetic aberrations have somewhat unpredictable effects on transcript abundance; for example, deletion of a region containing a gene often does not result in a corresponding decrease in the quantity of mRNA encoded by that gene.

CGH detects abnormalities by arraying sequences from one genome and detecting variations by measuring the fidelity of hybridization of another genome to the arrays. SNP arrays are structured to detect known single nucleotide polymorphisms in the human population. The resolution of array CGH and SNP genotyping arrays is determined by the coverage and density of the arrays. For example, using the Agilent 244K oligonucleotide CGH array, which has an average probe spacing of 6.4 kb (http://www.chem.agilent .com/Library/brochures/5989-5499EN_lo.pdf), the minimum size for a copy number alteration (identified by detection at three consecutive probes) would be ~20 kb. Using the Illumina 2.5 M SNP genotyping array, which has an average probe spacing of 600 bp, the corresponding minimum segment detected by three consecutive probes would be ~2 kb. These assays are performed on DNA extracted from large populations of cells, typically consisting of 10,000–100,000 cells. It has been shown that copy number alterations affecting ~20% of the cell population can be detected using SNP genotyping. However, genetic abnormalities that are present at lower frequencies would not be recognized. These array-based methods also do not map the genomic structures of copy number alterations. For example, using these methods, one would not be able to tell whether a duplicated region was adjacent to the original chromosomal location (e.g. in a tandem duplication) or was positioned on another chromosome.

Several data analysis tools are available for identifying copy number alterations from array CGH and SNP genotyping data. It has been reported that in most, but not all, cases the data from a given microarray platform was best analyzed using the software provided by the manufacturer. Here we provide a protocol for detecting copy number variations in hPSCs using an Infinium HD Illumina SNP genotyping array.

PROCEDURES

Infinium HD Super Assay

In the Infinium® HD assay, the whole genome is isothermally amplified up to 1000-fold using 200 nanograms of input genomic DNA and then enzymatically fragmented into an optimum length of 300–600 base pairs. The fragmented DNA is then precipitated with isopropanol and resuspended. Once purified, it is hybridized to the BeadChip overnight in a capillary flow-through chamber. The BeadChip is embedded with millions of 3-micron silica beads that are coated with hundreds of thousands of copies of a specific oligonucleotide. These locus-specific 50-mers are adjacent to the polymorphic nucleotide and act as a capture sequence for the fragmented DNA. Once any unhybridized or non-specific hybridized DNA is removed through washing, a single base labeled with either biotin (C and G nucleotides) or dinitrophenyl (DNP) (A and T nucleotides) is extended onto the probe using the hybridized DNA as a template. Once the hybridized DNA is washed away, green fluorescent Streptavidin and red fluorescent anti-DNP antibodies are bound to the labeled probe. By adding several rounds of fluorescent molecules and biotin-labeled and DNP-labeled antibodies, the fluorophore signal is amplified. The iScan high-resolution scanner then employs a dual color approach to identify the allele at each SNP site. The number of SNP loci examined are limited only to the number of markers on the microarray. For more information about the technology, see https://icom.illumina.com/Download/Summary/VXs_3jdtw068EuKI8TgSzA, http://www.illumina.com/technology/infinium_hd_assay.ilmn, and http://www.illumina.com/support/training/infinium_chemistry.ilmn). There are currently 5 million variant markers available for one sample (http://www.illumina.com/documents/products/datasheets/datasheet_gwas_roadmap.pdf). All steps of the Infinium HD protocol are followed according to Illumina's Infinium HD Assay Super Protocol Guide and they are summarized in Figure 14.1.

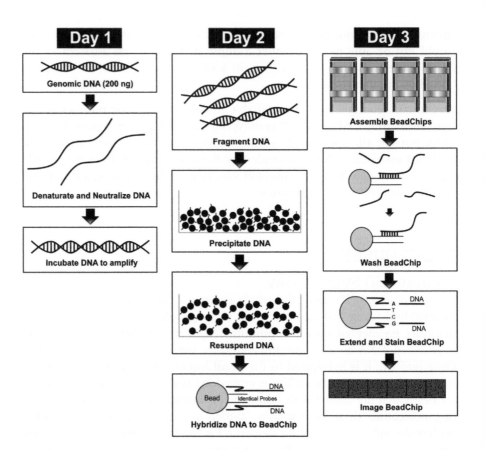

FIGURE 14.1

Infinium HD protocol.

Data Analysis

Copy number variation data obtained from an Illumina SNP genotyping microarray can be visualized and analyzed using Illumina's data analysis software, GenomeStudio®. The workflow for data analysis is summarized in Figure 14.2.

GenomeStudio is a modular software package provided by Illumina that allows users to analyze data generated from any Illumina platform. GenomeStudio® allows the user to analyze both SNP genotyping and Copy Number Variation (CNV). In order to analyze SNP genotyping data generated using Infinium and Golden Gate assays the Genotyping Module is required.

Creating a Project for Analysis in GenomeStudio

To begin the data analysis process it is first necessary to "create a project." A project allows the user to define and input all the parameters collected from

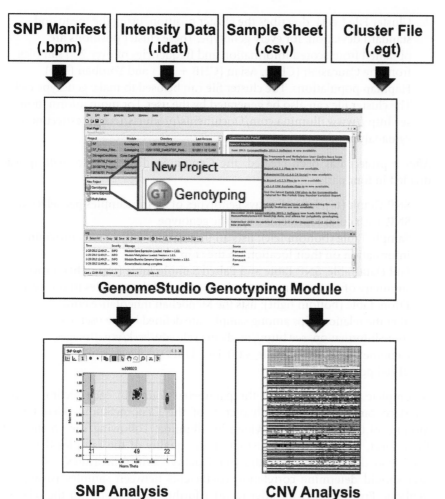

FIGURE 14.2
Workflow for data analysis using GenomeStudio®.

the assay for each sample. In order to create a project using GenomeStudio®, the user will need four files:

1. SNP Manifest (.bpm): describes the probe content of BeadChips and contains annotations.
2. Intensity Data (.idat): contains the intensities detected in the samples. These files contain the statistics (number of beads, intensity mean and standard deviation, etc.) for both colors of each bead in all the samples on the BeadChip. The user should have two.idat files per sample, one for the green channel and one for the red.
3. Sample Sheet (.csv): lists the locations (BeadChip bar code and position) of each sample that will be included in the project. This file is generated by the user.

4. Cluster File (.egt): describes the cluster position, using the mean and standard deviation, for every genotype. These cluster positions are obtained from scored information and frequencies of over 200 samples from the Caucasian (CEU), Asian (CHB + JPT) and Yoruban (YRI) HapMap populations. The cluster file can be used to make genotype calls. The cluster file can also be generated by the user. For more information, see http://www.illumina.com/Documents/products/technotes/technote_cytoanalysis.pdf.

After a project has been created, the GenomeStudio® interface will appear divided in four parts:

1. SNP Graph (top left): displays genotypes of all the samples per selected SNP. Each color represents a specific genotype.
2. Sample Table (bottom left): lists all the samples under study providing information on their location on the chip.
3. Full Data Table, SNP Table and Paired Sample Table (top right): provide a summary of all the parameters obtained for all the samples in the assay.
4. Errors Table (bottom right): lists the Mendelian inheritance errors, after the relationships among samples are defined by the user. For more information, see http://medicine.yale.edu/keck/ycga/Images/GenomeStudio_GT_Module_v1.0_UG_11319113_RevA_tcm240-21464.pdf.

A complete description of all the parameters found in each section of the interface can be found in Chapter 8 of the GenomeStudio, Genotyping Module v1.0 User Guide provided by Illumina. Following the creation of the project, the user should first check the quality of the samples that have been loaded into GenomeStudio (see Quality Control Measures). Next, the user should determine correlation coefficients between samples (using the Replicate Error Report) in order to infer further information on the relatedness of the samples and make sure samples were not switched while performing the assay (a reproducibility frequency between two samples of >0.85 indicates that the samples are related).

For SNP analysis it is useful to generate a final report that summarizes the genotype of each probe for each sample on the BeadChip. Illumina does not use a phasing method (the order in which the alleles are listed in the reports and tables does not refer to the same DNA strand) to report genotypes. Therefore, to correctly identify the nucleotide present in a sample it is necessary to check both the genotype (denoted as AA, BB or AB) and the Allele TOP 1 and Allele TOP 2. Allele TOP 1 indicates the nucleotide is on the top strand being read left to right, while Allele TOP 2 indicates that the nucleotide is on the bottom strand being read right to left. Allele A is the first nucleotide listed and Allele B is the second (see Converting A/B Genotype Calls to Nucleotide Calls,

Table 14.1 Definition of Values Associated with LRR and BAF

LRR	Aberration	BAF	Genotype
<0	Deletion	0	AA
0	Homozygosity or Heterozygosity	0.5	AB
>0	Duplication	1	BB

Technical note, Illumina and http://www.illumina.com/documents/products/ technotes/technote_topbot.pdf).

CNV analysis is performed with cnvPartition, a GenomeStudio CNV Analysis plug-in. The algorithm uses Log R Ratio (LRR) and B Allele Frequency (BAF) to identify aberrant regions of the genome. LRR is the ratio of observed to expected intensities in log scale, Table 14.1 (left columns). A LRR of 0 represents the diploid state, whereas a value of >0 indicates a duplication has occurred and a value <0 designates a deletion has occurred. The BAF represents the portion of hybridized samples that carry the B allele, Table 14.1 (right columns). The BAF ranges from 0 to 1 identifying AA and BB genotypes, respectively. Using a combination of the LRR and BAF values for each probe, the user is able to estimate the CNV state of a sample.

cnvPartition is able to identify 14 different genotypes. Additionally, cnvPartition includes an algorithm that identifies aberrant homozygous regions, known as Loss of Heterozygosity (LOH) (see http://www.illumina.com/ Documents/products/technotes/technote_cnv_algorithms.pdf and http://www .illumina.com/Documents/products/technotes/technote_cnv_algorithms.pdf). GenomeStudio contains a visual interface (Genome Viewer) that allows the user to graphically view the LRRs and BAFs for each chromosome of each sample. The combination of cnvPartition results and Genome Viewer are powerful tools that can be used to validate CNVs in samples of interest.

Steps in Making a Project

1. Prepare the Sample Sheet. Open Excel and beginning in cell A1 prepare a file as shown in Figure 14.3. The user should not edit the words in red font. Only the words in blue font should be edited according to the users' particular circumstance. No empty cells should be left between the different sections. Save the file as a .csv.
2. Open GenomeStudio.
3. In the bottom left "New Project" window of the GenomeStudio® start page, select the Genotyping Module.
4. A Genotyping Project Wizard window will open. Select "Next."
5. A new window will open. The user is now required to select the location where the project will be saved "Project Repository," and to give the

	A	B	C	D
1	[Header]			
2	InvestigatorName	PI		
3	ProjectName	SNP Genotyping Project		
4	ExperimentName	hESCs Analysis		
5	Date	January 1st 2012		
6	[Manifests]			
7	A	HumanOmni5-4v1_B		
8	[Data]			
9	Sample_ID	SentrixBarcode_A	SentrixPosition_A	
10	Sample 1	6285588028	R04C01	
11	Sample 2	...	...	
12	...			
13				

FIGURE 14.3

Excel file for sample sheet. Only edit the text in blue. The text in red should not change.

project a name "Project Name | Create." When all the selections have been made, press "Next."

6. A new window will open. Select "Use sample sheet to load sample intensities." Press "Next."

7. A new window will open. Select the location of the sample sheet, .csv file ("Sample Sheet"), the location where the .idats, the data generated from imaging, are stored ("Data Repository"), and the location where the manifest, the .bpm file for the BeadChip being used, is stored ("Manifest Repository"). When all the selections are made, press "Next."

8. A new window will open. Check the box called "Import cluster position from a cluster file." If the user is using the Cluster file (.egt) provided by Illumina, the file name should appear automatically in the panel under the checked box. If it does not, the Cluster file is usually located in the directory where the manifest is saved. Before pressing "Finish," check the box on the left of the window entitled "Pre-Calculate."

9. The project will be created and saved in the folder previously indicated by the user.

CNV Analysis

The first part of this protocol describes how to use the CNV Analysis plug-in and CNV Region Report plug-in of cnvPartition.

1. Download and install the cnvPartition CNV Analysis plug-in (available on the Illumina website).

2. From the menu bar of GenomeStudio pull down the "Analysis" menu and select "CNV Analysis."

3. A new window will appear. This window is divided into two panels (top and bottom). On the top panel, in the first pull-down menu select "cnvPartition 3.2.0" (or the version that was just downloaded). On the right, write a name to give to the partition. On the bottom panel, the user is able to edit the partition algorithm parameters. "Confidence threshold" refers to the score the algorithm gives to each call. "Detect Extended Homozygosity" is a LOH neutral detection algorithm. "Exclude Intensity Only" allows the user to exclude probes for which the BAF value will be ignored from the analysis. GC waves are variation in the hybridization intensity of certain probes and they originate from bad quantification of samples or from regions of the DNA rich in GCs (Diskin et al., 2008). "GC Wave Adjust" allows the user to adjust for these problematic regions using linear regression of LRR vs GC content. "Include Sex Chromosome" allows the user to decide whether or not to include X and Y chromosomes in the analysis. "Minimum Homozygous Region Size" indicates the minimum length in base pairs of a LOH. "Minimum Probe Count" refers to the number of consecutive probes that show the same aberration. After adjusting the above parameters, select "Create New CNV Analysis," located in the top panel.

4. The cnvPartition output is a visual summary of all the calls produced by the algorithm. Each call is color-coded (X stands for a generic allele):

Color	CNV Value	Aberration	Genotype
Red	0	Deletion	00
Orange	1	Deletion	X0
Green	2	LOH	XX
Blue	3	Duplication	XXX
Purple	4	Duplication	XXXX

5. To export the results from the partition in text format, use the CNV Region Report plug-in found in the "Analysis" drop-down menu under select "Reports | Report Wizard."

6. A new window will appear. Select the "Custom Report" and then pick "CNV Report 2.1.1 by Illumina, Inc from Illumina, Inc." Press "Next" and a new window will open. Decide whether or not to include "zeroed SNPs" and "invisible SNPs." When the appropriate choice is made, press "Next." Next, browse to the location where the report will be saved ("Output path") and name the new file. Press "Finish." GenomeStudio will generate a .txt report containing information about each call.

7. Save the project before closing it, otherwise the cnvPartition results will be lost.

CNV Validation

The following protocol describes how to use Genome Viewer and Chromosome Browser to validate the CNV calls.

1. In order to open Genome viewer (the visual interface in GenomeStudio) the user needs to select "Tools | Show Genome Viewer" from the top, drop-down menu bar.

2. All the samples in the project, and their parameters, will be listed in a new window. Select the sample(s) that need to be checked in the top part of the window. From the list of parameters on the bottom of the window select the two that are used to determine the presence of a CNV, LRR and BAF. When the selection has been made, first press "Add to Favorite" and then select "OK."

3. The first chromosome of the first sample selected will appear in a new window. To change the number of chromosomes visible in the window, the user can select "Chromosome Slide Show Mode" and then choose the number of chromosomes to be visualized.

4. To view one chromosome or one specific region of a chromosome in multiple samples, pick "View | Chromosome Browser." This operation might require some time to finish. When it has completed, the initial area of the chromosome will be highlighted in all the samples initially selected. In order to change the displayed area, just move the red rectangle over the chromosome.

5. To simplify the validation process it might be useful to map the cnvPartition results over the entire genome. To accomplish this, select "View | CNV Analysis as Bookmarks" and select one of the saved cnvPartition files in the newly opened window. After the partition file has been selected, press "OK." The chromosomes will be highlighted in the regions where a CNV is called using the same color scheme as the cnvPartition output (Figure 14.4).

■ Note

cnvPartition is not the only way to analyze CNV data. There are other algorithms that can be used for the same purpose such as PennCNV, Nexus, CLCbio, and R. However in our experience they have not worked as well as cnvPartition. ■

ALTERNATIVE PROCEDURES

SNP genotyping has much higher resolution than conventional methods such as karyotyping, and has the advantage of providing whole-genome data without prohibitive cost.

As the cost of next-generation whole genome sequencing drops, it may become a viable option for routine detection of genomic changes in hPSC

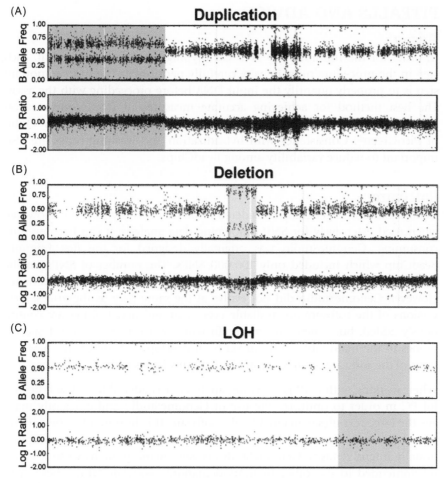

FIGURE 14.4
Examples of aberrations.
Genome viewer is
bookmarked with the results
from cnvPartition.
(A) Example of duplication.
(B) Example of deletion.
(C) Example of LOH.

populations. Sequencing has the advantages of producing quantitative data at single base-pair resolution. If whole-genome paired-end sequencing is performed at sufficient depth (standard sequencing runs provide ~30× coverage), both point mutations and copy number alterations can be detected. The publication of improved alignment and variant calling pipelines (VarScan, SamTools, GATK, Dindel) have made the analysis of whole genome data more accessible.

Targeted sequencing strategies, such as sequencing the 1% of the genome that is in exons, is a less expensive approach to obtaining high resolution information about changes in the hPSC genome. But while SNP genotyping surveys the whole genome, "exome" sequencing does not query the intronic and intragenic regions that contain regulatory sequences.

PITFALLS AND ADVICE

Assay

In order for the Infinium HD SNP Genotyping assay to be successful, each step in the procedure must be very carefully performed. The most important step is to properly quantify the input DNA before proceeding with the assay. The best method for achieving accurate measures of DNA concentration is to use a fluorescence-based method such as the Life Technologies' Qubit or PicoGreen®. Consistency with hybridization and washing times are also important to reduce variability among BeadChips.

Software

GenomeStudio® is generally considered to be user-friendly software. However, when used extensively for SNP genotyping data analysis, it can become time consuming. GenomeStudio was originally designed for the first Infinium BeadChip which included only 100,000 SNPs. The number of SNPs on the arrays has increased over the years until to the current 500,000 variants. However, GenomeStudio did not evolve at the same pace as the arrays. New versions of the software are available every year and new features are continuously added, but a new and more efficient way to filter and select specific probes in the main tables has not yet been developed. Additionally, new versions of the software often result in compatibility issues with old plug-ins.

When working with SNP genotyping analysis, it is advisable to use a powerful PC in order to optimize the use of GenomeStudio and efficiently manage the large text files generated by the software. It is important to be patient, especially when operating GenomeStudio's visual tools or running computationally intensive tasks. GenomeStudio is very memory-intensive so it is not recommended to use other memory-demanding software while these tasks are being executed. If the user has software/plug-in incompatibility problems, he/she should check the availability of newer versions of the plug-ins on the Illumina website. In the event that the problem is not resolved after updating all plug-ins to their most recent version, contact Illumina technical support. Frequently, technical support is aware of the problem and often they will promptly send an updated and functional version of the plug-in.

Analysis

After scanning through samples using the chromosome browser, users may notice the presence of clear aberrations (deletions or duplications) that are not called by the cnvPartition algorithm. Does that mean that they are not real? If they are real, why aren't they called? The answer to these questions is simple. These aberrations are often real and they are not called because either the algorithm is not sensitive enough, or the area where the aberration occurred is too noisy. It has been demonstrated that the algorithm can detect aberrations (duplications specifically) only if they are present in at least 80% of the population. Therefore, in order to gain further information on non-called duplications, we recommend using the BAF distance method to evaluate the percentage of the population carrying the duplication.

EQUIPMENT AND SUPPLIES

Assay

There are many university core facilities that provide Illumina genotyping services. A list of Illumina core labs can be found at http://www.illumina.com/services/corelabs.ilmn. To set up Infinium genotyping, the iScan, Infinium Starter Hardware Kit (catalog # WG-15-304), and HumanOmni BeadChips must be purchased from Illumina (www.illumina.com). Materials and reagents that must be supplied separately are listed below.

Item	Source/Supplier	Catalog Number
DNeasy Blood and Tissue Kit	Qiagen	69504/69506
Qubit 2.0 Fluorometer	Life Technologies	Q32866
Qubit dsDNA BR Assay Kit	Life Technologies	Q32850
96-well 0.2 mL skirted microplates	Bio-Rad	MSP-9601
0.8 mL storage plate (MIDI plate), conical well bottom (2 per run)	Abgene	AB-0765
Heat sealing foil sheets, Thermo-Seal	Abgene	AB-0559
Top-loading BeadChip rack	Illumina	
Vacuum Desiccator	PGC Scientific	L-198593
Disposable pipetting troughs	VWR	21007-970
Compressed air can (Whoosh-Duster)	VWR	16650-027
0.1 N NaOH (Sodium hydroxide) (65 mL)	Sigma-Aldrich	S0899
100% Ethanol	Sigma-Aldrich	E7023
100% 2-Propanol	Sigma-Aldrich	I9516
OmniPur Formamide	EMD Chemicals	4610
0.5 M EDTA (100 mL)	Sigma-Aldrich	E7889
0.5% Sodium Hypochlorite (10% bleach)	Sigma-Aldrich	425044

Data Analysis

Item	Source/Supplier
PC (CPU Speed: 2.0 GHz or more; Processor: 2 or more cores; Memory: 8 GB or more; Hard Drive: 250 GB or more; Operating System: Windows XP, Vista or 7)	Any
Genome Studio (v2011.1 or most updated)	Illumina
cnvPartition CNV Analysis plug-in (v3.2.0 or most updated)	Illumina
CNV Region Report plug-in (v2.1.1 or most updated)	Illumina
Microsoft Excel	Microsoft
Genome Studio Genotyping Module v1.0 User Guide	Illumina

QUALITY CONTROL METHODS

The Infinium HD assay contains internal controls in all steps of the protocol. Once data are generated, these controls must be examined in the software to verify that all steps were successful. As previously mentioned, proper quantification of input DNA using a fluorescence-based method such as Life Technologies' PicoGreen is critical. High quality DNA is also important for whole genome amplification.

In order to determine the quality of the samples and the efficiency of the assay, the first step is to understand the scores and the controls Illumina has placed in the assay itself. There are two types of controls: sample-independent and sample-dependent controls. The first type is designed to evaluate the performance of the assay, while the second set is used to determine the input sample quality. GenomeStudio provides a controls panel, where the user can evaluate the controls for both the assay and the samples at the same time. The Non-Polymorphic control graph should be assessed first as it describes both the performance of the assay and the quality of the sample. In the panel (Figure 14.5) there are two graphs: red dots represent A and T and green dots represent C and G. The controls should cluster together as shown in Figure 14.5. If the controls present in the same graph show a split, it means that there is a problem.

However, from these two graphs it is not possible to determine if the assay failed or if the DNA quality was poor. To gain further information it is necessary to evaluate the other controls, starting with the sample-dependent controls. Stringency and Non-Specific Binding controls evaluate the hybridization ability of the fragmented DNA depending on wash stringency and on non-specific binding, respectively. If the samples are of high quality, the intensity for the Perfect Match probe (PM) will be high while the intensity

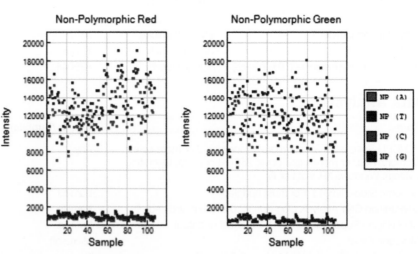

FIGURE 14.5
Non-Polymorphic Controls. This control assesses both the assay and the sample but does not distinguish between the two. Samples on both the red and green graphs should cluster together as shown.

for the MisMatched probe (MM) will be low as shown in Figure 14.6A on the Stringency control graph. At the same time, the intensity for non-specific binding will be low as shown in Figure 14.6B. In a situation in which the samples are of low quality, the intensity for the MM probe will be higher than the one from the PM probe and the intensity for non-specific binding will be high. These two sample-dependent controls are correlated, thus they should be evaluated together.

In the event that the sample-dependent controls are inconclusive, the remaining controls should be evaluated. The Staining controls evaluate the signal detection reaction. The Staining Red graph (referred to as DNP) should display high intensity only for DNP (High) while the Staining Green graph (referred to as biotin) should have a high intensity only for Biotin (High) (Figure 14.7A). The Extension controls evaluate the extension reaction and, like

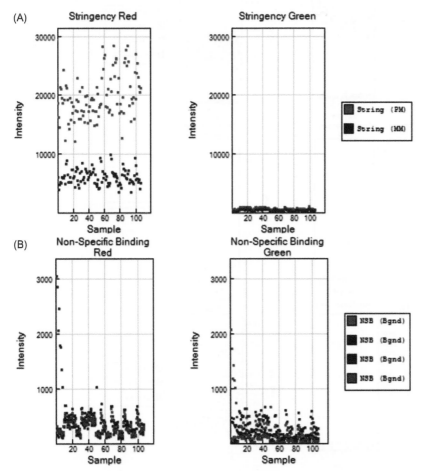

FIGURE 14.6
Sample-dependent controls. Stringency and Non-Specific Binding controls examine the ability of the fragmented DNA to bind to target sequences on the BeadChip. (A) Stringency control graph: in high quality samples, the intensity for the Perfect Match (PM) probes will be high, while the MisMatched (MM) probe intensity will be low. (B) Non-Specific Binding should be low.

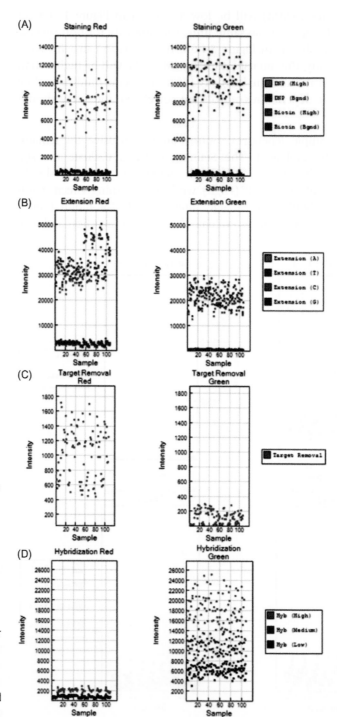

FIGURE 14.7

Sample-independent controls. Staining, Extension, Target Removal and Hybridization examine the performance of the assay. Graphs give examples of a properly performed assay. (A) Staining control graph: DNP (High) and Biotin (High) should be high for Staining red and green controls respectively. (B) Extension control graphs: samples in red and green should cluster together. (C) Target Removal control graphs: intensities should be low for both red and green graphs. (D) Hybridization control graphs: high, medium and low probes will be displayed from high to low intensities.

the Non-Polymorphic control, should have the two sets of controls clustered together (Figure 14.7B). The target removal control determines the efficiency of the stripping reaction. The intensities should be low in both the channels (Figure 14.7C). In the event that the intensity values are high, it can be assumed that the targets were not efficiently removed from the BeadChip. Finally, the hybridization controls evaluate the hybridization reaction itself. The graph should display the High, Medium and Low probes ordered from high to low intensities (Figure 14.7D). If this is not the case, hybridization was not successful.

It is also important to check the Call Rate for all the samples. This value can be found in the Sample Table. Each time the samples are re-clustered (for example, when one sample is removed from the analysis) the Call Rate should be recalculated. The Call Rate represents the percentage of SNPs with genotype call reliability higher than a pre-determined threshold. The values are displayed in decimal scale thus they range from 0 to 1. A good Call Rate is greater than 0.99.

READING LIST

FDA on Stem Cell Therapies
Fox, J.L., 2008. FDA scrutinizes human stem cell therapies. Nat. Biotechnol. 26, 598–599.

Karyotyping
Trask, B.J., 2002. Human cytogenetics: 46 chromosomes, 46 years and counting. Nat. Rev. Genet. 3, 769–778.

Karyotyping Aneuploidies in hPSCs
Baker, D.E., Harrison, N.J., Maltby, E., Smith, K., Moore, H.D., Shaw, P.J., et al., 2007. Adaptation to culture of human embryonic stem cells and oncogenesis in vivo. Nat. Biotechnol. 25, 207–215.

Draper, J.S., Smith, K., Gokhale, P., Moore, H.D., Maltby, E., Johnson, J., et al., 2004. Recurrent gain of chromosomes 17q and 12 in cultured human embryonic stem cells. Nat. Biotechnol. 22, 53–54.

Imreh, M.P., Gertow, K., Cedervall, J., Unger, C., Holmberg, K., Szoke, K., et al., 2006. In vitro culture conditions favoring selection of chromosomal abnormalities in human ES cells. J. Cell Biochem. 99, 508–516.

Maitra, A., Arking, D.E., Shivapurkar, N., Ikeda, M., Stastny, V., Kassauei, K., et al., 2005. Genomic alterations in cultured human embryonic stem cells. Nat. Genet. 37, 1099–1103.

Mitalipova, M.M., Rao, R.R., Hoyer, D.M., Johnson, J.A., Meisner, L.F., Jones, K.L., et al., 2005. Preserving the genetic integrity of human embryonic stem cells. Nat. Biotechnol. 23, 19–20.

Peterson, S.E., Westra, J.W., Rehen, S.K., Young, H., Bushman, D.M., Paczkowski, C.M., et al., 2011. Normal human pluripotent stem cell lines exhibit pervasive mosaic aneuploidy. PLoS One 6, e23018.

Teratocarcinomas
Atkin, N.B., Baker, M.C., 1982. Specific chromosome change, i(12p), in testicular tumours? Lancet 2, 1349.

Rodriguez, E., Houldsworth, J., Reuter, V.E., Meltzer, P., Zhang, J., Trent, J.M., et al., 1993. Molecular cytogenetic analysis of i(12p)-negative human male germ cell tumors. Genes Chromosomes Cancer 8, 230–236.

Skotheim, R.I., Monni, O., Mousses, S., Fossa, S.D., Kallioniemi, O.P., Lothe, R.A., et al., 2002. New insights into testicular germ cell tumorigenesis from gene expression profiling. Cancer Res. 62, 2359–2364.

SNP, CGH, Sequencing of hPSCs

Gore, A., Li, Z., Fung, H.L., Young, J.E., Agarwal, S., Antosiewicz-Bourget, J., et al., 2011. Somatic coding mutations in human induced pluripotent stem cells. Nature 471, 63–67.

Hussein, S.M., Batada, N.N., Vuoristo, S., Ching, R.W., Autio, R., Narva, E., et al., 2011. Copy number variation and selection during reprogramming to pluripotency. Nature 471, 58–62.

Laurent, L.C., Ulitsky, I., Slavin, I., Tran, H., Schork, A., Morey, R., et al., 2011. Dynamic changes in the copy number of pluripotency and cell proliferation genes in human ESCs and iPSCs during reprogramming and time in culture. Cell Stem Cell 8, 106–118.

Lefort, N., Feyeux, M., Bas, C., Feraud, O., Bennaceur-Griscelli, A., Tachdjian, G., et al., 2008. Human embryonic stem cells reveal recurrent genomic instability at 20q11.21. Nat. Biotechnol. 26, 1364–1366.

Mayshar, Y., Ben-David, U., Lavon, N., Biancotti, J.-C., Yakir, B., Clark, A., et al., 2010. Identification and classification of chromosomal aberrations in human induced pluripotent stem cells. Cell Stem Cell 7, 521–531.

Narva, E., Autio, R., Rahkonen, N., Kong, L., Harrison, N., Kitsberg, D., et al., 2010. High-resolution DNA analysis of human embryonic stem cell lines reveals culture-induced copy number changes and loss of heterozygosity. Nat. Biotechnol. 28, 371–377.

Spits, C., Mateizel, I., Geens, M., Mertzanidou, A., Staessen, C., Vandeskelde, Y., et al., 2008. Recurrent chromosomal abnormalities in human embryonic stem cells. Nat. Biotechnol. 26, 1361–1363.

SNP Resolution

Toujani, S., Dessen, P., Ithzar, N., Danglot, G., Richon, C., Vassetzky, Y., et al., 2009. High resolution genome-wide analysis of chromosomal alterations in Burkitt's lymphoma. PLoS One 4, e7089.

Yu, S., Bittel, D.C., Kibiryeva, N., Zwick, D.L., Cooley, L.D., 2009. Validation of the Agilent 244K oligonucleotide array-based comparative genomic hybridization platform for clinical cytogenetic diagnosis. Am. J. Clin. Pathol. 132, 349–360.

Copy Number Variation Software for Microarrays

Halper-Stromberg, E., Frelin, L., Ruczinski, I., Scharpf, R., Jie, C., Carvalho, B., et al., 2011. Performance assessment of copy number microarray platforms using a spike-in experiment. Bioinformatics 27, 1052–1060.

Kresse, S.H., Szuhai, K., Barragan-Polania, A.H., Rydbeck, H., Cleton-Jansen, A.M., Myklebost, O., et al., 2010. Evaluation of high-resolution microarray platforms for genomic profiling of bone tumours. BMC Res. Notes 3, 223.

Sequencing Software Methods

Albers, C.A., Lunter, G., Macarthur, D.G., McVean, G., Ouwehand, W.H., Durbin, R., 2011. Dindel: Accurate indel calls from short-read data. Genome Res. 21, 961–973.

DePristo, M.A., Banks, E., Poplin, R., Garimella, K.V., Maguire, J.R., Hartl, C., et al., 2011. A framework for variation discovery and genotyping using next-generation DNA sequencing data. Nat. Genet. 43, 491–498.

Koboldt, D.C., Chen, K., Wylie, T., Larson, D.E., McLellan, M.D., Mardis, E.R., et al., 2009. VarScan: Variant detection in massively parallel sequencing of individual and pooled samples. Bioinformatics 25, 2283–2285.

Li, H., Handsaker, B., Wysoker, A., Fennell, T., Ruan, J., Homer, N., et al., 2009. The Sequence Alignment/Map format and SAMtools. Bioinformatics 25, 2078–2079.

Other Topics

Atkin, N.B., Baker, M.C., 1982. Specific chromosome change, i(12p), in testicular tumours? Lancet 2, 1349.

Fox, J.L., 2008. FDA scrutinizes human stem cell therapies. Nat. Biotechnol. 26, 598–599.

Gore, A., Li, Z., Fung, H.L., Young, J.E., Agarwal, S., Antosiewicz-Bourget, J., et al., 2011. Somatic coding mutations in human induced pluripotent stem cells. Nature 471, 63–67.

Rodriguez, E., Houldsworth, J., Reuter, V.E., Meltzer, P., Zhang, J., Trent, J.M., et al., 1993. Molecular cytogenetic analysis of i(12p)-negative human male germ cell tumors. Genes Chromosomes Cancer 8, 230–236.

Skotheim, R.I., Monni, O., Mousses, S., Fossa, S.D., Kallioniemi, O.P., Lothe, R.A., et al., 2002. New insights into testicular germ cell tumorigenesis from gene expression profiling. Cancer Res. 62, 2359–2364.

CHAPTER 15

Analysis and Purification Techniques for Human Pluripotent Stem Cells

Carmel O'Brien, Jack Lambshead, Hun Chy, Qi Zhou, Yu-Chieh Wang, and Andrew L. Laslett

EDITOR'S COMMENTARY

The tremendously diverse differentiation power of human pluripotent stem cells (hPSCs) makes them uniquely valuable for therapeutic and research applications that require specific human cell types. But pluripotency is also a challenge: pluripotent cells must be guided along specific differentiation pathways in order to obtain the desired cell type. As populations of cells differentiate they lose synchronicity; even when the best directed differentiation methods are used, a single culture of 10 million cells might contain millions of cells in variable states, or even types, of differentiation. How can heterogeneous populations of differentiating cells be characterized and purified? This chapter describes methods for flow cytometry, fluorescence-activated cell sorting (FACS) and magnetic-activated cell sorting (MACS) that have made it possible to distinguish, and separate, many of the different types of cells that develop in hPSC cultures.

OVERVIEW

High-fidelity antibody and lectin-based technologies are important tools for the accurate identification and enrichment of starting hPSC populations and desired end–point cell types. The technologies are also important for the subsequent removal of potentially tumor-forming remnant pluripotent cells and off-target cell types resulting from differentiation protocols.

Flow cytometry provides an accurate and efficient technology with which to identify, quantify and separate, at a single cell level, distinct subpopulations of cells contained within complex cell mixtures. Given the heterogeneity typically found in hPSC cultures, flow cytometry is a useful tool for sorting pluripotent cells from both support feeder fibroblast cells and those hPSCs that are undergoing early stage differentiation. On a cell-by-cell basis,

J.F. Loring & S.E. Peterson (eds): Human Stem Cell Manual, Second edition.
DOI: http://dx.doi.org/10.1016/B978-0-12-385473-5.00015-1

FIGURE 15.1

Dot plot of the forward light scatter (FSC-A) and side light scatter (SSC-A) characteristics of a hPSC population. Two major clusters of cell subpopulations can be observed in this example. Typically, data for the region of lowest forward scatter FSC_{Lo} and lowest side scatter SSC_{Lo} (blue arrowhead) is not collected as it contains predominantly, but not solely, cell debris.

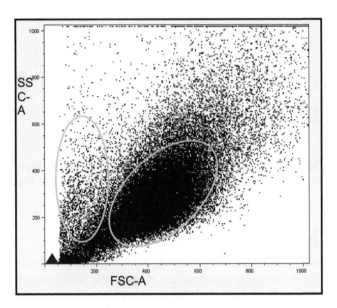

separation and analysis can be achieved by exploiting the differences in basic optical properties, specific reactivity to antibodies or lectins, singly or in combination, or using their physiological or biochemical properties.

Flow cytometric techniques characterize individual cells in a flowing suspension by exposing them to a light source, typically a laser beam, and recording how the light scatters as a cell flows past the source. Forward light scatter (FSC) yields information about cell size while side scatter (SSC) characterizes cellular complexity (Figure 15.1). Single cell suspensions can be labeled with fluorophore-tagged antibodies or lectins or they can be loaded with fluorescent dyes that emit fluorescence as cells pass by a wavelength filtered light source. Similarly, fluorescent proteins such as green fluorescent protein (GFP) driven from the promoter of an endogenous gene can be detected in cells of interest. Detection of fluorescence from cells yields information about cellular parameters such as viability, metabolic activity, nucleic acid content, and the expression of cell surface-bound and intracellular antigens. With the use of multiple light sources and detectors along one focused stream of a single cell suspension, data for a number of cell parameters can be simultaneously and rapidly acquired, identifying distinct cell types and giving an insight into the biology of the cells in question.

Flow cytometric analysis of multiple cell parameters on individual cells at a very high speed enables the detection of subtle changes in small cell populations, such as the changes occurring as hPSCs exit self renewal and commence early differentiation and the subsequent divergence in cell fates. Modern analyzers have up to 7 lasers and are capable of measuring more than 20 fluorescent and 2 scatter parameters per cell at speeds of greater than 50,000 cells per second, generating millions of event data sets in under 1 minute. The attachment of high throughput units to analyzers further augments the possible rate of analysis across multiple samples.

An extension of flow cytometry technology is the ability to physically fractionate and enrich or deplete for cells of different phenotypes based on their detected scatter and/or fluorescence profiles using a fluorescence-activated cell sorter (FACS). In this way, cell populations with predetermined parameters can be collected into tubes or wells of a tissue culture plate for subsequent analysis or continued culture. FACS instruments use high analysis and sorting speeds to enable the enrichment of small cell populations, or the elimination of contaminating subpopulation cells, and can achieve defined parameter samples at purities >99.5% from mixed cell populations.

Magnetic-activated cell sorting (MACS) is an alternative approach to isolate desired groups of cells from heterogeneous cell populations. Through the reaction with micron- or submicron-sized magnetic beads covalently coupled to specific probes (typically antibodies or lectins), reactive cells become paramagnetic and can be gently collected in the presence of an externally-applied magnetic field. Although the number of selective probes that can be simultaneously or sequentially applied to any heterogeneous cell population for cell separation is more limited in MACS, it has advantages over FACS in maintaining the viability of the isolated cells and being easy to scale up.

This chapter describes single- and multiple-color flow cytometry protocols that use fluorescently-labeled monoclonal antibodies or lectins to detect antigen(s) present on the cell surface of dissociated live hPSCs. We describe an alternative protocol for the detection of intracellular antigens, which requires prior fixation and permeabilization of the cell membranes. The latter protocol necessarily precludes the ability to culture cells post-flow cytometric analysis or sorting, but is a useful tool for the detection of canonical transcription factors expressed in hPSCs. Both protocols have worked well in our hands and are applicable to hPSCs cultured in a variety of conditions. As a word of caution, it should be noted that single cell dissociation and FACS separation of hPSC subpopulations following immunostaining for extracellular antigens will inevitably lead to some loss of cell viability when cells are to be retrieved for further culture. We also describe an optional alternative to the protocols for flow cytometry analysis of cells using a high throughput format in 96-well plates. Finally, we provide a protocol using antibody- or lectin-mediated MACS for the isolation of viable hPSCs.

PROCEDURES

Antibodies and Lectins for Detection of Undifferentiated hPSCs

Antibodies that are reactive with hPSC surface epitopes or transcription factors associated with pluripotency are important tools for the identification, isolation, characterization, and enrichment of both hPSCs and their derivative cell types. Many of the antibodies that are conventionally used to characterize human embryonic stem cells (hESCs) and now human induced pluripotent stem cells (hiPSCs) were originally developed for epitopes on human embryonal carcinoma (EC) cell lines, and there have been only a few additional markers developed in more recent years (Table 15.1). Many of the epitopes detected by antibodies directed against the cell surface of hPSCs are not fully characterized either at the gene or protein level. Indeed, many are not proteins, but complex carbohydrate and lipid moieties for which a corresponding gene is not yet identified. There remains a paucity of good monoclonal antibodies available that specifically detect hPSCs with the markers conventionally used for characterizing hPSCs also observed to be immunoreactive in embryonic tissues, or in later stage more mature cell types (e.g. PODXL is also expressed in glomerular podocytes and POU5F1/OCT4 in germ cells). Within the discrete context of stem cell commitment and differentiation, however, the available markers are useful for the characterization and comparison of different hPSC lines and culture regimes, and enable the isolation of defined populations of live cells.

Many of the pluripotency-associated antigens on the hPSC surface are glycoproteins or glycolipids, and it has recently been reported that specific glycosylation patterns are a hallmark of undifferentiated hPSCs (Tateno et al., 2011; Wang et al., 2011). Lectins, which are glycan-binding proteins, are a useful tool to specifically recognize the carbohydrate moieties present on hPSCs. Several recent studies have confirmed the feasibility of using lectins to identify and isolate hPSCs (Tateno et al., 2011; Wang et al., 2011). Lectins offer certain advantages over antibodies in some cases. Lectins can be easily removed from cells without using a harsh stripping procedure and are significantly less expensive than antibodies. In addition, the reactivity of lectins to pluripotent stem cells appears to cross multiple species, in contrast to many antibodies, which are often species-specific (Wang et al., 2011).

While there is a group of markers that typically characterize hPSC cultures, it is important to bear in mind that there will be subsets of cells within a culture that will express varying levels of these markers. Regarding the heterogeneity displayed in hPSC cultures, we have previously determined that the use of single surface markers to detect and separate hPSCs from the early stage differentiation cell types is not as efficient as using a combination of cell surface markers (Laslett et al., 2007).

Table 15.1 Antibodies and Lectins that are Reactive with Cell Surface Markers and Transcription Factors Associated with hPSCs

Antibody/ Lectin	Isotype	hPSC Marker	Target	Source/Supplier	Literature Reference
TG343	IgM	Cell surface epitope	Keratin sulphate proteoglycan (KSPG)-protein core	Millipore MAB4346 http://www.millipore.com	Cooper et al., 2002.
TRA-1-60	IgM	Cell surface epitope	KSPG-carbohydrate side chain	Developmental Studies Hybridoma Bank http://dshb.biology.uiowa.edu	Andrews et al., 1984; Cooper et al., 2002.
TRA-1-81	IgM	Cell surface epitope	KSPG-carbohydrate side chain	Developmental Studies Hybridoma Bank http://dshb.biology.uiowa.edu	Andrews et al., 1984; Cooper et al., 2002.
GCTM-2	IgM	Cell surface epitope	KSPG-protein core (detects the same antigen as the TG343 antibody)	—	Laslett et al., 2003; Pera et al., 2003.
SSEA-3	IgM	Cell surface epitope	Globoseries glycolipid	Developmental Studies Hybridoma Bank http://dshb.biology.uiowa.edu	Kannagi et al., 1983a.
SSEA-4	IgM	Cell surface epitope	Globoseries glycolipid	Developmental Studies Hybridoma Bank http://dshb.biology.uiowa.edu	Kannagi et al., 1983b.
P1/33/2 (TG30)	IgG2a	Cell surface epitope	25-kDa tetraspanin CD9	Millipore MAB4427 http://www.millipore.com	Laslett et al., 2003; Pera et al., 2003.
PHM-5	IgG1	Cell surface epitope	Podocalyxin (PODXL); CD34 family member.	—	Kerjaschki et al., 1986.
20-202S	IgG2a	Cell surface epitope	70-kDa heat shock protein 8 HSPA8	—	Shin et al., Son et al., 2005.
OCT4 (C-10)	IgG2b	Transcription factor	OCT4/POU5F1	Santa Cruz Biotechnology sc-5279 http://www.scbt.com	Laslett et al., 2003; Pera et al., 2003.
OCT4 (C10H11.2)	IgG1	Transcription factor	OCT4/POU5F1	Millipore MAB4401 http://www.millipore.com	Zuk et al., 2009.
NANOG	IgG	Transcription factor	NANOG	R&D Systems AF1997. http://www.rndsystems.com	Hyslop et al., 2005.
UEA-I (biotinylated)	Lectin	Cell surface glycans	Fucosylated glycans	Vector Laboratories http://www.vectorlabs.com/	Wang et al., 2011

The pluripotency-associated antibodies and lectins listed in Table 15.1 fall into five main categories:

1. The stage-specific embryonic antigens (SSEA) 1, 3, and 4 are globoseries glycolipid cell surface markers recognized by monoclonal antibodies

(MAbs) originally raised to determine early stages in mouse development. hPSCs express SSEA-3 and SSEA-4 but express SSEA-1 only upon differentiation.

2. A second group of antibodies mark a set of antigens associated with a pericellular matrix keratin sulfate/chondroitin sulfate proteoglycan found on the surface of hPSCs and initially raised to human EC antigens (TRA-1-60, TRA-1-81, GCTM-2, and TG-343). These antibodies do not bind to mouse cells.

3. Prior to differentiation, hPSCs also express the 25-kDa tetraspanin molecule CD9 (P1/33/2, TG30 antigen), podocalyxin (PODXL) and heat shock 70-kDa protein 8 Isoform 1 (HSPA8).

4. Specific transcription factors are associated with the maintenance of pluripotency in hPSC and with early embryogenesis such as POU5F1/OCT4 and NANOG.

5. The fucose-specific lectin, UEA-I, recognizes fucosylated glycans that are preferentially present on the surface of pluripotent stem cells and absent in most types of differentiated human cells with few known exceptions, including erythrocytes, leukocytes, and endothelial cells. Human, mouse, drill monkey, and rhinoceros pluripotent stem cells all react with UEA-1.

Antibodies to Differentiation Markers

The most defining property of pluripotent stem cells is their ability to differentiate into all cell types in the body inclusive of the germ tissues. Differentiation involves changes in gene expression at the RNA and protein level, as well as alterations in the epigenetic marks that regulate cell function and are as yet not well understood. While gene expression parameters can be monitored as evidence of hPSC differentiation to defined lineage cell types, complete validation requires functional studies. A number of antibodies have been used to provide evidence for differentiation of hPSCs into many cell types, and indeed to achieve enrichment of such cell types. While a complete listing of antibodies that have been used to identify hPSC-differentiated cell types is beyond the scope of this chapter, comprehensive reviews of the subject can be found in Pera and Trounson (2004) and Hoffman and Carpenter (2005).

Single-Color Analyses Using Flow Cytometric Analysis and Cell Sorting
Preparation of Single Pluripotent Stem Cell Suspensions

hPSC cultures are maintained on mitotically inactivated mouse embryonic fibroblast feeder cells (MEFs) and in basal hESC medium supplemented with 20% v/v KnockOut Serum Replacement (KSR medium, see Reagents and Supplies), using standard hPSC bulk culture passaging techniques and enzymatic dissociation (refer to Chapter 1). hPSC cultures that have been maintained by manual passaging will require careful expansion and adaptation to enzymatic dissociation methods prior to preparing single cell suspensions for flow cytometry (see Chapter 3). To obtain a high yield of single, viable hPSCs for labeling and flow cytometry, enzymatically-adapted cultures are harvested with TrypLE™ Express (Life Technologies), a recombinant enzyme that is similar to porcine trypsin. With this and other enzymatic treatments, care must be taken to avoid damage to protein epitopes on the cell surface prior to immunostaining.

1. Aspirate the KSR medium from 75 cm^2 bulk culture flasks.
2. Wash cells twice with 10 mL calcium- and magnesium- free phosphate-buffered saline (CMF-PBS), aspirating after each wash.
3. Add TrypLE Express enzymatic solution to the flask (1 mL/75 cm^2 flask) and incubate for 5 minutes at 37°C.
4. Gently tap the side of the flask to dislodge hPSCs and check by phase contrast microscopy to ensure sufficient incubation such that most cells have lifted and begun dissociating into a single cell suspension.
5. Inactivate the enzyme solution with the addition of 20% Fetal Bovine Serum (FBS) v/v in DMEM/F12 medium (FD12 buffer) and pipette to wash the flask wall and dislodge hPSC colonies. Carefully dissociate the cell suspension by triturating against the flask wall, repeating the process a few times.
6. Collect the cell suspension in a tissue culture tube (Becton Dickson, 15 mL or 50 mL according to number of flasks being harvested) and repeat Step 5, washing the flask wall again and adding the second wash to the same tube.
7. Centrifuge the cell suspension at 500 g for 2 minutes, 4°C. Carefully aspirate supernatant and gently resuspend cells in 10 mL FD12 buffer.
8. Repeat wash as for Step 7 and resuspend cells in 10 mL FD12 buffer.
9. Take a 20–50 μL aliquot of suspended single cells, incubate for 5 minutes at room temperature with trypan blue vital stain in an appropriate dilution, (usually 2–5-fold), and estimate total viable single cell count using a hemocytometer chamber. Typically 15–20 × 10^6 hPSCs will be recovered from one 75 cm^2 flask. Note that the flow cytometry analyzer will also give a viability read out following labeling when cells are co-stained with the nuclear dye propidium iodide (PI).
10. Deliver required cell aliquots equally for labeling and control samples into round bottom (or V-bottom when preparing samples with very low cell numbers) 5 mL polypropylene tubes (Becton Dickson, Greiner Bio-One). We routinely use 2 × 10^5 cells/test sample in a final volume of 200 μL–2 mL in FD12 buffer.
11. Centrifuge the cell suspensions at 500 g for 2 minutes, at 4°C. Carefully aspirate supernatants and gently resuspend each pellet in 200 μL–2 mL cold FD12 buffer. These cells are now ready for labeling.

■ **Note**

For staining with UEA-I, the FBS in FD12 buffer should be replaced with ultra-purified Bovine Serum Albumin (BSA) at a final concentration of 5% to avoid the potential reaction between UEA-I and serum glycoproteins, such as immunoglobulins.   ■

■ **Note**

Handle cells gently and keep tubes on ice during the following labeling procedure and prior to flow cytometry analyses.   ■

Fluorescent Labeling of hPSCs with Extracellular Markers SSEA-3, GCTM-2, TG30 (CD9) or UEA-I for Flow Cytometry

We routinely use these antibodies and lectins singly and in combination to identify the subpopulations of cells found in hPSC cultures. Include unstained cell samples, and for antibodies, include isotype matched and secondary antibody controls for single-color analysis.

1. To a single cell suspension of hPSCs in 200 μL–2 mL of cold FD12 buffer add the primary antibody, either mouse anti-human GCTM-2 (IgM) to a final dilution typically of 1:100, mouse anti-human TG30 (IgG2a) to 1:1000, or mouse anti-human SSEA-3 (IgM) to 1:100. Alternatively, use biotinylated UEA-I at a final concentration of 6.5 μg/mL. For isotype class-matched (mouse IgM or mouse IgG2a) control antibodies the protein concentration of the specific immunoglobulin used should be equivalent to the primary antibody immunoglobulin.
2. Mix cell suspensions gently with a 1000 μL pipettor tip and incubate tubes on ice for 30 minutes to 1 hour, ideally with tubes placed horizontally on a rocker platform at slow speed.
3. Remove the primary antibody, isotype control or lectin by gentle centrifugation at 500 g for 2 minutes at 4°C and wash cells twice with 1 mL of cold FD12 buffer.

■ **Note**

When using lectins, use a buffer containing 5% ultra pure BSA in Hank's Buffered Saline Solution (HBSS) for all washes and streptavidin dilution to avoid cross-reactivity.   ■

4. After the repeat wash, incubate the cell suspension in a manner similar to Step 2, for 30 minutes in the dark (inside ice box) in 200 μL–2 mL of fluorescently-labeled appropriate secondary antibodies diluted typically to 1:500 in FD12 buffer. Secondary antibody-only controls can also be included, i.e. cell samples that are processed in the same way but not stained with the primary antibody, to determine any background

non-specific fluorophore staining. When using lectins, use fluorescently labeled streptavidin at 1:500 in HBSS with 5% ultra pure BSA.

5. Remove secondary antibodies or streptavidin by centrifugation and wash twice as described in Step 3. Resuspend cells in 300 µL–3 mL of FD12 buffer (for antibodies) or HBSS with 5% BSA (for the lectin). Keep labeled cells stored on ice prior to analysis.

6. Just prior to performing flow cytometry analysis add PI (propidium iodide) to a final concentration of 0.3 µg/mL and incubate for 10 minutes at room temperature to discriminate dead cells by their nuclear staining.

7. To remove clumps of cells, filter the final cell suspension through a 40 µm filter mesh into a new 5 mL tube using a 1000 µL pipettor tip with repeated pipetting to assist flow through the filter. Pre-wetting the filter with 50–200 µL FD12 buffer will assist with flow of the cells through the filter. Note that some loss of cell numbers will occur at this step but it is important for avoiding blockage of the flow cytometer aspiration nozzle due to cell clumps that have formed during the staining process.

8. Store on ice until analysis using an appropriate flow cytometer (see Equipment).

If plating sorted hPSC subpopulations for ongoing culture post-analysis, centrifuge cells at 240 g for 5 minutes, resuspend in KSR medium and plate cells onto mitotically inactivated MEFs as per standard hPSC conditions, supplementing with bFGF as required for the maintenance of pluripotency.

Immunostaining of hPSCs with the Intracellular Marker POU5F1 (OCT4) for Flow Cytometry (Cell Permeabilization Required)

The OCT4 transcription factor is perhaps the most widely used marker for determining the undifferentiated hPSC state (Boyer et al., 2005). Detection of the intracellularly expressed protein requires the prior fixation and permeabilization of the hPSC membrane to enable reactivity of the nuclear protein with the primary antibody. This treatment exposes all cytosolic proteins to both the primary and secondary antibodies and requires the use of a blocking agent as well as care with setting compensation levels on the flow cytometer (see Controls and Compensation) to minimize the resulting increase in background non-specific staining that can occur.

1. Prepare a single cell hPSC suspension in the same way as described above, resuspending cells in 50 µL of CMF-PBS and pipette slowly into 1–5 mL of 2% v/v paraformaldehyde in CMF-PBS in a 15 mL centrifuge tube.

2. Incubate cells in the fixative at room temperature (RT) for 30 minutes, with occasional gentle mixing, then centrifuge at 170 g for 5 minutes, RT.

3. Wash cells twice in 1–5 mL of 10% goat serum v/v in CMF-PBS (blocking buffer) and centrifuge at 170 g for 5 minutes each time at RT.

4. To permeabilize, resuspend fixed cells in 1–5 mL of 0.1% v/v Triton X-100 in CMF-PBS solution and leave at RT for 5 minutes.
5. Wash cells twice as in step 3 with the blocking buffer.
6. To block non-specific antibody binding, resuspend cells in 1–5 mL blocking buffer and incubate for 30 minutes at RT.

■ Note

Cells are now ready for intracellular immunostaining. Use the blocking buffer in all subsequent antibody incubation steps. ■

7. Resuspend fixed, permeabilized and blocked cells in 200 µL–2 mL cold blocking buffer adding the primary mouse anti-human POU5F1/OCT-4 antibody (we routinely use MAB4401, Millipore) (IgG1) to a final dilution typically of 1:200 or isotype mouse IgG1 control to an equivalent concentration.
8. Mix antibody-cell suspensions gently with a 200 µL pipettor tip and incubate tubes at 4°C for 30 minutes to 1 hour, ideally with tubes placed horizontally on a rocker platform at slow speed.
9. Remove the primary antibody or isotype control by gentle centrifugation at 500 g, 2 minutes at 4°C and wash cells twice with 1 mL of cold blocking buffer.
10. After washing, incubate the cell suspension as previously described for 30 minutes in the dark on ice in 200 µL–2 mL of an appropriate secondary antibody, e.g. AF488 or AF647 goat anti-mouse IgG1 (Life Technologies) diluted to 1:500 in the blocking buffer. As for the extracellular staining, the use of both isotype and a secondary antibody-only control will also aid in determining non-specific fluorophore binding.
11. Remove secondary antibodies by centrifugation and wash twice as described above and resuspend in 300 µL–3 mL of FD12 buffer. Keep immunostained cells stored on ice prior to analysis.

■ Note

Just prior to performing flow cytometry analysis filter the final cell suspension as described for the extracellular staining protocol. Note that as cells are now permeabilized, PI uptake cannot be used to determine cell viability. ■

■ Note

The cells are now ready for intracellular flow cytometry analysis. ■

Flow Cytometry Detection and Analysis of GCTM2, SSEA3, TG30, UEA-I or POU5F1 (OCT4) Reactivity with hPSCs

■ As the cells are aspirated through the cytometer nozzle and flow in a single cell stream past the light source, the forward (FSC) and side (SSC) light scatter are the first parameters to be established (Figure 15.1). Center the population in the dot plot by adjusting the FSC and SSC gains.

■ Run control unstained cells next (i.e. cells that have been subject to the same treatment but not labeled and used to eliminate any background autofluorescence), adjusting the FL1 photomultiplier tube (PMT) voltage until the background fluorescence is situated in the first decade (if PI is also in this tube, adjust the FL3 PMT voltages until the PI negative cells are located in the first decade of the FL3 plot.

■ Run the labeled cells through the cytometer next, collecting at least 10,000 positive events (cells) to re-analyze later as a control check following FACS sorting of cell populations.

■ To analyze the data, cells should be initially gated according to FSC and SSC light scatter, excluding the debris in the bottom left hand corner (FSC_{Lo}, SSC_{Lo}) and the cell clumps on the right hand boundary (FSC_{Hi}, suggesting large size), (Figure 15.1).

■ The PI negative events are then gated to select the subpopulation of viable cells prior to collection of appropriate emission signals for the tagged fluorophore used.

■ The unstained cell control is analyzed first as it is used to assess the amount of non-specific staining of the isotype control (Figure 15.2).

■ Analysis of the appropriate antibody isotype matched control samples (see Controls and Compensation) is used to set gates for non-specific fluorescence so that a positive or negative immunoreactive status can be correctly assigned for the cells being analyzed in the test samples.

■ The boundary of the gate should be set so the background of the isotype control readout is approximately 0.5%. Run the labeled file through these gates and read off the percentage positive cells. Representative immunostaining is shown in Figure 15.2.

Multiple-Color Analyses Using Flow Cytometric Analysis and Cell Sorting

Multiple-color flow cytometry allows assessment of a range of properties of individual cells in hPSC cultures by simultaneously detecting multiple markers, often using more than one primary antibody paired with an appropriate secondary, or a lectin in addition to an antibody. The protocol for multiple-color labeling is essentially the same as for the single marker labeling described above, except incubations for each marker are carried out simultaneously. An important exception to this is where one of the multiple antigenic targets is expressed intracellularly, in which case both the primary and secondary antibody immunostaining for the extracellular antigen(s) must be completed *prior* to fixation and permeabilization of the stained cells for the sequential intracellular staining protocol.

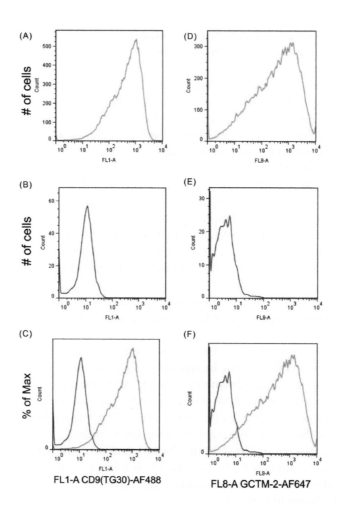

FIGURE 15.2

Multiple-color analysis of a hPSC population following two-antibody immunostaining with corresponding isotype controls. (A) Human PSC were stained with a mouse IgG anti-human CD9 antibody (TG30), which was detected with an Alexa Fluor 488 (AF488)-labeled goat anti-mouse IgG secondary antibody. The signal from the AF488 fluorophore is detected by photomultiplier tube FL1-A. An identical cell population (B) was stained with a non-specific IgG isotype control and the same secondary antibody as for A (IgG-AF488). Using the FlowJo software (version 7.2), an overlay plot (C) of A + B was created. The position of the green peak to the right of the red demonstrates that the CD9 epitope is detected in most cells in the population sample. The area under the green curve that is overlapping with the isotype control region (red) shows that a small proportion of cells are unstained for CD9 while detecting non-specific staining due to the class of antibody used. (D) hPSC were also stained with a rabbit IgM anti-human GCTM-2 antibody, which was detected with an Alexa Fluor 647 (AF647)-labeled goat anti-rabbit IgM secondary antibody. The signal from the AF647 fluorophore is detected by the photomultiplier tube FL8-A. Again, an identical population of cells (E) was stained with a non-specific IgM isotype control and the same anti-IgM antibody as for D. The overlay plot (F) of D + E shows cells that have stained specifically for GCTM-2 (green) while the area under the green curve overlapping with the isotype control region (red) shows the detection of some unstained cells and cells that have stained non-specifically for GCTM-2. In some cases unstained cells may be the result of weak immunostaining (see Pitfalls and Advice for help) but our work has shown that in this particular case there is a small population of cells that do not express CD9 or GCTM-2.

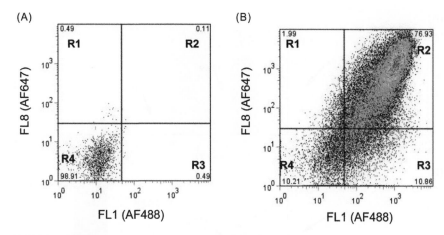

FIGURE 15.3

Alexa Fluor (AF) 488 vs AF647 staining of isotype controls (A) and antibodies (B). This is the same data set as Figure 15.2. The gates (quadrants) have been set on the isotype controls such that approximately 0.5% falls into each of the three positively-stained regions (R1: positive for AF647 only; R2: positive for AF647 and AF488, and R3: positive for AF488 only. The final region, R4, is negatively stained for both fluorochromes. Once the gates have been applied to the antibody data (B), it can be easily seen that most cells (77%) stain strongly positive for both markers. About half of the 20% AF647-negative cells are stained positive for AF488. Only a small number of cells stain with AF647 and not AF488 (R1). Proportions are automatically calculated by the FlowJo software once the gates have been set.

When using antibodies to differentiate between individual markers, a combination of primary and secondary antibodies must be used. Primary antibodies can belong to different classes (e.g. IgG1, IgG2a, IgG3, IgM) and/or be raised in different animals (mouse, rabbit, donkey, etc.) and these are detected by matched fluorophore-tagged secondary antibodies that bind the different antibody isotypes and are themselves bound to spectrally distinct fluorochromes enabling specific identification of two or more markers simultaneously on single cells by flow cytometry (Figure 15.3). For example, a combination of mouse IgG and mouse IgM primary antibodies could be distinguished by anti-mouse IgG-FITC and anti-mouse IgM-AF647 secondary antibodies. Or a combination of a rabbit IgG and a mouse IgG primary antibody could be deciphered by the use of secondary antibodies specific to rabbit and mouse IgGs, respectively.

For multiple color analyses using primary antibodies that are of the same isotype, an alternate approach is to conjugate these antibodies directly to differing fluorochromes by either using a conjugation antibody labeling kit or purchasing commercially available preconjugated antibodies (such as those available from Molecular Probes: http://www.invitrogen.com/site/us/en/home/References/Molecular-Probes-The-Handbook.html).

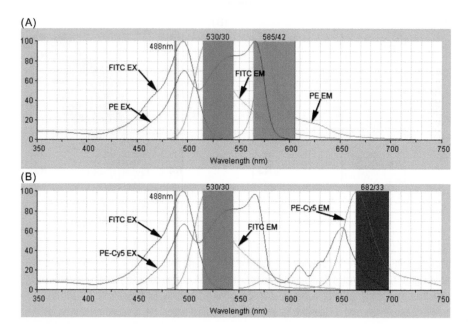

FIGURE 15.4

Choice of fluorophores has a significant impact on the complexity of data analysis of multicolour staining. (A) The choice of FITC and PE, historically one of the most common staining pairs used, shows spectral overlap: the FITC emission spectrum (FITC EM) overlaps the PE emission spectrum (PE EM) to such an extent that no appropriate filter can be chosen to eliminate the overlap (the 530/30 and 585/42 emission filter sets and FITC and PE, respectively, are the most common ones used). The advantage of using FITC and PE is that the same laser excitation source, 488 nm, may be used to excite both fluorophores simultaneously (see FITC EX and PE EX for their respective excitation spectra). (B) One solution to the problem, by using the PE-Cy5 fluorophore, which has the same excitation spectrum (PE-Cy5 EX) as PE but an emission spectrum (PE-CY5 EM) well outside that of FITC, one can still use 400 nm to simultaneously excite both fluorophores but also use a different filter set, 682/33, that will pick up PE-Cy5 emission but not FITC emission. Both panels are screen prints of data generated at the BD Biosciences website, an extremely useful site for visualizing spectra characteristics of many different fluorophores. Alternatively, computerized compensation can use proportions to calculate the amount of signal coming from undesired fluorophores one by one and subtract it/them from the total, leaving only signal from the fluorophore of interest. It is good practice to use spectrally distinct fluorophores where possible to minimize reliance on such compensation techniques.

Multiple conjugated primary antibodies are incubated simultaneously with cells, or the antibodies can be used for sequential immunostaining and wash steps. Where incubation is simultaneous, controls for single-color fluorescence must be included to ensure no cross-reactivity has occurred.

Spectrally distinct fluorochromes are required for multi-color staining (Figure 15.4), however the flow cytometry software can compensate for a small degree of spectral overlap. For the accurate analysis of all multiple-color immunostaining flow cytometry data, single-color control cell samples must always be included to determine appropriate compensation levels for each channel being used for the combined analysis.

High Throughput Screening (HTS) Unit

Allowing the adaptation of 96-well plates to the flow cytometer analyzer, high throughput screening (HTS) units will save time when a large number of cell samples (>50) are required to be analyzed sequentially using the same parameters. The BD Biosciences HTS unit requires a 15-minute wash procedure before and after use regardless of how many plates are run, and can analyze a full 96-well plate in approximately 15 minutes. This protocol is best suited to rapid analysis (not cell sorting) of a large number of hPSC samples for detection of pluripotency markers following experimental variation of culture conditions, detection of reactive hybridoma supernatants or in screens using reporter hPSC lines to determine pluripotency or differentiation outcomes.

Cell Preparation

1. Prepare a single cell hPSC suspension in FD12 buffer as described in the above protocols.
2. Aliquot 2×10^5 cells in a 200 μL volume into each well of 96-well U-bottom plate(s) using a multipipettor. Technical replicates are recommended for each sample being tested. Control sample wells should also be included for cells known to react with the primary antibody(ies), as well as the unstained cells, and matched isotype control samples.
3. Centrifuge the cell plate at 1000 g for 2 minutes at RT.
4. Immediately remove the supernatants by quickly flipping the plate over the sink. Grasp the plate in one hand from beneath and briefly turn it upside-down with a jolt. This action can be slightly unnerving at first, but cell pellets should remain intact at the bottom of the wells.

■ Note

The longer cells are left prior to removal of supernatant (>1–2 minutes), the more cells will be lost in the flipping action. If need be, recentrifuge the cells for an additional minute to avoid loss of cell pellets. ■

1. Resuspend hPSCs in 50 μL of primary antibody(ies) as per experimental design (e.g. mouse anti-human TG30 and mouse anti-human GCTM2) diluted with 10–20% FBS v/v in CMF-PBS (flow buffer). Leave the cell plate to incubate for 15–30 minutes at RT. Lectins can also be used, see notes and protocols above.
2. To wash and remove the primary antibody(ies): add 150 μL of FD12 buffer into each well using a multipipettor, then centrifuge the plate at 1000 g for 2 minutes.

3. Remove the supernatant by a quick plate flip as above.
4. Resuspend cells in 50 μL secondary antibody (e.g. AF647 goat anti-mouse IgG2a and AF488 goat anti-mouse IgM) diluted with flow buffer typically to 1:500. Incubate plate in the dark for 15–30 minutes at RT.
5. Wash cells as described in step 6.
6. Remove the supernatant by a quick plate flip as above.
7. Resuspend the cells into 150 μL flow buffer containing 0.30 μg/mL PI. The cells are now ready for flow cytometric analysis.

■ Note

The intracellular immunostaining protocol described earlier may be able to be adapted for the HTS protocol if required although we do not have specific experience with this in our laboratory.   ■

The HTS Hardware (Starting Up)

Instructions are given here for use of the BD HTS unit connected to a BD LSRII flow cytometer (Figure 15.5), although other cytometers and HTS attachments are available.

Ensure that the sheath water tank is full and the waste tank is less than half-full. Set the flow rate to "low" and the fluidics system to "standby" and remove the tube of deionized water that keeps the probe wet. Place the HTS unit onto the brace attached to the front of the cytometer (be careful, the unit is quite heavy for its size) and connect the HTS unit to the cytometer. There are five connection points between the cytometer and the HTS attachment, these are the sheath line (white), the waste line (orange), the power cable, the communication cable and the sample coupler (see Figure 15.5).

1. Once the HTS unit has been attached and turned on (switch on the right side of the unit), set the cytometer fluidics to "run" and "low."
2. Flick the switch at the side of the cytometer interface from the tube image to the plate image.
3. Prime the HTS unit (software menu: HTS→Prime)
4. Wash the HTS unit. Obtain a clean 96-well plate and fill 2 × 4 wells (A1–A4, B1–B4) with 2% bleach (*wells need to be completely full as 200 μL will be used for each wash*), and 2 × 4 wells (C1–C4, D1–D4) with deionized water.
5. Place the plate on the stage in the correct orientation with A1 in the top-right corner. Software menu: HTS→Wash→Daily Clean. This wash cycle will take approximately 15 minutes; the user may wish to also wash with detergent.

The HTS Software (Sample Acquisition)

■ The plate layout grid dictates the selection and sequence of wells to be analyzed. Wells can be annotated as "controls" or "samples." Wells that are allocated neither title will not be sampled from.

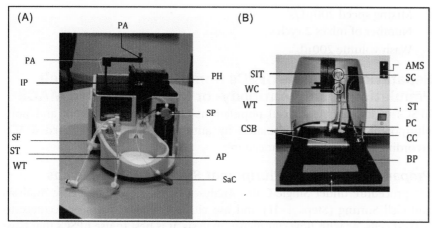

FIGURE 15.5

Adaptation of the BD LSRII flow cytometer for a 96-well plate readout format using the BD High Throughput Sampler (HTS) unit. The HTS unit (A) is supported in the LSRII port (B) by the base plate (BP) and the cytometer support bracket (CSB). From the LSRII port the power (PC) and communication cables (CC) control the electronics of the HTS unit. The 96-well plate sits atop the mobile plate holder (PH) and samples are acquired by the probe assembly (PA) and transported to the injection port (IP). The secondary pump (SP) then pushes the samples through the sample coupler (SaC) into the sample injection tube (SIT) of the cytometer. Sheath fluid is filtered by the sheath filter (SF) as it travels from the sheath connector (SC) of the cytometer to the HTS unit along the clear sheath tubing (ST). Waste is transported back to the waste connector (WC) of the cytometer through the orange waste tubing (WT). When the HTS is turned on, the acquisition mode switch (AMS) should be set to plate mode. The absorbent pad (AP) on the HTS unit catches any drips from the SIT during set-up and should not be required during regular use. The sheath and waste tubes are present in both images but there should only be one of each. *Adapted from the BD High Throughput Sampler Users Guide for the BD LSRII. http://facs.stanford.edu/sff/doc/BDLSRIIUserGuide.pdf*

- While using this grid, volume and flow rates for individual wells can also be determined, for both sample acquisition and washes.
- There are two modes of acquisition to choose from, "Standard Mode" and "High Throughput Mode." The standard mode has a flushing step between samples so it is slightly slower. Using the HTS acquisition mode poses a risk of well-to-well contamination from the probe, but in our experience this has not been a problem.

We use high throughput mode with the following volumes and flow rates, however sample volumes as low as $2\,\mu L$ can be analyzed:

- Sample flow rate $1\,\mu L/s$
- Sample volume $5\,\mu L/s$
- Mixing volume $50\,\mu L$

- Mixing speed 200 μL/s
- Number of mixes 2 cycles
- Wash volume 200 μL.

Separation of Viable hPSCs from Heterogeneous Cell Populations Using Antibody- or Lectin-Mediated MACS

hPSCs in heterogeneous cell populations containing pluripotent and non-pluripotent cells can be isolated by antibody- or lectin-mediated MACS according to the following procedure.

Preparation of Single Pluripotent Stem Cell Suspensions

See procedure under Single-Color Analyses Using Flow Cytometric Analysis and Cell Sorting (Steps 1–11) and use either TrypLE Express or accutase to harvest cells. As with flow cytometric analysis, it is best to use hPSCs that have been adapted to enzymatic passage (see Chapter 3).

Labeling of hPSCs with Extracellular Markers TG343 or UEA-I for MACS

We have used these probes singly to identify pluripotent cells found in hPSC cultures. Include unstained cell samples, isotype matched and secondary antibody controls for MACS with any of these markers:

1. To a single cell suspension of hPSCs in 200 μL–2 mL of cold FD12 buffer add the TG343 antibody to a final dilution typically of 1:500. For isotype class-matched (mouse IgM or mouse IgG2a) control antibodies the concentration of the specific immunoglobulin used should be equivalent to the primary antibody immunoglobulin with respect to protein concentration. For biotinylated UEA-I, use at a final concentration of 6.5 μg/mL in DMEM/F12 with 5% ultra purified BSA.
2. Mix antibody (or lectin)-cell suspensions gently with a 1000 μL pipettor tip and incubate tubes on ice for 45 minutes, ideally with tubes placed horizontally on a rocker platform at slow speed.
3. Remove the primary antibody, isotype control or biotinylated UEA-I by gentle centrifugation at 500 g for 2 minutes at 4 °C. Wash cells twice with 1 mL of either cold FD12 buffer (for antibodies) or cold HBSS with 5% ultra pure BSA (for the lectin).
4. React cells labeled with TG343 (or control antibodies) from Step 3 with biotinylated anti-mouse IgM Fab fragments on ice for 45 minutes, ideally with tubes placed horizontally on a rocker platform at slow speed. UEA-1 labeled cells do not need to be reacted with biotin because the lectin is already biotinylated. After the reaction, centrifuge samples at 500 g for 2 minutes at 4 °C and wash cells twice with 1 mL of HBSS with 5% ultra-purified BSA to avoid the potential reaction between streptavidin and free biotin in the FD12 buffer.

■ **Note**

> If co-staining with UEA-I and TG343 we recommend staining with UEA-I first and then with TG343.   ■

5. Incubate the cell suspension with anti-biotin microbeads (Miltenyi Biotec) according to the manufacturer's instructions.
6. Prepare the MACS LS Magnetic Columns (Miltenyi Biotec) for harvesting the beads from the cell suspension according to the manufacturer's instructions.
7. Pour the cell suspension through the magnetic columns. The cells captured by the magnetic beads will remain trapped in the columns until the columns are released from the magnet. Wash the columns with HBSS with 5% ultra pure BSA three times.
8. Release the columns from the magnet and flush the cells out of the columns with 5 mL of hESC medium.
9. Store on ice until cells are ready for analysis or subculture.

ALTERNATIVE PROCEDURES

Alternative Fixing Techniques

0.2% Glutaraldehyde or 4% paraformaldehyde (PFA) can also be used for fixing cells, however over-fixing with PFA can result in cross-linking that may obscure important epitopes and may also compromise ultrastructural integrity. For some intracellular antigens, cells may be able to be fixed and permeabilized adequately with cold methanol or ethanol, but each individual MAb will have to be validated for use with this type of fixation. Note that fixing hPSCs in 4% PFA is compatible with UEA-I staining.

Use of Reporter hPSC Lines

Sorting or analyzing genetically modified hPSCs that express a fluorescent reporter gene under the control of a pluripotency associated promoter (e.g. Oct4-eGFP cells) negates the need for any staining techniques, or can be used in conjunction with additional markers. Such cells can be simply dissociated to a single cell suspension and fed through the cytometer, using an appropriate spectral wavelength to gate fluorescing cells.

Alternative High Throughput Systems

Beckman Coulter sells a similar device to the BD HTS unit, the Hypercyt plate loader, designed to attach to their flow cytometers. Specifications list an ability to sample up to 40 wells per minute and compatibility with 384-well plates and the Beckman Coulter analysis software (http://www.coulterflow.com/bciflow/instruments16.php). We have not used this system in our lab.

PITFALLS AND ADVICE

Background Staining

A high level of background staining tends to be the result of non-specific antibody binding to the hPSCs. This is particularly pronounced with intracellular staining (e.g. POU5F1/OCT4) following the fixation and permeabilization of the cell membrane and exposure of all cytosolic proteins to the antibodies (both primary and secondary). The following approaches can be used to counter excessive background staining.

- Keep antibodies stored appropriately according to manufacturers' instructions and use freshly prepared dilutions on the day of the experiment. Over time, antibodies will degrade and increase the incidence of background and non-specific staining.
- Centrifuge the antibodies in a microfuge and carefully remove supernatant to avoid precipitates in aliquots. Alternatively, antibodies may be filtered prior to dilution.
- Optimize the concentrations of both primary and secondary antibodies for your experiment – you may need to lower concentrations for either or both antibodies.
- Increase the concentration of FBS in the wash and incubation buffers up to 20–30%. We use a high concentration of 20% especially for protection of FACS sorted cells intended for continuing culture, but usually concentrations down to 2–4% can be used for flow cytometry hPSC analyses. Too high a serum or protein concentration can lead to interference in the flow cytometry readout.
- Increase the dilution of either or both antibodies.
- Change the type of blocking serum (e.g. serum from secondary antibody host species). We have also used goat serum at 4% in CMF-PBS for the blocking buffer for many of our immunostaining protocols.
- Generally, where available, a monoclonal antibody will give greater specificity than a polyclonal antibody for the same antigen target.
- Ensure inclusion of secondary antibody-only control samples in experiments to assess and gate out background staining.
- Test a different clone of the antibody and different suppliers for the same antigen of interest.

Weak Staining

- Always determine the activity and specificity of new antibodies by testing on known positive and negative cell controls (e.g. characterized hPSC lines and MEF feeder cells).
- Test another antibody or batch of antibody to the same antigen. Titrate the primary antibody(ies) over a range of dilutions and/or increase the

secondary antibody concentration. Use an older antibody batch, known to be specific, as a positive control for determining optimal concentration of a new batch.

■ Increase incubation time for the primary antibody.
■ Use a fluorochrome that will yield brighter fluorescence (e.g. AF647 or AF488 to replace FITC).
■ Agitation/gentle rocking of cell suspensions during immunostaining incubation periods will aid in preventing cell clumping (particularly when staining large numbers of cells) and optimize consistency in cell contact with the antibodies.
■ Increase the PMT voltage for the fluorescence detector so the unstained cells sit outside the first decade. This can have the effect of achieving the flow cytometer's best resolution sensitivity. If this is the case you should see an increase between the ratio of the positive and negative cells compared with the lower settings.

Over Staining
■ Reduce the primary or secondary antibody concentration.
■ Shorten/optimize the antibody incubation times.
■ Decrease the PMT voltage for the fluorescence detector to place the brightest cells on scale.

Lectins
■ Be sure that there are no glycoproteins in any of the reagents used other than your cells or else the lectin may bind non-specifically. Do not use FBS as a blocking agent for this reason. Instead, use ultra pure BSA.
■ If you are using streptavidin, be sure that the buffers you are using do not contain biotin. Note that DMEM/F12 and many other cell culture media do contain biotin.
■ When labeling cells with both lectins and antibodies, we recommend doing the staining sequentially with the lectin first, then the antibody.

Controls and Compensation
■ The unstained cell control sample is the first reference. The cells in this tube are subjected to the same immunostaining procedures (i.e. trituration, washes, centrifugations and incubation conditions) but without the addition of primary or secondary antibodies. The unstained cell control is used to set the level of background autofluorescence on the cytometer and enables determination of any non-specific staining in the subsequent isotype control sample.
■ The isotype-matched control is used to determine the amount of non-specific staining due to the class of antibody used. The concentration

of subtype specific immunoglobulins used should be equivalent to that used for the isotype matched primary antibody with respect to protein concentration.

- Compensation: when using the dyes FITC and PI, a FL1 vs FL3 dot plot should be used to monitor the amount of FITC spill over in the PI detector due to the proximity of the wavelengths. If the level of FITC fluorescence is high, it will bleed through to the PI detector, making the cells appear to be PI positive (i.e. non-viable). A small amount of FL3– %FL1 will bring the FITC-high events back into the viable gate.
- When performing multiple-color analysis, each fluorochrome used must have a separate single-color control tube using the same concentration of each primary antibody as in the multiple-staining sample. Compensation is applied to the neighboring channels as each sample is run through the cytometer to minimize or eliminate any spill-over.
- Where possible, choose fluorochromes to minimize or eliminate spectral overlap.
- Fixed gates for repeated experiments can be identically set using 8-peak Ultra Rainbow calibration beads (Spherotech™) to enable reliable comparison between experiments and between alternative FACS sorter instruments.

Advice for Use of the HTS Mode

- False positive staining can appear in wells adjacent to positive samples/ controls when using the 96-well plate mode if care is not taken with pipetting technique.
- Cell pellets may be lost during the "flipping" supernatant removal process if the plate is left to sit for too long after centrifugation. Be prepared for your next step before centrifuging the plate so this delay can be avoided.
- The LSRII software program will change the volumes of wells one-at-a-time. To change volumes for the whole plate be sure this is pre-set for all wells prior to starting analysis.

EQUIPMENT

Flow Cytometer

- There are many different flow cytometers equipped with multiple combinations of laser beams and detectors. The minimum requirement for the protocols listed here is a three-color laser machine with the following filters:
  - 530 nm for FITC
  - 580 nm for phycoerythrin (PE)
  - Red filter >610 nm for PI

- We routinely use the BD LSRII for flow cytometry analyses and for sorting the Becton Dickson FACSVantage Diva. For the protocols described here we routinely use FITC, PE, AF488 and AF647 fluorochromes. If there is too much spectral overlap between the commonly paired FITC and PE fluorochromes, a multi-laser cytometer with an additional 633 nm laser to detect fluorochromes such as APC and AF633 is required.
  - High Throughput Screening (HTS) attachment. We use a BD HTS unit that adapts to the LSRII instrument.
  - Class II Biohazard hood for tissue culture work.
  - Multipipettor for immunostaining in 96-well format.

REAGENTS AND SUPPLIES

Item	Supplier	Catalog No.	Alternative
TrypLE Express	Life Technologies	12604	
Dulbecco's Phosphate Buffered Saline without Mg++, Ca++ (CMF-PBS)	Life Technologies	14190	HyClone SH30028
DMEM/F12 1:1 (DMEM/F12)	Life Technologies	11320	
KnockOut Serum Replacement (KSR)	Life Technologies	10828	
L-Glutamine	Life Technologies	25030	Glutamax (Life Technologies)
Non-Essential Amino Acids (NEAA)	Life Technologies	11140	
β-Mercaptoethanol	Life Technologies	21985	
Goat Serum	Life Technologies	16210	
Fetal Bovine Serum (FBS)	SAFC®	12203C	JRH Biosciences Cat# 12003-500M
Dissociation Reagents TrypLExpress™	Life Technologies	12604	TrypLE Select or Cell Dissociation Buffer (Life Technologies)
Propidium Iodide (PI)	Sigma Aldrich	P4864	
Glass Pasteur pipettes (230 mm)	Pacific Laboratory Products	PP/900/250	
1 mL sterile serological pipettes	BD Biosciences	357521	
2 mL sterile serological pipettes	BD Biosciences	357507	
5 mL sterile serological pipettes	BD Biosciences	357543	
10 mL sterile serological pipettes	BD Biosciences	357551	
25 mL sterile serological pipettes	BD Biosciences	357525	
50 mL sterile serological pipettes	BD Biosciences	357550	
10–1000 μL sterile aerosol barrier pipette tips	Interpath Services	24100-24800	
15 mL sterile centrifuge tubes	Grenier Bio-one	188 271	
50 mL sterile centrifuge tubes	Grenier Bio-one	227 261	
5 mL sterile polystyrene tubes w/40 μm cell strainer caps	BD Biosciences	352235	
5 mL sterile polystyrene tubes	BD Biosciences	352035	

Item	Supplier	Catalog No.	Alternative
5 mL non-sterile polystyrene tubes	BD Biosciences	352008	
96-well U-bottomed plates for HTS	Grenier Bio-one	650-101	
MidiMACS Separator	Miltenyi Biotec	130-042-302	
LS columns	Miltenyi Biotec	130-042-401	
Anti-Biotin Microbeads	Miltenyi Biotec	130-090-485	

Working Solutions
FD12 and Flow Buffers

For harvesting the single cell hPSCs and for immunostaining protocols where cells will be returned to culture post-flow analysis, a working buffer solution of 20% v/v FBS in DMEM/F12 medium (FD12) is used. Alternatively, where hPSC integrity post-flow analysis is not critical, a buffer comprising 10-20% v/v FBS in CMF-PBS (flow buffer) can be substituted for harvesting cells, and for the immunostaining wash steps and antibody dilutions.

KnockOut Serum Replacement (KSR) Medium

- 70% v/v DMEM/F12 medium
- 20% v/v KSR
- 0.1 mM NEAA
- 2 mM L-glutamine
- 90 μM β-mercaptoethanol
- Supplemented daily with 10 ng/mL bFGF (Millipore) for the maintenance of pluripotent cultures.

Blocking Buffer

10% v/v goat serum in CMF-PBS

Antibodies and Lectins

Antibody/lectin	Type	Dilution	Source	Catalog #
Mouse anti-human TG30 (CD9)	IgG2a	1:1000	The Australian Stem Cell Centre (ASCC) or Millipore	MAB4427
Mouse anti-human GCTM-2 (~0.4 mg/mL)	IgM	Hybridoma supernatant 1:50–1:100*	Kind gift from Professor Martin Pera	
Mouse anti-human Oct4	IgG1	1:200	Millipore	MAB4401
Mouse IgG1 isotype control (1 mg/mL)	IgG1	Equivalent to matched primary MAb	BD Pharmingen	553447

Antibody/lectin	Type	Dilution	Source	Catalog #
Mouse IgM isotype control (0.5 mg/mL)	IgM	Equivalent to matched primary MAb	BD Pharmingen	553472
Mouse IgG2a isotype control (0.5 mg/mL)	IgG2a	Equivalent to matched primary MAb	BD Pharmingen	554121
Alexa Fluor 488-conjugated goat anti-mouse IgG	IgG (H + L)	1:500	Life Technologies	A11029
Alexa Fluor 647 goat anti-mouse IgG	IgG (H + L)	1:500	Life Technologies	A21236
Alexa fluor 647 goat anti-mouse IgM	IgM	1:1000	Life Technologies	A21238
Alexa Fluor 488 goat anti-mouse IgG2a	IgG2a	1:500	Life Technologies	A21131
R-phycoerythrin goat anti-mouse IgG1 conjugate	IgG1	1:2000	Life Technologies	P21129
Biotinylated Ulex Europaeus Agglutinin I (UEA-I)	Lectin	6.5 µg/mL	Vector Laboratories	B-1065
Alexa Fluor 488-streptavidin	N/A	1:500	Life Technologies	S11223

*as per titration of each new batch of purified hybridoma protein. Millipore's TG343 antibody (MAB4346) detects the same protein and can be used at 1:200 for flow cytometry.

Various other fluorophore combinations can be used for the isotyped antibodies, but the above list includes secondary antibodies that we routinely use.

All antibodies are diluted either in FD12 buffer or the blocking buffer where one or more intracellular antigens are to be stained.

READING LIST

Andrews, P.W., Banting, G., Damjanov, I., Arnaud, D., Avner, P., 1984. Three monoclonal antibodies defining distinct differentiation antigens associated with different high molecular-weight polypeptides on the surface of human embryonal carcinoma cells. Hybridoma 3, 347–361.

Boyer, L.A., Lee, T.I., Cole, M.F., Johnstone, S.E., Levine, S.S., Zucker, J.R., et al., 2005. Core transcriptional regulatory circuitry in human embryonic stem cells. Cell 122, 947–956.

Cooper, S., Bennett, W., Andrade, J., Reubinoff, B.E., Thomson, J., Pera, M.F., 2002. Biochemical properties of a keratan sulphate/chondroitin sulphate proteoglycan expressed in primate pluripotent stem cells. J. Anat. 200, 259–265.

Hoffman, L.M., Carpenter, M.K., 2005. Characterization and culture of human embryonic stem cells. Nat. Biotechnol. 23, 699–708.

Hough, S.R., Laslett, A.L., Grimmond, S.B., Kolle, G., Pera, M.F., 2009. A continuum of cell states spans pluripotency and lineage commitment in human embryonic stem cells. PLoS One 4, e7708.

Hyslop, L., Stojkovic, M., Armstrong, L., Walter, T., Stojkovic, P., Przyborski, S., et al., 2005. Downregulation of NANOG induces differentiation of human embryonic stem cells to extraembryonic lineages. Stem Cells 23, 1035–1043.

Kannagi, R., Levery, S.B., Ishigami, F., Hakomori, S., Shevinsky, L.H., Knowles, B.B., et al., 1983a. New globoseries glycosphingolipids in human teratocarcinoma reactive with the monoclonal antibody directed to a developmentally regulated antigen, stage-specific embryonic antigen 3. J. Biol. Chem. 258, 8934–8942.

Kannagi, R., Cochran, N.A., Ishigami, F., Hakomori, S., Andrews, P.W., Knowles, B.B., et al., 1983b. Stage-specific embryonic antigens (SSEA-3 and -4) are epitopes of a unique globoseries gangliosideisolated from human teratocarcinoma cells. EMBO J. 2, 2355–2361.

Kerjaschki, D., Poczewski, H., Dekan, G., Horvat, R., Balzar, E., Kraft, N., et al., 1986. Identification of a major sialoprotein in the glycocalyx of human visceral glomerular epithelial cells. J. Clin. Invest. 78, 1142–1149.

Kolle, G., Ho, M., Zhou, Q., Chy, H.S., Krishnan, K., Cloonan, N., et al., 2009. Identification of human embryonic stem cell surface markers by combined membrane-polysome translation state array analysis and immunotranscriptional profiling. Stem Cells 27, 2446–2456.

Laslett, A.L., Filipczyk, A.A., Pera, M.F., 2003. Characterization and culture of human embryonic stem cells. Trends Cardiovasc. Med. 13, 295–301.

Laslett, A.L., Grimmond, S., Gardiner, B., Stamp, L., Lin, A., Hawes, S.M., et al., 2007. Transcriptional analysis of early lineage commitment in human embryonic stem cells. BMC Dev. Biol. 7, 12.

Oka, M., Tagoku, K., Russell, T.L., Nakano, Y., Hamazaki, T., Meyer, E.M., et al., 2002. CD9 is associated with leukemia inhibitory factor-mediated maintenance of embryonic stem cells. Mol. Biol. Cell 13, 1274–1281.

Pera, M.F., Trounson, A.O., 2004. Human embryonic stem cells: prospects for development. Development (Cambridge, England) 131, 5515–5525.

Pera, M.F., Blasco-Lafita, M.J., Cooper, S., Mason, M., Mills, J., Monaghan, P., 1988. Analysis of cell-differentiation lineage in human teratomas using new monoclonal antibodies to cytostructural antigens of embryonal carcinoma cells. Differentiation 39, 139–149.

Pera, M.F., Filipczyk, A.A., Hawes, S.M., Laslett, A.L., 2003. Isolation, characterization, and differentiation of human embryonic stem cells. Methods in Enzymol. 365, 429–446.

Shin, B.K., Wang, H., Yim, A.M., Le Naour, F., Brichory, F., Jang, J.H., et al., 2003. Global profiling of the cell surface proteome of cancer cells uncovers an abundance of proteins with chaperone function. J. Biol. Chem. 278, 7607–7616.

Son, Y.S., Park, J.H., Kang, Y.K., Park, J.S., Choi, H.S., Lim, J.Y., et al., 2005. Heat shock 70-kDa protein 8 isoform 1 is expressed on the surface of human embryonic stem cells and downregulated upon differentiation. Stem Cells 23, 1502–1513.

Wang, Y.-C., Nakagawa, M., Garitaonandia, I., Slavin, I., Altun, G., Lacharite, R.M., et al., 2011. Specific lectin biomarkers for isolation of human pluripotent stem cells identified through array-based glycomics analysis. Cell Res. 21 (11), 1551–1563.

Yamanaka, S., 2009. A fresh look at iPS cells. Cell 137, 13–17.

Zuk, P.A., 2009. The intracellular distribution of the ES cell totipotent markers OCT4 and Sox2 in adult stem cells differs dramatically according to commercial antibody used. J. Cell Biochem. 106, 867–877.

CHAPTER 16

Immunocytochemical Analysis of Human Stem Cells

Jamison L. Nourse, Boback Ziaeian, Theo Palmer, Philip H. Schwartz, and Lisa A. Flanagan

EDITOR'S COMMENTARY

Immunocytochemistry, which is the process of labeling cells with antibodies, is currently the best way to determine whether a population of cells is homogeneous or heterogeneous with regard to a particular molecular marker. Immunocytochemistry allows for the visualization of individual cells within a colony or culture and thus provides an overall assessment of the distribution of expression of a particular marker throughout the culture under specific culture conditions. For this reason, it is a valuable tool to complement biochemical assays that cannot discriminate individual cells in a population (such as immunoblots, PCR, and microarrays). In contrast to flow cytometry, which detects levels of expression of antigens in a population of cells in suspension, immunocytochemistry reveals the subcellular localization of the antigen and takes into account cell morphology. The success of immunocytochemistry is dependent on the quality of the antibodies used as well as the technique. This chapter illustrates how well immunocytochemistry works when it works well, and describes solutions to common problems that all labs encounter when analyzing human pluripotent stem cells (hPSCs) and their derivatives.

OVERVIEW

Antibodies can recognize antigens that are protein, glycolipid (such as the SSEA-4 epitope), carbohydrate, small molecule, or DNA. Staining of cells by a particular antibody is often described as "weakly positive" or "strongly positive." If a new antibody or cellular sample gives only a weakly positive signal, it is essential to confirm the presence of the antigen using another method. If the antigen is a protein, it is usually sufficient to demonstrate the presence in the same cells of the specific coding transcript, using a method such as RT-PCR. Other methods used for confirmation of antibody staining include

J.F. Loring & S.E. Peterson (eds): Human Stem Cell Manual, Second edition.
DOI: http://dx.doi.org/10.1016/B978-0-12-385473-5.00016-3

the use of a second antibody that recognizes another epitope on the same molecule, and immunoblots (Western blots), in which molecular weight information revealed by SDS gel electrophoresis adds confirmation of antigen identity.

Immunocytochemistry for cultured cells uses an amplification technique to make submicroscopic molecules visible. Ideally, every experiment includes negative controls (such as no primary antibody) and positive controls (such as a cell type known to express the antigen) in order to assess the efficacy of staining.

This chapter will describe the most popular immunocytochemical method in the stem cell field – using fluorescently tagged secondary antibodies to detect the primary antibody that is bound to an epitope on the molecule of interest. Secondary antibodies recognize the heavy chain of the primary antibody's isotype. Generally, these methods employ a long incubation period for the primary antibody, a series of washes to remove unbound antibody, and a shorter incubation for the secondary antibody, followed by washes and preparation for microscopy.

Primary antibodies vary widely in their binding affinities and specificities and must be tested to determine whether they recognize the antigen when the specimen is prepared for immunocytochemistry. Antibodies bind to specific epitopes, which may be short stretches of amino acids in a protein, conformational characteristics like an exposed alpha helix, or structural elements of a small molecule. Polyclonal antibodies contain multiple antibodies that usually recognize several different epitopes on a molecule. In contrast, monoclonal antibodies are of a single antibody type and recognize a single epitope.

Key Considerations
- Primary antibody: epitope(s) recognized
- Secondary antibody: match to the isotype of the primary antibody
- Fixation method: affects the accessibility of the epitope for the primary antibody
- Permeabilization: if the epitope of interest is located inside the cell, the membrane must be permeabilized to allow antibody entry. If the epitope is on the cell surface, permeabilization may interfere with its detection.

PROCEDURES

The protocols described below are easy, routinely give publication-quality photos, and can be done by devoting only a short time each day to the protocol. If rapid analysis is desired, the alternative protocol can be used, with timing indicated at the end of each section.

Choosing the Right Antibodies

Most fluorescence microscopes have the ability to excite and discern several unique fluorochromes using various optical filters. In designing a plan for staining for more than one antigen, it is important to select primary antibodies of a unique species or subtype (e.g. mouse IgG, mouse IgM, rabbit IgG, goat IgG, chicken IgY, guinea-pig IgG, rat IgG). If the primary antibodies for different antigens are from the same species and subtype, secondary antibodies will indiscriminately bind to both markers.

Preparation of Samples
Growth of Cells on a Glass Surface

Several days prior to staining, passage the cells to chamber slides or sterile glass coverslips with the appropriate substrata (extracellular matrix such as laminin or feeder layer of cells) so that the cells will adhere to the surface and not wash off during the staining process.

■ Note

Plastic dishes scatter light and are often autofluorescent, so fluorescent antibody staining on plastic culture dishes is not advised. ■

Chamber Slides or Glass Coverslips

There are practical considerations that dictate whether to use chamber slides or coverslips in immunostaining. An advantage of employing glass coverslips instead of chamber slides is the reduced amount of antibody required, up to 25-fold less, which is important if the antibodies are expensive or limited in availability. In addition, coverslips can be grown in 12- or 24-well plates, which allows fixation of individual samples on different days by transferring coverslips to fixative in a separate plate. This is useful for experiments that involve a time course. In contrast, chamber slides require all chambers to be fixed on the same day. *Chamber slides are more expensive but come ready to use, whereas coverslips require some preparation prior to use.* Chamber slides are also easier to use when performing the incubation steps of the immunostaining protocol because they do not involve additional physical manipulation steps that the coverslip method requires.

Detection of Proliferation: BrdU or EdU Labeling

The proliferative ability of cells can be examined at the single cell level by incorporation of modified nucleosides during cell culturing, which are later detected and visualized with fluorescent reagents. BrdU labeling employs antibodies to detect BrdU, but requires a denaturing step, or treatment with HCl, to allow antibody access to the epitope. This step can interfere with other antibodies or perturb cell morphology. EdU incorporation is detected with "Click" chemistry (Invitrogen), which does not require a denaturing step and is compatible with the immunostaining protocol.

For proliferation experiments, BrdU or EdU ($10\,\mu M$ final concentration) should be incubated with the cells for 1–24 hours prior to fixation. In some cases it will be desirable to remove the nucleoside-containing medium and culture the cells in regular medium for a few days before fixation.

■ Note

BrdU-labeled cells should be treated with HCl (1N HCl for 20–30 minutes at $37\,°C$) after fixation but prior to blocking and antibody incubation. Wash well with PBS after HCl incubation.   ■

Fixation

Fixation with paraformaldehyde (PFA) allows retention of cellular structures and has lower autofluorescence than other fixatives. Generally a simple PFA solution is sufficient for many antibodies, however, additional ingredients can improve the staining results for some antibodies. Including $MgCl_2$ to stabilize microtubules, and EGTA to inhibit Ca^{2+} dependent proteases, helps maintain cytoskeletal structure. Sucrose preserves cell morphology by providing osmotic balance. Including saponin, a cholesterol disrupting detergent, in the fixative can be used to permeabilize the membrane without compromising some membrane proteins, and allows for dual staining of external and internal proteins. Finally, including fluorescently-labeled phalloidin, a stabilizer of actin filaments, improves retention of cytoskeletal structure and allows visualization of the actin cytoskeleton.

Immunostaining Procedure
Fixation of the Cells

1. On the day of staining, carefully aspirate the growth medium and rinse cells once with prewarmed PBS (optional). Importantly, the cells should never be allowed to dry out, so you should not completely aspirate all the liquid from the well and you should always have the next solution at hand to add immediately after aspiration.
2. Fix cells for 10 minutes at room temperature with 4% paraformaldehyde in PBS (see Recipes). Dispense the solution down the side of the well so that it slowly floods the well without disturbing the cell surface. Use this same technique any time you add solution to the wells. Use the same volume as used for growth conditions.

■ Note

For time course experiments or to fix cells on different days use coverslips to culture the cells. Transfer coverslips with bent forceps to a new dish containing fixative.   ■

3. Wash cells three times with PBS (approximately 5 minutes each wash).

4. For best results, stain fixed cells within 24 hours of fixation. However cells are good for up to at least a month when fixed cells are stored at 4°C in PBS containing 0.05% (w/v) sodium azide, which prevents bacterial growth.

Permeabilizing the Cells

1. For antigens that reside within the cell, permeabilization of the plasma membrane is required to allow antibody entry. For antigens on the cell surface, permeabilization is not required and may prevent optimal staining. For dual staining of internal and external proteins, omit the Triton X-100 step and instead add saponin in the fixative. Alternatively, some external antigens may withstand a lower concentration (0.1 %) of Triton X-100.
2. Incubate cells with 0.3% Triton-X 100 solution for 5 minutes.

■ Note

Omit this step for surface antigens that are sensitive to detergent treatment.   ■

3. Wash cells three times with PBS (approximately 5 minutes each wash).

Set Up Primary Antibody Incubation

1. Remove PBS and add blocking buffer. Incubate for 30 minutes to 1 hour at ambient temperature. Proteins in the blocking buffer adhere to non-specific "sticky" sites to reduce non-specific antibody binding. The protein component of blocking buffer can be provided by serum from the antibody host species or IgG-free bovine serum albumin (BSA).

■ Note

Incubate any BrdU-treated wells with HCl, then rinse with PBS prior to blocking (see notes on BrdU above).   ■

2. Design an antibody staining plan for each sample well. This should be written on a matrix representing the wells, as it is easy to make mistakes when pipetting reagents. Make certain that antibody isotypes do not overlap within a given well for samples stained with more than one antibody.
3. Antibody concentration: most manufacturers provide recommendations for antibody concentrations for specific applications (for example for immunocytochemistry (ICC) or immunohistochemistry (IHC)). When using an antibody for the first time, it is a good idea to try a range of concentrations around that provided by the manufacturer. For example, if the recommended concentration is 1:100, try a range from 1:10 to 1:1000.

4. Aliquot antibody dilution buffer (ADB) into microcentrifuge tubes for the set of cells to be stained.
5. Add an appropriate volume of primary antibody (or antibodies) to each tube with ADB and mix gently.
   - **Chamber slides:** If using eight-well chamber slides, you will need a final volume of 250 μL per well. For four-well chamber slides, use 400 μL per well (adjust volume per well accordingly for wells that are other sizes).
   - **Coverslips:** If using 18 mm coverslips, you will need a final volume of 20 μL per coverslip, if using 12 mm coverslips, use a final volume of 10 μL per coverslip.

■ Note

Secondary-only control wells should be incubated in ADB alone (no primary antibody) or with a control immunoglobulin (Ig) diluted in ADB. ■

6. Remove protein precipitates from the diluted primary antibody solution by spinning at 11,000 rpm for 5 minutes in a microcentrifuge (optional).
7. If the solutions were centrifuged, gently remove supernatants to new tubes, leaving a small amount of liquid at the bottom to avoid picking up the sediment. Keep diluted antibodies on ice until added to cells.
8. Remove blocking buffer and add the diluted primary antibodies to the wells or coverslips.
   - **Chamber slides:** Remove the covers from the multi-well slides and carefully seal the tops of the wells with parafilm. Parafilm is used since condensation on the multi-well slide cover increases the probability of cross-contamination across the wells. Place slides into a humidity-controlled bin (i.e. Tupperware with damp napkin).
   - **Coverslips:** Remove coverslips from wells and incubate coverslips with antibody dilutions within a parafilm sandwich. First, tape a sheet of parafilm on a glass plate, making sure it is taut and level. Place a 10 μL drop of the antibody dilution on the parafilm. Using forceps, transfer the 12 mm glass coverslip cell side down at a 45° angle onto the droplet, slowly, to avoid creating bubbles. Repeat for each coverslip. Carefully tape another sheet of parafilm over the coverslips. Label with Sharpie on the parafilm next to the coverslips, if desired. For a visual demonstration see http://www.jove.com/index/Details.stp?ID=267.
9. *Recommended method:* Incubate slides overnight at 4°C. *Alternate method:* Incubate slides 1–2 hours at room temperature.

■ Note

Overnight incubations generally give better results, but results vary depending on the primary antibody. ■

Table 16.1 Common Fluorophores

Fluorophore	Absorption Peak (nm)	Emission Peak (nm)
Blue: AMCA, Hoechst, DAPI	350	450
Green: FITC, Cy2, Alexa488	492	520
Red: TRITC, Cy3, Alexa555	550	570
Far red: Cy5	650	670

Secondary Antibody Incubation

1. Dilute secondary antibody (or antibodies) in ADB using the concentration as recommended by vendor or determined empirically to give the best results. We usually dilute secondary antibodies 1:250. See Table 16.1 for common fluorophores and their absorption and emission peaks.
2. Remove the primary antibody from each well or coverslip.

■ Note

Place a disposable pipette tip on the end of the aspirator pipette and replace tips for each aspiration. Using the same aspirator tip for multiple wells increases the likelihood of cross-contaminating adjacent wells. ■

- **Coverslips:** Before removing the coverslips from the primary antibody solution, add PBS for the wash step to the original culture dish. Then remove the top layer of parafilm from the coverslips. Carefully remove the coverslips from the parafilm, and place the coverslip, cell side up, into the original culture dish that contains PBS. To reduce tension that can result in the cells detaching from the coverslip it is best to remove the coverslips as follows:
- Remove tape from one side of the lower layer of parafilm and lift up that layer from the plate. Then, simultaneously peel the parafilm away from the coverslip with one hand while using forceps with the other hand to remove it from the parafilm. Transfer coverslips back to their original culture dishes with PBS.
3. Wash cells three times with PBS to remove unbound primary antibody. Replace aspirator tips after each use.
4. Spin secondary antibodies at 11,000 rpm for 5 minutes to remove protein precipitates (optional).
5. Incubate cells with secondary antibody dilution.
    - **Chamber slides:** Carefully add secondary antibodies to aspirated wells.
    - **Coverslips:** Make a fresh parafilm sandwich as above, except use $15\,\mu L$ secondary antibody dilution per 12 mm coverslip or $30\,\mu L$ per 18 mm coverslip.

6. *Recommended method:* Incubate for 1–2 hours at room temperature in the dark (to reduce photobleaching of fluorescent antibodies). *Alternate method:* Incubate cells overnight at 4°C in the dark. If using chamber slides incubate in the humidity-controlled bin as described for primary antibody incubation.

Visualizing Immunofluorescence

1. Wash cells three times with PBS for 5 minutes each wash. See coverslip notes in previous section for more information.
2. If desired, incubate cells with DAPI or Hoechst reagent to counterstain nuclei (Hoechst 33342, Molecular Probes/Invitrogen, 1 mg/mL in DMSO, stored at 4°C). Dilute Hoechst 1:500 in PBS, incubate cells for 1–5 minutes at room temperature, wash cells with PBS. Alternatively, use a mounting medium that contains DAPI or Hoechst.

■ Note

It is often useful to have a cellular counterstain if it does not interfere with another antibody being detected by a fluorophore in the same channel. A nuclear counterstain is helpful when evaluating the nuclear localization of an antigen (particularly in stem cells that have a high nucleus-to-cytoplasm ratio) or when quantitation of stained cells is desired.   ■

3. Prepare mounting medium (used to minimize photobleaching of fluorescence). Examples of mounting media are: Vectashield® (Vector Labs), Slow Fade® (Invitrogen/Molecular Probes), and ProLong® Antifade Reagent (Invitrogen/Molecular Probes). Some mounting media interfere with particular fluorophores, so be careful to take into consideration the fluorescent label on the secondary antibodies when choosing a mounting medium.
4. If using chamber slides:
   a. Aspirate wells.
   b. Snap off plastic wells.
   c. Carefully use a razor on one of the short ends of the gasket to cut it. Using fine tweezers peel back the gasket slowly.
   d. Pipette a line of the mounting medium along the long end of the slide. Be careful not to allow bubbles to form in the mounting medium. Gently lower a rectangular coverslip at a 45° angle on the slide. Allow the mounting medium to spread slowly over the cells.
   e. Using two fingers, very gently squeeze out the extra mounting medium and/or trapped air bubbles over a paper towel (to absorb excess liquid). Pressing too hard will displace and/or damage cells. Aspirate the extra medium from the edge of the slide.
5. If using coverslips:
   a. Carefully remove coverslip from well containing the PBS wash. This is best done with forceps bent for the purpose.

b. Rinse the back of the coverslip with water to reduce salt crystal formation from residual PBS that interferes with visualization.

c. Lower the coverslip, cell side down, at a 45° angle onto a small drop (7 μL for 18 mm or 3 μL for 12 mm coverslips) of mounting medium on a glass slide. Slowly lower the coverslip to avoid formation of bubbles in the medium.

6. Briefly allow samples to dry at room temperature in a dark place then proceed to the next steps. Note that the coverslips will still be able to move around and should be handled with care.

Observation

1. Remove excess mounting medium by gently wiping the side of the slide without cells or coverslips with 70% ethanol (use Kimwipes® or cotton swabs).

2. Seal slide with nail polish (e.g. "top coat"). Allow to dry. Sealing with nail polish prevents movement of the coverslip and subsequent shearing of cells; however, this step can be omitted. Alternatively, use hard setting mounting medium.

3. View slides on a fluorescence microscope. Afterward, store slides at −20°C (with desiccant for best preservation). Storage at −20°C can preserve the signal for months (depending on the sample, antibody, etc.).

Using the Microscope

1. Seat slide on microscope stage with the coverslip facing the objective lens.

▪ Note

If you are using an inverted microscope this means that the slide has to be flipped so that the coverslipped-side is down. ▪

2. Make sure the microscope shutter is closed. Turn on fluorescent light source and white light.

3. Using a phase contrast 20× objective, bring the sample into the focal plane.

4. Turn off phase contrast white light.

5. Open shutter for fluorescence and view cultures through the microscope's binocular lenses.

6. Scan through areas of interest while cycling through the various fluorescence filters.

7. Remember to limit the exposure of the slide to fluorescent light. Close shutter when not analyzing samples.

Summary of Immunostaining Procedure

Once you have become familiar with the detailed methods, you can use this summary to remind yourself of the steps.

1. Remove medium from cells, wash with PBS if desired.
2. Add fixative, 10 minutes, room temperature.
3. Wash PBS, 3 × 5 minutes
4. Add HCl if cells are BrdU-treated, 20–30 minutes, 37°C, wash PBS, 2 × 5 minutes.
5. Add 0.3% Triton X-100 to permeabilize, 5 minutes, room temperature.
6. Wash PBS, 3 × 5 minutes.
7. Add blocking buffer, 30–60 minutes, room temperature, remove.
8. Incubate with diluted primary antibodies, overnight, 4°C.
9. Wash PBS, 3 × 5 minutes.
10. Incubate with diluted secondary antibodies, 1–2 hours, room temperature, or overnight, 4°C.
11. Wash PBS, 3 × 5 minutes.
12. Add Hoechst (1:500 in PBS) or DAPI (1×), 1–5 minutes, room temperature.
13. Wash PBS, 1 × 5 minutes.
14. Mount and coverslip chamber slides or mount coverslips on a glass slide, seal with nail polish if desired.
15. View on microscope.

PITFALLS AND ADVICE

Background Staining

A sample may have a high level of background fluorescence or fluorescent debris. Here are some possible remedies for resolving this common problem and further discussion of a few specific causes of background staining that are particularly useful for tissue staining.

- Spin the antibodies to remove precipitates before adding the antibody to the sample.
- Use fresh antibodies. Over time antibodies will degrade and increase the incidence of background and non-specific staining.
- Reduce primary antibody and/or secondary antibody concentrations.
- Do not let cells dry out during rinses, fixation or antibody incubation.
- Increase PBS rinsing time or number of washes.
- Use correct blocking serum or longer blocking time. Also try blocking with IgG-free BSA rather than animal serum (use 5% w/v in PBS for blocking buffer and 1% w/v in PBS for antibody dilution buffer).

- Refine growing conditions to avoid stressful cell culture conditions during growth.
- Attempt to use a different antibody for the antigen (try to choose an antibody that recognizes a different epitope on the molecule).

Species Mismatch

- *Problem:* Same-species antibodies yield high background. For example, when mouse primary antibodies are used on mouse tissues, detection with anti-mouse secondary antibodies will detect all mouse immunoglobulins that are native to the mouse tissue.
- *Solution:* Use species-mismatched primary antibodies or block the endogenous antibodies by pre-incubating with an unconjugated secondary antibody. If blocking, it is necessary to use Fab fragments and important to use a Fab preparation that matches the conjugated secondary antibody that will be used for detection. Vendors often sell unconjugated Fab preparations that match the detecting secondary antibody for this purpose.

■ Note

Why use Fab fragments for blocking endogenous Ig? Whole Ig is multivalent and a block with a multivalent antibody will leave many Fab ends unbound. Subsequent treatment with the primary antibody will simply bind these exposed ends and aggravate the background problems.   ■

Fc Receptors in Sample

- *Problem:* Fc receptors expressed by cells non-specifically bind primary and secondary antibodies. This is particularly problematic for tissues that have been damaged and contain activated immune cells.
- *Solution:* Use Fab preparations for detection rather than whole antibodies or block using unconjugated Fc fractions that match both primary and secondary antibody preparations.

■ Note

When using Fab fragments for detection, the secondary antibody must be one that recognizes a Fab fragment. Typically, the secondary antibody used will recognize the light chain rather than heavy chain and one must take care to determine the class of light chain present in the Fab fragment (i.e. either kappa or lambda light chain).   ■

Endogenous Enzymatic Activity

- *Problem:* Tissues and cells express peroxidases, galactosidases, and phosphatases that will create non-specific staining when using enzymatic methods to detect bound antibody (for example, horseradish peroxidase (HRP)-conjugated secondary antibodies).

■ Red blood cells contain high peroxidase activity. Vascular cells express high levels of phosphatase and macrophage/monocytes express high levels of all three types of enzymes.

■ *Solution:* Most kits include instructions for minimizing or inactivating endogenous enzyme activity. In general, the kit protocols work extremely well and can be used without modification.

■ Note

Endogenous peroxidase activity is ablated by pre-incubating tissues with high concentrations of H_2O_2. If the H_2O_2 used is old, then this step will not work well. Tissues should visibly bubble when pre-incubated. If not, the H_2O_2 is probably old. Use a higher concentration or use a new stock.   ■

■ Note

Controls for enzymatic detection should include: (a) no primary or secondary antibody, (b) secondary antibody alone. Incubation of the control samples with enzyme substrate can be used to confirm specificity of staining and check for endogenous enzyme background staining.   ■

Generalized Background

■ *Problem:* Very high overall background.

■ *Solution:* Titrate antibodies (both primary and secondary) for optimum signal to noise ratio. Primary or secondary antibodies may recognize non-specific antigens. To determine if the problem is with the primary or secondary antibody, prepare one sample that is treated with secondary antibody alone. If background is low, then the problem is with primary antibody. If background is present in samples treated with secondary antibody alone, then the problem is with secondary antibody. In both cases, an alternate antibody should be tried (if available) or more aggressive means to improve specificity should be explored.

■ Note

Secondary antibody background can be reduced if the vendor provides unconjugated pre-immune serum from the same species (ideally collected from the same animal prior to immunization). This is used in the initial blocking step to bind all non-specific sites prior to final detection using the conjugated secondary antibody preps.   ■

Weak Staining
- Test the antibodies on known positive and negative controls.
- Try another antibody to the same antigen.
- Fixation: Check the literature for papers that have used the antibody (and have nice images of immunostained cells) and follow the protocol verbatim. Most antibodies are sensitive to the type of fixation used or the amount of fixation (it is possible to over-fix).
- Increase the concentration of primary and/or secondary antibody.
- Increase the time of the primary antibody incubation.
- If positively staining slides have faded over time, be certain the nail polish sealant on slides is intact and that the slides are being stored in a desiccated environment.

Too Much Staining
- Reduce primary antibody concentration.
- Reduce primary or secondary antibody incubation period.
- Attempt to use different clone of antibody for the same antigen.
- See notes on blocking in Background staining section above.

Multiple Antibody Staining
The basic method described above is also used for staining with more than one antibody simultaneously on the same sample. Staining for more than one antigen involves use of multiple primary antibodies, each of a unique class or animal species, followed by use of multiple secondary antibodies, each specific for one of the primary antibodies and each carrying a unique enzyme or fluoro-chrome/fluorophore marker.

If using enzyme detection, sequential reaction in each substrate will be necessary. Order of application may be important (i.e. H_2O_2 used in the HRP reaction can oxidize other enzymes and reduce activity. Consider doing DAB last).

■ Note
Care should be taken to use secondary antibodies that are highly specific for the class and species of primary antibody that needs to be detected. Some vendors provide secondary antibody reagents that are validated to have minimal cross-reactivity to a wide spectrum of antibody classes and species (Jackson Immunoresearch is a reliable source).　■

■ Note
Histochemical reaction products are frequently opaque. If attempting to detect co-labeling in a given cell, care must be used in choosing the specific detection method. For example, if co-labeling for BrdU in the nucleus

along with a cytoplasmic marker to identify cell phenotype, use HRP-DAB for the BrdU to generate an opaque black nucleus. Use a more translucent marker for the cytoplasmic epitope (e.g. AP-vectorRed or AP-vectorBlue from Vector Labs). ∎

DIGITAL IMAGES OF FLUORESCENT CELLS

After immunostaining, cells are usually viewed on a fluorescence microscope and images of the stained cells captured with a digital camera. There are a variety of cameras and image capturing software packages available; therefore we will not go over the specific details of a particular program here.

Adobe Photoshop

Many scientists bring the captured images into Adobe Photoshop, to create output for publications. In the next section, therefore, we will describe how to use several features in Photoshop and briefly introduce a program available for image quantification (ImageJ).

Photoshop can open a wide variety of image files captured from a microscope-mounted camera, including ".tiff" and ".jpg" formats, and provides a variety of means to manipulate images. Here we will briefly describe how to set the color mode, alter the image size, create scale bars for an image, adjust the image brightness and contrast, and create color overlays of images.

Setting the Color Mode

Color digital images can either use RGB (red, green, blue) or CMYK (cyan, magenta, yellow, black) for color encoding. RGB images are more compatible for computer monitors or projectors, since they use an additive light system and printers rely on a subtractive light system.

Bright greens, reds, and blues cannot be reproduced in print as they can on a monitor, so prints of an RGB image may not convey the bright colors or fine detail visible on the computer monitor. For print purposes, it is best to convert an RGB image to CMYK. To convert to CMYK for printing, go to "Image"→"Mode" and select CMYK.

Adjusting the Image Size

Images captured by image acquisition software programs can come in a variety of sizes and resolutions. To find the size of your image, go to "Image"→"Image size." Images often are captured at 72 pixels/inch and are of fairly large dimensions (in terms of inches). Journals usually request photos at a resolution of 300 pixels/inch. The easiest way to do this is to change the resolution but not change the overall size of the file so that the dimensions (in inches) of the image are more suitable for printing or incorporating into a figure.

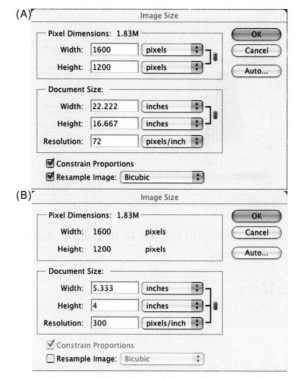

FIGURE 16.1
Adobe Photoshop "Image Size" window. (A) Before size adjustment. (B) After size adjustment.

To do this, make sure that the checkbox next to "Resample Image" is unchecked (as in Figure 16.1B) then adjust the resolution (see the examples above and note that the overall pixel dimensions (1.83M, 1600 × 1200 pixels) are the same for both while the document sizes (width, height, resolution) are different).

Scale Bars

Journals usually require scale bars for all microscopic images. One way to generate scale bars for your images and to make size/length determinations is to use a *scale micrometer*. A scale micrometer is a microscope slide that has lines etched a particular distance apart from each other. The micrometer can be placed on the microscope stage and an image taken using each of the microscope objectives. Because the images will be captured at the same width (in terms of pixels) as your photographs of cells or tissues, you can determine a conversion factor that will allow you to measure real distances on your images.

As an example, if an image taken with a camera on a particular microscope using a 20× objective has a total width of 580 μm (from the scale

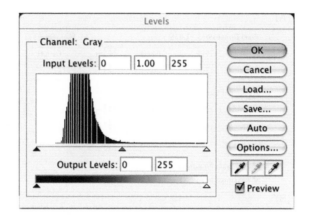

FIGURE 16.2
Adobe Photoshop "Levels" window.

micrometer) and 1600 pixels; this means that 100 μm would equal 276 pixels on that image.

■ Note

These measurements will be specific to the objective, microscope, and camera used, so attention must be paid to the conditions under which a particular image was captured in order to appropriately determine the scale. ■

For a scale bar on your photomicrograph, you can draw a line of a particular length (in pixels) in Photoshop by using the line tool (on the tool bar, which also contains the move tool, text tool, etc.) and watching the pixel location in the Navigator window ("Window"→"Navigator"; click on the "Info" tab in the Navigator window). The X and Y coordinates of the cursor location will be in pixels as long as the rulers for the image are set to "pixels" ("Preferences"→"Units and rulers").

Adjusting the Brightness/Contrast of an Image

There are multiple ways to adjust images in Photoshop, and most are found under "Image"→"Adjustments."

One straightforward way to adjust the brightness/contrast is to use the "Levels" option ("Image"→"Adjustments"→"Levels") and adjust the sliders under the histogram (Figure 16.2). The advantage of this option is that by viewing the histogram, you can more accurately adjust the intensity of the image without altering the data.

It is *imperative* when using any image adjustment for data images to be extremely careful to not alter the data with the adjustment. For example, decreasing the brightness should not remove signal and increasing the brightness should not create signal or expand its zone. See Figure 16.3 for images that have been appropriately and inappropriately adjusted.

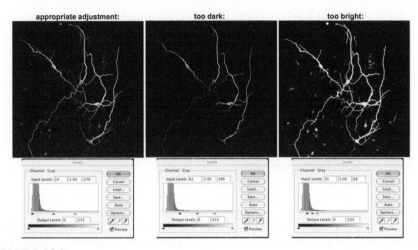

FIGURE 16.3

Adobe Photoshop "Levels" window and corresponding image changes. Moving the sliders on the "Levels" window can result in appropriate (left) image adjustment or images that are too dark (middle) or too bright (right).

Changing Grayscale Images to Color and Overlaying Color Images

Cells or tissues are often double- or triple-labeled with different fluorescent molecules to allow visualization of multiple signals. Photoshop can be used to convert captured grayscale images to color and overlay the color images so that all fluorescent signals can be visualized simultaneously.

In order to create a color overlay, the images of the different fluorescent channels are brought together into a single file. The separate images are maintained on individual layers and then assigned a different color.

1. To begin, select all of the image ("Select"→"All") and copy ("Edit"→"Copy").
2. Make a new file ("File"→"New") and the size, resolution, etc. will be identical to what you just copied.
3. In the window that opens and describes the new file, switch from "Grayscale" to "RGB" (or "CMYK" if the image is solely for print media).
4. Once the new file is created, paste in the copied image ("Edit"→"Paste"). Select all and copy the other images to be overlaid, then paste them into the new file.
5. Each image will automatically be pasted into a different layer ("Window"→"Layers").
6. To change the color of an image in a layer, open the "Levels" option ("Image"→"Adjustments"→"Levels") and use the tab marked "RGB" to select the Red, Green, or Blue channel.

7. Use the "Output levels" to alter the color: for a Red image, make the Green and Blue output levels 0 (change the number in the box on the right from 255 to 0), for a Green image, make the Red and Blue output levels 0, and for a Blue image, make the Red and Green output levels 0.
8. These steps can be repeated for different layers within the same document to create layers that are of different colors.
9. To overlay differently colored layers, position one colored layer directly above the other colored layer (in the "Layers" window) then change the button under the "Layers" tab from "Normal" to "Screen." You should now see both layers overlaid.

ImageJ

ImageJ is a free program available for download that can be used to quantify a wide variety of parameters in an image. In addition to the basic features of ImageJ, there are macros that others have created (or you can write yourself) that expand the functionality of the program. For details and downloads see http://rsbweb.nih.gov/ij/.

Counting in ImageJ

1. Open (jpeg or tif) files in ImageJ. If the image was saved as separate files for each channel, then open each file in ImageJ.
2. Using "Image" in menu bar choose "Color" from the drop down menu→"Merge Channels." Select the desired color for each file.
3. Open the Channels tool window ("Image"→"Color"→"Channels Tool") and click on a channel. Respond "yes" to the prompt to "convert to multichannel composite image."
4. Open the counter tool under "Plug-in"→"Analyze." Choose initialize, then choose a counter number and click on each cell that will be counted.
   a. Deselect all channels except the nuclear stain, and proceed to count nuclei.
   b. To count antibody stained cells, select another channel in the channels tool (and deselect the nuclear stain channel, if desired). Choose a second counter number and begin counting.
5. If desired, save your counted image by choosing the "Export Image" icon in the counter tool window. Choose "Save as" and select a file type.

Measurements in ImageJ

1. In order to measure in ImageJ, you must first know the scale of your picture in real dimensions (see Scale bar section above). An easy way to convert this information to a scale in ImageJ is to draw a line across the entire width of your image (use the straight line tool on the toolbar).

2. Once you have drawn the line, go to "Analyze"→"Set scale" and set the known distance to the numerical value and unit of length for your image width (for example, the width of the image described in "scale bars" above would be 580 μm).
3. Keep the Pixel Aspect Ratio as 1 and use "um" for "μm."
4. If you are analyzing multiple images that were taken under the same conditions and thus have the same scale, you can check "Global" in the "Set scale" window and the scale will be automatically applied to all the images. After setting the scale, the length of any line drawn and measured will be given in the desired units.
5. To measure an element in your image, you can draw a line (straight, segmented, or freehand) and then click on "Analyze"→"Measure." "Analyze"→"Set measurements" allows you to decide what parameters will be measured.

■ Note

You can also choose other types of shapes (other than a line) and measure parameters such as area. ■

EQUIPMENT

- Fume hood for working with paraformaldehyde.
- Inverted or upright microscope equipped for fluorescence.
- Objectives: 10×, 20×, 40×, and perhaps 60× or 100× objectives.
- Filter cubes appropriate for secondary antibody fluorophores. It is important to make sure that the cubes will give maximal signal for one fluorophore but not allow bleedthrough excitation of another fluorophore.

REAGENTS AND SUPPLIES

Item	Suppliers
Primary Antibodies	Various commercial vendors; colleagues
Secondary Antibodies	Jackson Immunoresearch, Life Technologies
Mounting Media	Vectashield (Vector Labs-#H-1000)
	Life Technologies (Molecular Probes)
Chamber Culture Slides	Lab-Tek II, Nalgene, Nunc
Cover slips, no. 1 thickness	Fisher (Carolina Biological Supply) range for high mag objectives
Microfuge tubes, 1.0 and 0.65 mL	Fisher, VWR
Pipette tips	Various
Conical tubes 15 mL, 50 mL	Fisher, Corning

Item	Suppliers
Nail polish "clear" top coat	Drug store
Pipettes, 1, 5, 10, 25 mL	Fisher, Corning
12- or 24-well tissue culture plates	Fisher, Corning
Bovine Serum Albumin (IgG free)	Jackson Immunoresearch (001-000-162)
Ethylenebis(oxyethylenenitrilo) tetraacetic acid (EGTA)	Fisher (Acros Organics-409910250)
Hoechst 33258	Fisher (AC22989-100)
Hydrochloric Acid (HCl)	Fisher (A144-500)
Magnesium Chloride (MgCl$_2$)	Fisher (M33-500)
Paraformaldehyde	Fisher (T353-500)
Phalloidin (fluorescently labeled)	Sigma or Life Technologies
Phosphate Buffer Saline (PBS), 10×	Fisher (Cellgro-46-013-CM)
Triton X-100	Sigma (T-8787)
Saponin	Sigma (S4521)
Sodium Azide (NaN$_3$)	Sigma (71289)
Sodium Hydyroxide (NaOH)	Fisher (BP-359-500)
Sucrose	Fisher (BP-220-1)

RECIPES

Paraformaldehyde 4% (100 mL)

Component	Amount
4% Paraformaldehyde	4 g
H$_2$O	85 mL
10× PBS	10 mL of 10×
5 mM MgCl$_2$	500 µL of 1 M
10 mM EGTA	2 mL of 0.5 M
Sucrose	4 g
Optional	
0.1% Saponin	5 mL of 2% in H$_2$O
250 nM Phalloidin	100 µL of 250 µM

In fume hood:

1. Add 4 g of paraformaldehyde to 80 mL of dH$_2$O.
2. Stir until dissolved.
3. Add a few drops of 10 N NaOH until solution is clear (it will not completely dissolve or clear without the addition of NaOH).
4. Add 10× PBS, MgCl$_2$, EGTA and Sucrose.
5. Stir until sucrose is dissolved. The MgCl$_2$ will come out of solution at this point, but will go back in once pH is brought to 7.5.
6. Use 6N HCl to bring pH to around 8.5–9.0. pH will change rapidly at this point so use 1N HCl to bring to pH 7.5.

7. Bring to 100 mL with H_2O. Filter with 0.45 or 0.22 µm filter unit.
8. Store at 4°C up to 1 week (alternatively store aliquots at -20°C).

Paraformaldehyde 4% (100 mL); Alternative Recipe

Component	Amount	Final Concentration
Paraformaldehyde	4 g	4%
H_2O	50 mL	
0.2 M Phosphate buffer pH 7.4	50 mL	0.1 M, pH 7.4

1. Dissolve paraformaldehyde in 50 mL H_2O as described above using NaOH.
2. Filter (0.2 or 0.45 µm) and add 50 mL of 0.2 M phosphate buffer, pH to 7.4 (recheck pH and adjust to pH 7.4).

Important Notes About Paraformaldehyde

- We recommend that paraformaldehyde be prepared fresh whenever possible. It will make a noticeable difference in the quality of immunofluorescence.
- Solubilization of paraformaldehyde powder is often accomplished with heat. If using heat, take care not to heat above 55–60°C and add just enough base to dissolve the paraformaldehyde. Under high heat or high pH, paraformaldehyde will iso-convert to formaldehyde which degrades rapidly to formic acid and water.
- If the solution temperature goes over 65°C during preparation, do not use it as it will produce a strong autofluorescence in cells or tissues.

Blocking buffer 1 (50 mL)

Component	Amount
PBS	48.5 mL
Serum from secondary antibody host species: rat, mouse, goat, donkey, etc. Final concentration: 3%	1.5 mL

Blocking Buffer 2 (5%BSA) (50 mL)

Component	Amount
PBS	45 mL
Bovine Serum Albumin (IgG free)	2.5 g

1. Rotate until BSA is dissolved, bring to 50 mL with PBS
2. Filter to remove particulates (0.45 µm or 0.22 µm).

Antibody Dilution Buffer 1 (50 mL)

Component	Amount
PBS	49.5 mL
Serum from secondary antibody host species: rat, mouse, goat, donkey, etc. Final concentration: 1%	0.5 mL

Antibody Dilution Buffer 2 (1% BSA) (10 mL)

Component	Amount
PBS	8 mL
5% BSA (from above)	2 mL

READING LIST

Research Articles with Antibody Staining of Human Stem Cells

Flanagan, L.A., Rebaza, L.M., Derzic, S., Schwartz, P.H., Monuki, E.S., 2006. Regulation of human neural precursor cells by laminin and integrins. J. Neurosci. Res. 83 (5), 845–856. Multiple examples of stained human and mouse neural stem cells are found in this paper.

Marchenko, S., Flanagan, L.A., 2007. Immunocytochemistry: Human neural stem cells. J. Vis. Exp. (7) (September) <http://www.jove.com/index/Details.stp?ID=267>. Video protocol demonstrating immunostaining of human neural stem cells.

Reubinoff, B.E., Itsykson, P., Turetsky, T., Pera, M.F., Reinhartz, E., Itzik, A., et al., 2001. Neural progenitors from human embryonic stem cells. Nat. Biotechnol. 19, 1134–1140. One of the first papers on using immunohistochemistry to identify specific derivatives of hESCs.

Schwartz, P.H., Bryant, P.J., Fuja, T.J., Su, H., O'Dowd, D.K., Klassen, H., 2003. Isolation and characterization of neural progenitor cells from post-mortem human cortex. J. Neurosci. Res. 74, 838–851. This paper has several examples of staining of individual cells with multiple markers.

Antibody Laboratory Manuals

Harlow, E.W., Lane, D., 1999. Using Antibodies, A Laboratory Manual. Cold Spring Harbor Laboratory Press, Cold Spring Harbor, NY.

Javois, L.C., 1999. Immunocytochemical Methods and Protocols. Methods in Molecular Biology, vol.115. Humana Press, Totowa, NJ.

Informational Websites

Antibody Cross-reactivity resource. <http://www.keithbahjat.com/abcxr/>.

B.D. Biosciences. <http://www.bdbiosciences.com/support/resources/protocols/immuno_microscopy.jsp>.

Millipore Antibody Learning Center. <http://www.millipore.com/antibodies/ab/abhome>.

Molecular Probes. <http://www.invitrogen.com/site/us/en/home/brands/Molecular-Probes.html>.

Protocol Online. <http://www.protocol-online.org/prot/Immunology/>.

Stem Cell Markers and Attributes. <http://stemcells.nih.gov/info/scireport/appendixe.asp>.

The Antibody Resource. <http://www.antibodyresource.com/educational.html>.

Vector Laboratories. <http://www.vectorlabs.com/>.

General References

Loring, J.F., Wesselschmidt, R.L., Schwartz, P.H. (Eds.), 2007. Human Stem Cell Manual. Elsevier.

CHAPTER 17

Analysis of Genome-Wide Gene Expression Data from Microarrays and Sequencing

Kristopher L. Nazor, Rathi D. Thiagarajan, and Louise C. Laurent

EDITOR'S COMMENTARY

Patterns of gene expression underlie fundamental differences that define cell type and function, and transcription of genes in human pluripotent stem cells (hPSCs) and their derivatives has become an essential part of their analysis.

Analysis of mRNA expression has evolved considerably over the past two decades, and changed radically in 1995, when research laboratories first introduced high-density arrays created by spotting cDNA libraries made from cell lines and tissues onto microscope slides. Genes expressed in experimental samples were detected by labeling cDNAs made from the experimental samples with fluorescent tags, hybridizing the labeled samples to the microarrays, then using optical methods to identify which spots hybridized to the arrayed sequences. This is essentially the same method that is used today with more sophisticated arrays. Incyte Genomics commercialized the first full-length cDNA array technology developed at Stanford, and Affymetrix took another strategy, using photolithography methods to synthesize multiple short probes representing cDNAs. Currently, many companies, including Affymetrix, Illumina, and Agilent, offer whole genome gene expression microarrays. Illumina and Agilent use longer oligonucleotides as probes; Illumina uses a bead-based technology.

It is important to note that the initiation of gene expression microarrays preceded by 5 years the publication of the first drafts of the sequence of the human genome. After the human sequence was deciphered and coding sequences were identified, the content and design of human gene microarrays was greatly improved. Improved versions of array content continue to be released by the principal suppliers every few years.

Over the past 10 years, as technical issues have been resolved, hybridization and imaging of microarrays have become standard techniques for core

271

J.F. Loring & S.E. Peterson (eds): Human Stem Cell Manual, Second edition.
DOI: http://dx.doi.org/10.1016/B978-0-12-385473-5.00017-5

laboratories, although some researchers prefer to control all of the aspects of microarray analysis. This means that researchers who want to analyze their stem cell lines can concentrate on the quality control of the input – the cells to be analyzed; and the output – the statistical analysis of the data. This chapter focuses on these two aspects of microarray experiments, and extends the advice about design of experiments and interpretation of data to next-generation RNA sequencing, which is becoming more popular as the cost decreases and the bioinformatic methods improve.

OVERVIEW

Gene expression profiling using high throughput, genome-wide platforms has fundamentally transformed the way that we approach problems in the biomedical sciences by expanding the classic single-gene approach toward comprehensive analysis of gene expression networks. To date, gene expression array data from nearly 400,000 human samples have been made publicly accessible via the Gene Expression Omnibus (GEO) as a resource from the NIH funded National Center for Biotechnological Information (NCBI).

As with all technologies, the microarray approach has both advantages and limitations that should be considered. Among the most distinct advantages are that array analysis can be performed quickly (2–3 days), at low cost (~$US150/sample), and with relatively small amounts of material (~50–100 ng total RNA). Additionally, analysis of the array output data can be easily performed with user-friendly open-source software or Microsoft Excel. The popularity of microarrays stems from their ability to profile all well-annotated genes in the human genome at once. However, because the content of microarrays is pre-defined, this approach lacks the potential to identify novel transcripts and isoforms.

An unexpectedly high level of transcriptional complexity has been revealed by initiatives such as the Functional Annotation of Mouse cDNA (FANTOM) and the Encyclopedia of DNA Elements (ENCODE). For example, it has been reported that an average of six to seven different mRNA isoforms arise from a single active locus. Given that most microarray platforms include between three and six probes per gene, the standard gene expression microarray is not capable of assessing this level of transcriptional variation. Exon arrays can identify alternative splicing events by using probes that span exon–intron junctions of known genes. Like most microarrays, however, the design of these platforms is dependent on existing annotations of known transcripts, which precludes the discovery of novel splicing events. Genomic tiling arrays employ a high density of partially overlapping probes that span across part or all of the genome. Although tiling arrays enable discovery of novel transcribed regions, they are expensive, and their construction still relies on existing genome annotations.

Next generation sequencing (NGS) technology was first applied to transcriptional analyses in 2008. Thereafter, RNA sequencing (RNA-Seq) applications have been shown to overcome many of the limitations imposed by pre-annotated platforms and hybridization-based methods. The RNA-Seq transcriptional profiling approach involves deep sequencing of total RNA populations to generate millions of short read sequences (35–150 bp) to characterize actively transcribed regions, allowing for the detection of novel splice variants as well as small and long non-coding RNAs. Despite the unprecedented discovery potential of RNA sequencing, many analytical obstacles have yet to be resolved.

In this chapter, we explain the basics of cDNA microarray and RNA-Seq technologies and provide guidelines for designing experiments in a way that permits the discovery of biologically and statistically significant results. We will highlight a few statistical concepts fundamental to data analysis and suggest some excellent options for data analysis packages/tools.

EXPERIMENTAL DESIGN

Recurrent genetic, epigenetic, and transcriptional variation and instability has been documented in hPSC cultures by our lab and several others (Bock et al., 2011; Laurent et al., 2011; Mekhoubad et al. (in press), Nazor et al (in press); Pomp et al., 2011), highlighting the need for frequent in-depth analysis of cell lines. Independently-derived cell lines may have different propensities to differentiate even when using the same protocols, and cultures of the same cell line have been shown to change during extended culturing. To avoid misinterpretation of experimental results, it is critically important to appreciate this sort of widespread variability during the design of experiments. We focus in this chapter on experimental design guidelines that we have developed to safeguard from variation-based confounders. Specifically, we will highlight four concepts that we believe to be most important in the design of every hPSC experiment using global profiling: *diversification, characterization, replication*, and *foresight*.

Diversification

Always work with multiple cell lines. Here are a few tips to consider when choosing hPSC lines for experimentation:

1. *Work with at least three independent cell lines per experimental group.* In the case of hESCs, we would consider lines to be independent if they were derived from different embryos, ideally, from different sources. For hiPSCs, independent lines would be would be minimally defined as independent clones generated in a single reprogramming experiment. Ideally, independent lines would come from independent reprogramming

experiments using different donor cell populations and/or different reprogramming techniques.

2. *Choose cell lines at relatively early passages.* As described above, hPSCs accumulate irreversible genetic and epigenetic aberrations over time in culture. Choose cells from relatively early passages whenever possible to avoid confounding passage-induced genetic and epigenetic changes.

3. *Know each cell line's history.* Because of the growing interest in the possible sources of variation among hPSC lines, peer reviewers are demanding more documentation of the culture history of hPSC lines. Chapters 1 and 3 described methods commonly used to culture hPSCs. We suggest that you make sure that you can answer the following questions when applicable. What lab or person cultured this cell line up to the point that it was used for your experiments? Have the cells always been cultured on feeder layers? Have they ever been cultured under feeder-free conditions? If so, for how long? Were the cells passaged enzymatically or mechanically? Was enzymatic passaging done with trypsin or Accutase™ to achieve single cells or with collagenase or dispase to passage intact colonies?

Characterization

Work with cell lines that have been well characterized.

1. *Know the cell lines' recent karyotype or genotype.* We emphasize recent karyotype or genotype, because this can change dramatically over 5–10 passages.

2. *Profile defined intermediate as well as starting and end-point populations.* hPSC differentiation protocols often pass through developmentally relevant intermediates. In the example of neuronal differentiation, many protocols first establish neural progenitor cells (NPCs), which subsequently will be directed to differentiate into neurons. Profiling of these intermediate stage NPC samples may lead to the discovery of specific genes or networks that are only *transiently* perturbed in the NPCs, but may be upstream and causal of any observed differences in expression in the neuronal cells.

3. *Assess the purity of your intermediate and endpoint populations.* The differentiation of hPSCs is an imperfect process, giving rise to heterogeneous populations comprised of a cell type of interest and "off-target" cell types. It is absolutely critical to understand the composition of your sample; as sample purity decreases so does the ability to make meaningful conclusions from the data. We suggest that at least 80% of the population have the same phenotype, defined by markers or other criteria. However, caution must be used in interpretation of low level transcripts if the sample is only 80% pure.

Replication and Foresight

Statistically supported data analysis necessitates the inclusion of *biological replicates*, which allow for the assessment of biological variation in a sample group. Biological replicates should not be confused with *technical replicates*, which come from a single biological sample and allow for the assessment of technical variation in a particular assay. At the most basic level, samples from a single cell line that were collected from different wells of a culture dish can be considered biological replicates. The selection of biological replicates must be tailored to the experiment, which in some cases may require multiple levels of biological replication.

Example

Let's design an experiment to determine if there are statistically significant differences *between* normal and disease-specific hiPSCs, or *among* the individual hiPSC lines (Figure 17.1). In this experiment, we are actually asking two distinct questions that can be easily addressed simultaneously with an adequately replicated experimental design. To test for differences *between* the two groups, we would need at least one biological replicate from each of the three non-affected cell lines (the normal group) and at least one biological replicate from each of the three disease-specific cell lines (the disease group)(Figure 17.1F). Notice that for each cell line to be used in this analysis, we suggest collecting three additional biological replicates for orthogonal assays (Western blots, RT-PCR, bisulfite sequencing, SNP genotyping, etc.) that may be used in future experiments to test any conclusions arising from transcriptional profiling. Additionally, we suggest cryopreserving two vials of cells in case the experimental results necessitate additional experiments using the transcriptionally profiled samples.

In the case that we are also interested in identifying differences *among* the individual hiPSC lines, we would have to expand the sample collection shown in Figure 17.1F, by collecting two additional biological replicates from an additional six-well plate for each cell line. In total, this would provide three biological replicates for transcriptional profiling, three biological replicates for orthogonal assays, and two cryopreserved vials of cells for each cell line (Figure 17.1G).

CRITICAL STATISTICAL CONCEPTS

The first step in achieving statistical significance is adequate sampling (see Diversification, above) for the experimental groups of interest. For analysis of gene expression, the law of large numbers would state that the average gene expression obtained for a number of samples from a given experimental group (population) more accurately estimates the actual average gene expression of the experimental group as sample size increases. For normally distributed data, the central limit theorem states that as sample size (n) increases,

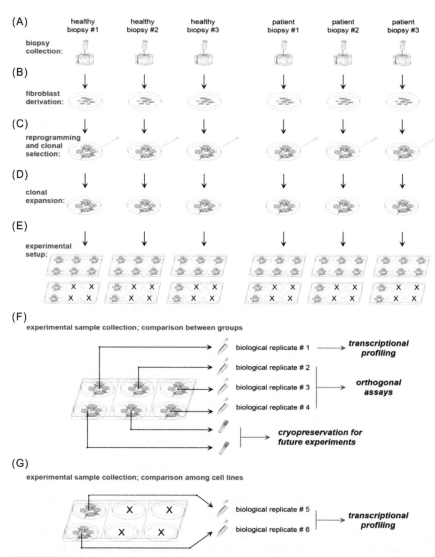

FIGURE 17.1

Experimental design and sample collection. (A) Normal and patient-derived (disease-specific) biopsies are collected from six individual sources (six different people). (B) From each biopsy, fibroblast cultures are derived, expanded and banked. (C) Fibroblast cultures are reprogrammed to pluripotency and individual clones are picked for subsequent expansion. (D) Individual hiPSC clones are expanded and banked prior to experimentation. (E) Individual hiPSC clones are passaged to enough wells to generate biological replicates for transcriptional profiling, orthogonal assays and cryopreservation. (F) In the comparison between groups, six biological replicates per cell line are sufficient. (G) In order to also perform comparisons among the individual cell lines, an additional two biological replicates would be needed.

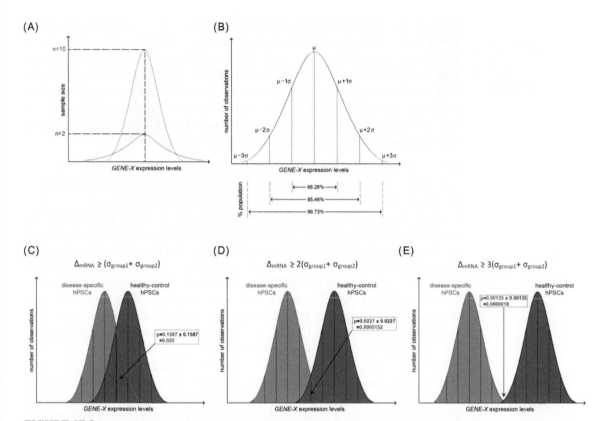

FIGURE 17.2

Critical statistical concepts. (A) As sample size (n) increases, the sample distribution more closely approximates the actual population distribution. Specifically, the sample mean ($\overline{X}$) more closely approximates the actual population mean (μ), and the sample standard deviation (s) more closely approximates the actual population standard deviation (σ). (B) Assuming a normal distribution, we find that 68.26% of the population falls within 1 standard deviation from the mean, 95.46% of the population falls within 2 standard deviations from the mean and 99.73% of the population falls within 3 standard deviations from the mean. (C) There is a joint probability of 2.5% (p = 0.025), that differences greater than 1 standard deviation from each population mean would be observed due to chance. (D) There is a joint probability of 0.05152% (p = 0.0005152), that differences greater than 2 standard deviations from each population mean would be observed due to chance. (E) There is a joint probability of 0.00018% (p = 0.0000018), that differences greater than 3 standard deviations from each population mean would be observed due to chance.

the sample distribution more closely approximates the actual population distribution. Specifically, the sample mean ($\overline{X}$) more closely approximates the actual population mean (μ), and the sample standard deviation (s) more closely approximates the actual population standard deviation (σ). Therefore, by simply increasing the number of diversified samples in an analysis, we can more accurately estimate average gene expression for a gene as well as its variance and have greater confidence in the validity of our results (Figure 17.2A).

One of the most common reported measures of differential expression between two sample groups is the fold change, which is a ratio of the mean expression values for a gene between two independent groups. While the fold change is a convenient way to describe the extent of differential expression between two groups (e.g. disease-specific hPSCs have a two-fold higher expression of GENE-X than healthy control hPSCs), it does not reflect the statistical significance of the observed difference. Therefore, we have to rely on statistical methods to determine which differences are significant.

While not all data are normally distributed, this is a common and fairly reasonable first assumption in the analysis of gene expression data. In the normal distribution, we can easily describe the probability that a sample's expression level for a given gene, GENE-X, will fall within a range of values according to the standard deviation for GENE-X (Figure 17.2B). Therefore, we can use the standard deviation of GENE-X for each experimental group in a comparison (e.g. disease-specific hPSCs vs healthy-control hPSCs) to calculate the joint probability of observing expression of GENE-X at a particular level in both populations. In Figure 17.2C–E, we observe a higher expression level of GENE-X in healthy control hPSCs compared to disease-specific hPSCs. From Figure 17.2B, we would expect that 15.87% $((100 - 68.26)/2)$ of the samples from the healthy control hPSCs would be expressed at levels less than the $\mu - 1\sigma$, and would expect that the same proportion of samples from the disease-specific hPSCs would be expressed at levels greater than the $\mu + 1\sigma$, giving a joint probability of 2.5% (Figure 17.2C). Therefore, it would be unlikely ($p = 0.025$), that differences of greater than 1 standard deviation from each population mean would be observed due to chance. As we increase the number of standard deviations between the means of two populations, we have decreasing probability that the observed difference was due to chance (Figure 17.2D–E).

Additional non-parametric statistical measures, including the Mann–Whitney–Wilcoxon statistical test, can be used for both normally distributed and non-normally distributed data. While it is out of the scope of this chapter to discuss the repertoire of statistical tests that can be applied to gene expression data, we refer you to the *Intuitive Biostatistics* textbook (Chapter 37), which provides excellent guidelines for choosing a statistical test. A convenient outline of how to choose a test is also available online, at http://www.graphpad.com/www/Book/Choose.htm.

MICROARRAY DATA ANALYSIS

Quality Control and Data Normalization

As the microarray platform has evolved over the past 15 years, extensive quality control measures have been engineered into the various microarray platforms. The most popular gene expression microarray manufacturers, Affymetrix and Illumina, provide excellent technical support, tutorial videos and detailed manuals to guide beginner-level microarray users through the quality control process. For each platform there are a number of R/Bioconductor packages that can be easily used for data normalization, including affy for Affymetrix data and lumi for Illumina data. Additionally, RMAExpress is an incredibly user friendly and open-source software for normalization of Affymetrix data. In any case, it is critical to identify and eliminate outliers prior to normalization and subsequent analysis. The effects of outlier samples on data normalization are exemplified by comparing data that were normalized in the presence or absence of outlier samples (compare Figure 17.3B and Figure 17.3D). In these plots, it should be immediately evident that the presence of outlier samples has drastic effects on the distribution of data following normalization.

Data Analysis

Once normalized data is in hand, it is important to perform a global level clustering analysis to identify how the individual samples relate to one

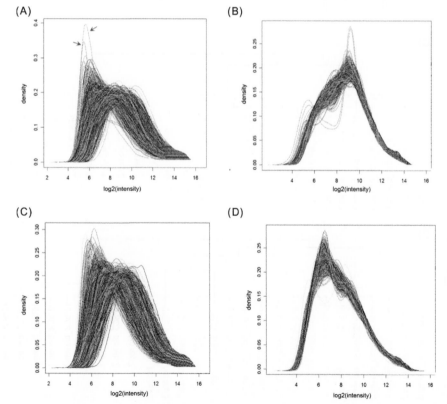

FIGURE 17.3
Quality control and data normalization. (A) Two outlier samples (red arrows) have a skewed probe density towards a lower intensity range. (B) Probe density plot for all samples in Figure 17.3A following Robust Spline Normalization (RSN) using the lumi package in Bioconductor. (C) Probe density plot after removal of outlier samples. (D) Probe density plot following RSN without the outlier samples.

another. There are a number of clustering methods that can be applied for this purpose, including principal component analysis (PCA), non-negative matrix factorization (NMF), and hierarchical clustering. For an excellent review of how clustering works, please refer to D'Haeseleer (2005).

Example

Hierarchical clustering of samples from a multistage directed differentiation of hPSCs into neurons may show that samples cluster together in a way that agrees with our knowledge of the samples (Figure 17.4A). Alternatively, the samples may not cluster as we would have expected (Figure 17.4B). In this

FIGURE 17.4

Identification of sample groups using global gene expression data. (A) Hierarchical clustering of nine samples from three sample groups cluster in accordance with our knowledge of the samples. (B) Hierarchical clustering of nine samples from three sample groups cluster in discordance with our knowledge of the samples. (C–D) Inclusion of multiple biological replicates from each line at each stage of the differentiation experiment permits the identification of potential reasons for unexpected clustering results. In some cases these may be explained by outlier samples (C), while others may be explained by biological differences among cell lines (D).

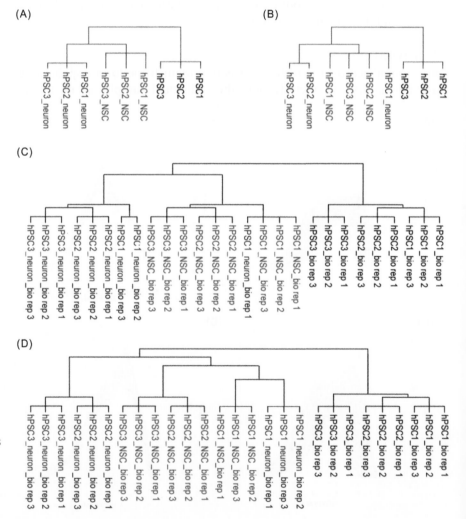

case, had we used our knowledge of the samples *a priori* to identify sample groups, any comparison among the three groups (e.g. differential expression) would have been compromised by the outlier NPC and neuron samples derived from a single cell line, hPSC1. This outlier example also highlights the importance of choosing at least three replicates per group. If we had only included samples from hPSC1 and hPSC2, the results from the global clustering would have been too confusing to move forward with the analysis. However, because we also included samples from hPSC3, we can determine that hPSC1 was the outlier sample and move forward with the analysis of samples derived from hPSC2 and hPSC3, after elimination of the hPSC1 samples. Finally, had we collected multiple biological replicates from each cell line at each stage of differentiation, we could potentially identify the source of this variation (Figure 17.4D, and 17.4C). According to the clustering results presented in Figure 17.4C, we could conclude that the hPSC1_neuron_bio_rep_1 sample is an outlier and should be removed from the analysis. In contrast, if the clustering results were as shown in Figure 17.4D, we might hypothesize that hPSC1 contained some defect that prevented terminal neuronal differentiation. In order to identify potential causes of this defect, we could perform the following statistically supported differential expression analyses:

1. hPSC1 vs hPSC2 and hPSC3 (test for baseline differences)
2. hPSC1_NSC vs hPSC2_NSC and hPSC3_NSC (differences in the NSC population)
3. hPSC1_neuron vs hPSC2_neuron and hPSC3_neuron (differences in the neuronal population)

After sample groups have been identified using global measures, a number of open-source software packages can be used for identification of differentially expressed genes, enriched pathways, enriched transcriptional regulatory motifs, over-represented gene ontologies, and a number of curated datasets. Here we provide a short list of data analysis packages for gene expression data:

- Cluster: http://rana.lbl.gov/EisenSoftware.htm
- Java treeview: http://rana.lbl.gov/EisenSoftware.htm
- Genepattern: http://www.broadinstitute.org/cancer/software/genepattern
- Expander: http://acgt.cs.tau.ac.il/expander
- Genomic Regions Enrichment of Annotations Tool (GREAT): http://great.stanford.edu
- Gene Set Enrichment Analysis (GSEA): http://www.broadinstitute.org/gsea/index.jsp
- Cyctoscape: http://www.cytoscape.org
- Database for Annotation, Visualization and Integrated Discovery (DAVID): http://david.abcc.ncifcrf.gov

- Limma: http://bioconductor.org/packages/release/bioc/html/limma.html
- Lumi: http://www.bioconductor.org/packages/2.0/bioc/html/lumi.html
- RMAExpress: http://rmaexpress.bmbolstad.com
- XCluster: http://www.stanford.edu/group/sherlocklab/cluster.html
- Human Protein Reference Database (HPRD): http://www.hprd.org
- JASPAR: http://jaspar.cgb.ki.se
- The Gene Ontology: http://www.geneontology.org

RNA SEQUENCING (RNA-Seq)

RNA sequencing provides quantitative analysis of gene expression and differential expression, as well as assessment of dynamic transcript architecture (such as alternative promoter usage and exon splicing) and expression of non-coding RNAs (such as microRNAs and long non-coding RNAs). This technique is ideal for analysis of a time course of differentiation or comparing diseased and normal cells. RNA-Seq has been used across a broad range of applications including transcript discovery and annotation, allele-specific expression using SNPs, mutation discovery, identification of gene fusion events and assessment of RNA editing. Current next-generation sequencing (NGS) datasets are archived and hosted in the Short Reads Archive (SRA) (equivalent of GEO) through a partnership among NCBI, the European Bioinformatics Institute (EBI), and the DNA Database of Japan. This section provides an overview of RNA-Seq experimental design, data generation, and analysis, as well as the current challenges associated with this maturing technology.

The RNA-Seq workflow involves isolation of the RNA population of interest (total RNA, polyA, etc.), followed by construction of a sequencing library (which involves conversion to cDNA, fragmentation, and addition of sequencing adaptors), and finally sequencing of millions to billions of short read sequences. Once the short-reads are mapped to a reference genome, mapped sequences can be visualized within a genome browser to assess transcriptional complexity and can be quantified to measure transcript abundance.

General Considerations Prior to Sequencing

To obtain highly informative datasets, the typical aims of RNA-Seq are to obtain:

1. *High coverage and depth*. The aim is to generate enough short-read tags to sample all RNAs, including rare and low-expressing transcripts. Recent reports suggest that ~700 million reads are necessary to identify >95% of the expressed transcripts in a mammalian sample.

2. *Long read length and accuracy with low bias.* Recent developments in library construction methods have been aimed at generating unbiased libraries from small amounts of starting material using rapid and reproducible protocols. The emphasis at the sequencing stage is to generate the greatest quantity of reads at the longest possible read lengths without resulting in sequencing artifacts and ambiguous base calls. High read lengths and low error rates both facilitate accurate alignment of the reads to the genome.

Specific Considerations Prior to Sequencing

To ensure that the desired information is obtained from sequencing output, it is important to map out a detailed experimental design, including an outline of the aims and desired outcomes. The following are several important issues and guidelines to consider prior to sequencing and during experimental design:

1. *RNA-Seq application.* RNA-Seq is highly versatile with many different applications that can be used to sample specific properties of the transcriptome. These various applications include GRO-Seq (Core et al., 2008), deep/nanoCAGE (Plessy et al., 2010; Valen et al., 2009), single molecule sequencing (Tang et al., 2009), and direct RNA sequencing (Ozsolak et al., 2009) (Table 17.1).

2. *Sequencing platform.* There are three major commercial sequencing platforms used for RNA-Seq, each with unique properties (for detailed review please see Metzker (2010): Genome Analyzer IIx (GAIIx)/HiSeq from Illumina (sequencing by synthesis; clonal amplification), SOLiD system from Life Technologies (sequencing by ligation; emulsion PCR),

Table 17.1 Properties of Various RNA-Seq Applications

Name	Description
mRNA-Seq	Sequencing of polyadenylated transcripts
ssRNA-Seq (Strand-specific)	Preserves transcriptional directionality/orientation
Small RNA	Size selected fraction of total RNA (<200 nucleotide (nt) fraction); ideal for microRNA (miRNA), short interfering RNA (siRNA) or other distinct classes of small RNA (~18–40 nt)
Single cell	Deep sequencing of single cells; ideal for low quantity samples and rare cell types
GRO-Seq (Global Run On)	Nascent RNA associated with transcriptionally engaged polymerase
Direct RNA sequencing	Single molecule capabilities; direct amplification of RNA molecules without cDNA conversion
DeepCAGE; nanoCAGE (Cap Analysis of Gene Expression)	Maps precise transcriptional start sites (TSS), 5′ ends, and promoters
Allele-specific expression	Used to identify differential expression of maternally and paternally derived alleles

Table 17.2 Primary Differences Across Major Sequencing Platforms

	GS FLX(454)	GAIlx(Illumina)	AB SOLiD
Read length	~350–400 bp	36,75, or 106 bp	50 bp
Single read	Yes	Yes	Yes
Paired-end reads	Yes	Yes	Yes
Long-insert (several Kbp) mate-paired reads	Yes	Yes	No
Number of reads per run	500 K	>100 M	400 M
Date output (max)	0.5 Gbp	20.5 Gbp	20 Gbp
Run time to 1Gb	6 Days	<1 Day	<1 Day
Base calling	Flow space	base space	Color space
Library preparation	Emulsion PCR	Solid phase amp	Emulsion PCR

and Roche 454 Genome Sequencer FLX (GS FLX) (pyrosequencing; emulsion PCR) (Table 17.2). The primary difference between the SOLiD and GAIIx/HiSeq systems is the sequencing outputs, which are in "base-space" (as in traditional Sanger sequencing (ACTG)) and "color-space" (two nucleotides are read together as one color; where there are 16 possible dinucleotides (4 bases squared) and 4 dyes (red, green, blue, and yellow)) respectively. The Illumina platform is currently the most widely used platform, and therefore has more data analysis tools available than the SOLiD system. The 454 Genome Sequencer FLX differs from the other two, in that the read lengths are significantly longer. However, the 454 platform produces fewer reads per run, which, overall, results in lower depth and coverage. Platform selection determines which library preparation protocols will be used.

3. *Starting sample material*. Starting material quantity and quality are critical for RNA-Seq. Typically, most RNA-Seq experiments require 0.05–2 μg of total RNA, depending on the sequencing application. Solutions to enable analysis of low quantity RNA samples include single molecule sequencing and direct RNA sequencing.

4. *cDNA library properties*. Depending on the experimental questions to be addressed, different RNA populations can be used for sequencing. For example, polyA selection can be used to purify mRNA transcripts, or size selection can be used to enrich for small RNA populations, such as microRNAs. Total RNA is ideal for surveying the complete transcriptional landscape of any given sample. For most sequencing runs, the ribosomal RNA (rRNA) population, which represents around 90% of total RNA in a cell, is depleted using kits such as RiboMinus™ from Life Technologies. Strand-specific RNA-Seq kits such as Encore® Complete RNA-Seq Library System from NuGen are useful for characterization of sense-antisense

transcripts and maintaining strand directionality. There are many commercial kits available to prepare samples for any of these purposes.

5. *Paired-end vs single-end sequencing.* If the purpose of the RNA-Seq experiment is to identify novel events in the transcriptome, such as alternative splicing or insertions and deletions, then paired-end sequencing (vs single-end) is the preferred sequencing method. The paired-end module is used to perform sequencing from both ends of the adapter-ligated fragments. Paired-end tags are also ideal for *de novo* transcriptome assembly (see Figure 17.5E).

Procedure

RNA-Seq workflow can be divided into three stages: *library preparation, sequencing,* and *data analysis* (Figure 17.5). Please note that the procedure outlined below serves as a general overview, as there are sequencing platform-specific differences within these steps. It is important to always consult the workflow suggested by the manufacturer.

Library Preparation (1–3 days)

1. *RNA isolation.* RNA extraction for RNA-Seq is performed using similar methods as are used for microarrays. Either ribosomal RNA-depleted total RNA or polyA RNA is extracted from the biological sample according to the desired sequencing output as described in the cDNA Library Properties section. Small RNA sequencing requires total RNA, which is then processed to enrich for the small RNA population. There are commercial kits available for RNA isolation (Qiagen, etc.) and also sequencing platform-specific kits such as the SOLiD total RNA-Seq.

2. *Fragmentation.* Larger RNA species need to be fragmented into smaller pieces (~200–250 bp); fragmentation is not necessary for the sequencing of small RNAs. RNA fragmentation methods include RNA hydrolysis or nebulization. Alternatively, cDNA can be fragmented using DNase I treatment or sonication. Most sequencing core facilities offer shearing services using either the Covaris E210 (a sonication-based instrument) or Hydroshear (which subjects the DNA to shear forces in a fluid stream).

3. *cDNA conversion.* The RNA fragments are converted to cDNA by reverse transcription primed by random hexamers or by oligo(dT).

4. *Sequence adaptor ligation and size selection.* Platform-specific sequencing adaptors are ligated to the ends of each fragment, which are then amplified using primers complementary to the adapters and subsequently purified. Fragment size selection is usually implemented by gel purification or selective binding to paramagnetic beads.

5. *Multiplexing.* "Indexing" or "barcoding" during library preparation allows for multiplexing of several samples in a single lane of a flowcell. This is

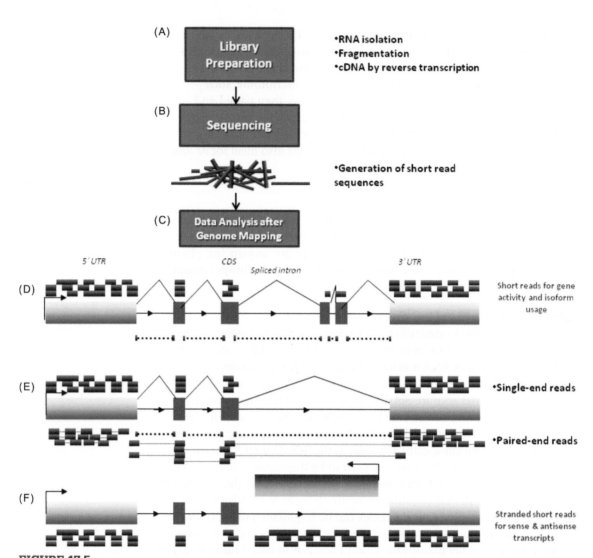

FIGURE 17.5

RNA-Seq workflow. There are three stages for an RNA-Seq experiment: (A) Library preparation; (B) Sequencing to generate short read sequences; (C) Data analysis, which includes mapping of short reads to the reference genome and gene annotations or through de novo assembly (not shown). Once short reads are assigned to a gene (faded pink boxes represent exons and the lines with arrows represent introns), visualization of read alignment to the genome can (D) provide information for gene abundance (blue tags aligned to exons/ coding sequences (CDS)); (E) reveal critical gene architecture information such as alternative splicing events or novel isoforms using reads spanning exon-exon junctions (paired-end) (dotted lines) – examples of single end and paired-end sequencing reads are displayed above and below the gene model, respectively; and (F) provide information on overlapping sense and antisense transcript expression for stranded cDNA libraries.

achieved by synthesizing short index sequences as part of the adaptors that are ligated onto the cDNA fragments; each sample in a given lane will have a unique index, such that the sequencing reads for the different samples can be distinguished. Multiplexing is a cost-effective solution when sequencing multiple samples.

Sequencing (2–10 days)

Each cDNA fragment is sequenced to produce millions of short reads. The sequencing technology varies for each platform (for a detailed review please see Metzker, 2010). During this step, the user specifies the desired read length and whether single reads or paired-end reads are desired.

Data Analysis

A typical RNA-Seq run generates large volumes of data, ranging from tens to hundreds of gigabytes (GB). Analysis of the data, therefore, requires significant computational infrastructure and informatic skills. The output sequencing files (FASTQ for GAIIx/HiSeq and CSFASTA for SOLiD) include annotation regarding the sequencing instrument used to generate the data, read tag properties (single vs paired-end), raw sequence reads, quality scores, and other systematic identifiers. Nearly all data analysis pipelines have the first four phases in common: *filtering, mapping, visualization*, and *expression counting*.

1. *Filtering*: Low quality reads, duplicate reads introduced from PCR bias during library preparation, and residual rRNA sequences are removed. If multiplexing was used, reads for the individual samples are identified and sorted. Adaptor sequences are then trimmed to permit mapping.

2. *Mapping*: For species such as human and mouse, with well-annotated and high-quality genome sequences available, a reference-based assembly is most commonly used. There are many mapping tools that align reads to the reference genome and map splice junctions, such as TopHat, Cufflinks and Scripture. Each tool has a unique algorithm for alignment strategies, speed, and assembly parameters, and all have detailed manuals. Alternatively, *de novo* assemblers such as Trans-ABySS and Trinity do not require reference genomes and are ideal for transcriptome discovery. Most mapping algorithms are open-source and require a Linux-based environment on 64-bit processors. The standard outputs of most mapping algorithms are in the Sequence Alignment Map (SAM) and/or the compressed binary version of SAM (BAM) (see http://samtools.sourceforge.net/) format.

3. *Visualization*: Mapped sequence reads can be viewed in genome browsers such as the University of California Santa Cruz (UCSC) Genome Browser and the Integrative Genomics Viewer (IGV) from the Broad Institute (Figure 17.6). Although the UCSC Genome Browser hosts many genome

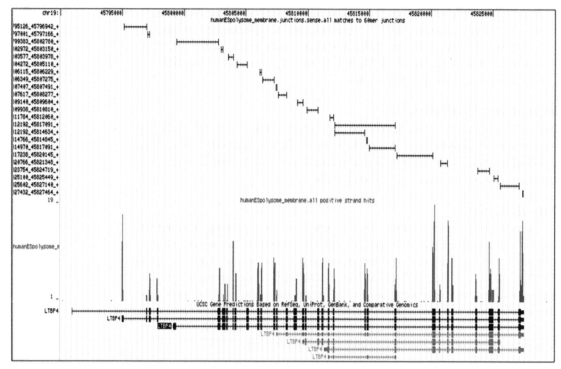

FIGURE 17.6

Screenshot of a typical RNA-Seq experiment visualized in UCSC Genome Browser. RNA-Seq dataset of human embryonic stem cell membrane polysome-associated RNA fraction (dataset from Kolle G et al., 2011). The browser shows a region from chromosome 19 (hg19) for the gene LTBP4. The bottom blue track represents UCSC Gene annotations with 3 known isoforms (dark blue) and 4 predicted transcript models (light blue). The red track represents RNA-Seq short read alignment after mapping. The red peaks pertain to the transcribed regions and highlight transcript specific expression in the 5′ end. The black track (top) represents exon–intron junction reads and highlights novel exon junctions.

annotation tracks that can be viewed along with RNA-Seq datasets, the public version of the browser does not support viewing of large data files. Local versions of the UCSC genome browser can be acquired, but require high levels of computational infrastructure. Alternatively, IGV supports installations that are compatible with most operating systems and can integrate annotation tracks from the UCSC Genome Browser.

4. *Expression counting*: Transcript abundance is calculated based on the number of reads per kilobases per million (RPKM). RPKM is defined as:

$$\text{RPKM} = \frac{\text{total exon reads}}{\text{mapped reads (millions)} \times \text{exon length (KB)}}$$

1 RPKM roughly translates to 1 transcript copy per cell. It is recommended to use reads with unique genomic coordinate start sites for calculating transcript abundance and to avoid bias from PCR artifacts. While RPKM can be measured against any gene annotation database, RefSeq is considered to be the gold standard. Alternatively, the Ensembl and UCSC databases contain more extensive isoform annotations, and include predicted transcript isoforms.

Differential Expression Analysis

A growing number of differential expression analysis packages for RNA-Seq are available, including Bioconductor tools such as DESeq, edgeR, and DEGSeq, all of which are implemented in the R statistical language framework. The output from such programs can be entered into gene ontology functional annotation analysis tools, such as GOSeq (Bioconductor), and commercial software packages, such as Ingenuity Pathway Analysis.

Integrative Data Analysis Platforms and Pipelines

The computational infrastructure and programming expertise required for RNA-Seq data analysis is the biggest challenge faced by most biologists interested in analyzing RNA-Seq data. Galaxy is an open source, web-based platform that has facilitated NGS analysis for researchers by providing an intuitive interface for many sequence analysis tools. Additionally, several commercial software packages, such as Partek, CLC Bio, and GeneSpring, provide integrative analysis platforms similar to microarray gene expression analysis tools.

PITFALLS AND ADVICE

The pitfalls of RNA-Seq are mostly associated with the ongoing evolution of the technology. Many of these issues are being addressed as an ongoing effort.

Standards for Presenting and Exchanging RNA-Seq Data

The Minimum Information About a Microarray Experiment (MIAME) guidelines describe the standards for recording and reporting microarray-based gene expression data to facilitate data access and interpretation from public repositories such as GEO. For RNA-Seq data, similar efforts are currently underway. The ENCODE consortium has released its *RNAseq Standards V1.0* data standards guidelines; although not exhaustive, they emphasize current practices and guidelines to guide experimental design and will be routinely updated.

Instrumentation

Since RNA-Seq instrumentation and computing infrastructure to host NGS datasets can be very costly (hundreds of thousands of dollars), NGS services are typically outsourced to large-scale sequencing and data analysis facilities.

Investment in Computing Power

For laboratories performing RNA-Seq on a regular basis, investments in computing power and storage are essential. An alternative to setting up expensive computing clusters is utilizing cloud computing on platforms such as the Amazon EC2 clusters. Many RNA-Seq analysis tools are being developed for the cloud, including personalized versions of Galaxy.

Sequencing Bias

Several steps during cDNA library preparation can result in uneven coverage along gene bodies. For example, cDNA fragmentation is biased towards the 3′ end versus the 5′ end of transcripts. RNA fragmentation offers more even coverage across the gene body; however, it is not ideal for 5′ and 3′ ends. High GC content and reverse transcription using oligo(dT) or random hexamer primers can also influence the level of coverage along gene bodies. Many of the known sequencing biases are currently being addressed during data analysis through statistically-based bias correction methods (for examples see Roberts et al. (2011) and Oshlack et al. (2009)).

RNA-Seq Summary

RNA-Seq is a relatively recently developed technology that offers additional benefits compared to microarrays for comprehensive profiling of the transcriptome beyond the quantification of gene expression. The benefits of RNA-Seq can be summarized as follows:

1. A relatively unbiased snapshot of the transcriptome, allowing *de novo* identification of transcript isoforms, alternative splicing events, and noncoding RNAs.
2. A large dynamic range of expression that accurately represents transcript levels in cells.
3. Single nucleotide resolution.

READING LIST

Bock, C., Kiskinis, E., Verstappen, G., Gu, H., Boulting, G., Smith, Z.D., et al., 2011. Reference maps of human ES and iPS cell variation enable high-throughput characterization of pluripotent cell lines. Cell 144, 439–452.

Brennand, K.J., Simone, A., Jou, J., Gelboin-Burkhart, C., Tran, N., Sangar, S., et al., 2011. Modelling schizophrenia using human induced pluripotent stem cells. Nature 473, 221–225.

Core, L.J., Waterfall, J.J., Lis, J.T., 2008. Nascent RNA sequencing reveals widespread pausing and divergent initiation at human promoters. Science 322, 1845–1848.

D'Haeseleer, P., 2005. How does gene expression clustering work? Nat. Biotechnol. 23, 1499–1501.

Dimos, J.T., Rodolfa, K.T., Niakan, K.K., Weisenthal, L.M., Mitsumoto, H., Chung, W., et al., 2008. Induced pluripotent stem cells generated from patients with ALS can be differentiated into motor neurons. Science 321, 1218–1221.

Eiges, R., Urbach, A., Malcov, M., Frumkin, T., Schwartz, T., Amit, A., et al., 2007. Developmental study of fragile X syndrome using human embryonic stem cells derived from preimplantation genetically diagnosed embryos. Cell Stem Cell 1, 568–577.

Israel, M.A., Yuan, S.H., Bardy, C., Reyna, S.M., Mu, Y., Herrera, C., et al., 2012. Probing sporadic and familial Alzheimer's disease using induced pluripotent stem cells. Nature 482, 216–220.

Laurent, L.C., Ulitsky, I., Slavin, I., Tran, H., Schork, A., Morey, R., et al., 2011. Dynamic changes in the copy number of pluripotency and cell proliferation genes in human ESCs and iPSCs during reprogramming and time in culture. Cell Stem Cell 8, 106–118.

Marchetto, M.C., Carromeu, C., Acab, A., Yu, D., Yeo, G.W., Mu, Y., et al., 2010. A model for neural development and treatment of Rett syndrome using human induced pluripotent stem cells. Cell 143, 527–539.

Metzker, M.L., 2010. Sequencing technologies - the next generation. Nat. Rev. Genet. 11, 31–46.

Oshlack, A., Wakefield, M.J., 2009. Transcript length bias in RNA-Seq data confounds systems biology. Biol. Direct. 4, 14.

Ozsolak, F., Platt, A.R., Jones, D.R., Reifenberger, J.G., Sass, L.E., McInerney, P., et al., 2009. Direct RNA sequencing. Nature 461, 814–818.

Plessy, C., Bertin, N., Takahashi, H., Simone, R., Salimullah, M., Lassmann, T., et al., 2010. Linking promoters to functional transcripts in small samples with nanoCAGE and CAGEscan. Nat. Methods 7, 528–534.

Pomp, O., Dreesen, O., Leong, D.F., Meller-Pomp, O., Tan, T.T., Zhou, F., et al., 2011. Unexpected X chromosome skewing during culture and reprogramming of human somatic cells can be alleviated by exogenous telomerase. Cell Stem Cell 9, 156–165.

Roberts, A., Trapnell, C., Donaghey, J., Rinn, J.L., Pachter, L., 2011. Improving RNA-Seq expression estimates by correcting for fragment bias. Genome Biol. 12, R22.

Sheridan, S.D., Theriault, K.M., Reis, S.A., Zhou, F., Madison, J.M., Daheron, L., et al., 2011. Epigenetic characterization of the FMR1 gene and aberrant neurodevelopment in human induced pluripotent stem cell models of fragile X syndrome. PLoS One 6, e26203.

Tang, F., Barbacioru, C., Wang, Y., Nordman, E., Lee, C., Xu, N., et al., 2009. mRNA-Seq whole-transcriptome analysis of a single cell. Nat. Methods 6, 377–382.

Valen, E., Pascarella, G., Chalk, A., Maeda, N., Kojima, M., Kawazu, C., et al., 2009. Genome-wide detection and analysis of hippocampus core promoters using DeepCAGE. Genome Res. 19, 255–265.

Recommended online links

ENCODE RNA-Seq Guidelines: <http://encodeproject.org/ENCODE/protocols/dataStandards/ENCODE_RNAseq_Standards_V1.0.pdf>.

Galaxy: <http://main.g2.bx.psu.edu>.

Tuxedo Suite for mapping (Bowtie, TopHat, and Cufflink):<http://tophat.cbcb.umd.edu/>

Bioconductor: <http://www.bioconductor.org>

SeqAnswers (discussion forum for NGS technologies): <http://seqanswers.com>

UCSC Genome Browser: <http://genome.ucsc.edu/index.html>

IGV Browser: <http://www.broadinstitute.org/igv/>

CHAPTER 18

PluriTest Molecular Diagnostic Assay for Pluripotency in Human Stem Cells

Johanna E. Goldmann, Bernhard M. Schuldt, Michael Lenz, and
Franz-Josef Müller

EDITOR'S COMMENTARY

Pluripotent cells have a unique phenotype that is defined by continuous self-renewal along with the ability to differentiate into mature cells representing the three germ layers (mesoderm, ectoderm, and endoderm). For mouse pluripotent stem cells, the definitive assay for pluripotency is the germ line transmission test; in the absence of such a clear assay for human cells, other tests must be used to confirm the pluripotency of human pluripotent stem cells (hPSCs). Induction of teratomas in immunodeficient mice (Chapter 21) is a common method for showing that a human cell line can differentiate into representative cell types from all three germ layers. The tissues that develop in experimentally induced teratomas are remarkably complex and resemble the structure of benign spontaneous clinical teratomas (Chapter 22). As an efficient assay for pluripotency, experimental teratomas have several shortcomings. First, there is no generally accepted set of standards for generating and analyzing teratomas; laboratories use a variety of methods, and many publications on new cell lines do not even report teratoma assays, making comparisons among studies problematic (Müller et al., 2010). Second, the teratoma assay is both technically challenging and resource intensive, requiring a mouse facility and skilled technicians. With the rapid increase in the generation and use of human ESCs and induced pluripotent stem cells (iPSCs), and concerns in some countries about the use of animals in research, there is a need for a fast, cost-effective, and animal-free alternative to the teratoma assay. We had developed a database of global gene expression profiles of hundreds of hESC and iPSCs. To model the expression characteristics we selected several hundred cell lines that were known to form teratomas, and used their gene expression profiles to model the key characteristics of hPSCs. The result is a molecular diagnostic test for pluripotency that we call "PluriTest". It is not a marker-based test; rather, it takes into account both the similarities and variations among cells that are proven to be pluripotent.

J.F. Loring & S.E. Peterson (eds): Human Stem Cell Manual, Second edition.
DOI: http://dx.doi.org/10.1016/B978-0-12-385473-5.00018-7

This chapter describes PluriTest in detail, provides information about how to use it and how to troubleshoot the results, and explains future improvements.

OVERVIEW

To provide an alternative to the teratoma assay for testing pluripotency in human stem cells, we developed a bioinformatic alternative assay called PluriTest (Müller et al., 2011). PluriTest assesses a cell's pluripotency based on global gene expression data as determined by microarray analysis. As a web-based assay, PluriTest provides easy access for stem cell researchers to a comprehensive alternative to the teratoma assay. PluriTest can be used as a standard measure of pluripotency as well as a resource providing information beyond what the teratoma assay reports. In this chapter we will outline how to use the PluriTest as a routine measurement of pluripotency. Furthermore we will provide some explanations concerning the biological and computational principles underlying PluriTest and provide some context-specific interpretations of PluriTest results.

All data entered into PluriTest is run through a standardized data processing and analysis pipeline. Briefly, raw microarray data in the .idat file format (Illumina) is normalized and transformed following published methods (Du et al., 2008). The processed data are then projected against a data-based model of pluripotency derived from the transcriptional profiles of hundreds of validated pluripotent and somatic cell lines. This projection allows for the direct comparison of your submitted sample data to hundreds of previously characterized datasets.

PluriTest reports back the following possible results:

- Whether the uploaded samples express a signature unique to pluripotent stem cells
- Whether the uploaded samples have a great similarity to know pluripotent stem cells but also express gene signatures that cannot be explained by a "pure" pluripotent signature
- If the uploaded samples' transcriptomic signature is significantly different from the global transcriptional make-up of human pluripotent stem cells.

The ability of PluriTest to assess pluripotency within a sample is based on the construction of a highly validated, computational ("machine learning") model of pluripotency based on gene expression profiles. Previously reported results have clearly demonstrated that gene expression profiles can be used as a unique identifier of a cell type or cellular properties such as pluripotency (Müller et al., 2008). The current PluriTest model was constructed using the gene expression data from several hundreds of well-defined and characterized

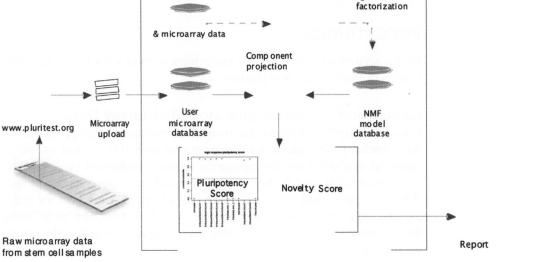

FIGURE 18.1
PluriTest workflow.

hESC and hiPSC samples, as well as approximately the same number of somatic, non-pluripotent cells. For an in-depth description of the mathematical model derivation, optimization, and bioinformatic processing steps please refer to our study describing the generation of the PluriTest model (Müller et al., 2011).

Using PluriTest for the identification of the pluripotent phenotype consists of a sequence of steps involving lab work and data processing through a web interface (see below). An overview of the PluriTest workflow is depicted in Figure 18.1.

These steps are discussed in this chapter:

1. Generation of gene expression microarray data
2. Uploading microarray data to www.pluritest.org
3. Data processing and bioinformatic analysis by PluriTest
4. Receiving a PluriTest report
5. Interpretation of the PluriTest results.

PluriTest determines two related but not identical metrics for any given sample: a *Pluripotency Score* and a *Novelty Score*. The *Pluripotency Score* indicates if a sample has shown up-regulation of pluripotency associated genes while the *Novelty Score* detects novel and potentially biologically relevant characteristics in the sample that may significantly diverge from most pluripotent stem cells. For real world examples for both scores, please see the *Output* section.

PROCEDURES

In this section we will cover the procedures necessary for use of the online version of PluriTest at www.pluritest.org.

Generation of Gene Expression Microarray Data

Currently PluriTest supports the analysis of microarray data generated with the Illumina gene expression microarray platforms HT12v3 and v4. We recommend that you have the microarray analyses performed by your institution's microarray core facility or a commercial microarray service provider. If there is not considerable experience and advanced equipment for extracting and especially quality controlling RNA in your lab (such as an Agilent Bioanalyzer), we also recommend that you outsource this task to the microarray core laboratory.

For mRNA extraction we have tested and recommend use of Ambion (Mirvana) or Qiagen extraction kits. We also recommend use of at least one million snap-frozen cells per sample, as we have observed that while lower cell numbers might still yield enough mRNA for microarray analysis, lower cell numbers frequently result in suboptimal RNA quality as measured, for example, by the RNA Integrity Number (RIN) by Agilent Bioanalyzers. We recommend not considering any RNA samples below a RIN of 9 for microarray and subsequent PluriTest analysis.

When evaluating your samples we further recommend running your samples in duplicates. PluriTest results are very robust against technical variation and a single array is statistically sufficient to compute reliable Pluripotency and Novelty Scores, yet the only way to detect possible sample and/or sample labeling mix-ups during the upstream wet lab processing steps is to include biological replicates.

Uploading Microarray Data to www.pluritest.org

In order to use PluriTest online, a user has to sign up for an account at the website www.pluritest.org and log in with their credentials (user name and password). Next, the microarray data are uploaded through a web interface (Figure 18.2). Depending on your computer setup, it may be necessary to install the Microsoft Silverlight 4 plug-in on your computer (see also Pitfalls and Advice: Frequently Asked Questions).

FIGURE 18.2
The data upload dialog box
at http://www.pluritest.org

PluriTest uses the raw microarray data which are stored in the .idat format. The .idat format is a proprietary file format created by Illumina, similar to Affymetrix's proprietary .CEL-file format. These files contain raw, gene probe-specific fluorescent intensity data from a hybridized array chip read by an Illumina array scanner (for more information regarding the format and why using the .idat file is critical please see the Pitfalls and Advice: Frequently Asked Questions). Using the .idat files allows PluriTest to normalize the transcriptomic data sets both efficiently and consistently.

To get these files from a microarray service provider, ideally inform them ahead of the planned experiment for this data output requirement. If you want to include samples that were previously analyzed on Illumina HT12v3 arrays, it is best to ask the core personnel to go back to the scanner output and have them send you the files that were used to generate the conventional GenomeStudio/BeadStudio output tables (the same .idat files are used to generate the table output for your GenomeStudio/BeadStudio analysis). When uploading your data please be aware you can only upload and analyze between 4 and 12 samples at a time.

Data Processing and Bioinformatic Analysis by PluriTest

After you have uploaded the .idat files, PluriTest will automatically extract the raw data and perform a variance stabilizing transformation (VST) (Du et al., 2008), robust spline normalization (RSN) (Du et al., 2008) and perform the PluriTest projection (Müller et al., 2011). This process should take no longer than 5–10 minutes.

Receiving a PluriTest Report

For inspection of a specific PluriTest, select your analysis run and either click the "View Results" icon or download the PluriTest Scores by clicking the "Results Tables" icon (Figure 18.3). The former will open an HTML web page with bitmap images, which can be easily copied and pasted into a word processor (e.g. Microsoft Word) for addition to a manuscript as supplementary material.

Interpretation of the PluriTest Results

The first question addressed with the PluriTest results is whether or not the sample has expression levels of genes characteristic of pluripotent cells and thus can be considered potentially pluripotent. This is the first part of the PluriTest report and is formalized as the *Pluripotency Score.*

Second, PluriTest can also indicate if an uploaded sample contains any novel characteristics that are generally not found in pluripotent samples with a so-called *Novelty Score*. In the next section we will describe the interpretation of both metrics on the basis of some real world examples.

PLURITEST OUTPUT

Pluripotency Score

The Pluripotency Score is designed to distinguish cells containing a pluripotent cellular gene expression profile from a non-pluripotent sample. It is optimized to robustly distinguish between pluripotent and non-pluripotent samples while allowing PluriTest to be flexible against experimental and technical variations. The PluriTest report will include graphs depicting the PluriTest results for up to 12 arrays at once.

The Pluripotency Score is a parameter that is based on all samples (pluripotent cells, somatic cells and tissues) in a stem cell model data matrix (Müller et al., 2011). Samples with positive Pluripotency Score values are more similar to the pluripotent samples in the model matrix than to all other classes of

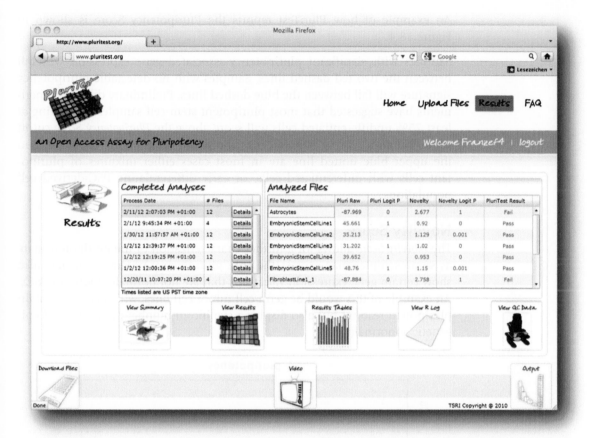

FIGURE 18.3
PluriTest reports page.

samples in the matrix. The area between the red lines indicates the range that contains approximately 95 % of the pluripotent samples tested (Figure 18.4). The Pluripotency Score gives an indication if a sample contains a pluripotent signature, but not necessarily if the cell preparation is a normal, bonafide hESC or iPSC. Partially differentiated pluripotent cells, teratocarcinoma cells or karyotypically abnormal embryonic stem cells may also have a high Pluripotency Score. The blue lines indicate those scores that we have observed in approximately 95 percent of the non-pluripotent samples (Figure 18.4).

▪ Note

We found it useful when submitting manuscripts for peer review to include reference samples into the PluriTest analysis to demonstrate, for example, the relative position of the studied iPSC line with fibroblasts and widely used human embryonic stem cell lines (such as WA01 or WA09). ▪

An example of how PluriTest reports the Pluripotency Score is shown in Figure 18.4.

Samples with significant pluripotency-associated gene expression will fall between the two red dashed lines. Samples with no detectable pluripotency signature will fall between the blue dashed lines. Preliminary dilution experiments have suggested that most pluripotent stem cell samples containing at least 75% undifferentiated cells will score above the *Pluripotency Score* cutoff value. Samples scoring between the cutoffs indicated by the lower red and the upper blue dotted line are in most cases either mixtures of pluripotent cells with differentiated cells, or early differentiating PSC-derived cell populations.

Novelty Score

The *Novelty Score* allows for a further characterization of the pluripotent state as some samples may express the core pluripotency program but may also show certain gene expression variations that are not normally found in PSCs. An example of this would be germ cell tumor cell lines, which clearly contain pluripotent features but also display a significant number of genetic and epigenetic abnormalities.

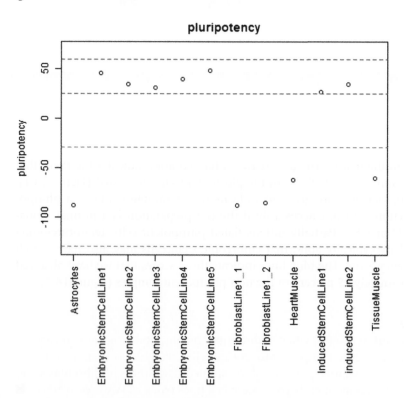

FIGURE 18.4
Pluripotency Score.

The Novelty Score is based on well-characterized pluripotent samples in a stem cell model data matrix (Müller et al., 2011). A low Novelty Score indicates that the test sample can be easily categorized based on existing data from other well-characterized iPSC and ESC lines. A high Novelty Score indicates that there are patterns in the tested sample that cannot be explained by the currently existing data from well-characterized, karyotypically normal pluripotent stem cells. Partially differentiated pluripotent cells, teratocarcinoma cells or karyotypically abnormal embryonic stem cells may have a high pluripotency score but cannot be easily categorized with data from well-characterized, normal pluripotent stem cells and thus are expected to have a high Novelty Score.

■ Note

Preliminary results indicate that the Novelty Score is more sensitive to early differentiation events and epigenetic as well as genetic variation among PSC samples than the Pluripotency Score. Surprisingly, the Pluripotency Score often counter-intuitively increases at early time points in PSC differentiation experiments (suggesting an "increase" in pluripotency). The Novelty Score also increases early on during differentiation, suggesting that differentiation-specific transcriptional programs are "booted up" while most of the PSCs are dominated by a pluripotent signature.   ■

An example of how PluriTest reports the Novelty Score is shown in Figure 18.5.

Density Plot

The density plot combines Pluripotency Scores and Novelty Scores with an empirical density distribution of pluripotent and somatic samples from the underlying stem cell model data matrix (Müller et al., 2011). This graph is best suited for use as supplementary material for peer reviewed stem cell studies (see Figure 18.6).

■ Note

The density plot can easily be exported to a graphic editor (such as Omnigraffle or Adobe Illustrator) and the samples indicated by black circles in the plot color coded and labeled for facilitated visual inspection by manuscript or grant application reviewers.   ■

Quality Control Outputs

The online PluriTest reports two basic microarray quality control plots to the user.

The first is a box plot (Figure 18.7) generated by the Lumi package after the samples were transformed with VST and before RSN. Outlier arrays with too much technical variation might be spotted if they show a different probe

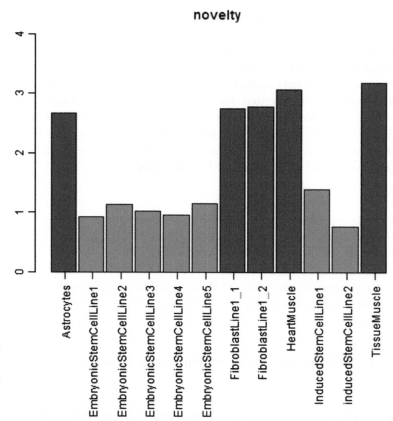

FIGURE 18.5
Novelty Score. Samples are color-coded green (pluripotent), orange (not shown in this example), red (not-pluripotent) based on the probabilities given from the logistic regression model. Orange and red samples are more dissimilar to the pluripotent samples in the model matrix than the other pluripotent samples in the matrix.

intensity distribution pattern in the box-plots when compared to the other arrays on the same chip or when compared to arrays on other chips.

The second is a simple clustering dendrogram drawn from the uploaded samples (Figure 18.8). If the user has included reference samples (such as fibroblasts and hESC samples), inferences can be drawn from the dendrogram. The hierarchical clustering plot is generated by the Lumi package after the samples are transformed with VST and before RSN. Outlier arrays with too much technical variation can be spotted if they do not cluster with their respective technical or biological replicates from the same sample or sample type. For example, if an array hybridized with the RNA from a pluripotent cell line clusters with fibroblasts on the same chip, but not with other pluripotent samples, something might be wrong with technical aspects of your experiment. Fibroblasts should never cluster with PSC samples. If this should be the case, either sample mix-ups or microarray hybridization issues should be considered first before Pluripotency and Novelty Scores are interpreted.

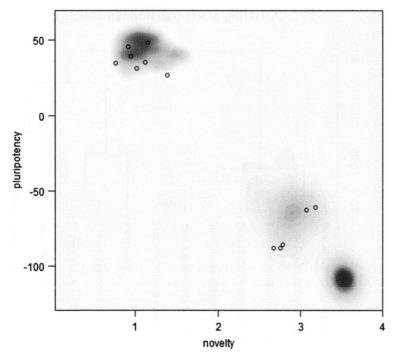

FIGURE 18.6
The density plot. The density plot combines the *Pluripotency Score* on the y-axis with the *Novelty Score* on the x-axis. The red and blue backgrounds illustrate the empirical distribution of the pluripotent (red) and non-pluripotent samples (blue) in the underlying stem cell model data matrix.

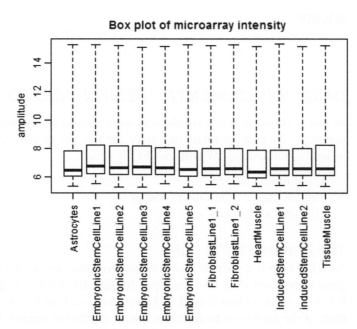

FIGURE 18.7
Sample box plots.

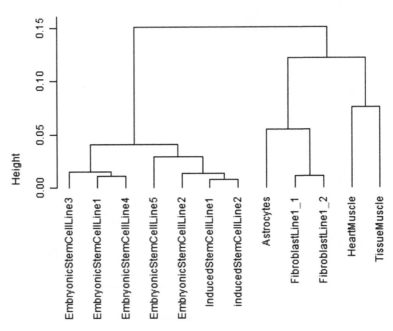

FIGURE 18.8
Hierarchical sample clustering plot.

■ **Note**

For details please see Du, Kibbe, and Lin, 2008 and the vignette to the lumi package on the BioConductor website. ■

DATA, EXPERIMENTAL DESIGN, AND FORMATS

In this section we will provide you with additional information regarding materials, data and formats necessary for the use of PluriTest and possible experimental designs. We will specifically cover:

1. The Illumina HT12 Chip
2. The "challenge experiment" design
3. The .idat file format.

The Illumina HT12 Chip

An HT12 chip is a glass slide with dimensions that are similar to a microscope glass slide. Each chip has a unique 10-digit ID number (e.g., 449359422). It has 12 areas etched into the glass surface, each of which is a microarray. Each of the arrays can be hybridized with a different sample, hence the name

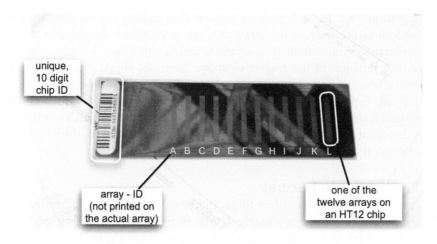

unique, 10 digit chip ID

A B C D E F G H I J K L

array - ID (not printed on the actual array)

one of the twelve arrays on an HT12 chip

FIGURE 18.9
Anatomy of an Illumina HT12 chip.

H(igh) T(hroughput) 12(arrays), see Figure 18.9. More information on microarrays is provided in Chapter 17.

Each array is treated separately with biological material, yet all the hybridization steps, incubation, etc. are done on all 12 arrays in parallel. This design is unlike the original Affymetrix system, in which each GeneChip cartridge contains one array, and more like the recently introduced 24- or 96- Affymetrix array plates. Each slot/array on a chip is identified by a letter from A to L. Thus the unique ID for each array consists of the 10-digit chip ID plus the letter denoting the array/slot on the HT12 array (e.g. 449359422E).

The "Challenge Experiment" Design

A specific experimental design based on the anatomy of an HT12 chip can be used for PluriTest experiments. We internally have often referred to this design as a "challenge experiment." For example, if a researcher wants to test four novel iPSC lines, these can be run on a single chip in biological duplicates. The remaining four arrays on a chip then could be hybridized with two samples from a well-characterized iPSC or hESC line maintained in the same culture conditions as the novel iPSC lines. Two more samples could be hybridized with samples from somatic cells (e.g. fibroblasts or keratinocytes) that were used to generate the novel iPSC lines.

Similarly, a researcher could design an experiment around a differentiation time course with two biological replicates representing the undifferentiated PSC stage, two biological replicates representing the somatic equivalent of the cell type which the researcher wants to obtain (e.g. human hepatocytes or CD34+ hematopoietic stem cells) and then 8 samples from time points spanning the differentiation time course.

It is important to note that we suggest distributing the samples in a near-random fashion across the array, as it has been demonstrated that there is a systematic signal variation on single Illumina chips that is dependent on the relative position of the respective arrays. Basically, signal intensities decrease from the right to the left side on a chip and if this variance is co-varying with different experimental conditions (e.g. by hybridization of sequential differentiation steps in a time course experiment with the same order as the arrays A–L on an Illumina HT12 chip), can cofound statistical differential gene expression analysis (Kitchen et al., 2010).

The .idat File Format

The .idat files are present in a folder that is written out by the Illumina microarray scanner. Its size is roughly 47 MB all together. For an example of how such a folder and its files normally look, please see Figure 18.10. Some core services produce a much larger, more detailed readout (called "bead-level data") by changing specific settings on their microarray scanner. In such cases the folder can be up to 1 GB in size. We currently do not access or use this information, although for internal quality control reasons we have generated bead-level data for all chips that were used for PluriTest model construction. This folder is named with the same 10-digit ID number (e.g. 449359422) as

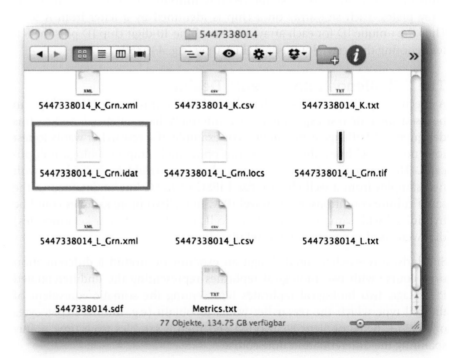

FIGURE 18.10

.idat files in a typical Illumina array scanner folder (marked with a red border).

the HT12 chip on which your experiments were run. It usually contains five file types; three of them are specific for each of the 12 arrays on the HT12 chip. The three array-specific file formats are .idat, .xml, and .jpg. The .jpg is a picture of the actual array, the .xml contains information pertaining to the probe content and the .idat file is, as mentioned, a binary container for the most relevant data for the specific array.

The naming convention for each array is the 10-digit ID number of the chip, followed by a letter ranging from A to L denoting one of the 12 array positions present on an HT12 array followed by the letters "Grn" (for the color used for labeling the chips, which in the case of the HT12 single color arrays is Cy3 ("green-ish")).

For using PluriTest, just navigate to the chip folder after clicking upload .idat files and only the PluriTest-relevant .idat will be selectable. Select the .idats you want to analyze, and click on upload.

PITFALLS AND ADVICE: FREQUENTLY ASKED QUESTIONS

Q: What is an .idat file?

A: This is a proprietary file format created by Illumina, similar to the proprietary .CEL-file format from Affymetrix. These files contain raw, gene probe-specific fluorescent intensity data read from a hybridized array chip in an Illumina array scanner. These files can only be read using Illumina's proprietary data extraction and analysis software, BeadStudio/GenomeStudio. This software translates the data into a ".txt" format that can be easily read and manipulated for compatibility with most available data analysis programs. In most cases, microarray core facilities use GenomeStudio to extract data from .idat files to provide their customers with .txt files that can be more broadly manipulated. We have incorporated the .idat-reader module from Illumina into PluriTest. Most importantly, the .idat file is completely raw and unmodified data.

Q: My core facility always gives me table format files from my microarray experiments; they tell me that's all I need. What do I do now?

A: Please ask your microarray facility to go back to the scanner output and to give you the files they have used to generate the GenomeStudio/BeadStudio output. Your microarray core staff has used the same .idat files for generating the table output for your experiment in GenomeStudio. We have learned, while testing the feasibility of the web-based PluriTest approach, that some very high turnover cores do actually erase the raw output (.idat) files and only retain the preprocessed data. This is very unfortunate and makes it impossible for you to use the web version of PluriTest with your data. If you are planning a new experiment please get in touch with your core facility before they run your experiment and ask them specifically to RETAIN and GIVE the data folder containing the .idat files from your experiment to you.

If your .idat files were erased, we provide the R/Bioconductor script as a download from the *Nature Methods* website, which essentially performs the same data projection modeling task and can start from the GenomeStudio/BeadStudio output after preprocessing it with the Bioconductor lumi-package. This can give you the same results as we provide with this website. Please understand that we can't support or troubleshoot your results from BioConductor. Instead, we strongly encourage you to use PluriTest to avoid problems associated with R script errors. We highly encourage you to retain and save the whole Illumina scanner output folder for future reference.

Q: Why do you only allow .idat files as input for PluriTest?

A: Primarily, .idat files are completely unmodified data files that contain information required for normalization and probe filtering of your data. We have worked hard to find the very best means to normalize these datasets and we don't want you to have to worry about this. PluriTest consists of hundreds of steps that we have incorporated into a single automated pipeline that executes in minutes. Just to give you some perspective: for our first paper, four bioinformatics specialists worked 6 months to complete a single analysis (Müller et al., 2008). With PluriTest, we can complete this truly predictive model within minutes. Using .idat files in PluriTest allows us to process all the data in a standardized fashion for an accurate, reproducible, and user-friendly assessment of your samples.

More specifically, we have two important reasons for using the .idat files:

1. The output from GenomeStudio is usually not complete: these output files usually do not contain critical information about the probes, which is designed to inform the user about quality criteria pertinent to each chip. While this information can be extracted from GenomeStudio, one has to go through several error-prone steps.
2. GenomeStudio offers some preprocessing functions, which, if used with PluriTest, would severely impair its predictive capabilities. We can't go into the details here, but there is plenty of room for error, a wrong checkbox here, a misguided click there and the whole thing just doesn't give the right answer. We have a well-trained and experienced team, some of whom have been working with Illumina arrays since before they were even commercially available, while others have been working as supporting scientists in the analysis of theses arrays and as bioinformatics core specialists for many years. All of the steps automated by PluriTest.org

can be performed using R/BioConductor and can be executed following a documented data transformation and normalization procedure with the R scripts provided as supplementary information to our *Nature Methods* manuscript (Müller et al., 2011). If you are up to it, you definitely can do all of this yourself, and every step is well supported by the literature.

Q: Why are you restricting the upload to 12 arrays, and why should the samples be run on the same array?

A: PluriTest is using the 12 samples processed in parallel as a unit that underlies similar technical variation. We found that the key to biologically relevant, predictive results was to discern biological variation from technical variation in our model building process. In our modeling efforts we discovered that batch effects from the multiple arrays processed in parallel could be used as the lynchpin to understand the technical variation and improve the biological interpretation, including predictive signatures. For this, we treat the newly uploaded arrays as chip/batch (because all arrays were then processed in parallel). If you have data from arrays that were run on different chips, you can run them through PluriTest chip-wise, but not all at once, even if their overall number is below twelve.

If your .idat files are not all run on the same chip or you want to process a larger batch than 12 arrays at once, we provide the R/Bioconductor script as supplementary information with our *Nature Methods* paper, which essentially performs the same data projection modeling task and can start from the GenomeStudio/BeadStudio output after preprocessing it with the BioConductor lumi-package. Still, we do not believe that it is truly relevant to process more than twelve arrays at once, since even we in the Loring Lab have rarely more than 60 iPSC lines to test at any given time point, which could be tested together. Testing 60 iPSC lines, even with several biological replicates can be achieved in one afternoon. Considering that even with the most cutting edge reprogramming methods it most likely took many weeks to generate so many lines, we don't think there is currently a pressing need for a large batch processing function. If you think this is important for you, please drop us a line by email fj.mueller@zip-kiel.de or schuldt@aices.rwth-aachen.de.

Q: What is Silverlight? Why do I have to install it on my computer to run the online PluriTest? And why doesn't the PluriTest.org run on my Linux box?

A: To quote Microsoft's Silverlight page (http://www.microsoft.com/silverlight/what-is-silverlight/):

> Silverlight is a powerful development platform for creating rich media applications and business applications for the Web, desktop, and mobile devices. Silverlight is a free plug-in powered by the .NET framework

that is compatible across multiple browsers, devices and operating systems to bring a new level of interactivity wherever the Web works. With support for advanced data integration, multithreading, HD video using IIS Smooth Streaming, and built in content protection, Silverlight enables online and offline applications for a broad range of business and consumer scenarios.

In less technical terms, Silverlight is a way for us to deliver a completely automated, bioinformatics analysis pipeline via the web. Our main intent was to make PluriTest accessible to "the rest of us" (to quote Apple, and not Microsoft, this time), especially with stem cell biologists in mind who are not experts in bioinformatics and maybe even still think that "R" stands only for the 18th letter of the alphabet and have never heard of the powerful statistical programming language with the same letter as its name.

Silverlight4, which currently needs to be installed on a computer in order to access PluriTest.org, runs on every Mac and Windows computer with a recent operating system. We had to work within a Microsoft product environment (90% of the folks involved in PluriTest are actually Mac- and Linux-based!), because this was the only way to use Illumina's proprietary .idat-reader module with PluriTest's quintessential, controllable .idat-parsing functionality. If we had not decided to create PluriTest within the Microsoft product world, every user of PluriTest, would have been forced to use GenomeStudio, which is not free and usually only owned by microarray cores. We wanted to enable PluriTest users to be independent of GenomeStudio.

If you are a user, who:

- has no access to a Mac or Windows machine
- gets the GenomeStudio output and uses only a Linux machine for bioinformatics analysis

you won't be able to use the web-PluriTest, but you can use the R/Bioconductor script.

All of the steps automated by PluriTest.org can be performed using R/BioConductor and can be executed following a documented data transformation and normalization procedure with the R scripts provided with our *Nature Methods* paper. If you are up to it, you definitely can do all of this yourself, and every step is well supported by the literature.

Q: My lab has been working with Affymetrix chips since at least the time of the Boston Tea Party, why do you support only Illumina?

A: Key for the PluriTest model is a consistent, large dataset with many biological replicates and a good statistical grip on batch effects and technical variations.

We are not aware of an Affymetrix stem cell dataset that fulfills these requirements. In principle, it should not be difficult to adapt PluriTest to the Affy data. We show in our *Nature Methods* paper that the predictive model for pluripotency works reasonably well with published Affy data, but is very limited in further in-depth exploratory analyses. Our decision for not supporting Affy data with the web-PluriTest at this point is due to the following two reasons:

- The basis for our model is the StemCellMatrix2, which was run on Illumina. We learned in our previous study on the StemCellMatrix1 that perfect control over the data generation and model building is quintessential to be able to come up with truly predictive models and not to generate over-fit signatures. To deal with the "garbage-in–garbage-out" problem we had to first build the dataset and we had to use a widely accepted, but also inexpensive, microarray platform.
- Our software development team has very limited bandwidth. We have only two part-time programmers working on PluriTest and hence have to work in a very focused manner. Thus, PluriTest would have never become a reality if we took the time to implement an Affy upload tool.

We're currently working on porting PluriTest to other platforms and even other species; if you are interested in getting involved, please contact Jeanne Loring (jloring@scripps.edu) or Franzef Müller (fj.mueller@zip-kiel.de).

READING LIST

Bradley, A., Evans, M., Kaufman, M.H., Robertson, E, Bradley, A., Evans, M., et al., 1984. Formation of germ-line chimaeras from embryo-derived teratocarcinoma cell lines. Nature 309 (5965), 255–256.

Dolgin, E, Dolgin, E., 2010. Putting stem cells to the test. Nat. Med. 16 (12), 1354–1357.

Du, P., Kibbe, W.A., Lin, S.M, Du, P., Kibbe, W.A., Lin, S.M., 2008. Lumi: a pipeline for processing Illumina microarray. Bioinformatics (Oxford, England) 24 (13), 1547–1548.

Kitchen, R.R., Sabine, V.S., Sims, A.H., Macaskill, E.J., Renshaw, L., Thomas, J.S., et al., 2010. Correcting for intra-experiment variation in Illumina BeadChip data is necessary to generate robust gene-expression profiles. BMC Genomics 11, 134.

Müller, F.-J., Laurent, L.C., Kostka, D., Ulitsky, I., Williams, R., Lu, C., et al., 2008. Regulatory networks define phenotypic classes of human stem cell lines. Nature 455 (7211), 401–405.

Müller, F.-J., Goldmann, J., Löser, P., Loring, J.F, Müller, F.-J., Goldmann, J., et al., 2010. A call to standardize teratoma assays used to define human pluripotent cell lines. Cell Stem Cell 6 (5), 412–414.

Müller, F.-J., Schuldt, B.M., Williams, R., Mason, D., Altun, G., Papapetrou, E.P., et al., 2011. A bioinformatic assay for pluripotency in human cells. Nat. Methods.

Nagy, A., Rossant, J., Nagy, R., Abramow-Newerly, W., Roder, J.C, Nagy, A., et al., 1993. Derivation of completely cell culture-derived mice from early-passage embryonic stem cells. Proc. Natl. Acad. Sci. USA 90 (18), 8424–8428.

Peterson, S.E., Tran, H.T., Garitaonandia, I., Han, S., Nickey, K.S., Leonardo, T., et al., 2011. Teratoma generation in the testis capsule. J. Vis. Exp. 57, e3177.

Epigenetics: Analysis of Histone Post-Translational Modifications by Chromatin Immunoprecipitation

Elisabetta Soragni, Sherman Ku, and Joel M. Gottesfeld

EDITOR'S COMMENTARY

The identity and functionality of cells are determined by the specific sets of genes that are expressed in each cell type, which in turn are regulated by other factors that affect translation and modifications of the proteins coded by the genes. The regulation of genes at the DNA level is termed "epigenetics", a term that expresses the fact that the DNA sequence – the genetics – alone do not determine the phenotype of a cell. In the last few years a considerable amount of effort has been devoted to understanding the epigenetic underpinnings of pluripotency and differentiation of human pluripotent stem cells (hPSCs). Gene expression is mainly coordinated by two major types of epigenetic control: methylation of CpGs (as well as CpAs and perhaps others) in regulatory regions, which often results in silencing of expression; and the binding of specific forms of modified histones that can enhance or repress transcription of genes. Chapter 20 describes methods for studying DNA methylation in hPSCs.

OVERVIEW

In this chapter, methods are provided for studying chromatin dynamics, the binding of modified histones to specific genes in hPSCs and their derivatives. hPSC chromatin is thought to be more "open" and hPSCs are characterized by an expression program that is defined by changes in histone post-translational modifications (PTMs) as well as the action of chromatin-remodeling complexes (for a review see Young, 2011).

Histone N-terminal tails are modified post-translationally on different residues (mostly lysines) and the combination of such modifications – the so-called "histone code" (Strahl and Allis, 2000) – determines the transcriptional activity of the genomic locus with which they associate. The most well-studied histone PTMs are acetylation and methylation. Many studies have established that

313

J.F. Loring & S.E. Peterson (eds): Human Stem Cell Manual, Second edition.
DOI: http://dx.doi.org/10.1016/B978-0-12-385473-5.00019-9

acetylation is associated with gene activation, while methylation is associated with either repression or activation, although it is now known that the interplay among these different modifications constitutes a complex language rather than a simple code (Berger, 2007). Histone tail modifications can directly affect chromatin structure or act through the recruitment of effector proteins and are "written" and "erased" by specific chromatin modifiers such as histone acetylases and deacetylases, methyltransferases and demethylases (Gardner et al., 2011). In this chapter we focus on a subset of histone H3 and H4 PTMs that are routinely analyzed in our laboratory to study the chromatin composition of the *FXN* gene locus (Herman et al., 2006). Acetylation of histone H3 lysine 9 (H3K9ac), H3 lysine 14 (H3K14ac), H4 lysine 5 (H4K5ac), H4 lysine 8 (H4K8ac), H4 lysine 12 (H4K12ac), and H4 lysine 16 (H4K16ac) are associated with active genes. Conversely, di- and trimethylation of H3 lysine 9 (H3K9me2 and H3K9me3) and trimethylation of lysine 27 (H3K27me3) are associated with constitutive heterochromatin, silent and poised genes that are repressed but can be rapidly turned on under activating conditions. High levels of trimethylation of H3 lysine 4 (H3K4me3) are associated with the 5′ regions of active genes and trimethylation of H3 lysine 36 (H3K36me3) is a mark of transcriptional elongation. For a recent review on histone modifications see Suganuma and Workman (2011).

Chromatin structure and dynamics play a pivotal role in the maintenance of human stem cell pluripotency and differentiation. Here we provide a detailed protocol on how to detect and analyze histone post-translational modifications on a gene of interest in hPSCs and differentiated neuronal cells, using chromatin immunoprecipitation (ChIP).

ChIP is an experimental technique used to study the association between a specific sequence of DNA and a protein in the cell (Orlando, 2000; Solomon et al., 1988). It has been successfully used to localize histone posttranslational modifications on a specific gene, thus allowing the study of the chromatin composition of that locus. The procedure consists of a crosslinking step, a sonication step, the use of an antibody to pull down crosslinked protein-DNA complexes and the analysis of the immunoprecipitated DNA by PCR. Formaldehyde is routinely used as a crosslinker. It has the advantage of being cell-permeable, so that a snapshot of protein to gene association can be taken at any specific time and condition, and the crosslinking is easily reversible by heat. It is a so-called "zero-length" crosslinker; because of its structure it can only crosslink proteins that directly bind the DNA. A "native" alternative of ChIP (NChIP) (Hebbes et al., 1988; O'Neill and Turner, 2003), which omits the crosslinking step, can also be used when analyzing histones because of their high affinity and stable binding to DNA.

Shearing the DNA by sonication into fragments of appropriate size is essential for the resolution of the assay. An average length of 400–500 bp is suitable for most applications. Antibodies raised against different post-translational modifications of histones and, in most cases, validated specifically for ChIP assays, are commercially available. The introduction of real-time quantitative PCR (qRTPCR) to analyze the immunoprecipitated DNA allows a quantitative comparison among protein occupancies on different regions of DNA. For a review on the ChIP assay, its variations and different downstream detection methods see Collas (2010).

PROCEDURES

In this chapter we provide a protocol for detecting a subset of histone post-translational modifications bound to a gene of interest, using ChIP in hPSCs and early neuronal derivatives of hPSCs. Figure 19.1 provides an example of chromatin immunoprecipitation in hiPSCs.

The hPSCs and neurons require two sample collection procedures that are described in Steps 1a to 4a for hPSCs and in Steps 1b to 4b for early neuronal derivatives.

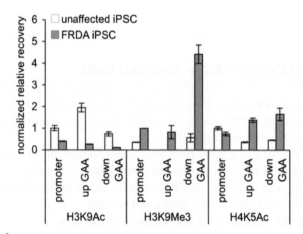

FIGURE 19.1

Chromatin marks on the Frataxin (FXN) gene in hiPSCs derived from fibroblasts of apparently healthy individuals and fibroblasts from patients with Friedreich's ataxia (FRDA). The unstable hyper-expansion of the triplet GAA·TCC in the first intron of the FXN gene causes a defect in expression of the essential protein frataxin and the neurodegenerative disease Friedreich's ataxia. Using ChIP we compared H3K9ac, H4K5ac and H3K9me3 on three regions of the FXN gene (the promoter, upstream and downstream of the GAA·TCC repeats) in hiPSCs derived from a Friedreich's ataxia patient and an unaffected individual. Lower levels of acetylation and higher levels of trimethylation of H3K9 indicate that the expanded FXN allele is silenced. DNA recovery after immunoprecipitation is expressed relative to total DNA and normalized to GAPDH coding region recovery. *Adapted from* Ku et al., 2010.

Chromatin Immunoprecipitation

The following protocol is for hPSCs (hESCs, hiPSCs) and early neuronal cells grown in one well of a 6-well plate. hPSC colonies are grown at a fairly high density (40–60 colonies per well) and early neuronal cells are plated at a concentration of 5×10^4 cells/cm^2.

Whole Cell Extract – hESCs and hiPSCs

1a. Add 54 µL of 37% formaldehyde to 2 mL of hPSC medium directly in the well (final formaldehyde concentration is 1%) and incubate for 10 minutes at ambient temperature with mild shaking.
2a. Add 125 µL of 2 M glycine (to a final concentration of 0.125 M) and incubate for 5 minutes at ambient temperature with mild shaking.
3a. Wash the cells with 2 mL of PBS, add 1 mL of fresh PBS and use a modified Pasteur pipette to cut, dislodge and collect the hPSC colonies from the well.
4a. Carefully transfer the hPSC colony fragments to a 1.5 mL microcentrifuge tube, briefly spin at 1500 rpm and resuspend in 350 µL of lysis buffer.

■ **Note**

Crosslinked hPSC colonies tend to stick to the walls of tips and plastic pipettes, so make sure to rinse the tips before discarding them.   ■

Whole Cell Extract – Early Neuronal Cells

1b. Add 54 µL of 37% formaldehyde to 2 mL of neural basal medium (or any other neuronal culture medium) directly in the well (final formaldehyde concentration is 1%) and incubate for 10 minutes at ambient temperature with mild shaking.
2b. Add 125 µL of 2 M glycine (to a final concentration of 0.125 M) and incubate for 5 minutes at ambient temperature with mild shaking.
3b. Wash the cells with 2 mL of PBS and add 130 µL of lysis buffer and scrape the cells off the well with a cell scraper.
4b. Transfer the lysate in a 1.5 mL microcentrifuge tube.

■ **Note**

For hPSCs one well can give enough extract to perform three to four pull-downs, while for early neuronal cells one well gives enough cells for one pulldown.   ■

Immunoprecipitation

5. Incubate the hPSC or neuronal cell lysate on ice for 15 minutes.
6. Sonicate four times for 15 seconds using a Brason sonifier at 9 W output, incubating the samples on ice for 1 minute after each round.

■ **Note**

Optimal shearing conditions must be determined for different sonicators. Avoid foaming during sonication; this will interfere with efficient shearing of DNA. Especially for early neuronal cells, it is advisable to pool the cell lysates from a few wells before the sonication step, to avoid foaming due to sonicating a very small volume of sample.   ■

7. Spin at 12,000 g for 10 minutes at 4°C and collect the supernatant. This is the whole cell extract.
8. Preclear the cell extract of immunoglobins by incubating with 60 μL of protein A-agarose beads for 1 hour at 4°C on a rotator.
9. Spin the sample for 30 seconds at 800 g to pellet the protein A-agarose beads and transfer the supernatant to a new tube. The supernatant is the whole cell extract to be used for ChIP.
10. Use 100 μL of whole cell extract per immunoprecipitation (IP) to be incubated with specific antibodies or with normal rabbit IgG as a negative control.
11. Set aside an additional sample of 10 μL and label as INPUT. The INPUT samples will be used to normalize the DNA analysis of the immunoprecipitated samples.
12. Dilute the 100 μL of extract with 0.9 mL of dilution buffer and add 2–4 μg of the desired histone antibody or normal rabbit IgG.
13. Incubate at 4°C overnight on a rotator.
14. The next day, add 60 μL of protein A-agarose beads to bind to the antibodies and incubate at 4°C for 2 hours on a rotator.

■ **Note**

All the antibodies listed in the reagent section can be efficiently bound by protein A. If using different antibodies from the ones listed below, check host and IgG isotype. Some IgG isotypes require protein G for efficient pulldown.   ■

15. Spin for 30 seconds at 800 g to pellet the agarose beads, and remove the supernatant.

■ **Note**

Remove as much supernatant as possible without aspirating the agarose beads. The use of 0.4 mm flat tips helps minimize the amount of the bead pellet aspirated with the supernatant.   ■

16. Wash the agarose beads to remove contaminants while retaining the protein–DNA complexes bound to the beads. Add 1 mL of low salt wash buffer to the beads and rotate for 5 minutes at 4°C. Spin to pellet the beads as above (30 seconds at 800 g) and remove the supernatant. Repeat two times, for a total of three low salt washes.

17. Perform a final wash with 1 mL of high salt wash buffer, spin as above and remove the supernatant.
18. Add 150 μL of elution buffer to the agarose bead pellet and incubate at 65°C for 10 minutes to elute the protein-DNA complexes. Spin as above (30 seconds at 800 g) and carefully collect the supernatant.
19. Repeat the elution step with another 150 μL of elution buffer and combine the two supernatants.
20. Add 300 μL of elution buffer to the 10 μL of INPUT previously set aside.
21. Add 3 μL of 20 mg/mL proteinase K to the eluates and the INPUT samples and incubate at 37°C for 30 minutes.
22. Reverse the crosslinks by incubating the samples at 65°C for 6–16 hours.

DNA Analysis

23. To purify DNA from immunoprecipitated samples:
    a. To each 300 μL sample, add 300 μL of phenol-chloroform-isoamyl alcohol (25:24:1).
    b. Vortex for 30 seconds and spin in a microcentrifuge at 12000 g for 5 minutes. The nucleic acids should collect in the upper aqueous phase while proteins will remain in the phenol/chloroform phase.
    c. Collect the upper aqueous phase and transfer it to a new tube.
    d. Precipitate the DNA by adding 600 μL of 100% ethanol, 30 μL of 3M sodium acetate pH 5.2, and 20 μg of glycogen.
    e. Cool the mixture by placing it on dry ice for 30 minutes.
    f. Centrifuge at 12,000 g at 4°C for 30 minutes to pellet the DNA.
    g. Remove the supernatant and wash the pellet with 500 μL of 75% ethanol.
    h. Spin for 10 minutes at 4°C, remove the ethanol and air-dry the pellet for 15 minutes at ambient temperature. Resuspend in 100 μL of water.
24. Analyze the immunoprecipitated (IP) samples and the INPUT samples for genes of interest using quantitative PCR.
    a. Prepare the following dilutions of the INPUT samples: 1:3, 1:9, 1:27, 1:81, 1:243 and 1:729
    b. Dilute each IP sample 1:4.
    c. Each 20 μL qPCR reaction contains: 4 μL of DNA (INPUT or IP), 10 μL of PerfeCTa SYBR® Green SuperMix and 10 pmoles of each primer.
    d. For each primer-set generate a standard curve using the undiluted INPUT sample and its dilutions and analyze each diluted IP sample in triplicate. Use the following conditions for the thermocycler: 95°C for 5 minutes, 40 cycles of 95°C for 10 seconds, 55°C for 10 seconds, and 72°C for 30 seconds. After cycling is complete, set the PCR machine to calculate dissociation curves from 70°C to 96°C.

■ **Note**

Cycling conditions depend on the thermocycler used; refer to your instrument's manual for optimal cycling conditions. The dissociation plot for every primer pair should be a single sharp peak.   ■

25. Visualize the amplification curve in a logarithmic scale and set the baseline from cycle 3 to cycle 9. Set the threshold in the linear phase of amplification and record the C_t value for each well. Average the C_t values for the three replicates of the IP samples and calculate the ΔC_t for each IP sample (normalized to the INPUT sample), using the standard curve generated with the C_t values of INPUT sample. Next calculate the $\Delta\Delta C_t$ ($\Delta\Delta C_t = \Delta C_t$ (GAPDH) $- \Delta C_t$ (target sequence)). For each IP sample, the relative recovery of the target sequence normalized to the relative recovery of the GAPDH coding region is $2^{-\Delta\Delta C_t}$.

■ **Note**

Make sure the C_t values obtained with IP DNA fall within the C_t values of the titration curve produced with INPUT DNA, to ensure accurate quantitation.   ■

ALTERNATIVE PROCEDURES

Alternative Protocol for DNA Purification with Chelex 100 Resin Prior to PCR

Chelex® 100 (BioRad) is a cationic exchange resin that is commonly used for preparing cellular samples for PCR. The following protocol is a much quicker alternative to the reversal of crosslinking and DNA purification steps provided above and has been used successfully in our laboratory. The DNA is not actually fully purified from the other cellular components, but the resin is used to remove some of the contaminants that can inhibit the subsequent PCR reaction. Two heating steps at 100°C seem to be sufficient to reverse the formaldehyde crosslinking.

1. Add 100 μL of Chelex suspension to the washed protein-A beads from Step 16 above (final wash after immunoprecipitation) and to 10 μL of saved INPUT. Vortex to mix resin and sample.
2. Incubate at 100°C for 10 minutes and let cool at ambient temperature.
3. Add proteinase K to a final concentration of 100 μg/mL and incubate at 55°C for 30 minutes with shaking.
4. Incubate at 100°C for 10 minutes and let cool at ambient temperature.
5. Centrifuge for 30 seconds at 800 g and transfer the supernatant to a new tube.
6. Add 100 μL of water to the beads, centrifuge as above, collect and combine the two supernatants. This DNA solution can be used directly in PCR.

Non-Quantitative PCR

If a real time PCR cycler is not available, PCR products from the INPUT and IP samples can be visualized on an agarose gel. In this case it is advisable to reduce the number of PCR cycles (if possible between 25 and 30 cycles depending on the initial amount of DNA) and produce a titration curve using varying amounts of INPUT DNA, to make sure the amplification is in the linear range. This type of analysis allows detecting qualitative differences in the amount of immunoprecipitated DNA but accurate quantitation is only achieved by quantitative methods like qPCR.

EQUIPMENT

- Tissue culture hood
- Tissue culture incubator, 37°C, 5% CO_2
- Inverted phase contrast microscope with a 4× objective
- Table-top centrifuge
- Heat block at 65°C
- 4°C room
- Rotator
- Sonicator (Branson Sonifier 150)
- MJ Research Chromo4 Thermal Cycler or similar real time PCR cycler

REAGENTS AND SUPPLIES

- 5 mL, 10 mL, 25 mL sterile disposable pipettes
- 6-well culture dishes
- 15 mL sterile conical tubes
- 50 mL sterile conical tubes
- Sterile 9″ pasture pipettes
- Pipette tips for Eppendorf or similar pipettor

Recommended Reagents

Item	Supplier	Catalog #	Note
DMEM/high glucose, L-glutamine (Dulbecco's Modified Eagle's Medium)	Life Technologies	11965-092	
KnockOut Serum Replacement (contains bovine products)	Life Technologies	108280-028	Use at 20%
L-Glutamine 200 mM	Life Technologies	25030-081	
MEM-Non-essential amino acids	Life Technologies	11140-050	

Item	Supplier	Catalog #	Note
2-Mercaptoethanol	Life Technologies	21985-023	
PBS (Dulbecco's Phosphate-Buffered Saline without Calcium and Magnesium)	Life Technologies	14190-144	
Recombinant human bFGF	Life Technologies	13256-029	Use at 20ng/mL
Pen/Strep (100x)	Life Technologies	15070-063	Optional
Neurobasal Medium A	Life Technologies	10888022	
B27 Supplement	Life Technologies	17504044	Use at 2%
N2 Supplement	Life Technologies	17502048	Use at 1%
ITS-A	Life Technologies	51300044	Use at 1%
Glycine	Sigma	410225	
Protease Inhibitor Cocktail	Roche	11873580001	
Protein A agarose beads	Millipore	16-157	
Proteinase K	Roche Diagnostics	03115828001	20mg/mL
Phenol-choloroform isoamyl alcohol (25:24:1)	EMD Chemicals	516726	
Glycogen	Roche Diagnostics	10901393001	
Chelex-100 Resin	Biorad	142-1253	10% w/v in water
PerfeCTa SYBR® Green SuperMix	Quanta Biosciences	95054	
Formaldehyde	Sigma	252549	Stock is 37%

RECIPES

- Lysis buffer: 1% SDS, 10 mM EDTA pH 8, 50 mM Tris pH 8, protease inhibitor cocktail (Roche Diagnostics)
- Dilution buffer: 1% Triton X-100, 150 mM NaCl, 2 mM EDTA pH8, 20 mM Tris pH 8, 1X
- Low salt wash buffer: 1% Triton X-100, 0.1% SDS, 150 mM NaCl, 2 mM EDTA pH 8
- High salt wash buffer: 1% Triton X-100, 0.1% SDS, 500 mM NaCl, 2 mM EDTA pH 8
- Elution buffer: 1% SDS, 100 mM NaHCO$_3$

ChIP ANTIBODIES

Antibody	Supplier	Catalog #
Rabbit anti-histone H3	Abcam	ab1791
Rabbit anti-histone H3K9ac	Millipore	07-352

Antibody	Supplier	Catalog #
Rabbit anti-histone H3K14ac	Millipore	07-353
Rabbit anti-histone H4	Millipore	07-108
Rabbit anti-histone H4K5ac	Millipore	07-327
Rabbit anti-histone H4K8ac	Millipore	07-328
Rabbit anti-histone H4K12ac	Millipore	07-595
Rabbit anti-histone H4K16ac	Abcam	ab61240
Rabbit anti-histone H3K4me1/2/3	Millipore	04-791
Rabbit anti-histone H3K36me3	Abcam	ab9050
Rabbit anti-histone H3K9me	Millipore	07-450
Mouse anti-histone H3K9me2	Abcam	ab1220
Rabbit anti-histone H3K9me3	Abcam	ab8898
Rabbit anti-histone H3K27me3	Millipore	07-449
Normal rabbit IgG	Santa Cruz Biotechnology	sc-2027
Normal mouse IgG	Santa Cruz Biotechnology	sc-2025

QUALITY CONTROL METHODS

Antibodies

A successful ChIP assay relies on the availability of good quality antibodies and on appropriate positive and negative controls. Always check antibody quality by western blotting. Although a successful western blotting experiment does not ensure that the antibody is suitable for a ChIP assay, western blotting gives a general idea of the quality and specificity of the antibody. Histone peptide arrays are also commercially available (Active Motif) to check the specificity of the histone PTM antibodies.

As negative controls, perform a "mock" pulldown using a normal IgG of the same isotype as the antibody used in the assay, and in the qPCR step use a gene that is known not to be occupied by the protein of interest. As a positive control we perform qPCR on GAPDH coding sequence for acetylated histone marks and satellite alpha DNA for heterochromatin marks like di- and trimethylated H3K9.

DNA Shearing

To determine the average size of sheared DNA, remove 20 µL of sheared cell extract and add 300 µL of elution buffer. Add proteinase K to a final concentration of 0.2 mg/mL and incubate at 37°C for 30 minutes and then from 6 to 16 hours at 65°C to reverse crosslinking. Add 300 µL of phenol-chloroform-isoamyl alcohol, vortex for 30 seconds and spin in a microcentrifuge at 12,000 g for 5 minutes. Collect the upper phase and transfer to a new tube. Precipitate DNA by adding 600 µL of 100% ethanol, 30 µL of 3M sodium acetate pH 5.2 and 20 µg of glycogen. Place on dry ice for 30 minutes and centrifuge at 12,000 g at 4°C for 30 minutes. Remove the supernatant and wash the pellet with 500 μL of 75% ethanol. Spin for 10 minutes at 4°C, remove the ethanol and air-dry the pellet for 15 minutes at room temperature. Resuspend in 10 μL of water and load onto a 1% agarose gel to check the fragment size. An average fragment size of 400–500 bp is recommended.

Primer Design

Primers for qPCR can be designed using a primer design software such as Primer3 (http://frodo.wi.mit.edu/primer3/). They should be between 18 and 25 nucleotides long, have a comparable GC content between 30 and 70%, have a T_m of 60°C or higher and give a product of 80 to 150 bp. Primers should also be checked for secondary structure and primer dimer formation.

READING LIST

Berger, S.L., 2007. The complex language of chromatin regulation during transcription. Nature 447, 407–412.

Collas, P., 2010. The current state of chromatin immunoprecipitation. Mol. Biotechnol. 45, 87–100.

Gardner, K.E., Allis, C.D., Strahl, B.D., 2011. Operating on chromatin, a colorful language where context matters. J. Mol. Biol. 409, 36–46.

Hebbes, T.R., Thorne, A.W., Crane-Robinson, C., 1988. A direct link between core histone acetylation and transcriptionally active chromatin. EMBO J. 7, 1395–1402.

Herman, D., Jenssen, K., Burnett, R., Soragni, E., Perlman, S.L., Gottesfeld, J.M., 2006. Histone deacetylase inhibitors reverse gene silencing in Friedreich's ataxia. Nat. Chem. Biol. 2, 551–558.

Ku, S., Soragni, E., Campau, E., Thomas, E.A., Altun, G., Laurent, L.C., et al., 2010. Friedreich's ataxia induced pluripotent stem cells model intergenerational GAATTC triplet repeat instability. Cell Stem Cell 7, 631–637.

O'Neill, L.P., Turner, B.M., 2003. Immunoprecipitation of native chromatin: NChIP. Methods 31, 76–82.

Orlando, V., 2000. Mapping chromosomal proteins in vivo by formaldehyde-crosslinked-chromatin immunoprecipitation. Trends Biochem. Sci. 25, 99–104.

Solomon, M.J., Larsen, P.L., Varshavsky, A., 1988. Mapping protein-DNA interactions in vivo with formaldehyde: evidence that histone H4 is retained on a highly transcribed gene. Cell 53, 937–947.

Strahl, B.D., Allis, C.D., 2000. The language of covalent histone modifications. Nature 403, 41–45.

Suganuma, T., Workman, J.L., 2011. Signals and combinatorial functions of histone modifications. Annu. Rev. Biochem. 80, 473–499.

Young, R.A., 2011. Control of the embryonic stem cell state. Cell 144, 940–954.

CHAPTER 20

Epigenetics: DNA Methylation

Marina Bibikova, Kristopher L. Nazor, Gulsah Altun, and
Louise C. Laurent

EDITOR'S COMMENTARY

Human pluripotent stem cells (hPSCs) are ideal for studying the epigenetic control of cellular phenotype. Epigenetic factors dynamically control what genes and regulatory molecules (including microRNAs and other non-coding RNAs) are expressed, which in turn determine what proteins are made by the cells. Both chromatin modification by histones (Chapter 19) and methylation of cytosines in DNA are involved in epigenetic regulation of cell phenotype. It is an interesting exercise to choose a gene that is characteristic of a specific differentiated cell type, such as tyrosine hydroxlyase in a dopamine neuron, and imagine what happens to that gene during the transition in culture from an undifferentiated hPSC to a mature cell type. It would be expected to be silenced in undifferentiated cells and activated at some stage during the differentiation process. But when, and how? What are the relative roles of chromatin modification and DNA methylation? And then, think about all of the other genes that define the phenotype of a dopaminergic neuron; how are they co-regulated?

The tools are now available to ask these questions. In this chapter we discuss using high density DNA methylation array technology to assess 450,000 potentially methylated sites in the genome. This method has less resolution than whole genome "methylome" mapping using bisulfite sequencing, but it gives more coverage of the genome than most other methods, and is ideal for surveying very large sets of samples to compare hPSCs with other cell types, and to determine how the methylome changes during expansion of hPSCs and their differentiation.

OVERVIEW

Increasing evidence suggests that epigenetic regulation is a key factor in the maintenance of the pluripotent stem cell state (Bhutani et al., 2010; Bock et al., 2011; Kim et al., 2011; Meissner, 2010). The best-studied epigenetic

325

J.F. Loring & S.E. Peterson (eds): Human Stem Cell Manual, Second edition.
DOI: http://dx.doi.org/10.1016/B978-0-12-385473-5.00020-5

modifications are those that affect transcription of genes: DNA methylation and chromatin modifications.

DNA methylation is the addition of a methyl group to a nucleotide base. In mammalian somatic cells, DNA methylation occurs almost exclusively on cytosine residues in the symmetrical CpG dinucleotide context. In human embryonic stem cells (hESCs), non-CpG methylation has been noted to preferentially occur in the CpA (Laurent et al., 2010) and CpHpG (Lister et al., 2009) contexts. Monitoring DNA methylation patterns is becoming a powerful method for characterizing human pluripotent stem cells (hPSCs) and their differentiated products *in vitro* (Meissner, 2010). In the past decade, many DNA methylation detection techniques have been developed, and a variety of methods enable quantitative DNA methylation profiling at a single-base resolution and at a large scale (Beck, 2010; Huang et al., 2010; Laird, 2010).

Bisulfite sequencing of genomic DNA is the highest resolution method for analyzing the methylation state of the entire genome (Eckhardt et al., 2006). While whole genome bisulfite sequencing (WGBS) can access every methylated site in the genome, the method is still quite expensive and is therefore limited to a low number of samples. A variation on this method, called reduced-representation bisulfite sequencing (RRBS), uses restriction enzyme digestion to fractionate the genome and targets regions of high CpG density (Meissner et al., 2005).

The microarray-based approaches are largely based on three major techniques: restriction enzyme digestion (Omura et al., 2008; Ordway et al., 2006; Schumacher et al., 2006), sodium bisulfite conversion of genomic DNA (Bibikova et al., 2009, 2011; Sandoval et al., 2011), and affinity-based assays (Robinson et al., 2011).

In this chapter, we describe a method that adapts SNP genotyping technology to the determination of DNA methylation of specific sites in the human genome using bisulfite converted genomic DNA (Bibikova et al., 2011).

Illumina DNA Methylation Assays

Illumina DNA methylation assays use Illumina's core BeadArray™ technology: arrays made of oligonucleotides linked to silica beads that are randomly assembled into microwells (Fan et al., 2006). The randomly distributed beads are then decoded to determine which bead is positioned in which well and to automate quality control for the array (Gunderson et al., 2004). Depending on the array density, 15–30 beads of each type (having the same oligonucleotide sequence) are present on each array, offering redundancy in analytical measurements.

Illumina has developed and commercialized several DNA methylation assays. The GoldenGate® DNA assay for Methylation (Bibikova et al., 2006) was a multiplexed assay using an indexing system that allowed interrogation of up to 1536 individual CpG loci per sample, with data read out by hybridization of the assay mixture to a universal array. The Infinium® assay uses a sample preparation method that enables interrogation of any number of CpG sites from one sample, limited only by the number of elements present on the microarray. Illumina has developed two Infinium DNA methylation arrays, which determine the methylation level for 27,000 and 450,000 loci per sample (Bibikova et al., 2009, 2011). The high density HumanMethylation450 BeadChip allows high throughput methylation profiling of over 450,000 CpG and CpHpG sites, and analyzes twelve samples in parallel. The innovative content includes coverage of 99% of RefSeq genes with multiple probes per gene, 96% of CpG islands from the UCSC database, CpG island shores and additional content selected from whole genome bisulfite sequencing data (e.g. Laurent et al., 2010) and input from DNA methylation experts.

The Infinium Assay

Figure 20.1A shows the workflow for the Infinium methylation assay. Genomic DNA is first converted with sodium bisulfite, whereby all unmethylated cytosines are converted to uracils and methylated cytosines remain unchanged – see Figure 20.1B (Clark et al., 2006; Frommer et al., 1992). Bisulfite converted DNA is then subjected to whole genome amplification (about 1000-fold amplification), fragmented, and hybridized to a BeadChip. The oligonucleotides attached to each bead type are complementary to a 50-base region adjacent to a query CpG site. A primer extension reaction

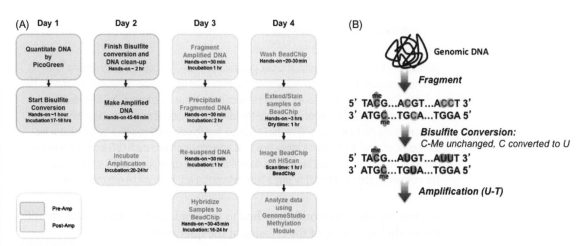

FIGURE 20.1

(A) Workflow for the Infinium Methylation Assay. (B) Illustration of bisulfite conversion.

(A) Infinium I

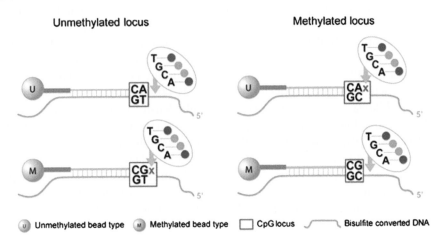

(B) Infinium II

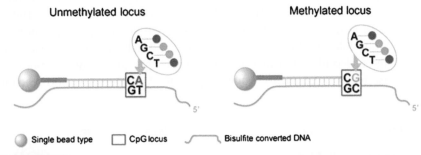

FIGURE 20.2
Two forms of the Infinium Methylation assay. (A) Infinium I assay. Two bead types correspond to each CpG locus: one bead type - to the methylated C, another bead type - to the unmethylated T state of the CpG site. The probe design assumes the same methylation status for adjacent CpG sites. Both bead types for the same CpG locus will incorporate the same type of labeled nucleotide, determined by the base preceding the interrogated "C" in the CpG locus, and therefore will therefore be detected in the same color channel. (B) Infinium II assay. One bead type corresponds to each CpG locus. Each probe can contain up to 3 underlying CpG sites, with a degenerate R base corresponding to C in the CpG position. The methylation state is detected by single-base extension. Each locus will be detected in two colors.

extends the oligonucleotide attached to the bead, and in this process identifies the methylation state of the cytosine at the query site.

Two forms of the assay are used (Figure 20.2). For the Infinium I assay, allele-specific primer extension (ASPE) is used to interrogate the CpG site, requiring two beads for each locus (Figure 20.2A) (Gunderson et al., 2005). For the Infinium II assay, one bead type is used, and the methylation status is scored by the single-base extension (SBE) reaction using labeled terminators (Figure 20.2B) (Steemers et al., 2006). The advantage of the Infinium assay is that it allows for a broad CpG locus choice and scales to a large number of CpGs in a single assay, limited only by the complexity of the array. However, each new set of CpG loci requires manufacture of a new array.

PROCEDURES

Bisulfite Conversion for Methylation Analysis

The EZ DNA methylation kit (Zymo Research, Orange, CA) is used for bisulfite conversion of the genomic DNA, according to the manufacturer's recommendations. Briefly, 500 ng of genomic DNA (1 µg for an automated process) is denatured by addition of Zymo M-Dilution buffer (contains NaOH) and incubated for 15 minutes at 37°C. The CT-conversion reagent (bisulfite-containing) is added to the denatured DNA and incubated for 16 hours at 50°C in a thermocycler and denatured every 60 minutes by heating to 95°C for 30 seconds. After bisulfite treatment, the remaining assay steps are identical to the Infinium genotyping assay.

Infinium Methylation

Infinium Methylation is conducted according to the Infinium HD Methylation Assay Guide (part # 15019519_B). First, whole genome amplification is used to amplify bisulfite-converted genomic DNA by a factor of $\approx 1000-2000\times$ in a relatively unbiased manner. The Infinium assay uses pre-formulated Whole Genome Amplification (WGA) reagents (RPM and MSM) for compatibility with downstream fragmentation and processing. Amplification proceeds at 37°C for 20 hours. After amplification, a fragmentation protocol is used to reduce the fragment size to 200–300 base pairs (bp). This fragmentation improves both resuspension and hybridization efficiency. After fragmentation, the reaction is stopped and the DNA is precipitated by addition of a precipitation reagent (PM1) and one volume of isopropanol. This step reduces the carry-over of dNTPs, and concentrates the DNA sample. The DNA pellet is resuspended in a formamide-containing hybridization buffer (RA1) by incubating at 48°C for 1 hour and vortexing for about 1 minute as specified in the user manual. The average yield from the WGA reaction is $\approx 1.5\,\mu g/\mu L$ and the final hybridization concentration is $5-6\,\mu g/\mu L$. After resuspension, the sample is denatured at 95°C for 20 minutes, allowed to cool on the bench for 5–10 minutes, and applied to the BeadChip array. The arrays are hybridized in the hybridization oven at 48°C for 16–24 hours.

After hybridization, the BeadChips are washed, primer extended, and stained on a Tecan Genesis/Evo robot using a GenePaint™ slide processing system. This "XStain" process involves pipetting of various reagents to the Te-Flow-Through Chambers placed on the GenePaint Te-Flow Chamber Rack, equilibrated to 44°C. These reagents are used according to the Infinium Methylation Assay Guide. Basically, the BeadChips are washed with hybridization buffer (RA1), blocked with protein (XC1), primer extended with a polymerase and labeled nucleotide mix (TEM), and stained with repeated application of STM (staining reagent) and ATM (anti-staining reagent). After staining is complete, the slides are washed with low salt wash buffer (PB1), coated to protect from photo bleaching (XC4), dried down, and imaged on Illumina's iScan instrument.

ALTERNATIVE PROCEDURES

Affinity-based DNA methylation profiling technologies have been developed to enrich the methylated fraction of the genome; these include methylated DNA immunoprecipitation (MeDIP) (Weber et al., 2005) or affinity chromatography over an MBD (methyl-binding domain) column (Ibrahim et al., 2006; Yamada et al., 2004). The affinity-based enrichment methods can be read out using tiling arrays (NimbleGen™ or Affymetrix™) or next-generation DNA sequencing.

Whole genome bisulfite sequencing offers the most comprehensive and unbiased analysis of the methylation state of every CpG site of the genome as well as non-CpG loci, although data analysis remains challenging due to the lower genome complexity after bisulfite conversion (Laurent et al., 2010; Lister et al., 2009). Several recent reviews compare available DNA methylation analysis methods and discuss the strengths and weaknesses associated with microarray and next-generation sequencing-based methods (Beck, 2010; Huang et al., 2010; Laird, 2010).

Profiling DNA Methylation in Undifferentiated and Specialized Cells

There have been a number of reports describing the DNA methylation state of embryonic stem cells (reviewed in Altun et al., 2010). Most of these studies were limited by small numbers of cell lines, the absence of somatic samples for comparison, and low assay resolution.

Recently, two more comprehensive manuscripts have been published. The first used RRBS and the HumanMethylation450 array to study 32 undifferentiated hPSC lines, 10 differentiated hPSC lines, and 10 somatic samples

(Ziller et al., 2011). The analysis of these data focused on non-CpG methylation, and concluded that non-CpG methylation is specific to undifferentiated hPSCs, with gain of non-CpG methylation during reprogramming and loss during differentiation. Non-CpG methylation was also correlated with expression of the *de novo* DNA methyltransferases DNMT3A and DNMT3B, and spatially correlated with CpG methylation, supporting roles for the DNMT3s in establishing pluripotency-specific DNA methylation patterns.

The second manuscript (Nazor et al., 2012) demonstrates how a high-resolution microarray-based method can be used to survey a large number of pluripotent and somatic samples. In this study, the HumanMethylation450 array was used to measure DNA methylation in the most comprehensive collection of hPSCs, primary cell lines, and tissue samples to date (193 undifferentiated hPSC samples and 80 samples from 17 tissue types). Analysis of these data revealed that somatic cell-type-specific genes were hypermethylated in hPSCs and that each tissue type was distinguished by unique patterns of DNA hypomethylation (Figure 20.3). In addition, widespread hPSC-specific epigenetic and transcriptional aberrations in genes subject to X chromosome inactivation (XCI) and genomic imprinting were identified (Figure 20.4). Finally, *in vitro* directed differentiation of hPSCs was found to recapitulate the unique patterns of DNA hypomethylation observed in tissues, but did not erase the hPSC-specific epigenetic aberrations in genes subject to XCI and genomic imprinting. Collectively, these studies show that tissue-specific DNA methylation patterns can be accurately modeled during directed differentiation of hPSCs, but caution is warranted in the interpretation of disease models in which the observed phenotype depends on proper XCI or imprinting. These studies highlight the immediate importance of identifying specific culture conditions that promote the stability of genomic imprints and XCI over long-term culture, and exemplify the importance of verification of baseline epigenetic status in hPSC-based studies.

Analysis of DNA Methylation Array Data

As with studies aimed at assessing genome-wide gene expression, the most critical feature of a successful experiment is the experimental design. General guidelines for experimental design for microarray-based experiments are presented in Chapter 17, so this section will focus on elements specific to DNA methylation analysis. The status of DNA methylation for a particular CpG using the Infinium HD Methylation Assay is most commonly represented by its β-value, which reflects the percent of CpGs that are methylated at a given position, and is calculated by the formula $\beta = $ Methylated/(Unmethylated + Methylated). For example, an unmethylated CpG would have a $\beta = 0$, a hemi-methylated CpG would have a $\beta = 0.5$, and a methylated CpG would have a $\beta = 1$.

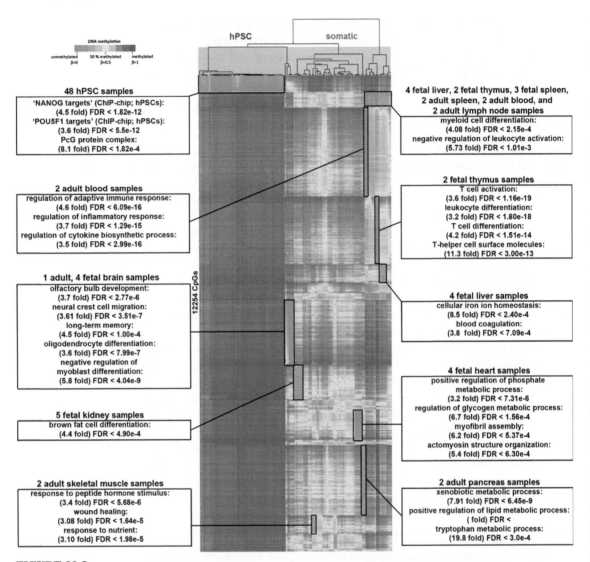

FIGURE 20.3

Tissue-specific patterns of DNA methylation. 12,254 CpGs on the 450K DNA Methylation array with uniquely hypomethylated CpGs in specific tissue types. Functional enrichments for tissue-specific hypomethylated clusters are identified with boxes. Samples are grouped according to hierarchical clustering and CpGs are rank-ordered for each tissue.

Logically, these would be the only three β-values that we would expect to see for any CpG. However, we frequently observe situations in which $0 > \beta < 0.5$ or $0.5 > \beta < 1$. So how could a particular CpG have a $\beta = 0.25$, when theoretically, this observation would be impossible in a homogenous population? Intermediate β-values can arise as a result of the specific hybridization

(A)

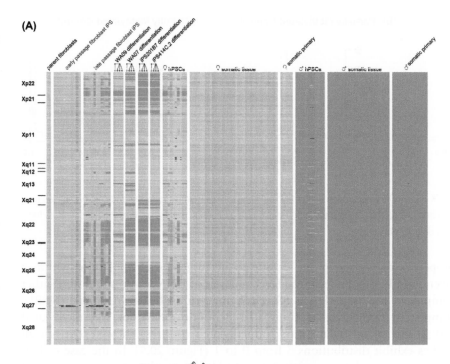

(B)

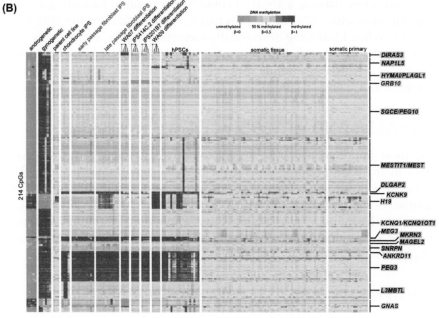

FIGURE 20.4
Analysis of DNA methylation of genes subject to XCI and genomic imprinting in hPSCs and somatic cells. (A) Heatmap of the 3279 XCI probes identified on the 450K DNA Methylation array for all samples analyzed. For the differentiation data, green dots above the heatmap indicate undifferentiated samples, red dots indicate differentiated samples and black arrows indicate direction of differentiation. (B) Heatmap of the 214 imprinted probes identified on the 450k array for all samples analyzed on the 450k methylation array. For the differentiation data, green dots above the heatmap indicate undifferentiated samples, red dots indicate differentiated samples and black arrows indicate direction of differentiation.

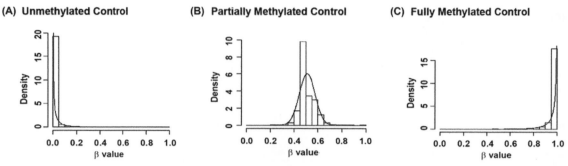

FIGURE 20.5

Distribution of control DNA on the 27K Human DNA Methylation BeadArray. (A) Fitting of data from the unmethylated control samples to a Beta distribution. (B) Fitting of data from the partially methylated control samples to a Gaussian distribution. (C) Fitting of data from the fully methylated control samples to a Beta distribution.

properties of a given probe or due to heterogeneous populations within a given sample. On the technical level, analysis of DNA controls that were either completely methylated (via SSI DNA methyltransferase treatment (NEB)), completely unmethylated (untreated genomic DNA), or half-methylated (50/50 mix of methylated and unmethylated controls) showed that not all probes exhibit distributions β from 0 to 1 (Figure 20.5). In the case of heterogeneous populations, a $\beta = 0.25$ could be the result of measuring a mixed population in which 50% of the cells had a $\beta = 0.0$ and the other 50% had a $\beta = 0.5$. When identifying differentially methylated CpGs, we generally filter for CpGs that have a difference in β of at least 0.2 between two experimental groups, in addition to meeting some statistical criteria as highlighted in the statistical and analytical guidelines in Chapter 17.

PITFALLS AND ADVICE

The most important factor to keep in mind for carrying out a successful DNA methylation profiling experiment is to be certain that adequate DNA is included in the assay. For this reason, accurate quantification of the DNA sample, using a reliable assay such as PicoGreen® (Life Technologies), is essential. The Infinium assay begins with a whole genome amplification step. Accurate representation of all parts of the genome during whole genome amplification requires high quality, intact DNA for best results.

EQUIPMENT, REAGENTS, AND SUPPLIES

All equipment and supplies for carrying out the Infinium Methylation assays are obtained from Illumina, Inc. (www.illumina.com). Illumina also provides DNA methylation services for large projects.

QUALITY CONTROL METHODS

HumanMethylation450 BeadChip includes internal controls that monitor each step of the assay. These may be used for trouble-shooting failed assays. As mentioned above, the most important factor for guaranteeing success is quantifying the DNA input with an accurate measure such as PicoGreen.

READING LIST

Altun, G., Loring, J.F., Laurent, L.C., 2010. DNA methylation in embryonic stem cells. J. Cell Biochem. 109, 1–6.

Beck, S., 2010. Taking the measure of the methylome. Nat. Biotechnol. 28, 1026–1028.

Bhutani, N., Brady, J.J., Damian, M., Sacco, A., Corbel, S.Y., Blau, H.M., 2010. Reprogramming towards pluripotency requires AID-dependent DNA demethylation. Nature 463, 1042–1047.

Bibikova, M., Lin, Z., Zhou, L., Chudin, E., Garcia, E.W., Wu, B., et al., 2006. High-throughput DNA methylation profiling using universal bead arrays. Genome Res. 16, 383–393.

Bibikova, M., Le, J., Barnes, B., Saedinia-Melnyk, S., Zhou, L., Shen, R., et al., 2009. Genome-wide DNA methylation profiling using Infinium® assay. Epigenomics 1, 177–200.

Bibikova, M., Barnes, B., Tsan, C., Ho, V., Klotzle, B., Le, J.M., et al., 2011. High density DNA methylation array with single CpG site resolution. Genomics 98, 288–295.

Bock, C., Kiskinis, E., Verstappen, G., Gu, H., Boulting, G., Smith, Z.D., et al., 2011. Reference Maps of human ES and iPS cell variation enable high-throughput characterization of pluripotent cell lines. Cell 144, 439–452.

Clark, S.J., Statham, A., Stirzaker, C., Molloy, P.L., Frommer, M., 2006. DNA methylation: bisulphite modification and analysis. Nat. Protoc. 1, 2353–2364.

Eckhardt, F., Lewin, J., Cortese, R., Rakyan, V.K., Attwood, J., Burger, M., et al., 2006. DNA methylation profiling of human chromosomes 6, 20 and 22. Nat. Genet. 38, 1378–1385.

Fan, J.B., Gunderson, K.L., Bibikova, M., Yeakley, J.M., Chen, J., Wickham Garcia, E., et al., 2006. Illumina universal bead arrays. Methods Enzymol. 410, 57–73.

Frommer, M., McDonald, L.E., Millar, D.S., Collis, C.M., Watt, F., Grigg, G.W., et al., 1992. A genomic sequencing protocol that yields a positive display of 5-methylcytosine residues in individual DNA strands. Proc. Natl. Acad. Sci. USA 89, 1827–1831.

Gunderson, K.L., Kruglyak, S., Graige, M.S., Garcia, F., Kermani, B.G., Zhao, C., et al., 2004. Decoding randomly ordered DNA arrays. Genome Res. 14, 870–877.

Gunderson, K.L., Steemers, F.J., Lee, G., Mendoza, L.G., Chee, M.S., 2005. A genome-wide scalable SNP genotyping assay using microarray technology. Nat. Genet. 37, 549–554.

Huang, Y.W., Huang, T.H., Wang, L.S., 2010. Profiling DNA methylomes from microarray to genome-scale sequencing. Technol. Cancer Res. Treat. 9, 139–147.

Ibrahim, A.E., Thorne, N.P., Baird, K., Barbosa-Morais, N.L., Tavare, S., Collins, V.P., et al., 2006. MMASS: an optimized array-based method for assessing CpG island methylation. Nucleic Acids Res. 34, e136.

Kim, K., Zhao, R., Doi, A., Ng, K., Unternaehrer, J., Cahan, P., et al., 2011. Donor cell type can influence the epigenome and differentiation potential of human induced pluripotent stem cells. Nat. Biotechnol. 29, 1117–1119.

Laird, P.W., 2010. Principles and challenges of genome-wide DNA methylation analysis. Nat. Rev. Genet. 11, 191–203.

Laurent, L., Wong, E., Li, G., Huynh, T., Tsirigos, A., Ong, C.T., et al., 2010. Dynamic changes in the human methylome during differentiation. Genome Res. 20, 320–331.

Lister, R., Pelizzola, M., Dowen, R.H., Hawkins, R.D., Hon, G., Tonti-Filippini, J., et al., 2009. Human DNA methylomes at base resolution show widespread epigenomic differences. Nature 462, 315–322.

Meissner, A., 2010. Epigenetic modifications in pluripotent and differentiated cells. Nat. Biotechnol. 28, 1079–1088.

Meissner, A., Gnirke, A., Bell, G.W., Ramsahoye, B., Lander, E.S., Jaenisch, R., 2005. Reduced representation bisulfite sequencing for comparative high-resolution DNA methylation analysis. Nucleic Acids Res. 33, 5868–5877.

Nazor, K., Altun, G.L.C., Tran, H, Aarness, JV, Slavin, I, Garitaonandia, I, et al., 2012. Recurrent variations in DNA methylation in human pluripotent stem cells and their differentiated derivatives. Cell Stem Cell (in press).

Omura, N., Li, C.P., Li, A., Hong, S.M., Walter, K., Jimeno, A., et al., 2008. Genome-wide profiling of methylated promoters in pancreatic adenocarcinoma. Cancer Biol. Ther. 7, 1146–1156.

Ordway, J.M., Bedell, J.A., Citek, R.W., Nunberg, A., Garrido, A., Kendall, R., et al., 2006. Comprehensive DNA methylation profiling in a human cancer genome identifies novel epigenetic targets. Carcinogenesis 27, 2409–2423.

Robinson, M.D., Stirzaker, C., Statham, A.L., Coolen, M.W., Song, J.Z., Nair, S.S., et al., 2011. Evaluation of affinity-based genome-wide DNA methylation data: effects of CpG density, amplification bias, and copy number variation. Genome Res. 20, 1719–1729.

Sandoval, J., Heyn, H.A., Moran, S., Serra-Musach, J., Pujana, M.A., Bibikova, M., et al., 2011. Validation of a DNA methylation microarray for 450,000 CpG sites in the human genome. Epigenetics 6, 692–702.

Schumacher, A., Kapranov, P., Kaminsky, Z., Flanagan, J., Assadzadeh, A., Yau, P., et al., 2006. Microarray-based DNA methylation profiling: technology and applications. Nucleic Acids Res. 34, 528–542.

Steemers, F.J., Chang, W., Lee, G., Barker, D.L., Shen, R., Gunderson, K.L., 2006. Whole-genome genotyping with the single-base extension assay. Nat. Methods 3, 31–33.

Weber, M., Davies, J.J., Wittig, D., Oakeley, E.J., Haase, M., Lam, W.L., et al., 2005. Chromosome-wide and promoter-specific analyses identify sites of differential DNA methylation in normal and transformed human cells. Nat. Genet. 37, 853–862.

Yamada, Y., Watanabe, H., Miura, F., Soejima, H., Uchiyama, M., Iwasaka, T., et al., 2004. A comprehensive analysis of allelic methylation status of CpG islands on human chromosome 21q. Genome Res. 14, 247–266.

Ziller, M.J., Muller, F., Liao, J., Zhang, Y., Gu, H., Bock, C., et al., 2011. Genomic distribution and inter-sample variation of non-CpG methylation across human cell types. PLoS Genet. 7, e1002389.

CHAPTER 21

Generation of Human Pluripotent Stem Cell-Derived Teratomas

Ha T. Tran, Trevor R. Leonardo, and Suzanne E. Peterson

EDITOR'S COMMENTARY

The teratoma assay has become a standard for proving pluripotency of human pluripotent stem cells (hPSCs). The main problem with claiming teratoma assays as a gold standard is that they are not standardized (Müller et al., 2010). There are alternatives to the teratoma assay to prove pluripotency, and two such alternatives are described in this book. Demonstration of three-germ layer differentiation can be done *in vitro*, using embryoid body methods (Chapter 23), and a molecular diagnostic test based on a gene expression database accurately correlates with teratoma results for cell lines (Chapter 18). The point made in the chapter describing the analysis of teratomas (Chapter 22) is that these tumors contain far more information than just indicating that the cells are pluripotent, so making teratomas will continue an important method for advanced analysis of hPSCs, even if alternative pluripotency assays gain prominence. The method described here is robust and has been used for generating teratomas from many hPSC lines, as well as from the first iPSC lines derived from endangered species (Ben-Nun et al., 2011).

OVERVIEW

In the teratoma assay, specific sites for hPSC injection are commonly used including the testis capsule, the kidney capsule, the liver and the leg muscle. After injection, mice must be monitored periodically for tumor formation, which can take between 5 and 12 weeks. Once a tumor is detected, it is fixed, sectioned and stained for analysis (see Chapter 22). In this chapter, we describe a reliable protocol for teratoma generation from hPSCs using the testis capsule as the site of injection. This protocol can easily be adapted

J.F. Loring & S.E. Peterson (eds): Human Stem Cell Manual, Second edition.
DOI: http://dx.doi.org/10.1016/B978-0-12-385473-5.00021-7

to other sites of injection with minor modifications (see Gertow, 2007 and Peterson, 2011). All procedures with animals should be performed under the appropriate institutional review and oversight.

PROCEDURES

We suggest that three people work together on this procedure. One person prepares the cells. A second prepares the mice for surgery and helps the surgeon. The third person does the actual surgery.

■ Note

See our JoVE article for a video showing some of the techniques (http://www.jove.com/details.php?id=3177). ■

Preparation for Surgery

1. Acquire 6-week old immunocompromised male mice such as CbySmn. CB17-Prkdc SCID/J mice.

■ Note

Other commonly used SCID mouse strains include NOD/SCID/γ_c^{null} (NOD/ShiJic-scid with γ_c^{null}) and SCID-beige mice (C.B-Igh-1^b GbmsTAC-Prkdc^{scid}-Lyst^{bg}N7). In addition, nude mice have also been used but may require greater numbers of injected cells. ■

2. Sterilize all surgical instruments, gloves, and gauze.
3. Dissociate hPSCs to be injected with Accutase™.

■ Note

Accutase should be used to dissociate hPSCs even if they are on feeder layers and not accustomed to enzymatic dissociation. We have found that mechanical dissociation of the cells does not work as well as dissociation with Accutase for teratoma generation. If cells are accustomed to enzymatic dissociation with Accutase, 2 minutes should be sufficient. If they are normally mechanically dissociated, they may require a longer incubation in Accutase, perhaps ~5 minutes. ■

4. Count cells and resuspend 1,000,000 cells per 20–30 μL in Matrigel™ diluted 1:1 in DMEM/F-12. Keep cells on ice until finished.

■ **Note**

It is recommended that the researchers do a run through of the experiment, injecting a dye, such as trypan blue, rather than cells, and using wild-type mice. This will help to identify any potential problems before any cells or immunocompromised mice are wasted.　■

Surgery

1. In the vivarium, use accepted institutional procedures to anesthetize your mouse. At our institution, we use an anesthesia machine with isoflurane. Mice are initially put in an induction chamber with 2 L/min oxygen and 3–4% isoflurane. Once anesthetized, a nose cone with 2 L/min oxygen and 2–3% isoflurane is used.
2. Use clippers to shave the abdomen, remove loose hair and clean the shaved area, starting from the center of the abdomen and working clockwise outward. First wash with Povidone-Iodine solution using a sterile swab. Next, wash with 70% ethanol. Repeat the procedure three times, changing swabs each time. Place the animal on a heating pad to keep the animal warm. Keep the heating pad and mouse inside a tissue culture or dissecting hood.
3. Just below the level of the hip joint, make an incision (approximately 1 cm) longitudinally along the midline of the abdomen through the skin and the abdominal wall using sterile surgical scissors.

■ **Note**

It is important to make an incision at the midline of the abdomen to avoid bleeding that may occur if the muscles are cut.　■

4. Take hold of the peritoneum using the Dumont Tweezers and reach down towards the right hip using the Graefe forceps. Pull the white fatty tissue out along with the attached testes.
5. Put the testes on sterile gauze.
6. Fill a tuberculin or Hamilton (1cc) syringe with the hPSCs to be injected.
7. Slowly inject the cells (20–30 μL) into the center of the testis capsule far from any major blood vessels. Stop if the testicular capsule begins to swell.
8. Take out the needle slowly to avoid reflux of the cells.
9. Using Graefe forceps, place the testes and fatty tissue back into its original location in the abdomen.
10. Close up the abdominal wall using the Olsen-Hegar Needle Holder and several reabsorbable sutures. Close the skin with wound clips using the wound clip applicator.

11. Keep the mouse warm until it recovers and give it some form of analgesic (see what is accepted in your institution) after surgery twice daily for 1–2 days.
12. Remove wound clips 7 days after the surgery.
13. Check the animal for tumor growth periodically for 5–12 weeks.
14. When the tumor is palpable and becomes approximately 5 mm in size, euthanize the mouse according to accepted procedures at your institution.
15. Remove the tumor and document it accordingly. Photograph it, measure the size, and weigh it.
16. Cut the tumor into smaller pieces and fix it in a 4% paraformaldehyde solution. Store the tumor in the fixative at 4°C until samples are sent to a pathologist for sectioning, staining and analysis. Alternatively, section the tumor and stain with Hematoxylin and Eosin. See examples from Chapter 22 in order to identify structures from each germ layer.

What to Expect

When this protocol is performed successfully and the cell line injected is pluripotent, a palpable tumor should form within 12 weeks. Established hPSC lines such as the WA09 hESC line typically generate tumors within 6 weeks. With some iPSC lines, tumors can take longer to form, often in the range of 8–10 weeks. It is very important to have a positive control for this assay in order to be sure that the procedure was performed appropriately. The best positive control is to inject an additional mouse with an hPSC line that is known to be pluripotent, such as WA09. Frequently, teratomas look very heterogeneous and have many attached cysts (Figure 21.1). Analysis of the tumor samples should reveal the presence of differentiated structures from all three germ layers (see Chapter 22).

PITFALLS AND ADVICE

- It is useful to go through the entire surgical procedure before using real hPSCs and immunocompromised mice, injecting only trypan blue into wild type mice, just to familiarize yourself with the procedure.
- It is very important to include a positive control of a well-established hPSC line when first performing this procedure to confirm that you are doing things correctly.
- Fewer cells can easily be used in this assay when injecting into the testis capsule or the kidney capsule. However, more cells may need to be used if they are injected into the leg.
- To ensure a successful surgical procedure, check mice for bleeding once they have regained consciousness and begin moving around their

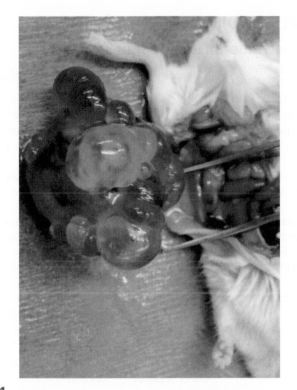

FIGURE 21.1

A hPSC-derived teratoma generated in the testis capsule. One million WA09 cells were injected into the testis capsule of a CbySmn.CB17-Prkdc SCID/J mouse. After 6 weeks, a palpable teratoma was observed. This is a very large tumor with cystic structures that expanded rapidly, increasing the size of the mass considerably in just a few days.

housing. If bleeding persists, re-anesthetize the mouse, identify the source of bleeding, and fix accordingly.

EQUIPMENT

- Animal balance
- Animal clippers (for shaving fur prior to surgery)
- Stereo dissecting microscope
- Fiber-optic lamp
- Heating pad, slide warmer, or heating lamp to warm cages for post-op recovery
- Hot bead sterilizer (or other method to sterilize surgical instruments during surgery)

REAGENTS AND SUPPLIES

Surgical Instruments and Supplies

Item	Supplier	Catalog #	Alternative
Micro Dissecting Scissors 4"	Roboz	RS-5882	Many
Operating Scissors 5"	Roboz	RS-6806	Many
Olsen-Hegar Needle Holder	WPI Inc.	500227-G	Many
Adson Forceps 12 cm 1 × 2 teeth	Roboz	RS-5233	Many
Graefe Forceps Slight Curve	Roboz	RS-5135	Many
Fine, blunt forceps "Thumb Dressing"	Roboz	RS-8102	Many
Dumon Tweezers	Roboz	RS-4905	Many
Silk suture size 5-0, with size 10 curved needle	Roboz	SUT-1074-21	Many
Wound clip applicator	Roboz	RS-9250	Many
Wound clips	Roboz	RS-9255	Many
Wound clip removing forceps	Roboz	RS-9263	Many
70% Ethanol	Henry Schein	1028715	Many
Tuberculin Syringe	BD	305945	BD 309623
Betadine Surgical Scrub	Henry Schein	6900581	
Paraformaldehyde	Sigma	P6148	Dust is very toxic
Matrigel	BD Biosciences	356230	
Accutase	Life Technologies	A1110501	
DMEM	Life Technologies	10566016	

READING LIST

Reviews

Gertow, K., Przyborski, S., Loring, J.F., Auerbach, J.M., Epifano, O., Otonkoski, T., et al., 2007. Isolation of human embryonic stem cell-derived teratomas for the assessment of pluripotency. Curr. Protoc. Stem Cell Biol. (Chapter 1: Unit 1B.4).

Przyborski, S.A., 2005. Differentiation of human embryonic stem cells after transplantation in immune-deficient mice. Stem Cells 23, 1242–1250.

Teratoma Production in Hind Leg Muscle

Thomson, J.A., Itskovitz-Eldor, J., Shapiro, S.S., Waknitz, M.A., Swiergiel, J.J., Marshall, V.S., et al., 1998. Embryonic stem cell lines derived from human blastocysts. Science 282, 1145–1147.

Xu, C., Inokuma, M.S., Denham, J., Golds, K., Kundu, P., Gold, J.D., et al., 2001. Feeder-free growth of undifferentiated human embryonic stem cells. Nat. Biotech. 19, 971–974.

Teratoma Production in Thigh Muscle

Richards, M., Fong, C.Y., Tan, S., Chan, W.K., Bongso, A., 2004. An efficient and safe xeno-free cryopreservation method for the storage of human embryonic stem cells. Stem Cells 22, 779–789.

Teratoma Production in Testis Capsule

Peterson, S.E., Tran, H.T., Garitaonandia, I., Han, S., Nickey, K.S., Leonardo, T.R., et al., 2011. Teratoma generation in the testis capsule. J. Vis. Exp. 57, e3177.

Rubenoff, B.E., Pera, M.F., Fong, C-Y., Trouson, A., Bongso, A., 2000. Embryonic stem cell lines from human blastocysts: somatic differentiation in vitro. Nat. Biotech. 18, 399–404.

Tissue Transplant in the Kidney Capsule

Hogan, B., Beddington, R., Constantini, F., Lacy, E., 1994. Manipulating the mouse embryo, second ed. Cold Spring Harbor Press, Cold Spring Harbour, NY.

Teratomas from Endangered Species

Ben-Nun, I.F., Montague, S.C., Houck, M.L., Tran, H.T., Garitaonandia, I., Leonardo, T.R., et al., 2011. Induced pluripotent stem cells from highly endangered species. Nat. Methods 8, 829–831.

Müller, F.J., Goldmann, J., Loser, P., Loring, J.F., 2010. A call to standardize teratoma assays used to define human pluripotent cell lines. Cell Stem Cell 6, 412–414.

CHAPTER 22

Characterization of Human Pluripotent Stem Cell-Derived Teratomas

Jessica Cedervall, Karin Gertow, Ivan Damjanov,
and Lars Ährlund-Richter

EDITOR'S COMMENTARY

Human pluripotent stem cells (hPSCs) have attracted much attention for their *in vitro* ability to differentiate into diverse specialized cell types. The differentiation of human PSC lines can be explored much further, however, by injecting them into experimental animals, where they can develop into remarkably organized tissues that resemble the structures found in fully developed organs.

Injections of PSCs with a normal karyotype have repeatedly been found to produce benign tissues. These tissues should be described as *experimentally-induced teratomas*, in analogy with the well-established clinical findings from patients spontaneously developing benign tumors in the ovary, testis, and extragonadal sites. These clinical teratomas are encapsulated and show, in most patients, limited growth potential, and typically include tissue or organ components resembling normal derivatives of all three germ layers: ectoderm, mesoderm, and endoderm. There are malignant versions of these tumors as well, but they are rare and are distinguishable from benign teratomas by their content of apparently undifferentiated cells.

In this chapter, the authors go far beyond the concept of using hPSC-derived teratomas as an assay for pluripotency and describe how they can be used to study human embryonic organ development.

OVERVIEW

Pluripotency refers to a stem cell's capacity to differentiate into cell types and tissues of all three developmental germ layers: ectoderm, endoderm, and mesoderm. This chapter focuses on the identification and analysis of various tissues arising from xenotransplantation of hPSCs into immunodeficient mice. The histology of a stem cell-induced mature teratoma is often

345

J.F. Loring & S.E. Peterson (eds): Human Stem Cell Manual, Second edition.
DOI: http://dx.doi.org/10.1016/B978-0-12-385473-5.00022-9

remarkable, with microscopic features equivalent to, or resembling, various tissues in developing or adult organs. Initial organoid arrangement of tissues is frequent, but advanced organogenesis is relatively uncommon. In this chapter we illustrate and describe the typical microscopic features of mature and immature tissues identified in these experimental teratomas. Usually, standard histology methods are sufficient to identify tissues in the teratomas, but in many situations, immunohistochemistry is needed as a complement to positively identify certain specialized cell or tissue types.

PROCEDURES

Based on our experience, and in consensus with the ISCI (International Stem Cell Initiative) working group on teratomas, which sought to agree on protocols for comparing the propensities of hPSC lines to differentiate (ISCI-3; meeting in September 2010 in Higher Disley, UK), we advocate the following outline for analysis of *in vivo* pluripotency:

- For experimental production of mature teratomas, hPSCs are typically transplanted to skeletal muscle, testes, or the kidney capsule of immunodeficient mice.
- For best engraftment of the human cells, it is recommended to use a genetically immunodeficient mouse strain that lacks mature T- and B-lymphocytes and NK cells. The NSG strain (NOD. Cg-PrkdcscidIl2rgtm1Wjl/SzJ; Jackson Laboratory) is ideal, but less immunodeficient strains such as SCID/Beige or NOD/Beige can be used.
- For a statistically relevant analysis, at least 10 mice per cell line should be injected and kept until mature teratomas have developed (~6 weeks).

For histology analysis and immunohistochemistry of the outgrowths, the teratoma should be cut into pieces (preferably 4–8). Randomly half of the fragments should be fixed in PFA and the other half snap-frozen in liquid nitrogen for use later in an RNA analysis if desired. A minimum of 10 sections of 5 microns at 3 separate levels should be analyzed after Hematoxilin-Eosin (HE) staining. The presence of germ layers should be recorded. To be scored as pluripotent, the teratoma should include at least one of the following easily identifiable tissues from each germ layer:

- Ectoderm: neuroectoderm/epithelium; pigmented epithelium corresponding to retina; ganglia.
- Mesoderm: cartilage, bone, fat cells, striated and/or smooth muscle.
- Endoderm: intestinal epithelium, bronchial epithelium.

Further detailed methods of the induction and analysis of hPSC-derived teratomas have been extensively described elsewhere (Gertow et al., 2007).

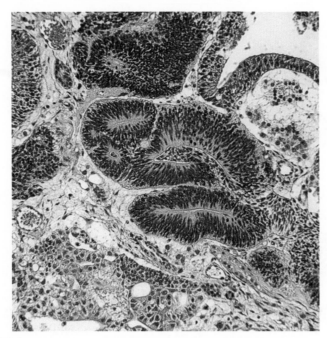

FIGURE 22.1
Immature teratoma composed of numerous fetal neural tube structures.

Ectodermal Tissues

Ectodermal tissues of both neural and non-neural character can be detected in hPSC-derived teratomas. Neural tissue is easiest to recognize when it is arranged into neural tubes or fetal neural rosettes that do not require immunohistochemistry (IHC) for identification as neural ectoderm (Figure 22.1). It is often multi-layered and shows considerable proliferative activity (Figure 22.2A,C). Condensation of fetal neural cells may be associated with the formation of glia-rich neuropil, indicative of maturation into adult-like cells (Figure 22.2C). Differentiation in such areas can be identified by positive staining for the neurite marker neurofilament protein (NFP) (Figure 22.2B,D). Also, functional synaptic activity can be indicated by the staining of synaptophysin in areas positive for NFP (Figure 22.3).

hPSC-derived teratomas frequently contain melanin-containing cells and (Figure 22.4, right side) retina-like epithelium (Figure 22.5). Although neural tissues often dominate the ectodermal content of teratomas, other types of ectoderm occur frequently. Among these, the most common are cysts lined by squamous epithelium. Periderm-like structures are somewhat less frequent (Figure 22.6) and the most rare are structures resembling skin (Figure 22.7), structures resembling the choroid plexus of the cerebral

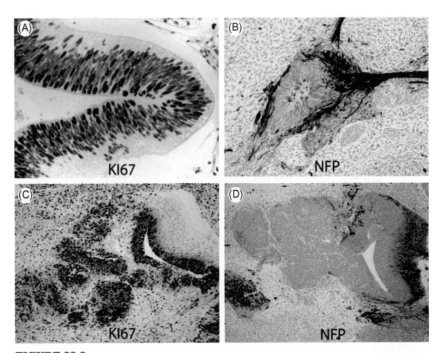

FIGURE 22.2
Suggestive neural epithelium (A, B) with proliferating cells (KI67 staining) and with staining for Neurofilament protein (NFP) (C, D) indicating presence of neurites. Original magnification A: 40×, B: 20×, C: 10×, D: 10×. *Reproduced with permission from Gertow et al., 2004.*

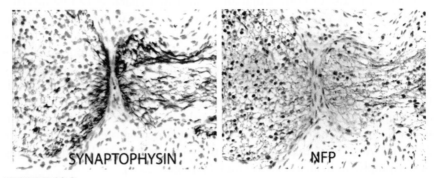

FIGURE 22.3
Synapses indicated in an area stained both for synaptophysin (left) and NFP (right). Original magnification 40×.

ventricles (Figure 22.8), and foci composed of primitive lens-forming cells of the eye (Figure 22.9).

Although mature teratomas with all three germ layers represented are the most common finding, sometimes hPSC-derived teratomas are composed

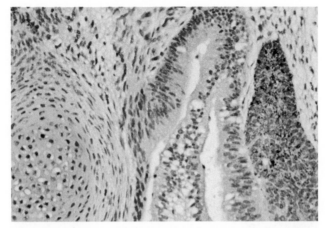

FIGURE 22.4
Teratoma composed of cartilage, primitive intestinal and neuroepithelial tissue, and a tube composed of pigmented retinal epithelium (right side).

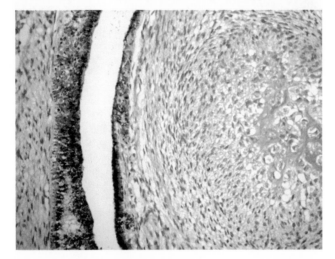

FIGURE 22.5
Specialized neural axis-derived cells such as pigmented, melanin-containing retinal cells may be seen focally. Original magnification 10×.

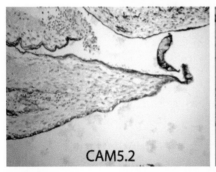
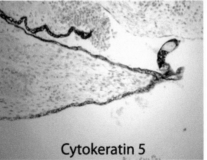

FIGURE 22.6
Periderm-like structures stained with cytokeratin 8/18 (CAM 5.2) (left) or cytokeratin 5 (right). Original magnification 10×.

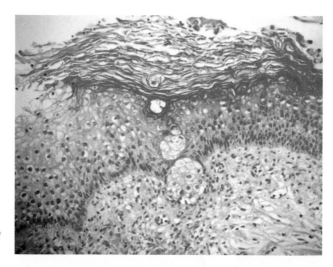

FIGURE 22.7
Ectodermal development of a cornified, stratified squamous epithelium including an excretory duct. Original magnification 10×. *Reproduced with permission from Gertow et al., 2004.*

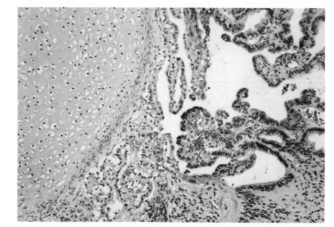

FIGURE 22.8
Teratoma composed of cartilage and choroid plexus-like tissue.

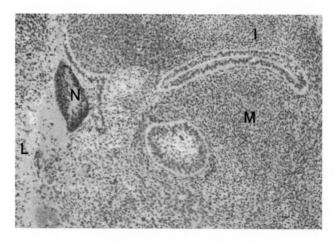

FIGURE 22.9
Teratoma composed of broad zones of mesenchymal stromal tissue (M), adjacent to fetal intestinal epithelium (I). Next to the neural tube (N) there is a small focus of lens-forming cells (L).

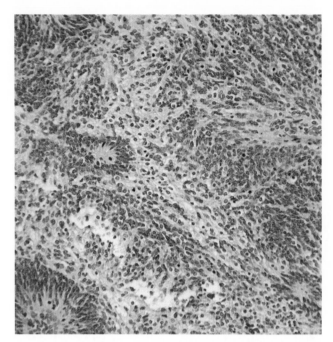

FIGURE 22.10
Immature teratoma composed of fetal neural tissue forming neural tubes or indistinct aggregates. The immature neural cells are surrounded by more differentiated neuropil that appears pink.

partly or almost entirely of fetal neural tissue of distinct or indistinct rosette aggregates, surrounded by more differentiated neuropil (Figure 22.10). Such immature neural cells cannot easily be distinguished microscopically from embryonic carcinoma cells (ECCs) and the tissues show considerable histological resemblance to teratocarcinomas (see below), as well as primitive neuroectodermal and neuroendocrine tumors commonly referred to as blastomas.

Endodermal Tissues

Endodermal structures typically include various glands and tubules lined by specialized epithelial cells. In some instances these tissues resemble fetal or adult intestine (Figure 22.11). The glandular structures can be surrounded by smooth muscle cells (Figure 22.12) and can indicate intestinal development if stained positively for mucin (Figure 22.13). Similar structures may also be respiratory epithelium (Figure 22.14). A simple rule of thumb is that these two common types of epithelia often can be identified in HE stained sections by the location of the nuclei in the epithelium. Respiratory epithelium is pseudostratified (nuclei are at different levels) while nuclei in the intestinal epithelium are generally lined up in a row. In other instances the glands cannot be positively identified (Figure 22.15). Sometimes nondescript mesenchymal cells surround the glands, and in such cases the juxtaposition of

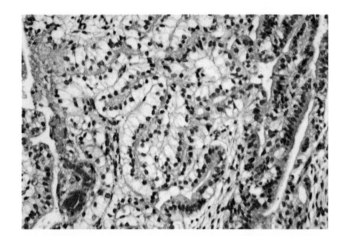

FIGURE 22.11
Teratoma containing clear cells resembling fetal intestine.

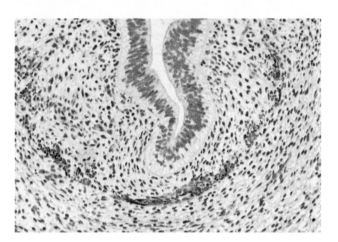

FIGURE 22.12
Teratoma containing an area of intestinal differentiation. The slide was immunohistochemically stained with antibodies to smooth muscle actin. With this approach one may see the layering of smooth muscle cells (brown) around the tubular intestinal epithelium, in a manner reminiscent of intestinal morphogenesis in the fetus.

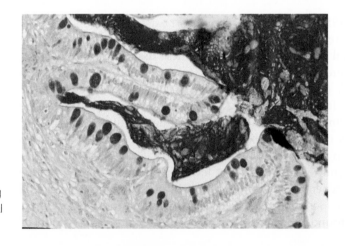

FIGURE 22.13
Teratoma containing intestinal epithelium. The slide was stained with the periodic acid-Schiff reaction, which outlines the mucus in the lumen (right upper quadrant) and individual mucus-secreting goblet cells in the intestinal epithelium.

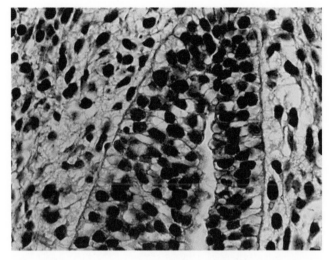

FIGURE 22.14
Ciliated and pseudostratified epithelium suggestive of respiratory epithelium surrounded by loose mesenchyme.

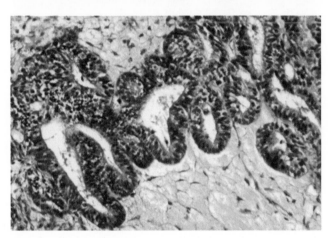

FIGURE 22.15
Teratoma containing glands that cannot be further characterized.

different cell types suggests complex organogenetic interaction of cells, similar to events that occur during fetal development (Figure 22.16).

Mesodermal Tissues

In hPSC-derived teratomas, loose or dense mesenchymal connective tissue, cartilage and early bone, fat cells, and smooth and striated muscle cells are the most frequently representatives of mesodermal tissues. Mesenchymal tissues may be composed of nondescript spindle cells, corresponding to fetal or adult connective tissue. Within such condensed mesenchymal tissue one may see focal areas of cartilage. Cartilage is readily recognized as it occurs in the form of discrete aggregates of chondroid cells surrounded by hyaline matrix (e.g. Figure 22.9). Such foci of cartilaginous differentiation are usually

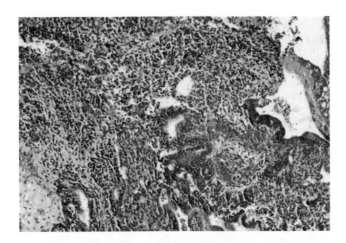

FIGURE 22.16
Teratoma composed of irregularly arranged nondescript glands and stromal tissue that cannot be further characterized on the basis of microscopy alone.

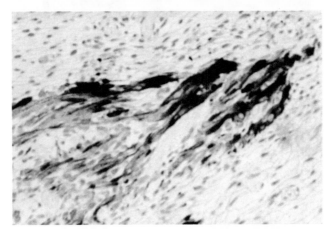

FIGURE 22.17
Teratoma containing striated muscle cells. These cells are immunohistochemically positive for desmin (brown).

distributed at random and may be adjacent to other mesenchymal derivatives, or ectodermal derivatives. By using antibodies to desmin one may see foci suggestive of striated muscle differentiation (Figure 22.17).

Mesenchymal condensations in association with epithelia reminiscent of glomeruli are good indications of kidney development. An example of very advanced nephron differentiation is indicated in Figure 22.18, and in this case, its identity is supported by *in situ* hybridization for kidney-associated *WT1* (Wilms tumor 1) and IHC for NCAM. Although the major part of the vascularization is comprised of mouse host vascularization, compound immature human vessels may be seen anastomosed with the host vessels.

Remaining Undifferentiated hPSCs: Analogies to Patient Immature Teratomas and Teratocarcinomas

Self-renewal, the ability to proliferate continuously in an undifferentiated state, is a requirement for hPSC culture. Clearly, *in vivo* injection into

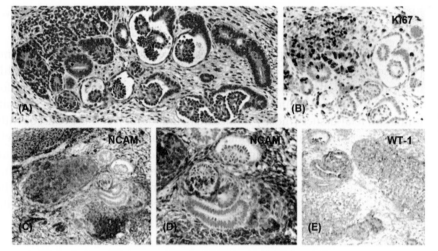

FIGURE 22.18
HE staining of an area suggestive of kidney development (A), showing proliferative cells (B, Ki67), and NCAM staining (C, D). *In situ* hybridization (ISH) using a probe specific for human *WT-1* (E). *Reproduced with permission from Gertow et al., 2004.*

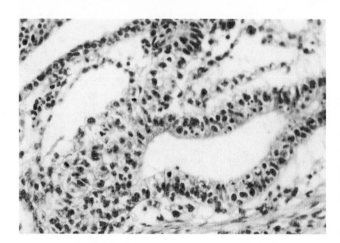

FIGURE 22.19
Immature teratoma containing undifferentiated hESCs. The cells are arranged into inter-anastomosing strands surrounding empty spaces. *Slide courtesy of Dr. PW Andrews.*

immunodeficient mice initiates differentiation, and the presence of undifferentiated cells in teratomas would indicate the persistence of proliferative pluripotent cells. In the largest study of experimental hESC-derived teratomas to date, histopathology analysis revealed that teratomas from only a very small minority of the cell lines analyzed contained undifferentiated cells (International Stem Cell Initiative, 2007). The rare foci of undifferentiated cells (Figures 22.19–22.21) have a high nucleus-to-cytoplasm ratio and their nuclei are slightly irregular, containing finely dispersed chromatin and prominent nucleoli. These cells are arranged into small groups, or in interconnected strands. Typically, the foci contain numerous apoptotic bodies and also scattered mitoses. It is not clear whether these cells represent mutated (potentially karyotypically abnormal) derivatives. Further systematic studies of hPSC-derived teratomas under standardized conditions are needed to answer this question.

Also, the significance of prolonged immaturity of some neural components (as described above for Ectodermal tissues) remains to be determined. This has a potential clinical parallel in that while mature teratomas account for most of the teratoid tumors in humans, some patient teratomas contain immature fetal tissues (mostly immature neural tissue), and are consequently denoted *immature teratomas*. Immature teratomas may grow indeterminately and spread to other organs and are thus considered potentially malignant. Intriguingly, in the hESC-xenograft model, the prolonged immaturity of some neural structures does not seem to cause the teratoma's encapsulation to be breached, and metastatic spreading has not been reported for hESCs with normal karyotypes.

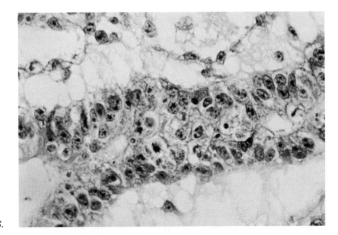

FIGURE 22.20

Immature teratoma containing undifferentiated hESCs. The hESCs are arranged into structures reminiscent of the early embryo and resemble embryoid bodies formed by hESCs *in vitro. Slide courtesy of Dr. PW Andrews.*

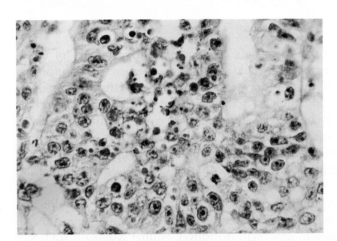

FIGURE 22.21

Immature teratoma containing undifferentiated hESCs. The hESCs are loosely arranged and many of them have initiated apoptosis. *Slide courtesy of Dr. PW Andrews.*

The malignant counterparts of human teratomas include *teratocarcinomas*, which appear in the testis (also known as non-seminomatous germ cell tumors or NSGCTs) as well as malignant mixed germ cell tumors, which are tumors that contain both teratoma and also seminoma (germ cell tumor) elements. All these tumors contain undifferentiated malignant stem cells, embryonal carcinoma cells (ECCs). It should be noted that ECCs are usually, similar to hPSCs, capable of differentiating into various somatic tissues, such as the usual derivatives of ectoderm, mesoderm and endoderm. However, some human ECCs can be developmentally "nullipotent," i.e. show no sign of differentiation and cannot differentiate into other cell types. Tumors composed of such nullipotent ECCs are called embryonal carcinomas, and account for approximately 10–15% of all human testicular NSGCTs.

hESC-derived teratocarcinomas have been observed only for xenografts of hESC carrying karyotypical mutations (Werbowetski-Ogilvie et al., 2009; Yang et al., 2008). However, the mere fact that *in vivo* injection of hPSCs initiates the generation of benign tissues analogous to clinical mature benign *teratomas* is a source of concern regarding the safety of using hPSCs in patient cell replacement therapies.

hiPSC-DERIVED TERATOMAS

Although the field is expanding rapidly, studies examining the capacity of human induced pluripotent stem cells (hiPSCs) to form teratomas are not yet as comprehensive as reports on hESC-derived teratomas. However, it is evident from the results of many groups, including the original report by Takahashi et al. (2007) that hiPSCs are capable of forming teratomas in a manner similar to hESCs. Examples of this are illustrated in Figures 22.22–22.25.

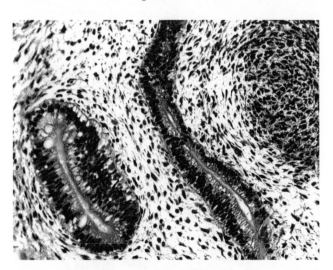

FIGURE 22.22
hiPSC-derived teratoma showing an area that appears to be respiratory epithelium surrounded by loose mesenchyme. The mesenchyme on the right side is more dense. *Photo courtesy of Dr. Ludovic Vallier.*

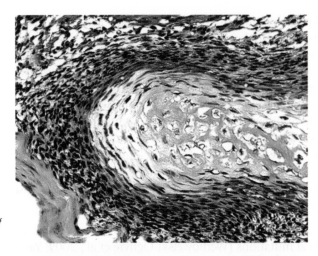

FIGURE 22.23
hiPSC-derived teratoma
showing area that
appears to be condensed
mesenchyme surrounding
an area of early bone
formation. *Photo courtesy of
Dr. Ludovic Vallier.*

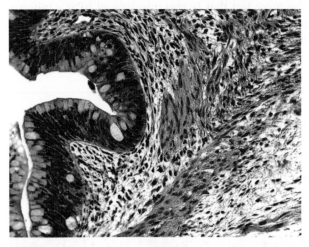

FIGURE 22.24
hiPSC-derived teratoma
showing area that appears
to contain respiratory
epithelium (left) and smooth
muscle surrounded by
loose mesenchyme. *Photo
courtesy of Dr. Ludovic
Vallier.*

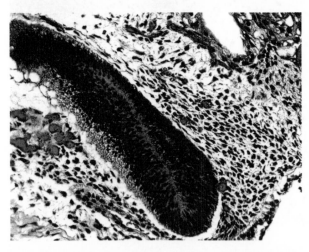

FIGURE 22.25
hiPSC-derived teratoma
showing an area of
loose and condensed
mesenchyme surrounding
an area of primitive
neuroectoderm formation.
*Courtesy of Dr. Ludovic
Vallier.*

READING LIST

Blum, B., Benvenisty, N., 2008. The tumorigenicity of human embryonic stem cells. Adv. Cancer Res. 100, 133–158.

Damjanov, I., Solter, D., 1974. Experimental teratomas. Curr. Top. Pathol. 59, 69–130.

Gertow, K., Wolbank, S., Rozell, B., Sugars, R., Andang, M., Parish, C.L., et al., 2004. Organized development from human embryonic stem cells after injection into immunodeficient mice. Stem Cells Dev. 13, 421–435.

Gertow, K., Przyborski, S., Loring, J.F., Auerbach, J.M., Epifano, O., Otonkoski, T., et al., 2007. Isolation of human embryonic stem cell-derived teratomas for the assessment of pluripotency. Curr. Protoc. Stem Cell Biol. (Chapter 1: Unit1B.4).

International Stem Cell Initiative, 85 authors, 2007. Characterization of human embryonic stem cell lines by the international stem cell initiative. Nat. Biotechnol. 25 (7), 803–816.

Lensch, M.W., Ince, T.A., 2007. The terminology of teratocarcinomas and teratomas. Nat Biotechnol. 25, 1211–1212.

Reubinoff, B.E., Pera, M.F., Fong, C.Y., Trounson, A., Bongso, A., 2000. Embryonic stem cell lines from human blastocysts: somatic differentiation in vitro. Nat. Biotechnol. 18, 399–404.

Takahashi, K., Tanabe, K., Ohnuki, M., Narita, M., Ichisaka, T., Tomoda, K., et al., 2007. Induction of pluripotent stem cells from adult human fibroblasts by defined factors. Cell 131, 861–872.

Thomson, J.A., Itskovitz-Eldor, J., Shapiro, S.S., Waknitz, M.A., Swiergiel, J.J., Marshall, V.S., et al., 1998. Embryonic stem cells lines derived from human blastocysts. Science 282, 1145–1147.

Werbowetski-Ogilvie, T.E., Bosse, M., Stewart, M., Schnerch, A., Ramos-Mejia, V., Rouleau, A., et al., 2009. Characterization of human embryonic stem cells with features of neoplastic progression. Nat. Biotechnol. 27, 91–97.

Yang, S., Lin, G., Tan, Y.Q., Zhou, D., Deng, L.Y., Cheng, D.H., et al., 2008. Tumor progression of culture-adapted human embryonic stem cells during long-term culture. Genes Chromosomes Cancer 47, 665–679.

4 PART

Differentiation

Spontaneous Differentiation of Human Pluripotent Stem Cells via Embryoid Body Formation

John P. Schell

EDITOR'S COMMENTARY

Embryoid bodies are aggregates of pluripotent cells that are induced to differentiate by a combination of a change in culture medium (removal of factors that support pluripotency) and allowing the cells to interact in three-dimensional structures. The initial interest in studying mouse embryonic stem (ES) cells by making them into embryoid bodies was what the name implies: creating structures that are embryo-like and could be used for studying complex developmental processes. Because the test for pluripotency for mouse ES cells was germ line transmission of the cells, embryoid bodies have never become an important method for demonstrating pluripotency of mouse cells. Embryoid bodies began to serve new purposes when human pluripotent stem cells (hPSCs) became widely used. For human cells, embryoid bodies are of less interest as models of to study embryonic development, and have taken on the roles of testing cells for pluripotency in vitro and generating differentiated cells. The term "embryoid body" is appropriate only when the cells are allowed to fully differentiate within the three-dimensional structures. As is pointed out in Chapter 26, aggregation of hPSCs is not necessarily used for making embryoid bodies. For many applications, the cells are aggregated to achieve a reproducible density as part of a directed differentiation protocol. It should also be noted that aggregation itself does not necessarily cause differentiaton; several laboratories have developed methods for maintaining hPSCs in an undifferentiated state while being grown as cellular aggregates in suspension culture (Amit et al., 2011). With that distinction firmly in mind, this chapter describes methods for generating embryoid bodies from hPSCs for the purpose of producing a variety of differentiated human cells.

OVERVIEW

The culturing of embryoid bodies (EBs) is a qualitative technique for testing pluripotency, as well as producing cells of multiple lineages. By creating a

363

J.F. Loring & S.E. Peterson (eds): Human Stem Cell Manual, Second edition.
DOI: http://dx.doi.org/10.1016/B978-0-12-385473-5.00023-0

suspension culture, dissociated colonies are allowed to form cellular aggregates under ultra-low attachment conditions. These aggregates coalesce to form sphere-like masses that eventually lose their pluripotency and begin down paths of differentiation. The initial transformation of cell commitment within an EB enables further differentiation based on localized cell signaling gradients from the emerging cell types.

EB formation and differentiation are frequently used to demonstrate the pluripotency of new hPSC lines. To demonstrate pluripotency, EBs derived from hPSCs must be able to differentiate into cells from the three germ layers: endoderm, ectoderm, and mesoderm. Many varieties of specialized cell types originate from each of these three germ layers.

Many researchers are using the initial stages of EB differentiation to enrich for derivatives of a particular germ layer, usually by blocking differentiation into undesired cell lineages and encouraging the survival and maturation of the desired cell type. This is usually achieved in several stages using a variety of small molecules, growth factors and substrates.

This chapter provides methods for generating EBs from hPSCs and using them to demonstrate spontaneous differentiation into the three germ layers.

PROCEDURES

Embryoid body formation has been accomplished using a variety of methods and culturing techniques, including hanging drop culture, methylcellulose culture, suspension culture, plating of cells in non-adhesive round-bottomed 96-well plates, and spin aggregates (Chapter 27). Figure 23.1 shows the three basic steps of using EBs for spontaneous differentiation assays: culture of hPSC colonies, formation and culture of EBs, and replating of EBs onto adhesive substrata to analyze differentiation.

Embryoid Body (EB) Generation and Spontaneous Differentiation into Cells From the Three Germ Layers

Summary of the procedure:

Enzymatically dissociate colonies from MEF feeder layer and plate in ultra low attachment dishes.	Day 1
Allow aggregates to form EBs in ultra-low attachment conditions.	Day 2–Day 7
Plate EBs onto 0.1% gelatin-coated dish.	Day 8
Allow differentiation of EBs, and immunostain for lineage markers.	Day 8–20

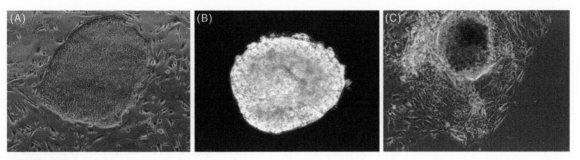

FIGURE 23.1
(A) hPSC colony being cultured on mouse embryo fibroblasts (MEFs). (B) Embryoid body in suspension. (C) Embryoid body differentiation after attachment to adhesive substrate.

Generating Embryoid Bodies from hPSCs Cultured on Feeder Layers (MEFs)

1. Begin with a confluent plate of hPSCs cultured on a feeder layer (MEFs). The cell culture should be at the density at which the cells would normally be passaged.
2. Prepare dispase at 2 mg/mL in D-MEM/F12 and sterilize with 0.22 μm filter.
3. Wash wells with PBS.
4. Aspirate PBS from wells and add enough dispase to cover the entire well, typically, 0.5 mL/well for a 6-well plate.
5. Incubate at room temperature watching under a phase-contrast microscope until the colonies fully dissociate from the feeder layer, approximately 7–10 minutes.

■ Note

The feeder layer should remain attached to the dish while the hPSC colonies should begin to lift off. Colonies at first become phase-bright around the edges and then eventually lift off the feeder layer. ■

6. Add 2 mL KOSR medium to each well to dilute the dispase, then gently pipette and transfer all floating hPSC colonies from an entire plate to a 15 mL conical tube.
7. Triturate cell chunks gently with the pipette 2–3 times to form smaller clumps roughly a quarter of the size of the original colony.
8. Allow all clumps to settle to the bottom of the 15 mL conical tube, then carefully aspirate off the medium.
9. Wash the cells once by adding 5–6 mL of fresh KOSR medium, then allow the clumps to settle once again; repeat Step 8.

10. Resuspend in KOSR medium without bFGF and plate on an ultra low attachment culture plate.

11. Incubate cells at 37°C, 5% CO_2.

Alternative Dissociation Methods

Use of a combination of Collagenase IV and Dispase (1 mg/mL Dispase + 1 mg/mL Collagenase IV) allows for slightly faster removal of hPSCs from their feeder layer. In addition, cells can simply be mechanically passaged (see Chapter 1) instead of passaged with enzymes. After mechanically passaging cells, transfer cells to a 15 mL conical tube and continue with Step 8 of the above procedure.

Feeding/Handling

Change the medium on the cells every other day as they form EBs. EBs in suspension culture require extra care so as to not damage or lose developing aggregates.

1. Tilt the plate to a 45° angle and allow EBs to settle in the bottom corner of each well.
2. Carefully aspirate or pipette off three-fourths of the media from the well without agitating EBs.
3. Gently add 1.5 mL KOSR media without bFGF back to each well.

Lineage Differentiation

Allow EBs to spontaneously differentiate.

1. Culture EBs in suspension in an ultra-low attachment dish for 7 days.
2. Coat an appropriately sized culture dish, chamber slide or coverslip with 0.1% gelatin.
3. Allow gelatin to incubate in the dish for 30–45 minutes at room temperature.
4. Aspirate gelatin completely.
5. Carefully remove EBs from suspension culture with a 5 mL pipette, and plate them onto the gelatin-coated dishes.
6. Allow EBs to settle and attach overnight.
7. Feed differentiating EBs DMEM with 20% FBS every other day, being mindful to not aspirate the attached cell masses.

■ Note

For the first few days after plating, the EBs are not well-attached and will come off if the medium is aspirated off too quickly. Gently pipette the medium off the cells until they become well attached to the substrate. ■

8. Observe differentiating cells migrate away from plated EBs. Allow EBs to mature and differentiate for 6–12 days (or longer if desired).
9. Fix and immunostain for early lineage markers.

Immunostaining: Endoderm, Mesoderm, Ectoderm

Table 23.1 lists some of the markers that are commonly used to detect derivatives from EBs of the three germ layers. Figure 23.2 shows examples of immunocytochemical staining of EB-derived cells. See Chapter 16 for detailed methods for immunocytochemistry.

1. Wash the cells with PBS twice. Fix the cells with 4% paraformaldehyde (PFA) in PBS for 10 min at 4 °C.
2. Wash the cells with PBS three times.
3. Block the cells with 10% serum in PBS containing 0.1% Triton X-100 to permeabilize the cells.

■ Note

It is recommended to use serum from the species in which your secondary antibody is derived to reduce non-specific binding. ■

Table 23.1 Examples of Germ Layer Derivative Markers; NCBI Nomenclature

Ectodermal	
MAP2	Microtubule-associated protein 2
NES	Nestin
NOG	Noggin
PAX6	Paired box 6
TUBB3	Tubulin, beta 3 class III
Mesodermal	
BMP4	Bone morphogenetic protein 4
INHBA (Activin A)	Inhibin, beta A
NODAL	Nodal homolog
SMA	Smooth muscle actin
TGFB1	Transforming growth factor, beta 1
WNT3A	Wingless-type MMTV integration site family, member 3A
Endodermal	
AFP	Alpha-fetoprotein
FOXA1	Forkhead box A1
FOXA2	Forkhead box A2
GATA4	GATA binding protein 4
GSC	Goosecoid homeobox
SOX17	SRY-box 17
HNF1B (TCF-2)	HNF1 homeobox B

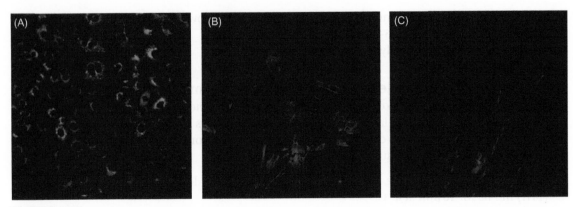

FIGURE 23.2
Examples of immunofluorescent staining for derivatives of the three germ lines. (A) Alpha-fetoprotein, endoderm; (B) Smooth muscle action, mesoderm; (C) Microbule associated protein 2, ectoderm.

■ Note

Cell surface antigen specificity may be lost if cells are treated with Triton X-100. See Chapter 16. ■

4. Incubate the cells for 1 hour to overnight at room temperature with the primary antibody.
5. Wash cells with PBS three times.
6. Incubate the cells for 1 hour at room temperature with an appropriate fluorescently-conjugated secondary antibody at room temperature in the dark.
7. Wash the cells three times with PBS.
8. Counterstain nuclei with DAPI.

ALTERNATIVE PROCEDURES

One of the major problems associated with differentiation of hPSCs via embryoid bodies is heterogeneity. Researchers often find significant variability in yields of cells differentiated using this technique. One of the easiest variables to control is EB size. This can be done using AggreWell plates (StemCell Technologies) that have a defined size and shape and contain one EB per well. hPSCs are counted and plated in the AggreWell plates, then spun down into aggregates that share relatively the same basic size and morphology with one another. This method controls for size and shape, while also eliminating any variability that may result from contact with other EBs. AggreWell plates are also often used in directed differentiation protocols that use an initial aggregation (not embryoid body) stage to control the cell density.

PITFALLS AND ADVICE

- It is very easy to accidentally aspirate (or even pipette off) EBs that have just been plated onto gelatin. Be very careful not to do so when EBs are first attaching to the dish.
- Not all "ultra-low attachment" dishes are created equally and sometimes EBs will attach to these dishes. This is often specific to particular lots and the manufacturers will often swap the bad lot for a different one. Be sure to report it if you get a bad lot.
- Don't neglect sterile technique when handling EBs. Eliminate the risk of contamination by always using proper aseptic technique and keeping a neat hood.
- Feed cells on a regular schedule, even down to the same hour each day.
- The first day after EB formation in suspension generally yields a large amount of cell death. You do not want dead cells incorporated into or sticking to your EBs. Eliminate the accumulation of dead cell debris by washing EBs in PBS before feeding.
- Differentiation is a stochastic event that can occur in a highly variable manner. Some cell types can be very proliferative, and have very short doubling times. Be careful not to let your culture overpopulate. If cells become too dense, visibility and space for further differentiation is sacrificed.

EQUIPMENT

- Tissue culture hood, Class II A/B3
- Tissue culture incubator, 37°C, 5% CO_2, with humidified air
- Low-speed centrifuge
- Water bath, 37°C
- Pipettors: p20, p200 and p1000
- Pipette Aid
- Inverted Phase contrast microscope with 4×, 10× and 20× objective
- Refrigerator at 4°C
- Freezers: −20°C and −80°C
- Fluorescence microscope

REAGENTS AND SUPPLIES

Supplies

- 6-well Corning® Cell Culture Surfaces and Ultra-Low Attachment Surface
- AggreWell 400 or 800 plates (Stem Cell Technologies)

- 5 mL, 10 mL, 25 mL sterile disposable pipettes
- 15 mL sterile conical tubes
- 50 mL sterile conical tubes
- Sterile 9″ pasture pipettes
- Pipette tips for Eppendorf or similar pipettor

Recommended Reagents

Item	Supplier	Catalog#
Basal Medium		
DMEM/high glucose, L-glutamine (Dulbecco's Modified Eagle's Medium)	Life Technologies	11965-092
DMEM (high glucose, GlutaMAX [dipeptide L-Alanyl-L-Glutamine])	Life Technologies	10566-016
KnockOut-DMEM (lower osmolarity and lower bicarbonate for optimal pH at 5% CO_2)	Life Technologies	10829-018
DMEM/F12 1:1 liquid (Dulbecco's Modified Eagle's Medium/Ham's F12). Contains HEPES buffer, L-glut, pryidoxol HCL	Life Technologies	11330-032
DMEM/F12 1:1 (Glutamax, no HEPES)	Life Technologies	10565-018
Serum Components		
Knockout Serum Replacement (contains bovine products)	Life Technologies	108280-028
Fetal Bovine Serum (FBS)	Life Technologies	10439-016
L-Glutamine 200 mM	Life Technologies	25030-081
MEM-Non-essential amino acids (100× = 10 mM)	Life Technologies	11140-050
2-Mercaptoethanol	Life Technologies	21985-023
D-PBS (Dulbecco's Phosphate-Buffered Saline without Calcium and Magnesium)	Life Technologies	14190-144
Basic FGF (FGF2)		
Recombinant Human bFGF	Gemini Bio-Products	300-112P
Recombinant Human bFGF	Life Technologies	13256-029

Recommended Collagenase IV and Dispase

Component	Catalog #	Supplier
Collagenase IV	17104019	Life Technologies
Dispase	17105041	Life Technologies

Recommended Primary Antibodies

Item	Supplier	Catalog #	Recommended Dilution
Alpha Fetoprotein mouse monoclonal	Millipore	2004189	1:100
Smooth Muscle Actin mouse monoclonal	Millipore	CBL171	1:100

Item	Supplier	Catalog #	Recommended Dilution
Nestin rabbit polyclonal	Millipore	AB5922	1:100
MAP-2 mouse monoclonal	Millipore	MAB3418-50UG	1:100
Brachyury rabbit polyclonal	Santa Cruz Biotechnology	sc-20109	1:250
GATA-4 mouse monoclonal	Santa Cruz Biotechnology	sc-25310	1:250

Recommended Secondary Antibodies

Item	Supplier	Catalog #	Recommended Dilution
Alexa Fluor 488 Goat anti-mouse IgG (H + L)	Life Technologies	A11001	1:750
Alexa Fluor 488 Goat anti-rabbit IgG (H + L)	Life Technologies	A11008	1:750
Alexa Fluor 555 Goat anti-mouse IgG (H + L)	Life Technologies	A21422	1:750
Alexa Fluor 555 Goat anti-Rabbit IgG (H + L)	Life Technologies	A21428	1:750
Mounting Media with DAPI	Vector Labs	H-1200	

RECIPES

Stock Solution Human Basic FGF (bFGF) 25 μg/ml

Component	Amount	Stock Concentration
Human bFGF	10 μg	25 μg/mL PBS with BSA 0.1%

1. Dissolve 10 μg in 400 μL of PBS containing 0.1% BSA.
2. Aliquot 25 μL per Eppendorf tube and store at −20°C.

■ Note

Avoid repeated freezing and thawing. ■

hPSC KOSR Medium 100 mL

Component	Amount	Final Concentration
DMEM/F12	77.8 mL	
KnockOut Serum Replacement	20 mL	20%
100× MEM-Non-Essential amino acids	1 mL	10 mM
100× L-Glutamine (If DMEM/F12 lacks glutamine)	1 mL	2 mM

Component	Amount	Final Concentration
55 mM 2-Mercaptoethanol or 0.1 M (1000×) stock	182 μL of 55 mM (0.1 mL of 0.1 M)	0.1 mM
Human bFGF (1 μg/ml)	20 μL	20 ng/mL

1. Prepare all media in the tissue culture hood using aseptic techniques.
2. Add basal medium to filter top of a 0.22 μm, low protein binding filter unit, then add serum and other supplements.
3. Store at 4°C for up to 4 weeks.

Embryoid Body Medium (KOSR medium – bFGF) 100 mL

Component	Amount	Final Concentration
DMEM/F12	77.8 mL	
KnockOut Serum Replacement	20 mL	20%
100× MEM-Non-Essential amino acids	1 mL	10 mM
100× L-Glutamine	1 mL	2 mM
55 mM 2-Mercaptoethanol or 0.1 M (1000×) stock	182 μL of 55 mM (0.1 mL of 0.1 M)	0.1 mM

1. Prepare all media in the tissue culture hood using aseptic techniques.
2. Add basal medium to filter top of a 0.22 μm, low protein binding filter unit, then add serum and other supplements.
3. Store at 4°C for up to 4 weeks.

Embryoid Body Differentiation Medium (DMEM with 20% FBS) 100 mL

Component	Amount	Final Concentration
DMEM/F12 (containing glutamine or GlutaMAX)	79 mL	
FBS	20 mL	20%
100× MEM-Non-Essential amino acids	1 mL	10 mM

1. Prepare all media in the tissue culture hood using aseptic techniques.
2. Add basal medium to filter top of a 0.22 μm, low protein binding filter unit, then add serum and other supplements.
3. Store at 4°C for up to 4 weeks.

READING LIST

Amit, M., Laevsky, I., Miropolsky, Y., Shariki, K., Peri, M., Itskovitz-Eldor, J., 2011. Dynamic suspension culture for scalable expansion of undifferentiated human pluripotent stem cells. Nat. Protoc. 6, 572–579.

Assady, S., Maor, G., Amit, M., Itskovitz-Eldor, J., Skorecki, K.I., Tzukerman, M., 2001. Insulin producing by human embryonic stem cells. Diabetes 50, 1691–1697.

Cerdan, C, Hong, SH, Bhatia, M., 2007. Formation and hematopoietic differentiation of human embryoid bodies by suspension and hanging drop cultures. Curr. Protoc. Stem Cell Biol. (Chapter 1:Unit 1D 2).

Itskovitz-Eldor, J., Schuldiner, M., Karsenti, D., Eden, A., Yanuka, O., Amit, M., et al., 2000. Differentiation of human embryonic stem cells into embryoid bodies comprising the three embryonic germ layers. Mol. Med. 6, 88–95.

Kehat, I., Kenyagin-Karsenti, D., Snir, M., Segev, H., Amit, M., Gepstein, A., et al., 2001. Human embryonic stem cells can differentiate into myocytes with structural and functional properties of cardiomyocytes. J. Clin. Invest. 108, 407–414.

Kurosawa, H, 2007. Methods for inducing embryoid body formation: *In vitro* differentiation system of embryonic stem cells. J. Biosci. Bioeng. 103, 389–398.

Levenberg, S., Golub, J.S., Amit, M., Itskovitz-Eldor, J., Langer, R., 2002. Endothelial cells derived from human embryonic stem cells. Proc. Natl. Acad. Sci. USA 99, 4391–4396.

Differentiation of Human Pluripotent Stem Cells into Neural Progenitors

Philip Koch, Franz-Josef Müller, and Oliver Brüstle

EDITOR'S COMMENTARY

Neural differentiation of human pluripotent stem cells (hPSCs) is one of the most important lineage specific differentiation trajectories in the stem cell field for disease modeling, drug development, and clinical applications. The neural progenitor stage is particularly useful because the cells are mitotic, can be cloned, genetically manipulated, cryopreserved and thawed for later use. They are multipotent, possessing the ability to differentiate into many kinds of neurons and glia, depending on the directed differentiation protocol that is used. This chapter describes a straightforward, reliable method for producing neural progenitor cells from hPSCs and differentiating them further into neurons and glia.

OVERVIEW

Numerous protocols have been published over the last decade with increasingly refined methodologies (Chambers et al., 2009; Greber et al., 2011; Koch et al., 2009; Watanabe et al., 2007; Zhang et al., 2001). Here we focus on a relatively simple and highly reproducible protocol that should help reliably obtain highly enriched populations of neural progeny even in the hands of non-experts. This protocol provides a platform for studying differences in differentiation propensities among diverse induced pluripotent stem cells (iPSCs) and human embryonic stem cell (hESC) lines as well as their terminal neural differentiation potential.

The protocol is based on aggregation cultures and does not require feeder-free cultivation of hPSCs or single cell suspensions. In our protocol, differentiation into neuroectoderm is guided by medium conditions and the dual inhibition of SMAD signaling (Chambers et al., 2009). We replaced noggin with the small molecule LDN-193189 ($0.5\,\mu$M), which allows cost-efficient differentiation of hPSCs towards neural progeny. Even without manual selection (see

375

J.F. Loring & S.E. Peterson (eds): Human Stem Cell Manual, Second edition.
DOI: http://dx.doi.org/10.1016/B978-0-12-385473-5.00024-2

Isolation of neural islands) this protocol typically gives rise to greater than 80% neuroepithelial cells within 15 days. A simple manual selection step described here allows further enhancement of neural progeny to 90–100%. We also provide instructions for single cell suspension of neuroepithelial islands and for terminal differentiation of neuroepithelia into neurons and glial cells.

PROCEDURES

The protocol described here consists of a series of steps in which the hPSCs are guided toward a neuronal lineage. The initial steps take about 15 days, and result in cultures that contain approximately 80% neural epithelial/neural progenitor cells. A second protocol can be applied at this stage, which involves enrichment by mechanical selection of patches of neuroepithelial cells. With this additional step, cultures may reach greater than 90% neural precursors. These cells can be cultured further and differentiated into functional neurons or other nervous system cell types.

Figure 24.1 shows a diagram of the protocols that are described in detail below.

Generation of hPSC Aggregates

hPSCs are maintained on irradiated mouse embryonic fibroblasts (MEFs) at 5% CO_2 in KSR medium (see Recipes), using standard hPSC bulk culture passaging techniques and enzymatic dissociation (Chapter 1). To generate

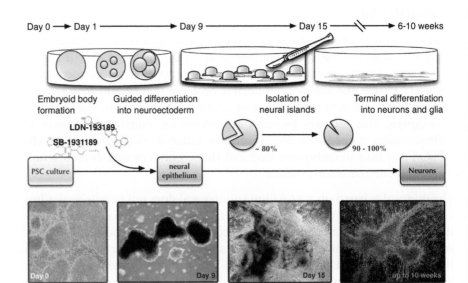

FIGURE 24.1

Overview of protocols for generating neuronal epithelium/neuronal precursors.

aggregates, hPSCs are incubated with collagenase IV, keeping the cells in small clumps. The volumes used here are all for one well of a 6-well plate (approximately 35 mm^2) and have to be adjusted accordingly when culturing hPSCs in other formats.

1. Aspirate the KSR medium from the culture dish.
2. Add 1 mL of collagenase IV (1 mg/mL in KSR medium) and place dish in the incubator at 37°C for 20 minutes.
3. Starting from 20 minutes, check every 5–10 minutes until hPSC colonies start to detach.
4. When about 10–20% of the colonies have detached and edges of most colonies started to curl, add 2 mL of KSR medium and wash the hPSC colonies from the plate using a p1000 pipetman.
5. Transfer the colonies to a 15 mL conical tube and let the colonies sink to the bottom of the tube for 5 minutes. Colonies from several 6-well dishes can be pooled into the one tube.
6. Aspirate the supernatant and add 3 mL of fresh medium.
7. Gently triturate the colonies by pipetting 3–5 times using a p1000 pipetman. The colonies should still be visible by eye.
8. Centrifuge colonies for 3 minutes at 200 g.
9. Aspirate the supernatant, resuspend cells in 5 mL of fresh KSR medium and transfer the colonies to a non-adhesive 60 mm Petri dish (use one 60 mm dish per 3 wells of a 6-well plate).
10. Incubate overnight at 37°C in 5% CO_2.

Guided Differentiation into Neuroectoderm

Neural differentiation is induced by sequentially changing the medium from KSR medium to N2 medium (see Recipes).

Inhibition of SMAD signaling using the ALK5 inhibitor SB-431542 (10 μM) in combination with the ALK2/3/6 inhibitor LDN-193189 (0.5 μM) is used to block mesendodermal differentiation and to direct the differentiation into neuroectoderm.

Day 1: The hPSC colonies will have formed aggregates, which float in the Petri dish.

1. Collect the medium and the aggregates with a 5 mL pipette and transfer into a 15 mL conical tube.
2. Wait 5 minutes until colonies have sunk to the bottom of the tube. Carefully aspirate the supernatant.
3. Add 5 mL of fresh KSR medium containing 10 μM SB-431542 and 0.5 μM LDN-193189.

4. Replate colonies on the same 60 mm dish and incubate at 37°C in 5% CO_2 for 2 days.

Day 3: Replace half the medium with N2 medium with inhibitors.

1. Transfer the medium and aggregates to a 15 mL conical tube.
2. Wait 5 minutes until colonies have sunk to the bottom of the tube. Carefully aspirate half of the supernatant.
3. Add 3 mL of fresh N2 medium containing 10 μM SB-431542 and 0.5 μM LDN-193189 and transfer colonies back on the 60 mm dish.

Days 5 and 7: Repeat the procedure from Day 3, replacing half the medium with N2 medium containing 10 μM SB-431542 and 0.5 μM LDN-193189. Using this procedure the medium will be sequentially changed from KSR medium to N2 medium.

Day 9: Repeat the procedure as described above but aspirate all the supernatant and resuspend the colonies in 6 mL N2 medium containing 10 μM SB-431542 and 0.5 μM LDN-193189.

1. Plate the colonies on poly-L-ornithine/laminin precoated plastic dishes (see precoating procedure below).
2. Use three 35 mm dishes or a 6-well plate per 60 mm Petri dish. Plate 2 mL of N2 medium containing neural aggregates per dish.
3. Change medium every 48 hours for another 6 days.

■ Note

The technique described above usually generates neuroectodermal cells with efficiencies of greater than 80%. The cells can be further differentiated in Neural Differentiation medium (see Recipes) or pre-purified before further differentiation by manual dissection of "neural islands" that appear in the cultures. This method is described below.   ■

Precoating Dishes with Poly-L-Ornithine and Laminin

1. Dilute poly-L-ornithine in sterile water to a final concentration of 0.1 mg/mL.
2. Add 1.5 mL of the poly-L-ornithine solution per well of a 6-well plate and incubate at 37°C for at least 3 hours.
3. Aspirate poly-L-ornithine and wash once with PBS (don't let the dishes dry out).
4. Dilute laminin in PBS to a final concentration of 10 μg/mL.
5. Add 1 mL of the laminin solution per 6-well plate and incubate at 37°C for at least 4 hours or at 4°C overnight.
6. When plating the cells, aspirate the laminin solution and directly plate the cells on the dishes (don't let the plates dry out).

Isolation of Neural Islands

Depending on the hPSC line, contaminating mesodermal or endodermal cells might also be present in the culture dish. The manual isolation of neural islands is a simple method to further enrich the cultures for neuroectoderm.

Day 15: Check your culture dishes for neuroectodermal and non-neuroectodermal differentiation.

Neuroectodermal islands are typically large and tend to grow three-dimensionally. They appear yellow–brown in color and contain rosette-like and neural-tube-like structures. Colonies comprising other germ layers are mostly flat with fibroblast-like cells, cobblestone-shaped cells or remaining pluripotent cells.

■ Note

These patches of neuroepithelium are generally larger and more complex than the "neural rosettes" that are described in Chapter 25.   ■

1. Scrape off all areas that appear non-neuroectodermal.
2. Gently wash dishes twice with pre-warmed PBS.
3. Add 2 mL of N2 medium and dislodge neural islands by pipetting with a p1000 pipetman. The larger neural islands detach easily. Flat cells and cells from other germ layers stick more tightly to the dish.
4. Check under the microscope to see how many neural islands remain and repeat the procedure if necessary.
5. Gently dissociate the islands into smaller clumps by pipetting the colonies up and down 10 times with a p1000 pipetman.
6. Collect the cells in a 15 mL conical tube.
7. Centrifuge the colonies for 3 minutes at 200 g
8. Plate cells on a Matrigel™-coated plastic dish in 2 mL of Neural Differentiation medium (see Recipes). The method for precoating the dishes with Matrigel™ is shown below.
9. Plate a portion of the cells on Matrigel™ -coated glass coverslips or multiwell slides; after the colonies have attached and cells have started to grow out, they can be checked with immunocytochemistry (see Chapter 16) with SOX2 and PAX6 (see Antibodies section for dilutions of primary and secondary antibodies). Both antibodies should show a nuclear staining pattern. More than 90% of the cells should express both neural progenitor markers.

Precoating Dishes with Matrigel™

1. Dilute Matrigel 1:30 in ice-cold DMEM/F12.
2. Coat the dishes with 800 µL Matrigel solution at 4°C overnight.
3. Aspirate the Matrigel solution immediately before plating the cells. Do not let the dishes dry out during this procedure.

Terminal Differentiation of Neuroectodermal Islands

To further differentiate the neural cells they are cultured in Neural Differentiation medium containing Neurobasal medium supplemented with B27 (see Recipes section).

1. Culture the cells in Neural Differentiation medium, changing the medium every other day.
2. Neuronal and glial differentiation will be apparent after 4 weeks of differentiation.
3. For electrophysiological assessment of the neurons you many need to differentiate the cells longer (8 weeks for the formation of spontaneous neuronal networks).
4. Check the efficiency of your differentiation by staining parallel cultures (on coverslips or multiwell slides) with the neuronal markers TUBB3 (beta III-tubulin) and MAP2 (microtubule-associated protein 2), the astrocyte marker GFAP (glial fibrillary acidic protein) and the oligodendrocytic antigen O4 (see Antibodies for dilutions of primary and secondary antibodies).

ALTERNATIVE PROCEDURES

After the cells have formed neural islands, the islands can be replated as single cells by digestion with trypsin. The cells are to be pre-incubated with the ROCK inhibitor Y-27632 to enhance their recovery.

1. After centrifugation resuspend the neuroectodermal islands in 5 mL of Neural Differentiation medium supplemented with 10 μM ROCK inhibitor Y-27632.
2. Incubate the cells for 20 minutes.
3. Centrifuge the islands at 200 g for 3 minutes.
4. Resuspend the islands in 1 mL of 0.05% typsin/EDTA.
5. Incubate the cells for 10 minutes at 37 °C.
6. Add 1 mL of Trypsin Inhibitor.
7. Dissociate by pipetting the islands up and down 10–20 times using a p1000 pipetman.
8. The remaining undissociated cells can be eliminated by filtering the solution through a cell strainer (optional).
9. Centrifuge the solution for 5 minutes at 300 g
10. Plate cells on freshly coated Matrigel dishes in 2 mL of Neural Differentiation medium per well of a 6-well dish.
11. Continue culture for several weeks until cells are fully differentiated (check with immunocytochemistry as above).

■ **Note**

Individual hPSC lines may show differences in survival when trypsinized. The optimal re-plating density must be determined for each cell line.   ■

PITFALLS AND ADVICE

Contamination with Neural Crest Cells

The neural crest is a migratory population of cells that arises at the dorsal edge of the neural tube as the tube closes. The generation of neural crest is not inhibited by the dual-SMAD inhibition protocol. In the aggregate-based protocol described here, the cellular density of the aggregates efficiently inhibits the formation of large numbers of neural crest cells.

However, plating dissociated neural islands or single neuroectodermal cells at too low density can cause neural crest cells to proliferate and expand. The propensity to give rise to neural crest cells varies among hPSC lines.

If after switching to Neural Differentiation medium, you mostly get flat cells and only a few neurons, neural crest cells are likely dominating your culture. Neural crest cell contamination can be identified by staining the cells with an antibody to CD271 (NGFR: NGF receptor p75). Other markers such as HNK-1 (B3GAT1: beta-1,3-glucuronyltransferase 1) may also distinguish neural crest cells. A small number of neural crest cells <5% is likely present in each preparation and cannot be completely eliminated using this protocol.

REAGENTS AND SUPPLIES

Item	Supplier	Catalog no.
PBS without Ca^{2+} and Mg^{2+}	Life Technologies	14190094
DMEM/F12 with Glutamine	Life Technologies	11320074
DMEM (high glucose) with Glutamine	Life Technologies	41965039
Knockout™ DMEM	Life Technologies	10829018
KnockOut Serum Replacement (KSR)	Life Technologies	10828-028
Neurobasal Medium	Life Technologies	21103049
L-Glutamine	Life Technologies	25030024
Pyruvate MEM	Life Technologies	11360039
Non-essential amino acids	Life Technologies	11140035
β-Mercaptoethanol	Life Technologies	31350010
Penicillin/Streptomycin	Life Technologies	15140122
Trypsin	Life Technologies	15400054
N2 Supplement	Life Technologies	17502-048

Item	Supplier	Catalog no.
B27 Supplement	Life Technologies	12587-010
Trypsin Inhibitor	Life Technologies	17075029
Polyornithine	Sigma Aldrich	P3655-500 mg
Laminin	Sigma Aldrich	L2020
5 mL sterile serological pipettes	Corning	4487
10 mL sterile serological pipettes	Corning	4488
25 mL sterile serological pipettes	Corning	4489
15 mL sterile centrifuge tubes	BD Falcon	352096
50 mL sterile centrifuge tubes	BD Falcon	352070
6-well plastic dishes	PAA	PAA30006X
60 mm Petri dishes	BD Falcon	351016
Gelatin	Sigma-Aldrich	G1890
cAMP	Sigma-Aldrich	D0627
Glucose	Sigma-Aldrich	G8270
bFGF (hPSC cultures)	Life Technologies	PHG0021
bFGF (neural cultures)	R&D Systems	234-FSE/CF
EGF	R&D Systems	236-EG-200
Matrigel	BD	354230
Rho-kinase inhibitor Y-27632	Stemgent	04-0012
SB-431542	Stemgent	04-0010
LDN-193189	Stemgent	04-0074

Antibodies

Antibody	Isotype	Dilution	Source	Catalog #
Mouse Anti-Sox2	IgG2A	1:600	R&D	MAB2018
Mouse Anti-Pax6	IgG	1:500	Developmental Studies Hybridoma Bank (concentrate)	Pax 6
Mouse Anti-Beta III-tubulin	IgG2A	1:3000	Covance	MMS-435P
Mouse Anti-CD271_FITC	IgG1	1:10	Miltenyi	130-091-917
Mouse Anti-MAP2ab	IgG1	1:300	Sigma	M1406
Mouse Anti-GFAP	IgG1	1:500	ICN Biomedicals	69110
Mouse Anti-O4	IgM	1:200	R&D	MAB1326
Mouse Alexa 555	IgG	1:1000	Life Technologies	A21424
Mouse Alexa 555	IgM	1:1000	Life Technologies	A21426
Mouse Alexa 488	IgG	1:1000	Life Technologies	A11001
Mouse Alexa 488	IgM	1:1000	Life Technologies	A21042

RECIPES

KSR Medium (500 mL)

Component	Amount (mL)	Final Concentration
KnockOut-DMEM	395 mL	
KnockOut serum replacement	100 mL	20%
Non-essential amino acids	5 mL	1%
β-mercaptoethanol	182 μL	0.1 mM
L-glutamine	2.5 mL	1 mM
FGF2	400 μL	4 ng/mL

N2 Medium (500 mL)

Component	Amount (mL)	Final Concentration
DMEM/F12	485 mL	
L-glutamine	5 mL	2 mM
L-glucose	0.8 g	1.6 g/L
Penicillin/Streptomycin	5 mL	1×
N2 supplement	5 mL	1×

Neural Differentiation Medium (500 mL)

Component	Amount (mL)	Final Concentration
DMEM/F12	241 mL	
L-glutamine	5 mL	2 mM
L-glucose	0.8 g	1.6 g/L
Penicillin/Streptomycin	5 mL	1×
N2 supplement	2.5 mL	0.5×
Neurobasal medium	241 mL	
B27 supplement	5 mL	0.5×
cAMP	1.5 μg	300 ng/mL

READING LIST

Chambers, S.M., Fasano, CA, Papapetrou, EP, Tomishima, M, Sadelain, M, Studer, L., 2009. Highly efficient neural conversion of human ES and iPS cells by dual inhibition of SMAD signaling. Nat. Biotechnol. 27, 275–280.

Greber, B, Coulon, P, Zhang, M, Moritz, S, Frank, S, Muller-Molina, AJ, et al., 2011. FGF signalling inhibits neural induction in human embryonic stem cells. EMBO J 30, 4874–4884.

Koch, P, Opitz, T, Steinbeck, JA, Ladewig, J, Brustle, O., 2009. A rosette-type, self-renewing human ES cell-derived neural stem cell with potential for in vitro instruction and synaptic integration. Proc. Natl. Acad. Sci. USA 106, 3225–3230.

Watanabe, K, Ueno, M, Kamiya, D, Nishiyama, A, Matsumura, M, Wataya, T, et al., 2007. A ROCK inhibitor permits survival of dissociated human embryonic stem cells. Nat. Biotechnol. 25, 681–686.

Zhang, SC, Wernig, M, Duncan, ID, Brustle, O, Thomson, JA., 2001. In vitro differentiation of transplantable neural precursors from human embryonic stem cells. Nat. Biotechnol. 19, 1129–1133.

CHAPTER 25

Dopaminergic Neuronal Differentiation of Human Pluripotent Stem Cells

Qiuyue Liu, Oliver Z Pedersen, and Xianmin Zeng

EDITOR'S COMMENTARY

One of the greatest challenges in the development of successful stem cell-based replacement therapies is determining which cell type and transplantation location is most likely to be successful in treatment of a specific disease. One of the problems with making these decisions is that there is very little history for cell replacement therapy for most degenerative diseases. Because there have been clinical trials using transplanted fetal dopamine-producing neurons for treatment of Parkinson's disease (PD) dating back to the 1980s, there is an unusual amount of historical precedent for cell therapy for this neurodegenerative disease. The clinical trials with fetal tissue had mixed results, but a great deal of information was obtained from both the successes and failures. The selective degeneration of dopaminergic neurons in the substantia nigra is associated with PD; these midbrain dopaminergic neurons play a fundamental role in the control of voluntary movement. Recently, efficient methods have been developed to generate dopaminergic neurons from human pluripotent stem cells (hPSCs), including human embryonic stem cells (hESCs) and induced pluripotent stem cells (iPSCs). These cells are useful in the short term as experimental disease models for studying the initiation and progression of PD. In the longer term, there is hope that these cells can serve as an unlimited source for cell replacement therapy.

OVERVIEW

The methods for differentiation of midbrain dopaminergic neurons are based on knowledge about how these cells develop *in vivo*. Development of mesencephalic dopaminergic neurons from neural stem/progenitor cells (NSCs) follows a number of stages that are regulated by signaling molecules and specific transcription factors. These stages include dopaminergic neuron induction, differentiation, specification, and maturation. Early proliferating dopaminergic progenitors are specified by two secreted signaling proteins, SHH (sonic

385

J.F. Loring & S.E. Peterson (eds): Human Stem Cell Manual, Second edition.
DOI: http://dx.doi.org/10.1016/B978-0-12-385473-5.00025-4

hedgehog) and FGF8 (fibroblast growth factor 8), which can induce the formation of dopaminergic neurons in the ventral midbrain. The transcription factor OTX2 (orthodenticle homeobox 2) is required for positioning control and regulation of SHH and FGF8 expression. Early midbrain markers such as PAX2 and PAX5 (paired box 2 and 5) and EN1 and EN2 (engrailed homeobox 1 and 2) are expressed at this stage and maintained by expression of the signaling molecule WNT1 (wingless-type MMTV integration site family, member 1). Dopaminergic progenitor cells stop proliferating and enter the differentiation process, expressing the neuron-specific class III beta-tubulin (TUBB3), which is recognized by the antibody TuJ1. The transcription factor genes LMXB1 (LIM homeobox transcription factor 1 beta) and NURR1 (NR4A2; nuclear receptor subfamily 4, group A, member 2) are involved in specification of mesencephalic dopaminergic neurons. NURR1 expression in the ventral midbrain is required for dopaminergic neuron development, and precedes transcriptional activation of the tyrosine hydroxylase gene (TH), which is the rate-limiting enzyme for dopamine synthesis. LMXB1 contributes to the specification of mesencephalic dopaminergic neuronal progenitors and regulates the expression of another transcription factor, PTX3 (pentraxin 3), which is crucial for survival and development of mesencephalic dopaminergic neurons.

The final steps involved in maturation of the dopaminergic phenotype include expression of genes directly involved in synthesis of dopamine and in development of the complete machinery for process extension and synaptic and neuronal function. Markers associated with the mature dopaminergic neural phenotype include TH, the enzyme aromatic-L-amino-acid decarboxylase (DDC; dopa decarboxylase), the vesicular monoamine transporter type 2 (SLC18A1; solute carrier family 18, vesicular monoamine, member 1), and the dopamine transporter (SLC6A3; solute carrier family 6, neurotransmitter transporter, dopamine, member 3). Morphologically, TH-expressing neurons exhibit multipolar morphology with fine axons and dendrites, and additional markers of synaptic function such as synaptic vesicle proteins and synapsins can be detected.

Mimicking the *in vivo* process of dopaminergic neuron development described above, we have developed a stage-specific process of generating midbrain dopaminergic neurons from hPSCs. Our protocols provide methods for generating homogenous NSCs from hPSCs and subsequent differentiation into midbrain dopaminergic neurons with high efficiency.

This chapter describes a process of generating functional midbrain dopaminergic neurons from hPSCs. This process includes four steps: 1) culture of hPSCs, 2) derivation of NSCs from hPSCs, 3) induction of dopaminergic precursors, and 4) differentiation and maturation of dopaminergic neurons. Each step uses defined substrates and media, and cells at intermediate stages can be stored, making the process scalable and amenable to good manufacturing practice (GMP).

The initial neural induction step via embryoid body formation of hPSCs involves the isolation of neural tube-like rosette structures that give rise to homogeneous NSCs that can differentiate into neurons and glia. Treatment of NSCs with SHH and FGF8 for 10 days followed by GDNF (glial cell derived neurotrophic factor) and BDNF (brain-derived neurotrophic factor) for 3 weeks results in an efficient differentiation to dopaminergic neurons with more than 30% of total cells expressing TH.

PROCEDURES

hESC Cultures under Defined Conditions

1. At least 1–2 hours prior to cell culture, coat culture dishes with a solution of Geltrex® substrate.
2. If differentiated colonies are present on a confluent dish of hPSCs, remove them by scraping off differentiated areas with a pipette tip, syringe needle, or a tool made of a sterile heat-stretched glass Pasteur pipette.
3. Aspirate hESC complete medium from the dish together with the differentiated cells scraped previously and wash once with pre-warmed DMEM/F12.
4. Add 3 mL of 1 mg/mL Collagenase buffer. Incubate for 50–55 minutes at 37°C. Wash with DMEM/F12 until colonies detach. Collect colonies in 15 mL tube. Centrifuge colonies for 2 minutes at 1000 rpm. Make sure the colonies are in a pellet, not floating.
5. Add 4–5 mL hESC complete medium and pipette all of the hESC colonies gently.
6. Transfer equal aliquots from the dish to 4–5 new dishes coated with Geltrex containing fresh hESC complete medium. Distribute hESC colonies evenly in incubator.
7. Change medium daily.

■ Note

Cells cultured under these conditions for >20 passages usually maintain a normal karyotype and remain pluripotent (Figure 25.1).   ■

Generation of NSCs from hPSCs
Making "embryoid bodies"

1. Aspirate hESC medium off hESCs and rinse cells with pre-warmed DMEM/F12. Wash 2–3 times with DMEM/F12.
2. Add 3 mL of 1 mg/ mL Collagenase buffer. Incubate for 50–55 minutes at 37°C. Wash with DMEM/F12 until colonies detach. Collect colonies

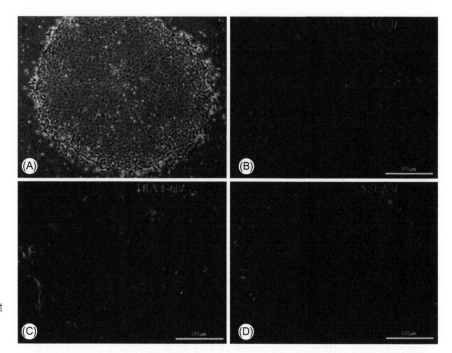

FIGURE 25.1
Generation of NSCs from hESCs adapted to defined medium. hESCs were adapted to a chemically defined medium, StemPro. (A–D) Morphology (A) and expression of the pluripotent markers Oct4 (B), Tra 1-60 (C) and SSEA4 (D).

in 15 mL tube. Centrifuge colonies for 2 minutes at 1000 rpm. Make sure colonies are in a pellet and not floating.
3. Add 3 mL embryoid body (EB) medium and pipette all of the hPSC colonies gently. Transfer the entire contents into a Petri dish (non-tissue culture treated or low-attachment dish) for embryoid body formation.
4. Culture EBs in suspension for eight days in EB medium with agitation. Change medium every other day (5 mL/60 mm dish) (Figure 25.2B).

■ Note
To form EBs, pipette the ESC colonies 3–5 times to break them into smaller clusters but do not overly triturate colonies. To change medium, transfer EBs to a 15 mL conical tube for sedimentation (do not centrifuge). Aspirate or pipette off medium and carefully resuspend EBs in fresh medium. Transfer EBs into the original Petri dish.   ■

Neural Induction
1. Change medium to neural induction medium (NIM) containing 20 ng/mL bFGF at day 8 and culture EBs in suspension for an additional 2 days.
2. Induce attachment by plating EBs on a Geltrex-coated 60 mm cell culture dish. When plating EBs, make sure there is enough space for colonies

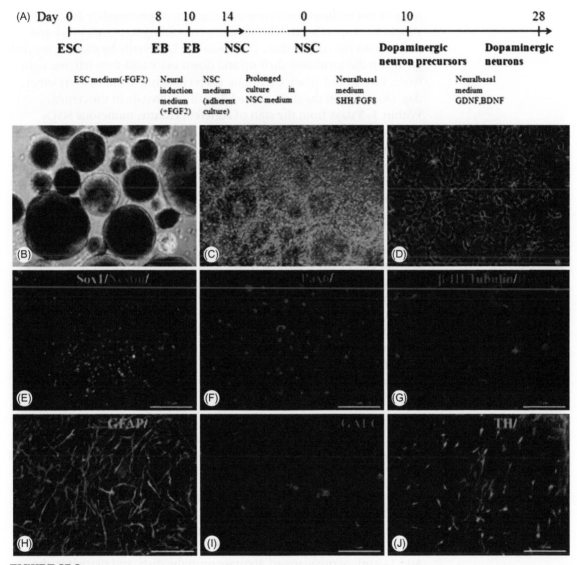

FIGURE 25.2

Generation of midbrain dopaminergic neurons from hESCs. (A) A schematic diagram of the differentiation process for generation of dopaminergic neurons in completely defined media. (B) EBs were formed in suspension culture from hESCs. Neural tube-like rosette structures (C) appeared in the center of the colonies after 12 days of differentiation. The rosettes were manually dissected and expanded. A monolayer of homogeneous NSCs (D) co-expressed SOX1, NESTIN (E), and PAX6 (F). NSCs can be expanded in NSC medium for prolonged periods while retaining the capacity to differentiate into neurons (G), astrocytes (H), and oligodendrocytes (I). For dopaminergic neural differentiation, hESC-derived NSCs were differentiated on poly-ornithine-laminin coated dishes in the presence of FGF8 and SHH for 10 days followed by 20–30 days in the presence of BDNF and GDNF. Approximately 30% of total cells expressed TH after 5 weeks of differentiation (J).

to grow out without contacting one another. Approximately 20–30 EBs should be suspended in NIM medium containing 20 ng/mL bFGF and plated onto the culture dish. Distribute the EBs evenly by shaking the dish gently on the incubator shelf up and down twice and then left and right twice. EBs should attach overnight. Keep changing medium every other day. Do not swirl the dish, as the EBs will congregate in the center.

3. Within 3–5 days from the start of adherent culture, numerous NSCs containing neural rosettes (appearing in patches with columnar morphology) will be formed (Figure 25.2C).

4. Isolate all patches of neural rosettes manually using a heat stretched glass Pasteur pipette tool and transfer them into a new culture dish coated with Geltrex (60 mm or 35 mm) in NIM medium. Avoid co-isolation of other cell types surrounding rosette patches.

5. Isolate 10–15 homogenous rosette patches, transfer them into a 1.5 mL Eppendorf tube and treat with Accutase (300–500 µL, 5 minutes at 37°C). Disrupt patches by pipetting with P-200 pipette. Centrifuge at 150 g for 5 minutes and plate cells onto a new Geltrex-coated dish containing NSC medium.

6. Expand NSCs in NSC medium as described below (Figure 25.2D).

7. Confirm NSC identity by immunocytochemistry using antibodies against markers characteristic of neural stem cells. They will stain positive for the definitive neural tube stage markers SOX1, NESTIN, and PAX6 (Figure 25.2E–F).

■ Note

It is recommended that NSCs be seeded at high density (approximately 50% confluence) in order to proliferate properly.   ■

In vitro Expansion and Cryopreservation of NSCs
Enzymatic Passage of NSCs

1. Carefully aspirate medium from a 60 mm dish containing confluent NSCs. Aspirate from the side of the dish and keep tilting it.

2. Add 1–2 mL of pre-warmed Accutase onto the dish, and place it in the incubator for 3–5 minutes until cells detach.

3. Wash off NSCs from the dish using a p1000 pipette and place cells in a 15 mL conical tube. Rinse the dish once again with 3 mL of pre-warmed Neurobasal medium and collect in the same tube.

4. Centrifuge cells at 180 g for 3 minutes.

5. Aspirate supernatant carefully and resuspend cells in 3 mL of NSC medium.

6. Aspirate Geltrex substrate solution from freshly coated dishes, add 4 mL of NSC medium into each plate and transfer 1 mL of NSCs (from step 5)

into each plate. Distribute the NSCs evenly by shaking the dish gently on the incubator shelf up and down twice and then left and right twice.

7. Incubate at 37°C, 5% CO_2 and change medium every other day.

■ Note

Cells from one confluent 60 mm dish can be split onto three new 60 mm dishes (split 1:3). Approximately 2–2.5 million cells per 60 mm dish is a desirable density.   ■

Cryopreservation of NSCs

1. Perform Steps 1–4 as described in Enzymatic passage of NSCs.
2. Resuspend cells in a small volume of NSC-FREEZE-A medium (1 mL or more if multiple dishes are combined) and count them.
3. Dilute cells using the same medium to obtain desired cell density (4–5 million cells/mL).
4. Carefully add an equal volume of NSC-FREEZE-B medium containing 20% DMSO under constant swirling (final cell density is about 2 million cells/mL). Mix gently 2–3 times by pipetting.
5. Aliquot into cryogenic vials (1 mL/vial).
6. Immediately freeze at −80°C using a "Mr. Frosty" isopropanol freezing device, and transfer vials into the liquid nitrogen tank the following day.

Dopaminergic Differentiation of NSCs in Defined Conditions

1. To start dopaminergic neural differentiation, evenly plate NSCs onto a 35 mm poly-ornithine/laminin coated dish at approximately 50% density in 2 mL of NSC medium and incubate overnight at 37°C, 5% CO_2.
2. The second day, aspirate off the NSC medium and replace it with 2 mL of DA1 medium containing FGF8 (100 ng/mL) and SHH (200 ng/mL). Differentiate NSCs in DA1 medium for 10 days with medium changes every other day.
3. Passage the cells when confluent.
4. 10 days later, change DA1 medium to DA2 (2 mL) medium containing BDNF (20 ng/mL) and GDNF (20 ng/mL). Continue dopaminergic neuronal differentiation for another 20 days. Change medium every other day. Supplement medium with ascorbic acid (200 μM) daily.

■ Note

Cells will proliferate in DA1 medium. Passage cells when necessary as described. Cells in DA2 medium may need further splitting. Medium should be changed slowly to avoid detachment of neurons. After 30 days of differentiation (10 days in DA1 medium and 20 days in DA2 medium), approximately 30% of total cells express TH (Figure 25.2J) by immunocytochemistry and the majority of TH$^+$ neurons also express GIRK2 (KCNJ6; potassium inwardly-rectifying channel, subfamily J, member 6) and VMAT (SLC18A1; solute carrier family 18, vesicular monoamine). GIRK2 and VMAT are markers for the subtype A9 dopaminergic neurons in the substantia nigra. A9 neurons have a distinctive morphology and degenerate preferentially in Parkinson's disease.   ■

Immunocytochemistry

We routinely use single, double, or triple labeling to assess various marker expression at different stages of dopaminergic differentiation. Below is an example protocol for double labeling with TuJ1 and TH.

1. Aspirate medium from dish and wash three times with PBS.
2. Add 4% paraformaldehyde to cover neurons for 10 minutes.
3. Wash the cells three times with PBS.
4. Add 10% goat serum in 0.1% Triton-X in PBS for 20 minutes for blocking.
5. Prepare primary antibodies in 8% goat serum in 0.1% Triton-X in PBS (Mouse anti-β III-Tubulin 1:1000, Rabbit anti-TH 1:500).
6. Aspirate blocking reagent, then add primary antibody solution overnight at 4°C.
7. Aspirate the primary antibody solution and wash three times with PBS.
8. Prepare secondary antibodies in 8% goat serum in 0.1% Triton-X in PBS (Alexa Fluor 488 goat anti-rabbit and Alexa Fluor 594 goat anti-mouse, both at dilution 1:1000).
9. Add secondary antibody solution for 1 hour in the dark at room temperature.
10. Wash three times with PBS.
11. Add 15 μL of ProLong® antifade reagent with Hoechst on the cell layer and carefully place a cover slip on the top avoiding the generation of bubbles. Keep in dark at 4°C until you need to view.
12. Observe the staining under a fluorescence microscope with appropriate filter combinations.

ALTERNATIVE PROCEDURES

■ EBs can be generated using special tools, e.g. AggreWell™ from STEMCELL Technologies. This product enables users to control the size and shape of the EBs allowing for optimization of differentiation protocols to specific lineages.
■ Dopaminergic neurons can be differentiated from NSCs in medium conditioned on the stromal cell line PA6 as efficiently as in the defined media described here.

- ESCs and NSCs can be cultured on a defined substrate, CELLStart™ (Life Technologies).
- We prefer the method described because of its reproducibility in our laboratory and others. There are alternative methods that do not use the EB/neural rosette steps, but we do not have sufficient experience with them to recommend them.

PITFALLS AND ADVICE

Manual Dissection of Neural Rosettes

The best way to isolate neural rosettes is to cut the cell layer along the edge line, detach it from the dish by scooping with the tool's edge and transfer into a new dish. Since co-isolation of other cell types is possible, this process may need to be repeated several times until homogeneous rosette patches are obtained.

Splitting Cells During Dopaminergic Differentiation

Examine the differentiating cells carefully every day and split the cells when necessary. We do not recommend splitting the cells after extensive processes are produced (14 days).

EQUIPMENT

- Biological safety cabinet
- Incubator
- Inverted phase contrast microscope
- Water-jacketed incubator
- Fluorescence microscope
- Centrifuge

REAGENTS AND SUPPLIES

Item	Supplier	Catalog No.
L-Glutamine Solution	Life Technologies	25030-081
60 mm polystyrene dish	Corning Inc.	430166
35 mm polystyrene dish	Corning Inc.	430165
Serological pipettes 5, 10, 25 mL	Costar	4487, 4488, 4489
60 × 15 mm Petri dish	VWR	25384-092
500 mL filter unit	Thermo	566-0010
100 mL filter unit	Thermo	565-0010

Item	Supplier	Catalog No.
Cryogenic vial	Corning Inc.	430488
GlutaMAX-I CTS	Life Technologies	A1286001
MEM-Non-Essential amino acids solution	Life Technologies	11140-050
KnockOut™ serum replacement	Life Technologies	10828-028
Dulbecco's modified Eagle's medium/ Ham's (DMEM/F12)	Life Technologies	11330-032
Dulbecco's modified Eagle's medium/ Ham's (DMEM/F12 with Glutamax)	Life Technologies	A10565
Geltrex®	Life Technologies	12760013
Neurobasal medium	Life Technologies	21103
2-Mercaptoethanol	Life Technologies	21985
B27 supplement	Life Technologies	17504
N2 supplement	Life Technologies	17502
StemPro hESC supplement	Life Technologies	A10006-01
Bovine serum albumin (BSA)	Life Technologies	A10008-01
Ascorbic acid	Sigma	A4544
Sonic hedgehog (SHH)	R&D	13-14-SH
Recombinant human FGF8	R&D	423-F8
StemPro Accutase	Life Technologies	A1110501
Basic fibroblast growth factor, human Rec (Rec Hu bFGF)	Life Technologies	PHG0314
Recombinant Human BDNF	R&D	248-BD
Recombinant Human GDNF	R&D	212-GD
Poly-ornithine	Sigma	P3655-50MG
Laminin from human	Life Technologies	23017-015
Dimethylsulfoxide	Sigma	D8418

Antibodies

Item	Supplier	Catalog No.
Rabbit anti-SOX1	Chemicon	AB5768
Mouse anti-Nestin	BD Transduction laboratories	611658
Mouse anti-Tra 1-60	eBioscience	14-8863
Mouse anti-SSEA4	eBioscience	14-8843
Rabbit anti-Oct4	Abcam	Ab19857
Rabbit anti-tyrosine hydroxylase	Pel-Freeze	P40101
Rabbit anti-β III-tubulin isotype III clone SDL.3D10	Sigma	T8660
Mouse anti-GFAP	DAKO	M0761
Alexa Fluor 488 goat anti-mouse	Molecular Probes	A-11001
Alexa Fluor 488 goat anti-rabbit	Molecular Probes	A-11008
Alexa Fluor 594 goat anti-rabbit	Molecular Probes	A-11012
Alexa Fluor 594 goat anti-mouse	Molecular Probes	A-11005

Item	Supplier	Catalog No.
Alexa Fluor 594 rabbit anti-goat	Molecular Probes	A-110080
Alexa Fluor 486 goat anti-rat	Molecular Probes	A-11006
Alexa Fluor 594 goat anti-rat	Molecular Probes	A-11007

RECIPES

Stock Solutions

Component	Amount	Stock Concentration
Recombinant human bFGF	Dissolved in sterilized PBS with 0.1% BSA	10µg/mL, aliquot and store at −20°C
SHH	Dissolved in sterilized PBS with 0.1% BSA	50µg/mL, aliquot and store at −20°C
Recombinant human BDNF, GDNF	Dissolved in sterilized PBS with 0.1% BSA	50µg/mL, aliquot and store at −20°C
Ascorbic acid	Dissolve 1 mg ascorbic acid in 1 mL PBS	200mM, aliquot and store at −20°C
Recombinant FGF8	Dissolved in sterilized PBS with 0.1% BSA	50µg/mL, aliquot and store at −20°C

Media for Maintenance of ESCs/iPSCs

Component	Amount	Final Concentration
DMEM/F-12+GlutaMAX™ (1×)	90.8mL	1×
Stempro hESC SFM Growth Supplement (50×)	2mL	1×
BSA 25%	7.2mL	1.8%
FGF-basic (10µg/mL)	80µL	8ng/mL
2-Mercaptoethanol	182µL	0.1mM

Sterile filter with a 0.22 µm filter, add FGF-basic, STEMPRO hESC SFM Growth Supplement, and 2-Mercaptoethanol just prior to feeding cells. Medium is stored at 4°C for up to 1 week.

EB Medium

Component	Amount	Final Concentration
DMEM/F12	79mL	
KnockOut serum replacement(KSR)	20mL	20%
Glutamax	2mL	1×
MEM-Non-Essential amino acids solution	1mL	4mM
2-Mercaptoethanol	182µL	0.1mM

Sterile filter with a 0.22 µm filter, add 4ng/mL bFGF just prior to feeding cells. Medium is stored at 4°C for up to 1 week.

Neural Induction Medium (NIM)

Component	Amount	Final Concentration
DMEM/F12	97 mL	
MEM-Non-Essential amino acids	1 mL	2 mM
GlutaMAX-I CTS	1 mL	2 mM
N2	1 mL	1×

Sterile filter with a 0.22 μm filter, add 20 ng/mL bFGF just prior to feeding cells. Medium is stored at 4°C for up to 1 week.

NSC Medium

Component	Amount	Final Concentration
Neurobasal medium	96 mL	1×
MEM-Non-Essential amino acids solution	1 mL	2 mM
GlutaMAX-I CTS	1 mL	2 mM
B27	2 mL	1×

Sterile filter with a 0.22 μm filter, add 20 ng/mL bFGF just prior to feeding cells. Medium is stored at 4°C for up to 1 week.

NSC Freezing Medium A (NSC-FREEZE-A)

Component	Amount	Final Concentration
Neurobasal medium	96 mL	1×
MEM-Non-Essential amino acids solution	1 mL	2 mM
GlutaMAX-I CTS	1 mL	2 mM
B27	2 mL	1×

NSC Freezing Medium B (NSC-FREEZE-B)

Component	Amount	Final Concentration
Neurobasal medium	78 mL	1×
MEM-Non-Essential amino acids solution	1 mL	2 mM
GlutaMAX-I CTS	1 mL	2 mM
DMSO	20 mL	20%

DA1 medium

Component	Amount	Final Concentration
Neurobasal medium	96 mL	
B27	2 mL	1×
MEM-Non-Essential amino acids solution	1 mL	2 mM
GlutaMAX-I CTS	1 mL	2 mM
SHH	400 µL	200 ng/mL
FGF8	200 µL	100 ng/mL

Sterile filter with a 0.22 µm filter, add SHH and FGF8 just prior to feeding cells. Medium is stored at 4°C for up to 1 week.

DA2 Medium

Component	Amount	Final Concentration
Neurobasal medium	96 mL	
B27	2 mL	1×
MEM-Non-Essential amino acids	1 mL	2 mM
GlutaMAX-I CTS	1 mL	2 mM
BDNF	40 µL	20 ng/mL
GDNF	40 µL	20 ng/mL
Ascorbic acid	100 µL	200 µM

Sterile filter with a 0.22 µm filter, add BDNF and GDNF just prior to feeding cells. Medium is stored at 4°C for up to 1 week.

READING LIST

Perrier, A.L., Tabar, V., Barberi, T., Rubio, M.E., Bruses, J., Topf, N., et al., 2004. Derivation of midbrain dopamine neurons from human embryonic stem cells. Proc. Nat. Acad. Sci. USA 101, 12543–12548.

This is one of the first reports of the generation of dopaminergic neurons from hESCs using a co-culture (on MS5 stromal cells) method.

Pruszak, J., Isacson, O., 2009. Molecular and cellular determinants for generating ES-cell derived dopamine neurons for cell therapy. Adv. Exp. Med. Biol. 651, 112–123.

This review summarizes current protocols used for differentiation of authentic dopaminergic neurons in vitro.

Swistowski, A., Peng, J., Han, Y., Swistowska, A.M., Rao, M.S., Zeng, X., 2009. Xeno-free defined conditions for culture of human embryonic stem cells, neural stem cells and dopaminergic neurons derived from them. PLoS One 4, e6233.

This study reports the first scalable process for generating dopaminergic neurons from hESCs under completely defined and xeno-free conditions.

Swistowski, A., Peng, J., Liu, Q., Mali, P., Rao, M.S., Cheng, L., et al., 2010. Efficient generation of functional dopaminergic neurons from human induced pluripotent stem cells under defined conditions. Stem Cells 28, 1893–1904.

This manuscript describes a scalable protocol for efficient generation of midbrain dopaminergic neurons from iPSC lines using a completely defined xeno-free system the authors have previously developed for hESC differentiation.

Yan, Y., Yang, D., Zarnowski, E.D., Du, Z., Werbel, B., Valliere, C., et al., 2005. Directed differentiation of dopaminergic neuronal subtypes from human embryonic stem cells. Stem Cells 23, 781–790.

This study reports a method of generating midbrain and non-midbrain dopaminergic neurons from hESCs using defined media.

Zeng, X., Cai, J., Chen, J., Luo, Y., You, Z.B., Fotter, E., et al., 2004. Dopaminergic differentiation of human embryonic stem cells. Stem Cells 22, 925–940.

This is one of the first reports of dopaminergic differentiation of hESCs.

Directed Differentiation of Human Pluripotent Stem Cells to Oligodendrocyte Progenitor Cells

Jason Sharp, Gabriel I. Nistor, and Hans S. Keirstead

EDITOR'S COMMENTARY

Oligodendrocytes form multilayered, concentric wraps of modified plasma membrane, called myelin, that insulate axons of the vertebrate central nervous system (CNS) and allow rapid and extended conduction of neuronal signals. Death of oligodendrocytes or loss of their function results in a loss of myelin, which leads to nerve cell dysfunction. Myelin loss is associated with a number of central nervous system disorders and injuries, including multiple sclerosis, Alzheimer disease, stroke, and brain and spinal cord injuries.

Oligodendrocyte development has been studied in mice and rats for over 20 years, and has led to the identification of specific markers that define phenotypic stages of the lineage from embryonic stem cells to oligodendrocyte progenitor cells (OPCs) to myelinating oligodendrocytes. Much of this research has been applicable to the development of methods to generate oligodendrocytes from human pluripotent stem cells (hPSCs). There is still a great deal to learn about the biology of oligodendroglial development and function, but hPSC research has made remarkable progress in understanding human oligodendrocytes and how their dysfunction underlies disorders of the human central nervous system.

OVERVIEW

In this chapter, we describe an efficient way to produce OPCs from hPSCs. The protocol was initially established using the WA07 hESC line (WiCell) but has since been used with iPSCs (Pouya, 2011). The derivation method is designed to recapitulate the spatiotemporal development of OPCs within the motor neuron progenitor domain in the developing CNS. Induction of hPSCs toward OPCs occurs through differentiation and restriction to multipotent, neural-lineage cells. Cultures are maintained in a novel glial restriction medium that includes epidermal growth factor (EGF). The presence of EGF is

J.F. Loring & S.E. Peterson (eds): Human Stem Cell Manual, Second edition.
DOI: http://dx.doi.org/10.1016/B978-0-12-385473-5.00026-6

important for amplification of the derivative population to counter the slow growth rate of human cells and permit a more cost-effective protocol. In addition, it appears that reduced concentrations of certain morphogens and spontaneous differentiation contribute to the purity and yield of OPCs.

Oligodendroglial lineage commitment is confirmed using a panel of biomarkers that include OLIG1 (oligodendrocyte transcription factor 1), SOX10 (SRY sex determining region Y-box 10), PDGFRα (platelet-derived growth factor receptor, alpha polypeptide), NG2/CSPG4 (chondroitin sulfate proteoglycan 4), and A2B5 antigen. It is important to note that a panel of biomarkers is required to definitively identify oligodendroglial lineage cells; individual markers such as A2B5 or NG2 are not unique to oligodendroglia and can label other cell populations in human tissue. We find that oligodendroglial lineage cells are prominent in the final population with biomarker-positive cells reproducibly representing more than 70% of the cells. The contaminating cells are about half GFAP (glial fibrillary acidic protein)-positive astrocytes and half β-III tubulin (TUBB3; tubulin, beta 3 class III)-positive neurons. In addition to these markers, the morphology of the oligodendroglial lineage-labeled cells is bipolar, reminiscent of OPCs.

PROCEDURES

The OPC derivation protocol consists of four stages outlined below.

hPSC Cultures for OPC Production

In the first stage, undifferentiated hPSCs are maintained in colonies on Matrigel™-coated flasks in the absence of an exogenous feeder layer (see Chapter 3). Ideal hPSC colonies have smooth, rounded edges and are surrounded by migrating fibroblastic "stromal" cells that emerge from the colonies (Figure 26.1A). Colonies that show severe multilayered areas or areas of spontaneous differentiation should be excised from the pool. hPSC colonies are expanded until the appropriate confluence for the differentiation protocol is obtained (see Chapters 1 and 3).

One Day Prior to hPSC Culture: Preparation of Matrigel-Coated Flasks

1. For each flask, add 28 mL KnockOut DMEM to each 2 mL 1:1 aliquot of diluted Matrigel (see Recipes) for a total volume of 30 mL of 1:30 diluted Matrigel. Coat T75 flasks by adding 1:30 Matrigel to flask. Swirl and/ or tap flask to ensure even spreading of solution on flask bottom and incubate for 3 hours to overnight at room temperature (RT) and then store at 4°C until use.

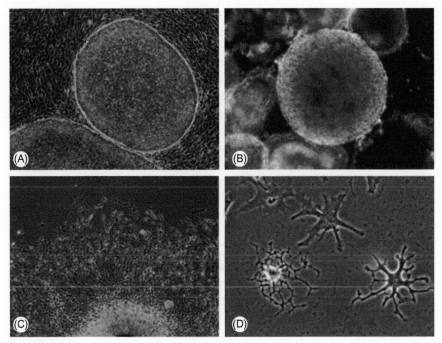

FIGURE 26.1

Derivation of OPCs from hPSCs. (A) An ideal culture for the start of the OPC differentiation protocol will show well-defined, rounded hPSC colonies without internal spontaneous differentiation, protrusions, or multiple layers (notable yellow coloration), with colonies separated by stromal cells. (B) When transferred to a non-adherent substrate and exposed to retinoic acid (RA), colonies take on a yellow, spherical morphology and become neuralized. Cell debris, individual cells, and small cell aggregates will be removed during media changes. (C) Plated spheres extend processes along which OPCs migrate from the spheres. OPCs continue to migrate from the spheres over time, so removal of the spheres is recommended to maintain desired cell concentration and maturation. Along these lines, inclusion of the spheres in the final product will lead to an increase in nestin-positive cells, early OPCs, and contaminants in the population. (D) When maintained at low density, without growth factors, and on an appropriate substrate, a subpopulation of OPCs will mature into oligodendrocytes with the characteristic morphology and ICC profile of cultured primary OPCs.

■ Note

Matrigel is stored at $-20°C$ and must be kept cold until ready to use. Undiluted Matrigel forms a gel at RT. Diluted aliquots (as used here) can be prepared as long as the solution remains cold/on ice. This coating procedure can be applied to other culture flasks and plates.   ■

2. Before use, incubate the flask for at least 1 hour in a tissue culture incubator to equilibrate the temperature and balance the pH.

Day 1: Thawing hPSCs and Initial Growth

1. Warm 40 mL of MEF-conditioned medium (MEF-CM) in a 37°C water bath (MEF-CM can be prepared in advance and frozen at −80°C for at least 1 month).
2. Thaw a 1 mL vial of hPSCs (approximately 1.5×10^6 cells) at 37°C for 90–120 seconds and add to 9 mL of warmed MEF-CM in a 15 mL conical. Spin cells down at 200 g for 3 minutes, aspirate supernatant, and resuspend in 5 mL of MEF-CM medium. If necessary, break up the pellet by gentle trituration through a wide bore pipette. Do not break up the cell clusters.
3. Aspirate the Matrigel coating solution, add the resuspended cells to the Matrigel-coated flask, and add 20 mL MEF-CM supplemented with 8 ng/mL of bFGF (add 0.8 μL of 10 μg/mL stock solution for each milliliter of medium). Place flask in incubator at 37°C with 5% CO_2.

Stage 1: Preparation of hPSCs
Days 2–14: Expansion and Passage of hPSC Cultures

1. Feed cells with 20–50 mL of new MEF-CM containing 8 ng/mL bFGF every day. Add 20 mL if the pH of the medium is neutral, more if the pH is acidic (judged by color or the phenol red in the medium).
2. Passage cells once per week at 1:4 to 1:6.
3. Before passaging, coat plates with Matrigel as described above, one day in advance.
4. On the day of passage, prepare 100–200 mL of warm MEF-CM and 20 mL of warm DPBS.
5. Lift hPSC colonies by adding 10 mL of 2 mg/mL collagenase IV in DPBS to the flask and place in the incubator for 2–10 minutes. Monitor dissociation every 2 minutes and as soon as colonies start to lift at edges, aspirate collagenase solution, wash with DPBS, add MEF-CM, and then scrape the flask bottom and colonies with a sterile rubber cell scraper.
6. Transfer cells to a sterile 15 mL conical tube, add 5 mL MEF-CM, and spin cells at 200 g for 3 minutes. Aspirate supernatant and resuspend in 5 mL of MEF-CM medium.
7. Collect the cells and distribute according to desired splitting ratio into Matrigel-coated flasks. Add 20–30 mL medium plus 8 ng/mL of bFGF to each flask.
8. Mark the passage number and return flasks to incubator.
9. Continue to feed the cells daily with fresh media plus 8 ng/mL of bFGF.
10. To start differentiation, grow culture(s) to 70–90% confluence.

Stage 2: Cellular Aggregates

In the second stage, neural lineage differentiation is initiated by removing the colonies from the adherent substrate and treating with a new medium ("transition medium"). Dissociated hPSC colonies will aggregate into clusters when transferred into ultra-low adherent culture flasks. The aggregates will initially be inconsistent in size and not perfectly spherical. Isolated floating cells are usually not viable and are discarded at feeding with gravity or low force centrifugation. These aggregates are not "embryoid bodies," as defined for mouse ESCs, although some published reports use this term for hPSC aggregates. Embryoid bodies are highly structured and have discernable layers such as Reichert's membrane. Consequently, we refer to the hPSC aggregates as "cellular aggregates."

Day 1: Transfer of hPSCs to Low-Attachment Flask in Transition Medium with bFGF

1. For each flask, prepare 30 mL of transition medium +4 ng/mL bFGF (use 0.4 μL of 10 μg/mL bFGF stock for each milliliter of medium). Pre-warm and pH balance the medium in the CO_2 incubator.
2. Lift hPSC colonies by adding 10 mL of 2 mg/mL collagenase IV in DPBS to the flask and place in the incubator for 2–10 minutes. Monitor dissociation every 2 minutes and as soon as colonies start to lift at the edges, aspirate collagenase media, wash with DPBS, add MEF-CM, and then scrape the flask bottom and colonies with a sterile rubber cell scraper.
3. Transfer cells to a 15 mL sterile conical tube, add 5 mL MEF-CM, and spin cells down at 200 g for 3 minutes. Aspirate supernatant and resuspend in 5 mL of MEF-CM medium.
4. Distribute cells evenly to low-attachment T75 flasks. Add 20–30 mL of transition medium + 4 ng/mL bFGF to each flask. Incubate at 37°C, 5% CO_2.

Day 2: Feed Cells with Transition Medium with bFGF, EGF, and RA

1. Prepare 30 mL of transition medium+4 ng/mL bFGF+20 ng/mL EGF and 10 μM RA for each flask (use EGF and RA stocks).

■ Note

Use minimal light during feeding since RA is light sensitive. For RA, use 0.5 μL of 20 mM stock solution per mL of medium. Discard the vial after use. ■

2. Collect aggregates from each flask in a 50 mL conical tube. Spin cells at 200 g for 3 minutes.
3. Aspirate the old medium and add 30 mL of new transition medium with supplements. Resuspend gently, without breaking up the cell clusters.
4. Add cells and fresh medium back to the flask. Return cells to incubator.

Stage 3: Yellow Sphere Formation: Neural Progenitors

In the third stage, the aggregates are maintained in glial restrictive medium (GRM) to induce OPC differentiation. Yellow spheres/neural progenitors form from the aggregate clusters, acquire a near-perfect spherical morphology, and are bright yellow (Figure 26.1B). There will be small and large yellow spheres surrounded by cellular debris that includes dying cells. The spheres should be homogenous without visible dark, necrotic centers. During this stage, more and more yellow spheres are produced and fewer contaminants seen. Establishment of healthy yellow spheres is critical for the next steps.

Days 3–10: Formation of Aggregates: Growth with RA

At the beginning of this period, small clusters of 20–50 cells can be seen floating in the medium. At the end of this period, yellow spheres containing neuralized cells will be clearly observed in the culture.

■ Note

The following procedure must be done every day for days 3–10. ■

1. Prepare 30–35 mL of GRM+20 ng/mL EGF+10 µM RA for each flask.
2. Collect suspended aggregates in a 50 mL conical tube and centrifuge at 200 g for 3 minutes.
3. Aspirate supernatant and add 30 mL of GRM+EGF+RA medium. Do not dissociate clumps.
4. Add cells and fresh media back to flask. Return cells to incubator.

Days 11–15: Medium Aggregates: Growth without RA

Yellow spheres are accumulating in the cultures and are apparent macroscopically. Darker clusters in the culture have an irregular shape and a loose composition. Individual floating cells are discarded at every feeding.

■ Note

Feed the cultures every other day (e.g. Monday, Wednesday, Friday). ■

1. Prepare 30 mL of GRM+20 ng/mL EGF for each flask.
2. Collect suspended aggregates in a 50 mL conical tube and let settle (do not centrifuge) for 5–10 minutes.
3. Aspirate supernatant and add 30 mL of GRM+EGF medium. Use caution as clumps are not as adherent to vial bottom as when centrifuged. Do not dissociate clumps.
4. Add cells and fresh medium back to flask. Return cells to incubator.

Days 16–27: Large Clusters

Yellow sphere growth continues as non-selected cells are discarded.

■ Note
Feed the cultures every other day (e.g. Monday, Wednesday, Friday).   ■

1. Prepare 30 mL of GRM + 20 ng/mL EGF for each flask.
2. Collect the cells in 50 mL conical tubes and let them settle by gravity (without centrifugation) for 2 minutes.
3. Aspirate the supernatant and add 30 mL of GRM + EGF medium. Gentle dissociation of large aggregates during pipetting is acceptable.
4. Add cells and fresh medium back to flask. Return cells to incubator.

Stage 4: Oligodendrocyte Progenitors
In the fourth stage, plating of yellow spheres/neural progenitors allows selection of viable cells, dissociation of the spheres, and further differentiation of neural progenitors into OPCs. During this stage, migrating cells can exhibit either an epithelial or a bipolar morphology with short thick branches (Figure 26.1C). Most importantly, they are positive for oligodendrocyte markers such as OLIG1 and NG2. Some plated yellow spheres will extend long processes first, followed by migration of the OPCs along the radial branches over the next few days.

Day 28: Plating Spheres on Matrigel
1. Prepare a Matrigel-coated T75 flask with 1:30 Matrigel in KnockOut DMEM 1 day before plating.
2. Prepare 30 mL of GRM + 20 ng/mL EGF, aspirate medium from flask, place the fresh medium in the coated flask, and equilibrate for at least 1 hour in the incubator.
3. Collect cells from each low-attachment flask in a 50 mL centrifuge tube. Let the spheres settle to the bottom for 3 minutes.
4. Aspirate supernatant and add a small amount (~5 mL) of pre-warmed GRM + EGF.
5. Place cells in the coated T75 flask with the equilibrated medium. Moving the suspended cells in a forward and reverse figure 8 motion can help distribute clumps as they settle.
6. Return to the incubator.

Days 29–34: Oligodendrocyte Progenitors Migrate out of Yellow Spheres
1. Change medium every Monday, Wednesday, Friday.
2. Prepare 30 mL of GRM + 20 ng/mL EGF for each Matrigel-coated flask.
3. Tilt flask and aspirate used medium.
4. Add 30 mL of new GRM + EGF medium to flask. Return to incubator.

Days 35–42: Purification, Replating, and Imaging

On day 35, the dissociated cells will go through a panning process in which adherent cells (astrocytes, fibroblasts) attach to tissue culture plastic and the less adherent OPCs are collected, replated, and maintained. At the same time, the cell population is sampled for ICC by plating them on laminin or Matrigel-coated imaging slides or glass coverslips.

1. Prepare Matrigel-(1:30) or laminin-($10\,\mu g/mL/cm^2$) coated imaging slides or coverslips in wells 1 day in advance.

2. Prepare the required Matrigel-coated T75 flasks with 1:30 Matrigel in KnockOut DMEM one day in advance. Prepare enough flasks for splitting the cells at a 1:2 ratio.

3. About 1 hour before use, replace the coating solution with GRM (without EGF) and place the slides in the CO_2 incubator for temperature and pH equilibration. Add 25 mL of GRM + 20 ng/mL EGF to each flask; pre-warm and equilibrate the pH of the medium in the incubator.

4. Aspirate medium from each cell-containing flask. Wash with 10 mL DPBS. Aspirate again.

5. Add 7 mL of warm TrypLE to each flask. Incubate at 37°C until dissociated, about 5–15 minutes.

6. Add 7 mL of medium to the flask. Collect the dispersed cells from each flask into a 15 mL conical tube.

7. Spin the cells at $250\,g$ for 5 minutes. For each conical tube, resuspend the cells in 5 mL GRM (without EGF).

8. Take a 50 μL sample of the cell suspension and mix with 50 μL of trypan blue. Count live (unstained) and dead (blue-stained) cells using a hemocytometer.

9. To pan for adherent cells, transfer the resuspended cells from all conical tubes to two T75 or one T150 uncoated tissue culture plastic flask(s) and incubate for 1 hour at 37°C. This step allows astrocytes and other adherent cells to attach to the plastic bottom, while the less adherent OPCs will not attach.

10. Collect medium containing OPCs with a gentle shake of the flasks and transfer to 50 mL conical tubes fitted with a 100 μm cell strainer (BD Biosciences). The strainer will allow single cells to pass into the tube and remove clumps.

11. Transfer panned, strained cells to Matrigel-coated T75 flasks prepared earlier, containing GRM+20 ng/mL EGF.

12. Change GRM+20 ng/mL EGF medium every Monday, Wednesday, Friday for days 35–42.

13. On day 42, cells are ready to be used for *in vitro* and *in vivo* experiments. Take a sample from the cell population for ICC. Plate cells at 50,000 cells/cm^2 on poly-D-lysine/laminin-coated imaging slides. After 2 days, plated cells are ready to be fixed and stained.

When plated at low density in a growth factor-free medium, some OPCs will acquire the morphology of mature oligodendrocytes with branches and sheets and will label positively for oligodendroglial markers (Figure 26.1D).

ALTERNATIVE PROCEDURES

The protocol provided above for generating OPCs from hPSCs was first described by Nistor and colleagues using the WA07 hESC line and was based on knowledge of the detailed spatiotemporal patterns of morphogens and serial induction that occur during development of the CNS. Hence, differentiation of hPSCs into OPCs is attained via the use of specific substrates and specialized media supplemented with specific morphogens at specific time points.

Although some variations on the protocol are tolerated by the cells, we recommend that trained personnel adhere to the original procedures from start to finish without deviations. This process is highly sensitive to the status of hPSCs prior to initiation of the differentiation and to the course of medium changes and maintenance. Once high-purity OPCs are produced, variations from the given protocol can be done with certain limitations. The cultures are not very tolerant to changes in the concentration of supplements, and single-factor changes can affect the outcome of OPC versus contaminant cell ratio and yield.

This protocol has been used successfully for other cell lines. Pouya et al. (2011) reported use of this protocol to derive functional OPCs from human induced pluripotent stem cells (hiPSCs), specifically the hiPSC lines Royan hiPSC1 and hiPSC8. Some modifications of this protocol have been successfully used. Izrael et al. (2007) modified this protocol to include the use of NOGGIN to produce increased O4 antigen-positive oligodendrocytes in culture and increase myelin formation *in vivo*.

PITFALLS AND ADVICE

Pluripotent Stem Cell Condition

The quality of pluripotent stem cell colonies prior to directed differentiation is a major factor in the successful differentiation of hPSCs into OPCs. In general, the growth rate of the hPSC cultures should allow for a 1:4–1:6 split per week; if the growth rate is different, check for contamination and confirm normal karyotype. Prior to initiation of differentiation, confirm the pluripotency of the culture. An ideal starting population is free of intracolony spontaneous differentiation (as often evident by a change in cell morphology) and/or any bias toward endodermal or mesodermal lineage. The colonies will be clearly delineated with smooth surfaces, and are surrounded by fibroblastic "stromal" cells that migrate from the colonies. Markers for pluripotency, such as OCT4/POU5F1 or SSEA4, should be evenly distributed across each colony. Please refer to Part III, and Chapter 16 in particular, for more information.

Microbial Contamination

Due to the protracted protocol (42 days) for OPC differentiation, contamination is a serious risk. The hPSCs must be free of bacteria, mycoplasma, and yeast prior to the start of differentiation. If contamination is found, discard the cells and start new hPSC cultures. It is not recommended to use antibiotic-treated cultures to start differentiation. If contamination is caught during the differentiation process, it might be possible to salvage the cultures using repeated washes with sterile medium at feeding and adding an appropriate antibiotic for 7–10 days. However, use of antibiotics can affect OPC differentiation. If contamination persists, discard the cultures, as they will not be of the same purity and yield described in this protocol.

Extensive Cellular Death at the Stem Cell Culture Dissociation Step: "Day 1: Transfer of hPSCs to Low-Attachment Substrate in Transition Medium with bFGF"

This is often caused by excessive confluence of the hPSC culture and/or bias toward endodermal or mesodermal lineages. To attempt a recovery, consider combining remnant clusters in a 6:1–4:1 ratio. It might be necessary to use individual wells of 6-well low-attachment plates to combine clusters to this ratio.

Yellow Spheres are not Forming During "Yellow Sphere Formation: Neural Progenitors" Stage

In our experience, cultures with a low abundance or absence of "stromal" cells tend to generate low yields of OPCs. Cultures with excessive spontaneous differentiation toward endodermal or mesodermal phenotypes and/or multilayered colonies tend to generate more contaminants. In these cultures, hollow cystic cell aggregates and darker, brown spheres form. These spheres can fail to differentiate and/or produce contaminant cells. It is possible to remove these spheres via pipette to improve OPC-to-contaminant ratio in the final product.

Retinoic acid quality can also affect differentiation. We recommend replacement of the RA stock every 3–6 months. Good, active RA is bright yellow, but is degraded by light. Degraded RA has a more faded yellow appearance.

Immature or Failed Terminal Differentiation

The plating density for terminal differentiation in the absence of growth factors is critical. If the density is high, cells in culture will continue to proliferate. If the density is too low, the culture will not survive. Characterization of day-44 OPCs on imaging slides often demonstrates a mixture of young and semi-mature OPCs with some cells displaying a more mature morphology.

REAGENTS AND SUPPLIES

- 15 mL and 50 mL conical centrifuge tubes
- 1.5 mL centrifuge tubes
- 1 L vacuum filters, 0.22 μm
- T75 and T150 flasks
- Low-attachment T75 flasks and/or 6-well plates
- Chamber Slides for imaging, Nunc Lab-Tek™, Permanox
- Cell scrapers, rubber
- Cell strainers, 100 μm

Item	Supplier	Catalog No.
Trypan blue	Sigma	T8154
Water for embryo transfer (WET)	Sigma	W1503
Dulbecco's PBS (DPBS)	Life Technologies	14190-144
Dulbecco's Modified Eagle's Medium (DMEM)	Life Technologies	12430-047
KnockOut DMEM	Life Technologies	10829-018
Matrigel, growth factor reduced	BD Biosciences	356231
TrypLE Select (1×)	Life Technologies	12563
Progesterone	Sigma	P6149
Putrescine	Sigma	P5780
Sodium selenite	Sigma	S9133
Transferrin (human)	Sigma	T1408
T3 (triiodo-L-thyronine)	Sigma	T2877
Insulin (bovine)	Sigma	I1882
Retinoic acid – all *trans*	Sigma	R2625
bFGF (human)	Chemicon	GF003
EGF (human)	Sigma	E9644
B27 supplement	Life Technologies	17504-044
Collagenase IV	Life Technologies	17104-019
Mouse embryonic fibroblast feeder (MEF) cells	Chemicon	PMEF-CF

RECIPES

Component	Amount	Stock Concentration
Progesterone	Dissolve 1 mg in 1 mL ethanol, add 15 mL DMEM	63 μg/mL
Putrescine	Dissolved 100 mg in 10 mL DMEM	10 mg/mL
Sodium selenite	Dissolve 1 mg in 10 mL WET, dilute 1 mL in 20 mL WET	50 μg/mL

Component	Amount	Stock Concentration
Transferrin (human)	Dissolve 100 mg in 2 mL sterile culture medium	50 mg/mL
T3 (triiodo-L-thyronine)	Dissolve 100 mg in 1 mL 1N NaOH, dilute 40 μL in 100 mL DMEM	40 μg/mL
Insulin (bovine)	Add 100 μL glacial acetic acid to 100 mg, bring to 10 mL with WET	10 mg/mL
Retinoic acid – all *trans*	Dissolve 50 mg in 8.3 mL DMSO	20 mM in DMSO
bFGF (human)	Dissolve 50 μg in 5 mL PBS/0.5% BSA	10 μg/mL in PBS/0.5% BSA
EGF (human)	Dissolve 0.2 mg in 10 mL 10 mM acetic acid/0.1% BSA	20 μg/mL in acetic acid/0.1% BSA
B27 supplement	Use as supplied	50×

MEF-Conditioned Medium

MEF-conditioned medium (MEF-CM) is produced by maintaining a monolayer of MEF cells in embryonic stem cell medium (see Chapter 1) without bFGF for 24 hours. The medium is then filtered and can be stored at −80°C. bFGF is added as a supplement just before use to maintain hPSCs.

Glial Restrictive Medium (GRM)

Component	Amount (mL)	Final Concentration
DMEM/F12 (with glutamine or GlutaMAX)	974 (or 1 L)	
Progesterone stock	20 mL	63 ng/mL
Putrescine stock	1 mL	10 μg/mL
Sodium selenite stock	1 mL	100 ng/mL
Transferrin stock	1 mL	50 μg/mL
T3 stock	1 mL	40 ng/mL
Insulin stock	1 mL	25 μg/mL
B27 supplement	1 mL	1×

Add all components to a vacuum filter cup and apply vacuum. GRM can be stored at 4°C.

Transition Medium

Mix GRM and MEF-CM at a 1:1 ratio.

Matrigel Aliquots

Thaw Matrigel at 4°C overnight. Pipette cold (4°C) KnockOut DMEM into thawed, cold Matrigel at a 1:1 ratio. Make 2 mL aliquots from this solution and store aliquots at −20°C.

■ **Note**

Undiluted Matrigel forms a gel at RT. Diluted aliquots can be prepared as long as the solution remains cold or on ice. ■

QUALITY CONTROL METHODS

We recommend the use of morphological, immunocytochemical, and molecular characterization to control the quality of the hPSC-derived OPCs. Quality control can save valuable time by preventing the continuation of aberrant cultures or experiments using low quality cells. In particular, we recommend that users examine a number of markers over different time points and establish inclusion/exclusion criteria (Table 26.1). These markers and time points can then be used to determine the quality of culture samples during subsequent differentiations. The following outline lists markers that can be labeled at each defined stage (see Procedures) following ICC methods described in Chapter 16.

Undifferentiated hPSCs Stage

■ SSEA-4: a glycolipid epitope that is used as a marker of many human pluripotent cells.
■ POU5F1/OCT3/4: a transcription factor characteristic of pluripotent cells.

Cellular Aggregates Stage

This is a transition stage. After approximately 3 days of being plated on an adherent substrate, cells will label positive for embryonic markers (SSEA-4,

Table 26.1 Immunocytochemical Characterization of Cultured OPCs. An Example of an ICC Profile of 3 Cultures Monitored at Days 10, 42, and 56. In Addition to These Markers, We Recommend the Use of Oct4, A2B5, Sox10, and Nkx2.2

Antigen (Day of Differentiation)	Culture 1 % Total Cells	Culture 2 % Total Cells	Culture 3 % Total Cells
Pax6 (Day 10)	98% ±2%	97% ±3%	96% ±3%
Pax6 (Day 42)	>1%	>1%	>1%
Olig1 (Day 42)	83% ±7%	84% ±6%	88% ±5%
NG2 (Day 42)	98% ±2%	97% ±3%	99% ±1%
GalC (Day 56)	95% ±4%	94% ±6%	97% ±2%
RIP (Day 56)	95% ±2%	95% ±5%	90% ±6%
04 (Day 56)	85% ±5%	82% ±7%	80% ±3%
GFAP (Day 56)	4% ±4%	3% ±3%	5% ±3%
Tuj1 (Day 56)	8% ±2%	7 ±2%	6% ±2%
BMP4 (Day 56)	0	0	0
SSEA4 (Day 56)	0	0	0

OCT4) or neural markers such as nestin and Pax6. Occasionally, neurogenic cores can be observed surrounded by non-labeled cells.

Yellow Spheres/Neural Progenitors Stage

Yellow spheres and/or spheres plated on adherent substrate will stain positive for:

- Pax6: a transcription factor indicative of neural commitment
- Nestin: intermediate filament often used as a marker for neural commitment
- A2B5: marker for early neural progenitors.

Oligodendrocyte Progenitors Stage

Typical markers for OPCs are:

- Olig1/2: transcription factor expressed during oligodendroglial development (located in both the nucleus and cytoplasm)
- A2B5: marker for early neural progenitors
- NG2: a chondroitin sulfate proteoglycan
- PDGFRα: growth factor receptor.

Markers of Contaminating Cells at Oligodendrocyte Progenitors Stage

- GFAP: used as an indicator for astrocytes (typically less than 10%).
- Neurofilaments (βIII-Tubulin (also known as Tuj1), MAP2, etc.): used as indicators for neurons (typically less than 10%).
- SMA (Smooth Muscle Actin): occasionally detected in single cells (less than 0.1%).

READING LIST

Izrael, M., Zhang, P., Kaufman, R., Shinder, V., Ella, R., Amit, M., et al., 2007. Human oligodendrocytes derived from embryonic stem cells: Effect of noggin on phenotypic differentiation in vitro and on myelination in vivo. Mol. Cell Neurosci. 34, 310–323.

Nistor, G.I., Totoiu, M.O., Haque, N., Carpenter, M.K., Keirstead, H.S., 2005. Human embryonic stem cells differentiate into oligodendrocytes in high purity and remyelinate after spinal cord transplantation. Glia 49, 385–396.

Pouya, A., Satarian, L., Kiani, S., Javan, M., Baharvand, H., 2011. Human induced pluripotent stem cells differentiation into oligodendrocyte progenitors and transplantation in a rat model of optic chiasm demyelination. PLoS One 6 (Epub e27925).

Sharp, J., Keirstead, H.S., 2007. Therapeutic applications of oligodendrocyte precursors derived from human embryonic stem cells. Curr. Opin. Biotechnol. 18, 434–440.

CHAPTER 27

Cardiomyocyte Differentiation of Human Pluripotent Stem Cells

Cheryl Dambrot, Cathelijne Van Den Berg,
Dorien Ward-van Oostwaard, Richard Davis, Stefan Braam,
Elizabeth Ng, and Christine Mummery

EDITOR'S COMMENTARY

The most dramatic cellular phenotype differentiated from human pluripotent stem cells (hPSCs) is cardiac myocytes; the spontaneous rhythmic contractions of these cells when viewed through a microscope are clear proof of their remarkable developmental abilities. In 2007, when he was governor of California, Arnold Schwarzenegger visited our laboratory, and although he probably doesn't remember any of the scientists or even the research institute, he must certainly remember looking through the microscope and seeing beating heart cells differentiated from human ES cells. Reliable production of cardiomyocytes from hPSCs remains a challenge, and, as is also the case for many other derivatives, most hPSC-derived cardiac cells resemble fetal stages more than adult tissues. hPSC-derived cardiomyocytes have many uses, such as for the study of cardiac disease *in vitro*, and for screening drugs for cardiac toxicity, as well as in potential cell replacement therapies. This chapter provides the currently most reliable methods for generating cardiomyocytes from hPSCs.

OVERVIEW

Human cardiomyocytes differ significantly from those of rodents, most obviously in their electrophysiological properties. For this reason, many potential biomedical applications of cardiomyocytes would benefit from using human rather than rodent cells. This is of growing interest as new opportunities arise for using induced pluripotent stem cells (iPSCs) carrying cardiac disease-associated mutations to understand the pathophysiology of heart disease and to develop new treatment modalities.

There is an obvious need for robust and efficient cardiac differentiation protocols for human pluripotent stem cells, preferably under defined, serum-free conditions. Although protocols that are effective on multiple pluripotent

413

J.F. Loring & S.E. Peterson (eds): Human Stem Cell Manual, Second edition.
DOI: http://dx.doi.org/10.1016/B978-0-12-385473-5.00027-8

cell lines are beginning to emerge, none has yet been shown to be universally applicable without individual optimization, and most still require a cell aggregation step to initiate differentiation. More than 10 years of research on human embryonic stem cells (hESCs) has, however, provided important clues as to which parameters are candidates for further improvement. Many of the recent methodological advances made using hESCs are proving to be transferable to human iPSCs with minor modifications. Thus, it is now possible to produce cardiomyocytes from most hESC and hiPSC lines, although the efficiencies with which individual cell lines differentiate can still be highly variable.

Most protocols mimic in some way the signals that take place during fetal heart development. Among the first was one that recapitulated signals from endoderm adjacent to the developing heart: mechanically passaged hESC colonies were co-cultured with a mouse visceral endoderm-like cell line (END2) in serum-free culture medium. This method is effective in most hESC and hiPSC lines and will yield 5–15% cardiomyocytes depending on the individual line. This technique provides a simple way to check for cardiomyocyte differentiation capacity of multiple hiPSC clones while they are still at very early (mechanical passage) stages of development. For producing larger numbers of cardiomyocytes from pluripotent cells already adapted to enzymatic passage on feeder cells, however, a more effective method is based on centrifuging the cells to force them to make aggregates, which are called "Spin embryoid bodies". Human pluripotent cells have less inclination to form aggregates spontaneously in suspension than mouse pluripotent cells and forced aggregation makes cells attach more firmly to one another. In the cardiomyocyte differentiation protocol, the aggregates do not become true embryoid bodies, since they are not cultured long enough to induce all three germ layers.

In this chapter we outline a basic protocol for the differentiation of hESCs into cardiomyocytes. The method is based on the formation of aggregates of a standard size using a fixed number of undifferentiated cells in each aggregate. The aggregates are formed by centrifugation of the cells into embryoid body (EB)-like structures and are referred to as "Spin EBs". Key to the procedure, and most difficult, is the prior adaption of the undifferentiated cells to a standardized single cell passaging bulk culture system in which cells are passaged enzymatically and grown on feeder layers.

The method we recommend for making spin EBs after bulk culture adaption is straightforward and has worked well in our hands for multiple hESC lines and several hiPSC lines. The method requires titration of growth factor concentrations for each line to optimize efficiencies. We also discuss alternative methods and reagents that work but are not routinely used in our laboratories. In addition, we discuss aspects of the protocol that could be further optimized. The key variables that we outline in this chapter are:

- Basal culture conditions
- Adaption to bulk passage
- Generating spin "embryoid bodies" (Spin EBs)
- Growth factor titration curves
- Dissociation of EBs and replating cardiomyocytes.

While optimizing and standardizing conditions, it is important to keep in mind that changing one reagent or parameter in the system may unexpectedly impact the outcome of the entire protocol.

PROCEDURES

Basal Culture Conditions

In our laboratory, many of our stock hESC lines (e.g. HES2, HES3, HES4, NL-HES1) are grown and maintained as mechanically passaged "cut and paste" colonies in KnockOut Serum Replacement-based medium on irradiated or mitomycin C-treated mouse embryonic feeder cells (MEFs). The colonies are cut into small pieces using glass needles and, using a pipette, are transferred to new feeder layers once weekly. See Chapters 1 and 2. For details see Mummery et al., 2007.

From Mechanical Colonies to Bulk Culture: Tips for Successful Adaption Prior to Generating Spin EBs

Several of the hESC lines we use are routinely passaged enzymatically, twice weekly on MEFs using TrypLE Select. These cells can be used immediately in the spin EB protocol for cardiomyocyte differentiation. For cell lines that are not enzymatically passaged, the spin EB protocol requires that the undifferentiated colonies be adapted to single cell enzymatic passage on feeder layers (referred to as bulk culture).

It is important to select hPSC lines for adaption that have been well-maintained with timely medium refreshment and regular mechanical passaging. The colonies should look healthy with little debris and few differentiated cells around the colony edges (Figure 27.1).

For best results use 80–100 mechanically-passaged colonies per line to start bulk adaption of a line.

MEFs are inactivated by prior irradiation at 3000 rads in suspension culture. Plate irradiated MEFs the day before bulk culture adaption to allow attachment and spreading.

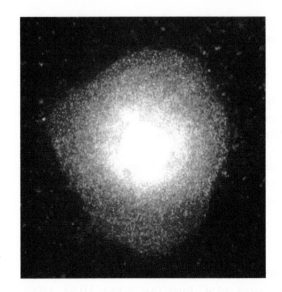

FIGURE 27.1
Example of a healthy
looking hESC colony
with a small amount of
differentiation in the center
of the colony.

Starting Bulk Cultures

1. Cut away any residual differentiated parts of the colonies. These are usually the flat cells around the edge and the cluster at the center.
2. Cut the colonies into extremely small square pieces in a grid (cut closely spaced lines first in one direction over the whole colony then similarly spaced lines in the perpendicular direction) with standard tungsten or glass needles. Note that the squares should be much smaller than those generated during routine mechanical passaging (Figure 27.2).
3. Lift the cut pieces from the dish using a p1000 pipette and place in a 15 mL tube.
4. Collect all of the pieces from a total of 80–100 colonies.
5. Centrifuge for 3 minutes at 250 g.
6. Aspirate medium and resuspend in HESC-medium using a p1000 to pipette colony pieces up and down gently several times.
7. Plate all of the pieces in a T25 flask seeded the day before with 2–3 × 10^4 irradiated MEFs (density of MEFs depends on mouse strain used to produce MEFs and should be optimized empirically) per cm^2 with the appropriate amount of HESC-medium.
8. Gently passage the cells in the flask 4 days later in a 1:1 ratio by incubation in TrypLE Select and resuspend in hESC-medium to spread the small colonies, which tend to pile up, over the surface of the flask (Figure 27.3).
9. The cultures should be observed and the medium should be changed daily. The cells should be passaged at a low ratio (1:2 or 1:3) on new feeder layers in HESC-medium when the islands of undifferentiated cells start to fill the flask (Figure 27.4).

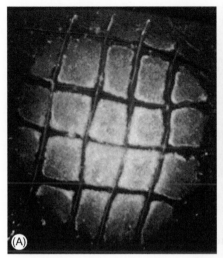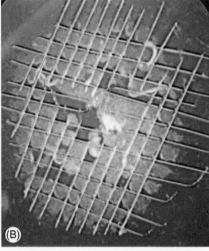

FIGURE 27.2
(A) Example of the size of pieces for mechanical passaging (pieces approximately 400–500 μm in length). (B) Example of grid made for bulk culture initiation (pieces approximately 100–150 μm in length).

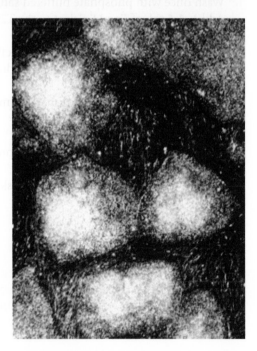

FIGURE 27.3
Example of a bulk culture after 4 days, just before the first passaging.

■ Note

This low split ratio should be continued until the cells adapt (possibly up to five or six passages). This depends somewhat on the individual cell line; some lines accept a higher split ratio after only two passages. The cells are then cultured according to the procedure below (see Maintaining Bulk Cultures).   ■

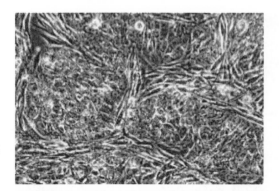

FIGURE 27.4
Example of bulk culture that has been adapted and is ready to be passaged.

Maintaining Bulk Cultures

Bulk cultures should be passaged approximately every 3–4 days as follows:

1. Wash once with phosphate buffered saline (PBS).
2. Add 1–1.5 mL of TrypLE Select to the T25 flask.
3. Place at 37°C for 5 minutes.
4. Tap the flask with fingers to dislodge cells.
5. Resuspend cells in 5 mL of culture medium containing 10% serum and place in a 15 mL tube.
6. Rinse the flask with an additional 3–5 mL of the same serum-containing medium and add to the same tube.
7. Centrifuge for 3 minutes at 250 g.
8. Gently resuspend the pellet in the appropriate amount of HESC-medium.
9. Plate down in an appropriate sized flask, plate, or dish pre-plated with 2–3×10^4 irradiated MEFs per cm^2.

■ Note

If the bulk cultures were started using T25 flasks, the flask size can easily be increased for cryopreservation or scaled down to 6-well plates for general maintenance. One well of a 6-well plate contains a sufficient number of cells for the next step of the procedure and reduces the number of MEF feeder cells required at each passage. ■

Generating Spin EBs
One Day Before Starting the Spin EB Differentiation Protocol

Deplete cultures of MEF feeder cells by passaging the bulk cultures and plating on to Matrigel-coated dishes or flasks (ratio about 1:2/1:3 or about 8×10^5 cells per well of a 6-well plate) (Figure 27.5). The following protocol uses one well of a 6-well plate.

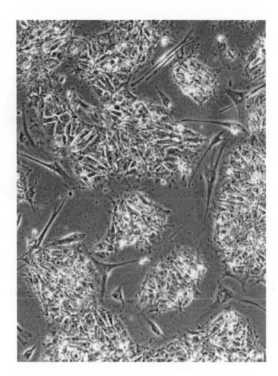

FIGURE 27.5
Bulk cultures one day after plating on Matrigel™ for MEF feeder cell depletion.

Day 0

1. Early morning on the day of differentiation: refresh the medium on the bulk culture in the well of the 6-well plate to maintain culture quality and wash away dead cells.
2. Prepare the growth factor solutions prior to harvesting the cells.
3. Midday on the day of differentiation, start harvesting the cells by washing the culture once with PBS.
4. Add 1 mL of TrypLE™ Select (this is for one well of a 6-well plate; adjust as necessary to the size of culture flask, dish, or well) and incubate for 5 minutes at 37 °C.
5. Add 3 mL (adjusting again as necessary to the size of culture flask, dish, or well) of serum-containing medium to collect cells.
6. Wash the flask again with 3 mL of serum-containing medium to collect remaining cells.
7. Centrifuge for 3 minutes at 250 g.
8. Resuspend cells in 2 mL of BPEL medium. BPEL medium is basal medium with BSA, Polyvinylalcohol (PVA), and Essential Lipids (see Recipes).
9. Centrifuge for 3 minutes at 250 g.
10. Resuspend cells in BPEL medium (1 mL per well of 6-well plate).

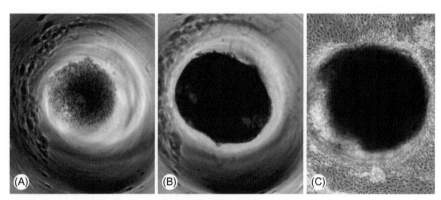

FIGURE 27.6
(A) Example of Spin EB in 96 well V-wells at day 0 after centrifuging. (B) Example of spin EB in V-wells on day 7 (C) Example of Spin EB after plating onto gelatin on day 8.

11. Count cells.
12. Plate cells at a final density of 3000 cells in 50 μL of BPEL medium plus growth factors in each V-shaped well of a 96-well plate from a stock solution containing the required number of cells in the correct volume for all wells to be seeded.

■ **Note**
For a complete plate only use the 60 inner wells and add 100 μL of PBS to the outer wells to compensate for any evaporation of culture medium.   ■

13. Centrifuge plate for 3 minutes at 450 *g* to form aggregates (Spin EBs) (Figure 27.6A).
14. Place in the incubator at 37°C for 3 days.

Day 3
15. Remove the BPEL medium containing growth factors and add 100 μL of growth factor-free BPEL medium.

Day 7
16. Plate the Spin EBs in BEL medium (BPEL without PVA) on 0.1% gelatin-coated 96-well plates with flat-bottomed wells (Figure 27.6B).

■ **Note**
For practical reasons BPEL can also be used but PVA is not needed after 7 days for the continuous differentiation of the EBs.   ■

Day 8
17. Begin checking for beating areas in the cell aggregates under a phase contrast microscope (20× objective) (Figure 27.6C).

■ Note

Beating may begin up to 20 days later, depending on the cell line. HES3 Spin EBs usually start to beat around day 8 and reach a maximum at day 12 (that is, all EBs containing cardiomyocytes are beating). Note that EBs that do not beat visibly do not usually contain cardiomyocytes if stained retrospectively with an antibody against sarcomeres. ■

18. After 14 days, refresh wells with BEL or BPEL medium once a week.

Growth Factor Titration Curves

Titration of Activin A and BMP4 is usually needed to determine the optimal concentrations for the induction of beating cells. Titration will be required for each new batch of growth factors (which are supplied on the basis of protein concentration and not specific activity) and each new cell line or clone:

	Final (ng/mL)	Stock (ng/μL)
BMP4	Determine using titration	30
VEGF	30	50
SCF	40	40
Activin A	Determine using titration	25

One way in which a titration can be done is to test new growth factor batches or cell lines in the middle 60 wells (6 wells by 10 wells) of a 96-well plate. Vary the concentration of Activin A (0–30 ng/mL) in one direction and BMP4 (0–30 ng/mL) in the other direction. Duplicate rows and columns so that there are multiple samples of each combination.

Constant concentrations of VEGF (30 ng/mL) and SCF (40 ng/mL) are present in the Spin-EB medium. PVA is an essential component and should not be omitted (Figure 27.7).

Dissociation of Beating Spin EBs for Experimentation on Single Cardiomyocytes

1. Wash Spin EBs with PBS.
2. Add TrypLE Select for 20 min at 37°C (this depends on the size of the EBs, days of differentiation, etc.). Break up the EBs by gentle pipetting.

■ Note

The time needed for dissociation increases with the age of the EBs and thus the time mentioned above should be monitored. ■

3. Neutralize the enzyme by resuspending cells in serum-containing medium and place in a 15 mL tube.

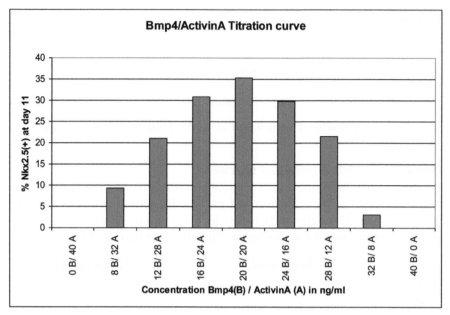

FIGURE 27.7

The outcome of a typical titration curve of BMP4 and Activin A, using HES3 cells, showing optimal BMP4/ Activin A concentrations of 20 ng/mL for each. Higher Activin A concentrations in HES3 lead to endoderm formation. Lower Activin A and higher BMP4 concentrations lead to other mesoderm derivatives like blood. Here, the number of Nkx2.5 + cells present was determined by FACS analysis of GFP (NKX2.5-GFP-Hes3 line [Elliot et al., 2011], but it may also be done using an Nkx2.5 antibody) in a single cell suspension (dissociated beating Spin EBs, see protocol below). For other cell lines, similar titration experiments showed concentrations as low as 5 ng/mL for these factors to be optimal for inducing cardiomyocyte differentiation.

■ Note

Pipette up and down vigorously (maximum three times) if clumps of EBs are still visible. ■

4. Centrifuge for 3 minutes at 250 g.

■ Note

For use in FACS, begin standard FACS staining protocol after this step. ■

5. Resuspend in BEL medium containing 5 μM Rock Inhibitor.
6. Plate dissociated cells on the appropriate 0.1% gelatin-coated cell culture treated dish, plate, or flask.

■ Note

5000 cells per cm² gives a nice spread of single cells although the ideal cell density depends on the subsequent experiments that will be carried out.

Coating dishes, plates, or flasks with other ECM materials (e.g. Matrigel or fibronectin) may improve attachment in some cases. ■

7. Refresh medium, removing the Rock Inhibitor 24 hours later.

ALTERNATIVE PROCEDURES

Use of APEL Instead of BPEL

The use of Bovine Serum Albumin (BSA) in the BPEL medium has some drawbacks, including batch-to-batch variation and the fact that it is an animal product. Replacing BSA with human recombinant albumin may decrease the batch-to-batch variation observed with BSA. Human recombinant albumin may also produce higher efficiencies of differentiation. However, human recombinant albumin is much more expensive than BSA. Stem Cell Technologies™ also supplies pretested APEL medium on a commercial basis.

PITFALLS AND ADVICE

Defined Media for hESC Cultures

The use of defined media for culture of hESCs is ideal for reducing the dependence on MEFs, eliminating the unknown components they produce and reducing variability in cultures over time. There are two commercial media, mTeSR and Nutristem, which we have used to grow hESC cultures under defined, feeder-free conditions for up to 20 or so passages. Long-term passage in either media may be more likely to cause earlier karyotypic abnormalities compared to MEF-based mechanically passaged cultures because it is thought that the total number of cell divisions in a given time period may be higher in defined media. In addition, the potential for subsequent differentiation using other protocols can vary depending on the media used, possibly due to the differences in media components. For example, higher bFGF levels in mTeSR have been found by some to impact the efficiency of differentiation to cardiomyocytes and endoderm derivatives.

Other Factors Influencing Differentiation Efficiency

While we have optimized the Spin EB method using specific growth factors, there are a number of factors which could still be added to help improve efficiency further. These include the addition of ascorbic acid or small molecular inhibitors of transforming growth factor β receptors. The presence of Wnts or Wnt inhibitors has also been reported to enhance cardiomyocyte differentiation if present during the earliest phases of mesoderm formation. Efficient nascent mesoderm formation is a prerequisite for subsequent patterning to cardiogenic mesoderm.

EQUIPMENT

- Tissue culture hood: Class II A/B3
- Tissue culture incubator, 37°C, 5% CO_2, in humidified air
- Inverted phase/contrast microscope with 4×, 10× and 20× objectives
- Centrifuge, 250 to 450 g
- Water bath, 37°C
- Pipettes, such as Eppendorf p2, p20, p200, p1000, multichannel
- Pipette Aid, automatic pipettor for use in measuring and dispensing media
- Aspirator in the hood, with flask
- Eight-channel plastic aspirator
- Roller bank
- Refrigerator, 4°C
- Freezers: −20°C, −80°C

REAGENTS AND SUPPLIES

Supplies

- 5 mL, 10 mL, 25 mL sterile disposable pipettes
- 6-well culture dishes
- T25 culture flask
- 96-well Greiner bioone V-shaped bottom low attachment tissue culture plates
- 96-well flat-bottomed tissue culture plates
- 15 mL sterile conical tubes
- 50 mL sterile conical tubes
- Sterile 9″ Pasteur pipettes (autoclaved)
- Pipette tips for Eppendorf or similar pipette (use blue and yellow low attachment polypropylene pipette tips from Corning)
- Hemocytometer
- 0.22 μm Stericup filtration unit
- Other filters and syringes
- Reagent reservoir, optional

Recommended Reagents

Item	Supplier	Catalog #
Basal medium		
DMEM/high glucose (Dulbecco's Modified Eagle's Medium)	Life Technologies	11960-044
DMEM/F12 1:1 (GlutaMAX, no HEPES)	Life Technologies	31331-028
IMDM, L-glutamine, 25 mM HEPES, no phenol red (Iscove's Modified Dulbecco's Medium)	Life Technologies	21056-023
F12 Nutrient Mixture (Ham), GlutaMAX	Life Technologies	31765-027

Item	Supplier	Catalog #
Serum components		
Knockout Serum Replacement (contains bovine products)	Life Technologies	10828-028
Fetal Bovine Serum (FBS) (batch selected)	Sigma-Aldrich	F7524
L-Glutamine 200 mM	Life Technologies	25030-024
GlutaMAX™-I Supplement 200 mM	Life Technologies	35050-038
MEM-Non-essential amino acids 100×	Life Technologies	11140-035
2-Mercaptoethanol 50 mM (1000×)	Life Technologies	31350-010
D-PBS (Dulbecco's Phosphate-Buffered Saline without Calcium and Magnesium)	Life Technologies	14190-094
Distilled H_2O	Life Technologies	15230-089
Basic FGF Recombinant human	PeproTech	100-18B
Albumin from bovine serum, powder	Sigma	A3311
Resin Beads	Bio-Rad	142-6425
Polyvinyl alcohol (PVA)	Sigma-Aldrich	P8136
Recombinant Human VEGF 165	R&D systems	293-VE
Recombinant Human SCF	PeproTech	300-07
Recombinant Human/Mouse/Rat Activin A	R&D systems	338-AC
Recombinant Human BMP-4, CF	R&D systems	314-BP
Fasudil, Monohydrochloride Salt, >99%	LC Laboratories	F-4660
BD Matrigel Matrix Growth Factor Reduced (GFR)	BD Biosciences	354230
Chemically Defined Lipid Concentrate	Life Technologies	11905-031
TrypLE Select (1×)	Life Technologies	12563-029
1-Thioglycerol	Sigma-Aldrich	M6145
Insulin-Transferrin-Selenium-X Supplement (100×)	Life Technologies	51500-056
Gelatin from porcine skin	Sigma-Aldrich	G1890
PFHM II: Protein free Hybridoma medium 1×	Life Technologies	12040-051
L-Ascorbic Acid 2-Phosphate	Sigma-Aldrich	A8960
Pen/Strep (100×)	Life Technologies	15070-063

RECIPES

■ Note

Glassware should be dedicated to tissue culture use only. If glassware is to be used instead of pre-sterilized plasticware, do not expose bottles to detergent. ■

Stock Solution Human Basic FGF (bFGF) 100 μL

Component	Amount	Stock Concentration
Human bFGF	10 μg	100 ng/μL

1. Dissolve 10 µg of human basic FGF in 10 µL of 5 mM Tris (according the product sheet).
2. Add 90 µL of PBS +0.1% BSA.
3. Aliquot in 5, 10 and 25 µL samples.
4. Store thawed aliquots at 4°C for up to 2 weeks.
5. Store frozen aliquots at −80°C (stable for up to 12 months).

Stock Solution 10% Deionized BSA 45 mL

Component	Amount	Stock Concentration
10% Deionized BSA	4.5 g	10% w/v in IMDM

1. Dissolve 4.5 g in 45 mL of IMDM at 37°C in a 50 ml tube.
2. Vortex to help mix.
3. After completely dissolved, add Bio-Rad Resin Beads in excess (>2 grams).
4. Set rolling at 4°C on a roller bank until beads turn yellow (beads are saturated).
5. Remove the supernatant and place in a new Falcon tube.
6. Add new beads and set rolling on a roller bank at 4°C.
7. Repeat until beads remain blue.

Stock Solution 5% PVA 50 mL

Component	Amount	Stock Concentration
PVA	2.5 g	5% PVA in dH$_2$O

1. Dissolve 2.5 g in 25 mL of dH$_2$O (tissue culture qualified).

■ Note

Add the powder to the dH$_2$O and move the tube while adding the PVA powder). ■

2. Mix slightly.
3. Set rolling at 4°C on a roller bank until completely dissolved and add dH$_2$O up to 50 ml (can take up to 2 days).

Stock Solution 1-Thioglycerol 10 mL

Component	Amount	Stock Concentration
1-Thioglycerol	130 µL	0.15 M

1. Add 130 µL of 1-Thioglycerol to 10 mL IMDM (making these larger volumes of 130 µL is more accurate than smaller volumes because the Thioglycerol is highly viscous).

Stock Solution BD Matrigel Matrix Growth Factor Reduced (GFR) 10 mL

Component	Amount	Stock Concentration
BD Matrigel Matrix Growth Factor Reduced (GFR)	10 mL	Undiluted

1. Thaw 10 mL stock overnight on ice and place in icebox at 4°C. Place Eppendorf tubes and pipette tips at −20°C.
2. Aliquot stock into 100, 200, and 500 µL samples on ice in ice-cold Eppendorf tubes and store at −20°C for up to 3 months.

For use in coating plates:

3. Thaw 100 µL aliquot on ice.
4. Using cold pipettes and medium add 100 µL of Matrigel to 10 mL of DMEM/F12 basal medium.
5. Using a cold pipette, add 2 mL of medium to each well of a 6-well plate.
6. Leave at room temperature for at least 45 minutes.
7. Plates can be used immediately or stored for 2 weeks at 4°C.

■ Note

Stored plates should be left out at room temperature for 45 minutes before use to allow them to warm up. ■

Stock Solution Recombinant Human VEGF 165 200 µL

Component	Amount	Stock Concentration
Recombinant Human VEGF 165	10 µg	50 ng/µL

1. Centrifuge prior to opening.
2. Dissolve in 200 µL of dH$_2$O (according to the product sheet).
3. Aliquot in 5 and 10 µL samples.
4. Store thawed aliquots at 4°C for up to 1 week.
5. Store frozen aliquots at −80°C (stable for up to 12 months).

Stock Solution Recombinant Human SCF 250 µL

Component	Amount	Stock Concentration
Recombinant Human SCF	10 µg	40 ng/µL

1. Centrifuge prior to opening.
2. Dissolve in 250 µL of dH$_2$O (according to the product sheet).
3. Aliquot in 5, 10, and 20 µL samples.
4. Store thawed aliquots at 4°C for up to 1 week.
5. Store frozen aliquots at −80°C (stable for up to 12 months).

Stock Solution Recombinant Human/Mouse/Rat Activin A 400 μL

Component	Amount	Stock Concentration
Recombinant Human/Mouse/Rat Activin A	10 μg	25 ng/μL

1. Centrifuge prior to opening.
2. Dissolve in 400 μL of PBS +0.1% BSA (according to the product sheet).
3. Aliquot in 5, 10, and 20 μL samples.
4. Store thawed aliquots at 4°C for up to 1 week.
5. Store frozen aliquots at −80°C (stable for up to 12 months).

Stock Solution Recombinant Human BMP4 330 μL

Component	Amount	Stock Concentration
Recombinant Human BMP-4, CF	10 μg	30 ng/μL

1. Centrifuge prior to opening.
2. Dissolve in 330 μL of PBS+0.1% BSA (according to the product sheet).
3. Aliquot in 5 and 10 μL samples.
4. Store thawed aliquots at 4°C for up to 1 week.
5. Store frozen aliquots at −80°C (stable for up to 12 months).

Stock Solution Fasudil, Monohydrochloride Salt, >99% 5 mL, ROCK Inhibitor

Component	Amount	Stock Concentration
Fasudil, Monohydrochloride Salt, >99%	8.15 mg	5 mM

1. Add 8.15 mg to 5 mL of dH_2O.
2. Mix until dissolved.
3. Filter using 0.22 μm.
4. Aliquot in 5 and 10 μL samples.
5. Store at −20°C (stable up to 3 months).

Stock Solution "Knockout Serum Replacement" (Invitrogen 10828-028)

This product can be kept (after thawing) at 4°C for 1 week.

1. After 1 week freeze aliquots into appropriate volumes in 50 mL tubes and store at −20°C.
2. Thaw at 37°C just prior to use.

Stock Solution L-Glutamine (200 mM) (Invitrogen 25030-081)

L-glutamine is unstable and must be stored frozen.

1. Thaw the bottle completely and aliquot in 10 mL tubes. Store at −20°C.
2. Thaw a tube just prior to use.
3. Do not refreeze tube, store at 4°C and discard unused glutamine after 2 weeks.

HESC Medium 100 mL

Component	Amount	Final Concentration
DMEM/F12 1:1 (GlutaMAX, no HEPES)	78 mL	
KnockOut Serum Replacement	20 mL	20%
100× MEM-Non-Essential amino acids	1 mL	10 mM
50 mM 2-Mercaptoethanol	200 μL	0.1 mM
Human bFGF (100 ng/μL)	10 μL	10 ng/mL
Pen/Strep (100×)	0.5 mL	1:200

1. Prepare all media in the tissue culture hood using aseptic techniques.
2. Combine the basal medium and all the components except bFGF in the filter top. Filter sterilize using a 0.22 μm Stericup filtration unit. Add 10 ng/mL bFGF after filtering.
3. Store at 4°C for up to 2 weeks.

BPEL Medium: Bovine Serum Albumin (BSA) Polyvinylalchohol Essential Lipids 100 mL

Component	Amount	Final Concentration
IMDM, L-glutamine, 25 mM HEPES, no phenol red (Iscove's Modified Dulbecco's Medium)	42.6 mL	
F12 Nutrient Mixture (Ham), GlutaMAX	42.6 mL	
PFHM II: Protein free Hybridoma medium 1×	5 mL	1:20
10% Deionized BSA in IMDM	2.5 mL	0.25% w/v
5% PVA in dH$_2$O	2.5 mL	0.125% w/v
Chemically Defined Lipid Concentrate	1 mL	1:100
Insulin-Transferrin-Selenium-X Supplement (100×)	100 μL	1:1000
0.15 M 1-Thioglycerol	300 μL	450 μM
5 mg/mL L-Ascorbic acid 2-phosphate	1 mL	0.05 mg/mL
200 mM GlutaMAX-I Supplement	1 mL	2 mM
Pen/Strep (100×)	0.5 mL	1:200

1. Prepare all media in the tissue culture hood using aseptic techniques.
2. Combine all components and filter sterilize using 0.22 μm filter.
3. Store at 4 °C for up to 2 weeks.

BEL medium: Bovine Serum Albumin (BSA), Essential Lipids 100 mL

Component	Amount	Final Concentration
IMDM, L-glutamine, 25 mM HEPES, no phenol red (Iscove's Modified Dulbecco's Medium)	42.6 mL	
F12 Nutrient Mixture (Ham), GlutaMAX	42.6 mL	
PFHM II: Protein free Hybridoma medium 1×	5 mL	5%
10% Deionized BSA in IMDM	5 mL	0.5% w/v
Chemically Defined Lipid Concentrate	1 mL	1:100
Insulin-Transferrin-Selenium-X Supplement (100×)	1 mL	1:100
1-Thioglycerol	300 μL	450 μM
5 mg/mL L-Ascorbic acid 2-phosphate	1 mL	0.05 mg/mL
200 mM GlutaMAX-I Supplement	1 mL	2 mM
Pen/Strep (100×)	0.5 mL	1:200

1. Prepare all media in the tissue culture hood using aseptic techniques.
2. Combine all components and filter sterilize using 0.22 μm filter.
3. Store at 4 °C up to 2 weeks.

Serum-Containing Medium 100 mL

Component	Amount	Final Concentration
DMEM/high glucose (Dulbecco's Modified Eagle's Medium)	87.5 mL	
FBS	10 mL	10%
100 × MEM-Non-essential amino acids	1 mL	10 mM
200 mM L-Glutamine	1 mL	2 mM
Pen/Strep (100×)	0.5 mL	1:200

1. Prepare all media in the tissue culture hood using aseptic techniques.
2. Store at 4 °C up to 4 weeks

QUALITY CONTROL METHODS

Lot-to-Lot Variability of Reagents

Growth factors are sold on the basis of weight/volume, not specific activity. Effective concentrations of growth factors required for growth and differentiation will then depend on how and how long the reagent has been stored. Thus, be careful to test each lot and record lot numbers.

READING LIST

Co-Culture

Mummery, C.L., Ward, D., Passier, R., 2007. Differentiation of human embryonic stem cells to cardiomyocytes by coculture with endoderm in serum-free medium. Curr. Protoc. Stem Cell Biol (Chapter 1:Unit 1F.2).

Spin EBs

Burridge, P.W., Anderson, D., Priddle, H., Barbadillo Munoz, M.D., Chamberlain, S., Allegrucci, C., et al., 2007. Improved human embryonic stem cell embryoid body homogeneity and cardiomyocyte differentiation from a novel V-96 plate aggregation system highlights interline variability. Stem Cells 25, 929–938.

Costa, M., Sourris, K., Hatzistavrou, T., Elefanty, A.G., Stanley, E.G., 2008. Expansion of human embryonic stem cells in vitro. Curr. Protoc. Stem Cell Biol. (Chapter 1:Unit 1C.1.1-1C.1.7).

Ng, E.S., Davis, R.P., Azzola, L., Stanley, E.G., Elefanty, A.G., 2005. Forced aggregation of defined numbers of human embryonic stem cells into embryoid bodies fosters robust, reproducible hematopoietic differentiation. Blood 106, 1601–1603.

Ng, E.S., Davis, R., Stanley, E.G., Elefanty, A.G., 2008a. A protocol describing the use of a recombinant protein-based, animal product-free medium (APEL) for human embryonic stem cell differentiation as spin embryoid bodies. Nat. Protoc. 3, 768–776.

Ng, E.S., Davis, R.P., Hatzistavrou, T., Stanley, E.G., Elefanty, A.G., 2008b. Directed differentiation of human embryonic stem cells as spin embryoid bodies and a description of the hematopoietic blast colony forming assay. Curr. Protoc. Stem Cell Biol. (Chapter 1).

Activin/BMP/Wnt signaling

Burridge, P.W., Thompson, S., Millrod, M.A., Weinberg, S., Yuan, X., Peters, A., et al., 2011. A universal system for highly efficient cardiac differentiation of human induced pluripotent stem cells that eliminates interline variability. PLoS One 6 (4), e18293.

Kattman, S.J., Witty, A.D., Gagliardi, M., Dubois, N.C., Niapour, M., Hotta, A., et al., 2011. Stage-specific optimization of activin/nodal and BMP signaling promotes cardiac differentiation of mouse and human pluripotent stem cell lines. Cell Stem Cell 8, 228–240.

Paige, S.L., Osugi, T., Afanasiev, O.K., Pabon, L., Reinecke, H., Murry, C.E., 2010. Endogenous Wnt/beta-catenin signaling is required for cardiac differentiation in human embryonic stem cells. PLoS One 5 (6), e11134.

Ren, Y., Lee, M.Y., Schliffke, S., Paavola, J., Amos, P.J., Ge, X., et al., 2011. Small molecule Wnt inhibitors enhance the efficiency of BMP-4-directed cardiac differentiation of human pluripotent stem cells. J. Mol. Cell. Cardiol. 51 (3), 280–287.

Directed Differentiation of Human Pluripotent Stem Cells into Fetal-Like Hepatocytes

Thomas Touboul, Nick Hannan, and Ludovic Vallier

EDITOR'S COMMENTARY

Of all of the cell types that researchers hope to obtain from human pluripotent stem cells (hPSCs), hepatocytes are arguably the cell type that has the greatest number of uses. Transplantation of immunologically matched iPSC-derived hepatocytes might someday help sustain the function of livers damaged by disease or metabolic disorders. Mechanisms of liver-associated diseases like hepatitis could be studied in iPSC-derived hepatocytes. Because the liver metabolizes pharmaceutical drugs, liver-derived cell lines are used for large-scale toxicity screening during the early stages of drug development. Because these hepatocyte-derived cell lines are generally of Caucasian origin, drugs that cause liver toxicity only in non-Caucasians are not identified in such screens; the genomics-based adverse effects are usually only discovered after marketing of the drug. Some examples are frontline HIV and tuberculosis treatments that have proved to be more toxic to people of Sub-Saharan origin. Because of this wide range of potential uses, generation of hepatocytes from iPSCs is an area of considerable interest. This chapter describes methods for directed differentiation of hPSCs along a hepatocyte lineage.

OVERVIEW

Hepatocytes represent the main cellular unit of the liver. This cell type displays a unique combination of functions such as protein secretion (albumin and alpha-antitrypsin), cholesterol uptake and glycogen storage. Hepatocytes are also the main detoxifying cell type of the body for drugs, alcohol, and other chemicals, and thus they are broadly used as a cellular model for toxicology screening and drug development. In addition, cell therapy using hepatocytes could be become an attractive alternative therapy to orthotopic liver transplantation for end-stage liver diseases and inherited metabolic disorders.

433

J.F. Loring & S.E. Peterson (eds): Human Stem Cell Manual, Second edition.
DOI: http://dx.doi.org/10.1016/B978-0-12-385473-5.00028-X

During the past two decades, a large number of groups have tried to grow primary hepatocytes *in vitro* with limited success. Indeed, culture systems currently available allow the maintenance of primary hepatocytes for only a few weeks; the cells do not proliferate, and their functional characteristics decrease over time. Because there is a shortage of organ donors, researchers are precluded from requesting large quantities of primary hepatocytes for experiments. Therefore, alternative sources of hepatocytes are urgently needed.

Human pluripotent stem cells (hPSCs) of embryonic origin (human embryonic stem cells or hESCs) or those generated by reprogramming somatic cells (human induced pluripotent stem cells or hiPSCs) could provide a solution to this major challenge. These cells are able to proliferate indefinitely *in vitro* while maintaining the capacity to differentiate into cell types of clinical interest including hepatocytes.

Of note, methods currently available allow only for the differentiation of hPSCs into fetal hepatocytes while production of fully mature hepatocytes remains problematic. Consequently, cells generated from hPSCs often display a limited level of functionality when compared to primary adult hepatocytes, including low levels of albumin secretion and the absence of adult cytochromes. Nevertheless, fetal hepatocytes generated from hiPSCs have been used to model liver diseases *in vitro* and could offer important benefits for cell therapy.

In this chapter we outline a three-step protocol to direct differentiation of hPSCs into fetal hepatocytes. This protocol relies on defined culture media devoid of feeder cells, serum, or complex extracellular matrices. This defined culture system drives differentiation of hPSCs by recapitulating the natural path of development, reminiscent of hepatic development in the mammalian embryo.

This protocol has been used successfully on 3 hESC lines and 18 hiPSC lines and additional lines are currently being validated. Nevertheless, solid expertise in hPSC culture and differentiation is required to apply this protocol. We also indicate alternative media and culture systems used by others, which may work more efficiently for specific hPSC lines.

The last part of the chapter describes techniques commonly used to characterize the functional activities of hepatocytes. We recommend that the same characterization experiments are performed on primary adult hepatocytes in parallel as a positive control.

Importantly, the culture system described here only works in defined conditions. The use of feeder layers, extracellular matrix (such as Matrigel™) or any other basal media can interfere with the efficacy of the protocol.

PROCEDURES

Tips for Successfully Differentiating hPSCs into Hepatocytes

Quality of the hPSCs

This method will only work efficiently on homogenous populations of hPSCs. Significant contamination by differentiated cells strongly interferes with endodermal differentiation. We recommend that at least 70% of the cells be positive for cell surface pluripotency markers, such as Tra-1-60/1-81 or SSEA-3/4.

hPSC Culture Conditions

This protocol is routinely used with hPSCs grown on feeder layers in the presence of Serum Replacer and bFGF or with hPSCs grown in chemically defined medium supplemented with Activin A and bFGF. Culture systems that use high doses of bFGF and Matrigel are not recommended because these factors disturb endoderm differentiation.

Extracellular Matrix (ECM) Compatible with hPSC Growth, Definitive Endoderm (DE) Differentiation and Hepatic Specification

Fibronectin can support hPSCs and DE cells but it is not compatible with hepatocyte growth, which relies on collagen. To solve this issue we coat plates with FBS and gelatin, which provides a complex and complete ECM that includes vitronectin, fibronectin, collagen, and other ECM components

Endoderm Differentiation

This step is the most important. Indeed, all the hPSC lines that differentiate efficiently into endoderm also differentiate efficiently into hepatocytes. Thus, this step must be carefully monitored by performing immunostaining for SOX17 expression and FACS analyses to define the percentage of cells expressing CXCR4. We recommend that at least 60% of the cells express CXCR4.

General Guidelines

- hPSCs can be plated at high density as long as the population is homogenously pluripotent.
- The entire process of differentiation occurs in the same well because endoderm cells and hepatocytes cannot be passaged without high levels of cell death.
- The size of the colony can influence endoderm differentiation and should contain between 500–1000 cells.
- Medium should be changed daily.

■ The morphology of differentiating cells needs to be monitored every day. See Recognizing changes in cellular morphology below.

Step 1: Endoderm Differentiation

In this part of the protocol hPSCs are differentiated into DE cells expressing SOX17 (SRY-box 17) and CXCR4 (chemokine receptor 4). Confluent hPSCs are first split onto a fresh plate in the absence of serum, feeder layers or Matrigel and left to recover for 48 hours. Then, hPSCs are grown for 3 days in the presence of high doses of Activin A (INHBA), BMP4 (bone morphogenic protein 4), bFGF (basic fibroblast growth factor), and the PI3 kinase (phosphoinositide-3-kinase) inhibitor LY294002.

hPSCs grown in these culture conditions will differentiate first into primitive streak-like cells marked by the expression of T (brachyury homolog), EOMES (eomesodermin), MIXL1 (Mix paired-like homeobox), and the absence of SOX2. Then, the cells will become positive for DE markers including SOX17, CXCR4, FOXA2 (forkhead box A2), and GATA4 (GATA binding protein 4) while being negative for the expression of pluripotency markers such as POU5F1/OCT4 and NANOG.

Day 1

Passage hPSCs using Dispase/Collagenase IV or an alternative method and plate them on a FBS/Gelatin coated plate in chemically defined medium (CDM) + Activin A (10 ng/mL) + bFGF (12 ng/mL).

■ Note

The passaging method does not affect the efficiency of differentiation as long as the size of the clumps is between 500 and 1000 cells. Avoid single cell suspensions.   ■

■ Note

One confluent well from a 6-well plate can produce one 12-well plate.   ■

Day 2

Add fresh medium: CDM + Activin A (10 ng/mL) + bFGF (12 ng/mL).

Day 3

Add differentiation medium: CDM + poly vinyl alcohol (PVA) + Activin A (100 ng/mL) + bFGF (20 ng/mL) + BMP4 (10 ng/mL) + PI3 kinase inhibitor LY294002 (10 μM) + GSK-3 inhibitor CHIR99021 (3 μM).

Day 4

Add differentiation medium: CDM + PVA + Activin A (100 ng/mL) + bFGF (20 ng/mL) + BMP4 (10 ng/mL) + PI3 kinase inhibitor LY294002 (10 μM).

Day 5

Add differentiation medium: RPMI + non-essential amino acids (NEAA) + B27 supplement + Activin A (100 ng/mL) + bFGF (20 ng/mL) + BMP4 (10 ng/mL) + LY294002 (10 µM).

■ Note

0.5 mL of medium per well is enough for a 12-well plate.   ■

■ Note

Withdrawal of insulin can increase differentiation of specific cell lines such as WA09. However, it can also induce a great deal of cell death and thus should be use cautiously.   ■

■ Note

At this stage 70% of the cells express CXCR4/SOX17 and 20% express PDGFRA (platelet-derived growth factor receptor alpha).   ■

Step 2: Ventral Foregut Specification

In this step, DE cells are differentiated into ventral foregut cells using Activin A. The resulting foregut cells express HHEX (hematopoietically expressed homeobox), HNF1B (HNF1 homeobox B), FOXA2, HNF4 (hepatocyte nuclear factor 4 alpha), GATA4, and GATA6.

Days 6–9

Add differentiation medium: RPMI + NEAA + B27 supplement + Activin A (50 ng/mL).

Step 3: Hepatic Specification

In this step, ventral foregut cells are differentiated into hepatic endoderm by adding BMP4 and bFGF, both of which are known to control hepatic specification during mammalian development. The resulting cells express HHEX, CK18 (KRT18; Keratin 18), CK19 (KRT19), FOXA2, and AFP (alpha-fetoprotein).

Days 10–12

Add differentiation medium: RPMI + NEAA + B27 supplement + BMP4 (10 ng/mL) and bFGF (50 ng/mL).

Step 4: Hepatic Maturation

In this step, hepatic endoderm cells are differentiated into hepatocytes using HGF (hepatocyte growth factor) and Oncostatin M (OSM). The differentiation process can take up to 14 days and markers specific to hepatocytes include albumin (ALB), AAT (SERPINA1: alpha-1 antiproteinase, antitrypsin), CYP (cytochrome P450) activity, glycogen storage, and cholesterol uptake.

Days 13–25
Add CMRL Culture Medium + Oncostatin M (10 ng/mL) + HGF (25 ng/mL)

■ Note
Hepatozyme-SFM can be used as basal medium instead of CMRL.   ■

■ Note
Proliferation strongly decreases during this step.   ■

■ Note
High level of apoptosis are observed toward the end of this protocol and the cells can rarely be grown for more than 25 days.   ■

■ Note
Change the medium every other day.   ■

Recognizing Changes in Cellular Morphology
Figures 28.1–28.4 show the changes that can be seen in cellular morphology.

ALTERNATIVE PROCEDURES
Endoderm Protocol
One alternative method commonly used to induce DE was first described by D'Amour et al. (2005). In this method, hPSCs are plated on Matrigel™ at high density. Cells are differentiated into mesendoderm in RPMI supplemented with Wnt3a and Activin A. The following days (days 1–2) cells are cultured in Activin A and 0.2% serum to induce DE. Other groups have reported similar differentiation efficiency using B27 supplement rather than serum (Hay et al., 2008; Si-Tayeb et al., 2009).

Liver Differentiation
Different methods allow specification of DE cells into hepatic progenitors. Cells can be directly exposed to various combinations of factors, such as BMP4 with bFGF (Si-Tayeb et al., 2009) or FGF4 (fibroblast growth factor 4) and HGF (Liu et al., 2011). The cells specified are then grown in maturation medium containing HGF and OSM (Liu et al., 2011; Si-Tayeb et al., 2009). Other methods

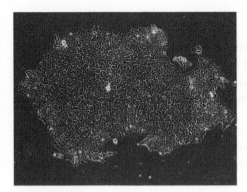

FIGURE 28.1

Day 2. Undifferentiated hPSCs cultured under feeder-free conditions. Cells are round, small and compact, with a small cytoplasma/nuclear ratio (magnification 10×).

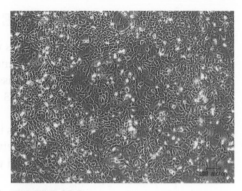

FIGURE 28.2

Day 6. Definitive endoderm. Cells are spreading from the colonies and differentiating. As cells differentiate their morphology becomes triangular and flat (magnification 10×).

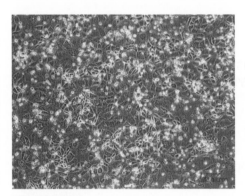

FIGURE 28.3

Day 10. Ventral foregut cells. Cells establish contact characteristic of epithelial cells. Foregut cells are larger than their DE counterparts and develop a square shape (magnification 10×).

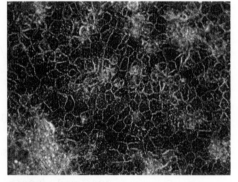

FIGURE 28.4

Day 25. Heptatocyte-like cells. The morphology of the cells resembles the cuboidal shapes typical of hepatocytes (magnification 20×).

use direct specification of DE by growing cells in a medium containing Serum Replacer and DMSO (Hay et al., 2008). We have also generated hepatic cells by inducing primarily anterior foregut with FGF10, and then hepatic progenitors with a combination of FGF10, retinoic acid, and SB431542 (an activin receptor-like kinase (ALK) receptor inhibitor) The hepatic progenitors are then matured in commercially available serum-free medium supplemented with HGF, EGF (epidermal growth factor) and OSM (Touboul et al., 2010).

PITFALLS AND ADVICE

- Variability among lines: hPSCs vary significantly in the ease at which they differentiate into endoderm, and the culture conditions may need to be adjusted to compensate. This issue may require increasing or decreasing the quantity of BMP4 or GSK-3 inhibitor CHIR99021, which are the main inducers of differentiation.
- Batch variation: some reagents including BMP4, Activin A and CHIR99021 can display variable activity between batches. Thus, differentiation quality should be carefully monitored when a new batch of these reagents is purchased.
- The confluence of the cells can also be an important factor. Some cell lines show higher mortality when plated at low confluence.
- It is also recommended to sometimes start the differentiation the day after plating, and not 2 days. Some cell lines will start differentiating when passaged from feeder layers to feeder-free conditions, affecting the yield of DE cells.

EQUIPMENT

- Tissue culture hood, Class II A/B3
- Tissue culture incubator, 37°C, 5% CO_2, in humidified air
- Low-speed centrifuge
- Water bath, 37°C
- Pipettors: p20, p200, and p1000
- Pipette Aid
- Inverted phase contrast microscope with 4×, 10× and 20× objective
- Refrigerator at 4°C
- Freezers: −20°C and −80°C
- Fluorescence microscope
- QPCR instrument

REAGENTS AND SUPPLIES

Recommended Reagents

Item	Supplier	Catalog #	Note
Basal Medium			
IMDM (Iscove's Modified s Dulbecco's Medium)	Life Technologies	12440053	
Ham's F-12 nutrient mix, GlutaMAX	Life Technologies	31765035	
RPMI (Roswell Park Memorial Institute 1640)	Life Technologies	11875-093	

Item	Supplier	Catalog #	Note
CMRL	Life Technologies	11530037	
HepatoZYME-SFM	Life Technologies	17705021	
Medium Components			
Poly(vinyl-alcohol) (PVA)	Sigma	P8136	
Chemically Defined Lipid Concentrate (100×)	Life Technologies	11905031	
1-thioglycerol	Sigma	M6145	
Insulin (100 mg)	Roche	11376497001	
Transferrin (20 mL)	Roche	10652202001	To be aliquoted and stored at −20°C upon arrival
B27 Supplement	Life Technologies	17504044	Use at 1× concentration in medium
MEM NEAA (Non-Essential Amino Acids)	Life Technologies	11140050	Use at 1× concentration in medium
Recombinant human Basic Fibroblast Growth Factor	BioPioneer	HRP-001-1	
Recombinant human Activin A	Stemgent	03-0001	
Recombinant human BMP4	R&D	314-BP	
Recombinant human HGF	Peprotech	100-39	
Recombinant human Oncostatin M	R&D	295-OM-010	
Collagenase, Type IV, microbial	Life Technologies	17104019	
Pen/Strep (100×)	Life Technologies	15070-063	Optional
CHIR 99021	Axon Medchem	Axon 1386	

RECIPES

Chemically Defined Medium (CDM) 500 mL

Componet	Amount
IMDM	250 mL
Ham's F-12 nutrient mix, GlutaMAX	250 mL
Chemically Defined Lipid Concentrate (100×)	5 mL
1-thioglycerol (11.55 M)	20 μL
Insulin (10 mg/mL)	350 μL
Transferrin (30 mg/mL)	250 μL
BSA or PVA	0.5 g/0.5 g

Stock Solution Human Insulin 10 mg/mL

Component	Amount	Stock Concentration
Human insulin	100 mg	10 mg/mL in sterile water

1. Dissolve 100 mg of insulin in 10 mL of sterile water.
2. Aliquot 700 μL in Eppendorf tubes and store at −20°C.

Stock Solution Human Basic FGF (bFGF) 25 μg/mL

Component	Amount	Stock Concentration
Human bFGF	10 μg	25 μg/mL PBS with BSA 0.1%

1. Dissolve 10 μg in 400 μL of PBS containing 0.1% BSA.
2. Aliquot 25 μL in eppendorf tubes and store at −20°C.

■ Note

As for any other growth factor, avoid repeated freeze–thaw cycles. ■

Stock Solution Human BMP4 10 μg/mL

Component	Amount	Stock Concentration
Human recombinant BMP4	10 μg	10 μg/mL PBS with BSA 0.1%

1. Dissolve 10 μg in a sterile 4 mM HCl solution.
2. Add 800 μL of PBS containing 0.1% BSA to obtain a 10 μg/mL stock solution.
3. Aliquot 10 μL in eppendorf tubes and store at −80°C.

Stock Solution Human Activin A 200 μg/mL

Component	Amount	Stock Concentration
Human recombinant Activin A	5 μg	200 μg/mL 10 mM HCL

1. Dissolve 5 μg in 25 μL of sterile 10 mM HCl solution.
2. Aliquot 5 μL in Eppendorf tubes and store at −80°C.

Stock Solution Human Hepatocyte Growth Factor (HGF) 50 μg/mL

Component	Amount	Stock Concentration
Human recombinant HGF	10 μg	50 μg/mL PBS with 0.1% BSA

1. Dissolve in 20 μL of sterile water (0.5 mg/mL).
2. Add 480 μL of PBS containing 0.1% BSA.
3. Aliquot 10 μL in Eppendorf tubes and store at −80°C.

Stock Solution Human Oncostatin M 10 μg/mL

Component	Amount	Stock Concentration
Human recombinant HGF	10 μg	10 μg/mL PBS with 0.1% BSA

1. Dissolve in 1 mL of PBS containing 0.1% BSA.
2. Aliquot 20 μL in Eppendorf tubes and store at −80°C.

QUALITY CONTROL METHODS

Here we describe different methods to monitor the efficiency of differentiation and characterize the functionality of the terminal differentiated cells.

RT-QPCR Analysis
1. Extract total RNA using the RNeasy Mini Kit.
2. Reverse transcribe RNA using the QuantiTect Reverse Transcription Kit.
3. Subject cDNA samples to QPCR amplification with DNA primers.
4. QPCR is performed using the Applied Biosystems TaqMan® Gene Expression master mix and the following program conditions: first step of 5 minutes at 94°C, 30 cycles for 39 seconds at 94°C; a 30 second annealing step at 60°C and 30 seconds at 72°C; and extension for 10 minutes at 72°C.

Immunostaining
1. Wash the cells with PBS twice. Fix the cells with 4% paraformaldehyde (PFA) in PBS for 15 min at 4°C.
2. Wash the cells with PBS three times.
3. Block the cells with 10% serum in PBS containing 0.1% Triton X-100 (use serum from the species in which your secondary antibody was derived to reduce non-specific binding).
4. Incubate the cells for 1 hour at room temperature with a primary antibody (i.e. AFP, ALB, HNF4, CYP3A4, AAT).
5. Wash the cells with PBS three times.
6. Incubate the cells for 1 hour at room temperature with an appropriate fluorescence-conjugated secondary antibody at room temperature in the dark.
7. Wash the cells three times with PBS.
8. Mount slides and counterstain nuclei with DAPI (Vector Laboratories).

Quantification of Albumin Secretion
1. Analyze albumin secretion using the Human Albumin ELISA Kit from Bethyl Laboratories.
2. Collect 24-hour supernatants and centrifuge in a benchtop centrifuge at 2000 rpm for 5 minutes to eliminate cell debris. Control medium should be collected and analyzed in parallel.

3. Prior to detection, samples should be diluted into $1\times$ Dilution Buffer C based on the expected concentration of the unknown sample.
4. Add $100\,\mu L$ of sample or standard solution to the designated well of the strips. Duplicates of each sample or standard should be done.

■ Note

The strips are placed on a plate frame to facilitate manipulation. ■

5. Incubate for 1 hour at room temperature.
6. Wash the plate four times with $1\times$ wash buffer. The solution can be discarded over the sink by flipping the plate. The remaining liquid can be eliminated by blotting the plate onto absorbent material.
7. Add $100\,\mu L$ of anti-albumin detection antibody to each well and incubate for 1 hour at room temperature in the dark (mix well by gently tapping the plate).
8. Wash the plate four times with $1\times$ wash buffer.
9. Add $100\,\mu L$ of HRP solution A to each well, and incubate 20–25 minutes at room temperature in the dark.
10. Stop the reaction by adding $100\,\mu L$ of TMB solution into each well. The reaction is performed at room temperature in the dark. The contents of the well should turn blue.

■ Note

Do not seal the plate using a plate sealer. ■

11. Stop the reaction by adding $100\,\mu L$ of stop solution. Mix the reagents by tapping the plate gently. The presence of albumin in the standard or sample will be revealed by a yellow color.
12. Measure the absorbance on an ELISA plate reader at $450\,nm$ within 30 minutes after the addition of the stop solution.

Alpha-1-Antitrypsin (AAT) ELISA

1. Coat high binding surface Costar 96-well plates overnight with affinity-purified rabbit polyclonal antibodies against $\alpha1$-antitrypsin (Abcam ab31657) at $2\,\mu g/mL$ in carbonate/bicarbonate buffer ($Na_2CO_3/NaHCO_3$, pH 9.5).
2. After washing (0.9% NaCl and 0.05% Tween-20), block plates for 2 hours in blocking buffer (PBS, 0.25% BSA, 0.05% Tween 20).
3. Dilute samples (culture medium or cells lysed in $50\,\mu L$ Nonidet lysis buffer) and standards (plasma purified M or Z $\alpha1$-antitrypsin) in blocking buffer.
4. Nonidet buffer: $150\,mM$ NaCl; $50\,mM$ Tris-Cl, pH 7.5; 1% [v/v] Nonidet P-40.
5. Add $50\,\mu L$ to each well and incubate for 2 hours.

6. Wash the cells and incubate with 9C5 monoclonal antibodies ($1\,\mu g/mL$ diluted in blocking buffer) for 2 hours.
7. Bound monoclonal antibodies are detected with rabbit anti-mouse IgG HRP-labeled antibody (1:20,000; Sigma-Aldrich) for 1 hour.
8. Develop the reaction with TMB liquid substrate (Sigma-Aldrich) for 10 minutes in the dark, and stop the reaction with $1\,M\ H_2SO4$.
9. Read absorbance at 450 nm on a ThermoMAX Microplate Reader (Molecular Devices).

Periodic Acid Schiff (PAS) Assay

1. Fix hPSC-derived hepatocytes with 4% PFA in PBS for 10 minutes at 4 °C.
2. Wash the cells three times with PBS.
3. Add 0.5% periodic acid (Dako) solution for 10 minutes and wash the cells three times with distilled water.
4. Stain the cells in Schiff's reagent for 10 minutes and wash them three times with distilled water for 3 minutes.
5. Counterstain the cells with Mayer's hematoxylin solution for 30 seconds to 2 minutes, and wash three times with distilled water for 3 minutes.

■ Note

Cells can be observed at this time during the experiment. ■

6. If mounting the preparation, dehydrate the cells with 95% alcohol and mount with a mounting solution (Southern Biotech).

Recommended Reagents

Item	Supplier	Catalog #	Note
RNeasy mini kit	Qiagen	74104	Other RNA isolation kits can be used
QuantiTect Reverse Transcription Kit	Qiagen	205310	
VECTASHIELD® with DAPI	Vector labs	H-1200	Other mounting medium can be used
Human Albumin Elisa Kit	Bethyl Laboratories	E88-129	
Rabbit anti-mouse IgG-HRP labeled antibody	Sigma-Aldrich	A9044	
TMB liquid substrate	Sigma-Aldrich	T0440	
Artisan Periodic Acid Schiff Stain Kit	Dako	AR165	
Fluoromount-G	Southern Biotech	0100-01	
TaqMan Gene Expression Master Mix	Applied Biosystems	4369016	

Antibodies

Item	Supplier	Catalog #	Note
OCT4: anti-human OCT-3/4 mouse monoclonal antibody (C10)	Santa Cruz Biotechnology	SC-5279	Dilution 1:100
Sox17: anti-human SOX17 goat polyclonal antibody	R & D Systems	AF1924	Dilution 1:100
FOXA2: anti-human FOXA2 goat polyclonal antibody	Abcam	Ab 5074	Dilution 1:50
GATA4: anti-human GATA4 goat polyclonal antibody (C-20)	Santa Cruz Biotechnology	SC-1237	Dilution 1:100
CXCR4: PE mouse anti-human CD184	BD Pharmingen	555974	20 μL per reaction
PDGFRα: PE mouse anti-human CD140a	BD Pharmingen	556002	20 μL per reaction
HNF1b: anti-human HNF1b, goat polyclonal	Santa Cruz Biotechnology	SC-7411	Dilution 1:100
HHEX: anti-HHEX polyclonal rabbit antibody	Abcam	Ab34222	Dilution 1:50
HNF4: anti-human HNF4 rabbit polyclonal antibody	Santa Cruz Biotechnology	SC-8987	Dilution 1:100
AFP: anti-human AFP polyclonal rabbit antibody	Dako	A0008	Dilution 1:100
ALB: anti-human ALB goat antibody	Bethyl Laboratories	A80-229A	Dilution 1:200
AAT: anti-human AAT goat polyclonal antibody (F16)	Santa Cruz Biotechnology	SC-31919	Dilution 1:100
CYP3A4: anti-human CYP3A4 goat polyclonal antibody (C17)	Santa Cruz Biotechnology	SC-27639	Dilution 1:100

READING LIST

Cai, J., Zhao, Y., Liu, Y., Ye, F., Song, Z., Qin, H., et al., 2007. Directed differentiation of human embryonic stem cells into functional hepatic cells. Hepatology 45, 1229–1239.

D'Amour, K.A., Agulnick, A.D., Eliazer, S., Kelly, O.G., Kroon, E., Baetge, E.E., 2005. Efficient differentiation of human embryonic stem cells to definitive endoderm. Nat. Biotechnol. 23, 1534–1541.

Hay, D.C., Fletcher, J., Payne, C., Terrace, J.D., Gallagher, R.C., Snoeys, J., et al., 2008a. Highly efficient differentiation of hESCs to functional hepatic endoderm requires Activin A and Wnt3a signaling. Proc. Natl. Acad. Sci. USA 105, 12301–12306.

Liu, H., Kim, Y., Sharkis, S., Marchionni, L., Jang, Y.Y., 2011. In vivo liver regeneration potential of human induced pluripotent stem cells from diverse origins. Sci. Transl. Med. 3 (82), 82ra39.

McLean, A.B., D'Amour, K.A., Jones, K.L., Krishnamoorthy, M., Kulik, M.J., Reynolds, D.M., et al., 2007. Activin A efficiently specifies definitive endoderm from human embryonic stem cells only when phosphatidylinositol 3-kinase signaling is suppressed. Stem Cells 25, 29–38.

Rashid, S.T., Corbineau, S., Hannan, N., Marciniak, S.J., Miranda, E., Alexander, G., et al., 2010. Modeling inherited metabolic disorders of the liver using human induced pluripotent stem cells. J. Clin. Invest. 120 (9), 3127–3136.

Rossi, J.M., Dunn, N.R., Hogan, B.L., Zaret, K.S., 2001. Distinct mesodermal signals, including BMPs from the septum transversum mesenchyme, are required in combination for hepatogenesis from the endoderm. Genes Dev. 15, 1998–2009.

Si-Tayeb, K., Noto, F.K., Nagaoka, M., Li, J., Battle, M.A., Duris, C., et al., 2009. Highly efficient generation of human hepatocyte-like cells from induced pluripotent stem cells. Hepatology.

Si-Tayeb, K., Lemaigre, F.P., Duncan, S.A., 2010. Dev. Cell. 18 (2), 175–189. (Review).

Thomson, J.A., Itskovitz-Eldor, J., Shapiro, S.S., Waknitz, M.A., Swiergiel, J.J., Marshall, V.S., et al., 1998. Embryonic stem cell lines derived from human blastocysts. Science 282, 1145–1147.

Touboul, T., Hannan, N.R., Corbineau, S., Martinez, A., Martinet, C., Branchereau, S., et al., 2010. Generation of functional hepatocytes from human embryonic stem cells under chemically defined conditions that recapitulate liver development. Hepatology 51, 1754–1765.

Vallier, L., Touboul, T., Chng, Z., Brimpari, M., Hannan, N., Millan, E., et al., 2009. Early cell fate decisions of human embryonic stem cells and mouse epiblast stem cells are controlled by the same signalling pathways. PLoS One 4, e6082.

Genetic Manipulation

Transfection of Human Pluripotent Stem Cells by Electroporation

Uma Lakshmipathy

EDITOR'S COMMENTARY

Transfection methods that work for human pluripotent stem cells (hPSCs) have a broad range of uses, such as making reporter lines with fluorescent or other markers for monitoring cellular state *in vitro* and *in vivo*, for introducing reprogramming or transdifferentiation factors, for inhibiting expression of genes using RNAi, for correcting mutations via gene targeting, and for strategies for introducing suicide genes to remove transplanted cells if they malfunction. The challenges for transfection are to deliver the nucleic acid (usually DNA, but possibly mRNA or miRNA) to the cells with sufficient integrity and efficiency. Development of methods for transferring DNA into cells dates back to the 1970s, and transfection tools now include chemical, viral, and physical methods. Several other chapters discuss methods for using viruses and lipids to introduce intact nucleic acids into hPSCs. This chapter focuses specifically on electroporation, which is the most common method historically used for mouse ES cells. The main challenge of adapting electroporation methods designed for mouse ESCs to human pluripotent cells has been overcoming the slower growth rate and greater fragility of hPSCs. The methods described here are the result of a great deal of trial and error in optimizing electroporation protocols for hESCs and hiPSCs.

OVERVIEW

This chapter outlines a protocol for DNA transfection into hPSCs via electroporation. The electroporation method we describe is designed for the Neon™ Transfection System, but is adaptable to other systems. This system uses an electronic pipette tip as a transfection chamber instead of the standard electroporation cuvette, which makes it more flexible for variations in cultures. We also provide alternative methods for other electroporation-based systems

J.F. Loring & S.E. Peterson (eds): Human Stem Cell Manual, Second edition.
DOI: http://dx.doi.org/10.1016/B978-0-12-385473-5.00029-1

and Nucleofection™. Although the work flow is common among the different electroporation methods, some variables are listed which can have an impact on the efficiency of transfection and survival of transfected cells:

- Quality of DNA
- Actively dividing, healthy cells
- Preferable absence of feeders
- Culture conditions that allow complete recovery of transfected cells

PROCEDURES

Preparation of DNA for Transfection

Prepare DNA and resuspend at a concentration of 1 mg/mL using an endotoxin-free DNA purification method.

■ Note

The presence of endotoxin in plasmid DNA preparations is known to result in poor transfection and low cell survival. Experiments with episomal vectors expressing GFP indicate a strong correlation between the level of endotoxin and the ability to obtain stable colonies. ■

Preparation of Cells for Transfection

1. Prior to transfection, culture hPSCs in 60 mm Geltrex®-coated dishes under feeder-free culture conditions, to eliminate feeder fibroblasts cells (see Chapter 3 for feeder-free culture acclimation methods).

■ Note

It is important to eliminate feeder fibroblasts since they can compete for the DNA during transfection. ■

2. Wash the plated hPSCs once with PBS and add 1 mL of TrypLE™ to the cells. After 2 minutes, add 1 mL of medium and gently pipette to resuspend cells using a 5 mL serological pipette.

■ Note

Pipette the cells gently after TrypLE to ensure a single cell suspension without damaging the cells. ■

3. Transfer cells to a 15 mL conical tube and add hPSC medium up to 5 mL.
4. Spin cells at 1000 rpm for 2 minutes at room temperature.
5. Aspirate the medium and cells are ready for assessment of viability and transfection.

■ Note

It is important to use healthy, actively diving cells with over 90% viability. Cell viability can be assessed using trypan blue or LIVE/DEAD® Cell Vitality Assay Kit. It is also important to ensure that the majority of the hPSC population used for transfection is pluripotent. This can be assessed by immunocytochemical staining of pluripotent surface markers such as SSEA4 and Tra-1-60 (see Chapter 16). Use of poor quality cells can lead to massive differentiation and loss of cultures during subsequent recovery. ■

Electroporation Using Neon System

1. Turn on the Neon Transfection System device and adjust parameters to 850 V, 30 ms, 1 pulse.
2. Rinse the hPSCs grown in feeder-free conditions with 2 mL PBS once.
3. Add 1 mL TrypLE per 60 mm dish, incubate for 3–5 minutes. After incubation, use a 5 mL pipette to gently dislodge cells.
4. Transfer cells to a 15 mL conical tube, wash the dish with at least 2 mL PBS and transfer to the same tube.
5. Save a small aliquot of cells for cell counting and for assessing viability using LIVE/DEAD assay kit or by trypan blue.
6. Centrifuge the remaining volume of cells at 1000 rpm for 4 minutes.
7. Remove supernatant and resuspend cells using "Neon Resuspension buffer (R buffer)" at a concentration of 1×10^7 cells/mL. Note that most hPSC lines yield $3–5 \times 10^6$ cells per 60 mm dish. Each electroporation uses 10^6 cells in 100 µL
8. Mix DNA (5 µg DNA per electroporation) and cells (100 µL per electroporation at a density of 1×10^7 cells/mL) in the 1.5 mL microcentrifuge tube.
9. Add 3 mL of "Neon Electrolytic buffer" into the Neon transfection tube, and place the tube in the Neon Transfection System device.
10. Use a "gold tip" to aspirate 100 µL of the DNA-cell mixture and place pipette and gold tip in the station.
11. Click Start and then the flashing arrow button to electroporate cells.
12. Immediately transfer electroporated cells to a 60 mm dish with mitotically inactivated mouse embryo fibroblasts (MEFs) and 5 mL of prewarmed MEF-conditioned medium (MEF-CM). Note that viability is better if hPSCs are seeded on inactivated MEFs.
13. Change medium the next day and observe cells daily to assess cell recovery and transfection efficiency. Expression of reporters such as GFP should be visible 24 hours post-transfection. Expression is robust when cells have recovered, typically by 48 hours (Figure 29.1, left panel). FACS analysis shows heterogeneous GFP expression in the transiently transfected cells (Figure 29.1, middle panel) with the majority of the

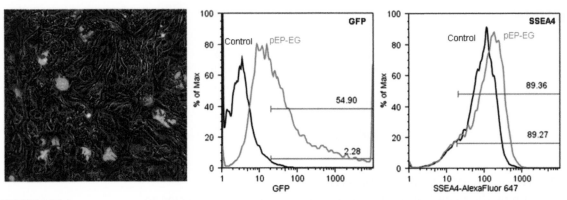

FIGURE 29.1

H9 hESC transfected with an episomal vector Ef1a-GFP plasmid and seeded on MEF and allowed to recover for 48 hours. Left panel: overlap of fluorescence and phase contrast images showing GFP+ cells in some of the recovering ESC colonies on MEFs. Middle panel: cells harvested at 48 hours post-transfection show heterogeneous expression of GFP by FACS analysis. Right panel: assessment of expression of the pluripotency marker SSEA4 shows comparable levels between transfected cells and untransfected controls.

cells expressing the pluripotent surface marker, SSEA4 (Figure 29.1, right panel).

14. Cells transfected with vectors that carry a drug resistance marker can be further subjected to drug selection with the appropriate drug to isolate stable clones. The overall workflow is represented as a schematic in Figure 29.2. A kill curve for each drug for each hPSC line needs to be carried out to determine the optimal drug concentration to be used for selection of cells stably expressing the exogenous transgene. Following 2–3 weeks of selection, drug-resistant colonies begin to appear and if the gene is a visual marker such as GFP, colonies can be monitored for the expression of the reporter. Individual clones can be picked and expanded for further analysis or pooled. Left and right panels (Figure 29.2) show generation of stable BG01V and WA09 hESC clones transfected with an episomal vector carrying Ef1a-GFP.

ALTERNATIVE PROCEDURES

Electroporation Using Standard Electroporator (e.g. BTX ECM630 Electrocell Manipulator)

1. Culture feeder-free hPSCs to 80–90% confluency.

■ Note

Two 60 mm dishes provide enough cells for one electroporation.

Work-flow	BG01V ESC	H9 ESC

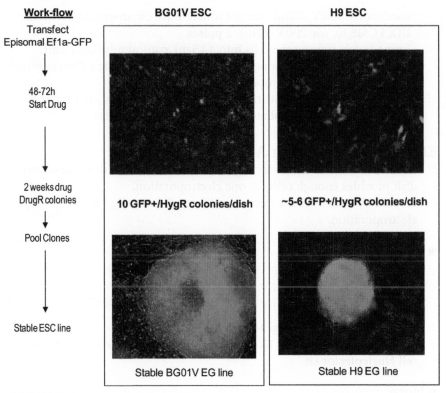

FIGURE 29.2

hESC transfected with episomal vectors can be cultured further in the presence of hygromycin to generate stable pooled clones. Schematic shows the basic workflow. BG01V hESC (left panel) and H9 hESC (right panel) transfected with an episomal vector carrying an Ef1a-GFP transgene. Cells were imaged to confirm GFP expression before initiating drug selection for 2 weeks until robust colonies expressing GFP appeared which were then pooled to create stable hESC clones.

2. Prepare DNA in a 1.5 mL microcentrifuge tube, usually 30 μg total DNA per electroporation.
3. Add 1 mL TrypLE to each 60 mm dish and incubate for 3–5 minutes until cells dislodge.
4. Collect cells and centrifuge to pellet them.
5. Resuspend cells in 1× PBS and recentrifuge. 1–2 cell clusters is optimal.
6. Collect cells in PBS and add to a 1.5 mL microcentrifuge tube that contains DNA.
7. Transfer the cell/DNA mixture to an electroporation cuvette with an electrode gap of 0.4 cm.
8. Electroporate the cells once using a standard electroporator (BTX ECM630, Bio-Rad Gene Pulser® II or Xcell™ System) and the following

conditions: 500 V, 250 μF. For the square wave electroporators such as BTX ECM830, use 200 V, 10 ms, 2 pulses.

9. Immediately transfer the cells into a 15 mL conical tube with 3 mL prewarmed MEF-conditioned medium (MEF-CM) using a Pasteur pipette and incubate at room temperature for 5 minutes.

10. Transfer the above cell mixture into a 60 mm dish with inactivated MEFs and 3 mL MEF-CM. Incubate at 37°C in a CO_2 incubator.

Electroporation Using Amaxa Nucleofector

1. Culture feeder-free hPSCs to 80–90% confluency. Note that one 60 mm dish provides enough cells for one electroporation.

2. Prepare DNA in a 1.5 mL microcentrifuge tube, usually 15–20 μ DNA per electroporation.

■ Note

An 80–90% confluent dish yields about 5×10^6 ESCs and it is optimal to use higher DNA amounts for efficient transfection. Add 1 mL TrypLE to the 60 mm dish and incubate for 3–5 minutes until cells dislodge.   ■

3. Collect cells and centrifuge to pellet them.

4. Resuspend cells in 100 μL Nucleofector solution provided in the mouse ES cell Nucleofector kit.

■ Note

Two human ESC-specific kits are also now available and can be used instead of the mouse ES cell Nucleofector kit (Lonza Cat#VPH-5012 and VPH-5022). Collect cells in Nucleofector solution and add to a 1.5 mL microcentrifuge tube that contains DNA.   ■

5. Transfer the cells and DNA mixture to the supplied cuvette and Nucleofect using the program A27.

■ Note

Programs A23 and B16 can also be used.   ■

6. Resuspend the electroporated cells in 1 mL prewarmed MEF-CM.

7. Transfer the above cell mixture into a 60 mm dish with inactivated MEFs and 3 mL MEF-CM. Incubate at 37°C in a CO_2 incubator.

PITFALLS AND ADVICE

It is important to monitor cell cultures and morphology before and after transfection. If stable transfection is desired, it is important to start with relatively low passage cells and ensure that the starting population of cells is karyotypically normal and truly pluripotent. Cultures also need to be pathogen- and mycoplasma-free. Contamination of transfected cells is a common issue especially since stem cells are cultured in the absence of antibiotics. This can be avoided by using clean DNA and good aseptic techniques.

EQUIPMENT

- Tissue culture hood: Class II A/B3
- Tissue culture incubator, 37°C, 5% CO_2, in humidified air
- Inverted phase/contrast microscope with 4×, 10×, and 20× objectives
- Low-speed centrifuge 300–1000 rpm
- Water bath, 37°C
- Pipettors, such as Eppendorf p2, p20, p200, p1000
- Pipette aid, automatic pipettor for use in measuring and dispensing media
- Hemocytometer
- Vortex mixer
- Aspirator in the hood, with flask
- Refrigerator at 4°C
- Neon Transfection System Device Life Technologies # MPK5000
- BTX ECM630 Electroporator
- Fluorescence microscope
- Freezers: −20°C, −80°C

REAGENTS AND SUPPLIES

Supplies

- 5 mL, 10 mL, 25 mL sterile disposable pipettes
- 6-well culture dishes
- 15 mL and 50 mL sterile conical tubes
- 1.5 mL sterile Eppendorf tubes
- Sterile 9″ pasture pipettes
- STEMPRO® EZPassage™ Tool Life Technologies Cat #23181-010
- Cell scraper Falcon Cat # 353085
- Neon™ transfection System kit Life Technologies Cat # MPK10096
- Neon transfection tubes Life Technologies Cat # MPT100
- Amaxa Nucleofector
- Pipette tips for Eppendorf or similar pipettor

Recommended Reagents

Item	Supplier	Catalog #	Notes
Dulbecco's Modified Eagles Medium	Life Technologies	10569-010	
Dulbecco's Modified Eagles Medium/F12 with GlutaMAX™	Life Technologies	10565-018	
Knock Out™ Serum Replacer (KSR)	Life Technologies	10828-028	
Fetal Bovine Serum (FBS), ESC Qualified	Life Technologies	16141-079	
MEM-Non-essential Amino Acids	Life Technologies	11140-050	
β-mercaptoethanol	Life Technologies	21985-023	
Collagenase Type IV	Life Technologies	17104-019	
Basic fibroblast growth factor (bFGF)	Life Technologies	13256-029	
DPBS with Calcium and Magnesium	Life Technologies	14040-133	
DPBS without Calcium and Magnesium	Life Technologies	14190-144	
GIBCO® Mouse Embryonic Fibroblasts (Irradiated)	Life Technologies	S1520-100	
Attachment Factor	Life Technologies	S-006-100	
TrypLE Select	Life Technologies	12563029	
LIVE/DEAD Cell Vitality Assay Kit	Life Technologies	L34951	
Trypan Blue	Life Technologies	15250061	
Mouse ES Nucleofector Kit	Lonza	VPH#1001	Alternatively, Human ESC nucleofector kits are also available
Bovine Serum Albumin	Life Technologies	A10008-01	
Geltrex™ hESC Qualified	Life Technologies	A10480-01	
Lipofectamine 2000	Life Technologies	11668027	

RECIPES

■ Note

Glassware should be dedicated to tissue culture only. If glassware is to be used instead of pre-sterilized plastic ware, do not expose bottles to detergent.   ■

Stock Solution Human Basic FGF (bFGF) 10 μg/mL 1 ml

Component	Amount	Stock Concentration
Human bFGF	10 μg	10 μg/mL PBS with 0.2% BSA

1. Dissolve 10 µg of human basic FGF in 1 ml PBS containing 0.2% BSA.
2. Aliquot in 50–100 µL samples.
3. Store thawed aliquots at 4°C for up to 2 weeks.
4. Store frozen aliquots at −20°C or −80°C.

■ Note

For all growth factors, pre-wet all pipette tips, tubes, and filters with PBS + 0.2% BSA to lessen the loss of the growth factor.   ■

Additional Information

■ Official Symbol: FGF2 and Name: fibroblast growth factor 2 (basic) [*Homo sapiens*]
■ Other Aliases: BFGF, FGFB, HBGH-2
■ Other Designations: basic fibroblast growth factor; basic fibroblast growth factor bFGF; fibroblast growth factor 2; heparin-binding growth factor 2 precursor; prostatropin
■ Chromosome: 4; Location: 4q26-q27.

MEF Medium 100 mL

Component	Amount	Stock Concentration
DMEM	89 mL	
FBS, ESC qualified	10 mL	
MEM-Non-essential amino acids	1 mL	10 mM
β-mercaptoethanol	182 µL	55 mM

ESC Medium 100 mL

Component	Amount	Stock Concentration
DMEM	79 mL	
FBS, ESC qualified	20 mL	
MEM-Non-essential amino acids	1 mL	10 mM
β-mercaptoethanol	182 µL	55 mM
Basic FGF	40 µL	10 µg/mL

Preparation of MEF Conditioned Medium (MEF-CM)

1. Plate 9.4×10^6 mitomycin C treated MEFs in a T175 flask coated with 0.1% gelatin in MEF medium.
2. The following day, replace the MEF medium with 90 mL ESC medium.
3. Collect MEF-CM from the flasks after 24 hours of conditioning for up to 7 days.

4. Filtered CM can be stored at $-20°C$ until use.
5. At time of use, freshly supplement CM with extra GlutaMAX (2 mM) and bFGF (4 ng/mL).

Preparation of Geltrex-Coated dishes

1. Thaw a 5 mL bottle of Geltrex (hESC Qualified) at 2–8°C overnight.
2. Dilute 1:1 with cold DMEM/F-12 to prepare 1 mL aliquots in tubes chilled on ice. These can be frozen at $-20°C$ or used immediately.
3. Dilute 1:100 with cold DMEM on ice.

■ Note

An optimal dilution of Geltrex may need to be determined for each line. Try dilutions from 1:30 to 1:200. ■

4. Cover the whole surface of each culture dish with the diluted Geltrex.
5. Incubate in a 37°C, 5% CO_2 incubator for 1 hour.

■ Note

Dishes can now be used or stored at 2–8°C for up to a week. Do not allow dishes to dry out. ■

6. Pipette out diluted Geltrex from the culture container and discard. Cells can be passaged directly into MEF CM on Geltrex coated containers.

■ Note

You do not need to rinse off the Geltrex from the culture container. ■

QUALITY CONTROL METHODS

DNA

Plasmid DNA used for transfection needs to be validated prior to use in stem cells. This includes preparation of high quality DNA with 260/280 of above 2, checking proper cloning of the gene of interest using restriction enzyme digests and transfection into a workhorse cell line such as HEK293 to determine functional expression of the gene.

Cells

Cells used for transfection must be handled carefully and if stable clones are being generated, it is critical to use lower passage number cells. Standard practices such as analysis to confirm normal karyotype and pathogen/myco-plasma-free cultures are important. Cells grown on feeders can be differentially harvested and directly used for transfection but since contaminating fibroblasts have greater transfection efficiency, excessive fibroblasts in hESC samples could compete for the DNA. It is best if cells are grown under feeder-free conditions for at least one generation. It is important to use cultures with over 90% pluripotent cells, based on surface marker expression. Use of poor quality hESC cells can lead to low transfection efficiencies, poor recovery or excessive cell death.

Transfection and Recovery Conditions
The electric parameters suggested here should work for most hESC and iPSC lines. It is always important to include controls.

1. No DNA, no transfection control: determine plating and quality of cells used for transfection
2. No DNA control: determine recovery of cells after transfection
3. GFP control: if using a target gene that cannot be visualized, use of GFP controls helps determine the efficiency of transfection. While comparing efficiencies, note that transfection efficiency is dependent on DNA size.

READING LIST

Gene Delivery Methods
Heiser, W.C. (Ed.). Methods in Molecular Biology, vol. 245. Gene Delivery to Mammalian Cells: nonviral Gene Transfer Techniques, Vol.1. Human Press Inc., Totowa, NJ.
  This book describes nonviral methods of gene delivery in detail.

Neon Electroporation System
The first paper described the Neon™ Transfection system in detail. The remaining two papers utilize this method to create stable clones in human embryonic stem cells.

Kim, J.A., Cho, K., Shin, M.S., et al., 2008. A novel electroporation method using a capillary and wire-type electrode. Biosen Bioelectron. 23 (9), 1353–1360.

Liu, Y., Thyagarajan, B., Lakshmipathy, U., Xue, H., Lieu, P., Fontes, A., et al., 2009. Generation of a platform human embryonic stem cell line that allows efficient targeting at predetermined genomic location.. Stem Cells Dev. 18 (10), 1459–1472.

Thyagarajan, B., Scheyhing, K., Xue, H., Fontes, A., Chesnut, J., Rao, M., et al., 2009. A single EBV-based vector for stable episomal maintenance and expression of genes in human embryonic stem cells. Regen. Med. 4 (2), 239–250.

Electroporation
Thyagarajan, B., Liu, Y., Shin, S., Lakshmipathy, U., Scheyhing, K., Ellerstrom, C, et al., 2008. Creation of engineered human embryonic stem cell lines using PhiC31 integrase. Stem Cells 26, 119–126.
  This paper describes creation of stable clones using phiC31 integrase using electroporation.

Nucleofection
Lakshmipathy, U., Pelacho, B., Sudo, K., Linehan, J.L., Coucouvanis, E., Kaufman, D.S., et al., 2004. Efficient transfection of embryonic and adult stem cells. Stem Cells 22, 531–543.
  This is one of the earlier papers that describe nucleofection of mouse and human embryonic stem cells.

CHAPTER 30

Lentiviral Vector Systems for Transgene Delivery

Pauline Lieu, Bhaskar Thyagarajan, Uma Lakshmipathy,
Jonathan Chesnut, and Ying Liu

EDITOR'S COMMENTARY

Methods for efficiently introducing DNA transgenes are vital for much of the future expansion of human pluripotent stem cell (hPSC) tools into broader areas of research and clinical applications. The challenges of introducing intact DNA into hPSCs are discussed in several chapters in this book (Chapters 29–31). Because of their high affinities for certain human cells and efficiency of infection, a great deal of effort has been invested in developing viral vectors for *in vivo* gene therapy. Viral vector methods were not used extensively for genetically modifying mouse ES cells, but they have been vital for introducing genes into human cells. For example, viruses have been particularly effective for delivering reprogramming factors to human (and animal) cells and continue to be the most reliable approach for making hiPSCs (Chapter 10). One of the most versatile viruses for introducing transgenes into hPSCs is replication-defective lentivirus. This chapter describes two approaches to insert transgenes into the genome. The first approach is to use viral vectors to deliver DNA that randomly integrates into the genome; this approach has the advantage of being relatively simple. The disadvantages of random integration are the possibility that the transgenes become epigenetically inactivated or that they may insert into coding regions and disrupt genes. The second approach, although more complicated, avoids the problems of random insertions by directing the transgene to a specific locus in the genome that is transcriptionally active.

Note that for this chapter, like several other chapters, we have asked professionals in this particular technology to outline their best techniques that they have experimentally developed and/or optimized. It is presented as a straightforward method that works well, which is the purpose of all of the methods described in this book. Because of the focus on optimized reagents and kits,

463

J.F. Loring & S.E. Peterson (eds): Human Stem Cell Manual, Second edition.
DOI: http://dx.doi.org/10.1016/B978-0-12-385473-5.00030-8

this chapter references products that are unique to a specific vendor and may contain proprietary components.

OVERVIEW

The powerful delivery capabilities of viral vectors have made them a prime tool in gene delivery for both basic research and therapy applications. Viral delivery tools combine both the cloning vector and delivery mechanism and they can target a wide variety of cell types both *in vitro* and *in vivo* with high efficiency (Park and Kay, 2001). The lentivirus, specifically, has become a favored gene therapy tool for its unique ability to transduce dividing as well as terminally differentiated, non-dividing cells (Naldini et al., 1996; Pandya et al., 2001; Zufferey et al., 1998) and to generate stable gene expression (Zhou et al., 2003). These features are especially critical when attempting to genetically modify stem cells, hematopoetic cells, neurons, and various tissues *in vivo*. Here we describe procedures to generate high titer lentivirus and transduce hPSCs. We then detail a method for site-specific integration of lentiviral vectors.

PROCEDURES

Cloning Gene of Interest into Lentiviral Expression Construct

These procedures are optimized for a specific set of reagents (Life Technologies), but can be adapted for reagents from other sources. This protocol is based on reagents from the ViraPower™ HiPerform™ Lentiviral Expression kits.

1. Amplify your gene of interest by PCR.
2. Clone your insert into Gateway® or TOPO® vectors of the lentiviral Expression Construct (Figure 30.1).

■ Note

TOPO vectors have topoisomerase I covalently bound to each 3′ phosphate. This enables the vectors to readily ligate DNA sequences with compatible ends. The Gateway vectors allow transfer of DNA fragments between different cloning vectors while maintaining the reading frame. They have a proprietary set of recombination sequences that are ligated to the PCR product, and the Gateway kit has proprietary enzyme mixes. ■

3. Generate large-scale prep of the pLenti-expression construct containing your gene of interest using an endotoxin free plasmid kit to enhance viral production.

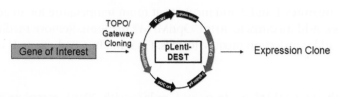

FIGURE 30.1
Generation of lentiviral Expression Construct. Clone gene of interest into the pLenti-expression construct via TOPO or Gateway vector. This construct can be used to produce a lentiviral stock.

4. Verify that your lentiviral plasmid has not undergone aberrant recombination by performing sequencing or restriction digests.

Producing Lentivirus in 293FT Cells

Before creating a stable transduced cell line expressing your gene of interest, you first need to produce a lentiviral stock by cotransfecting a packaging plasmid mix containing the pLP1, pLP2, pLP/VSVG plasmids and your lentiviral Expression Construct into the 293FT cell line.

■ Note

The 293FT line is a fast-growing variant of the original 293 line, and expresses SV40T antigen driven by the CMV promoter. ■

1. The day before transfection, plate 293FT cells in a 10 cm plate so that they will be 85–90% confluent the next day, in the absence of antibiotics. Incubate the cells overnight at 37°C in a humidified 5% CO_2 incubator.
2. Next day, replace the culture medium with 5 mL Opti-MEM® Medium containing 5% serum in the absence of antibiotics.
3. Prepare transfection mixture as described below:

Day 1
Per 10 cm plate combine:

Mixture 1:

- 500 µL of Opti-MEM
- 36 µL of Lipofectamine® 2000
- Incubate at room temperature for 5 minutes.

Mixture 2:

- 500 µL of Opti-MEM
- 3 µg of vector DNA (lentiviral Expression Construct)
- 9 µg Packaging mix.

Combine mixtures 1 and 2 and incubate at room temperature for an additional 15 minutes. Add mixtures to 10 mL OptiMEM® medium. Remove medium from cells, and add transfection mixture to cells. Final medium volume is 11 mL.

Day 2
Aspirate the OptiMEM® medium and replace with 20 mL complete medium: DMEM + 10%FBS, 1% pen/strep, 1× L-glutamine, 1× non-essential amino acids, and 1× sodium pyruvate.

■ Note
Expression of the VSVG glycoprotein causes 293FT cells to fuse, resulting in the appearance of large, multinucleated cells known as syncytia. This morphological change is normal and does not affect lentivirus production. ■

Day 3
1. Collect Lentivirus supernatants in 50 mL conical tubes.
2. Spin at 2000 rpm for 10 minutes to remove cell debris.
3. Filter the viral supernatants through a Millex R-HV 0.45 μm filter or an equivalent PVDF filter.
4. Aliquot into cryovials in 1 mL aliquots.
5. Store aliquots at −80°C.

■ Note
Repeated freeze–thaw of virus will reduce viral titers. ■

Concentrating and Titering Viral Stocks
It is possible to concentrate VSVG pseudotyped lentiviruses using a variety of methods without significantly affecting their ability to transduce cells. If your cell transduction experiment requires a relatively high Multiplicity of Infection (MOI), concentrate your virus before titering and proceeding to transduction. For details and guidelines on how to concentrate your viral supernatant by ultracentrifugation, refer to published reference sources (Yee, 1999). A guide to titering viruses is provided in the manual for the ViraPower HiPerform Lentiviral Expression kit.

Transduction of hPSCs
1. Coat your plate with Geltrex® (1:30 dilution), and leave plate for 1 hour at room temperature.
2. Split the hiPSCs from feeder plates using 1 mg/mL Collagenase IV solution by following standard protocols (see Chapter 1).

3. Aspirate the medium from the Geltrex plate and wash the wells once with PBS.
4. Plate the cells onto Geltrex-coated plates with 2.5 mL of medium designed for feeder free culture (e.g. Stempro® or TeSR®).
5. When hPSCs reach 50–80% confluence, add concentrated lentivirus ($\sim10^8$ transfection units: TU) into 2 mL of medium (for a 60 mm plate) in the presence of 4–8 µg/mL of polybrene.
6. Incubate the hiPSCs with viral supernatant for 4–6 hours in a tissue culture incubator.
7. Aspirate the virus-containing medium and add 2.5 mL medium.
8. Cells can be analyzed 3–5 days after transduction for transgene expression.

Spin-Inoculation Method

1. Coat your plate with Geltrex (1:30 dilution), and leave plate for 1 hour at room temperature.
2. Split the hiPSCs from feeder plates using 1 mg/mL Collagenase IV solution by following standard protocols.
3. Aspirate the medium from the Geltrex plate and wash the wells once with PBS.
4. Plate the cells onto Geltrex-coated plates (one 6-well plate) with 2 mL of medium designed for feeder-free culture (e.g. StemPro or TeSR). When hESCs reach 50–60% confluence, add concentrated lentivirus ($\sim10^8$ TU) into 1 mL of medium.
5. Centrifuge at 2000 g for 45 minutes using a bench-top centrifuge at room temperature.
6. After centrifugation, add equal volume of growth factor-containing media to the well to dilute the viruses and polybrene. Incubate overnight.
7. Next day, replace virus with feeder-free medium with growth factors.
8. Cells can be analyzed 3–5 days after transduction for transgene expression.

■ Note

The quality and concentration of DNA used plays a central role in the efficiency of transfection. It is crucial that the DNA is free of endotoxins. If viral titer is low, make sure the density of 293FT is at 90% confluence at the time of transfection. If you see low transduction efficiency, your viral supernatant is too diluted and needs to be concentrated. In addition, do not freeze/thaw virus more than three times. Virus should be stored at −80°C. ■

Site-Specific Integration

To target integration to specific sites in the genome, we have developed a site-specific integration platform, mediated by PhiC31 and R4 integrase. The process starts with the generation of a platform cell line with a single integration of the R4 *attP* site in the cell's genome, mediated by the PhiC31 integrase, and is followed by retargeting the gene of interest to the specific R4 *attP* site mediated by the R4 integrase (Figure 30.2). The PhiC31 integrase differs from the better-known recombinases such as Cre and Flp, in that it catalyzes recombination between two non-identical sites. In addition, unlike other recombinases, it lacks the excisionase enzyme, which makes the integration events catalyzed by PhiC31 unidirectional and virtually irreversible (Thyagarajan et al., 2001, 2008). This technology is described in detail in the Jump-In™ TI™ (Targeted Integration) information manual (http://tools.invitrogen.com/content/sfs/manuals/jump_in_ti_gateway_system_manual.pdf).

Generation of the R4 Platform Line

In order to target your gene of interest to a specific site in the genome, you first need to generate a platform cell line with a designated target sequence. The first step in generating the platform line is by co-transfecting the pJTI™ PhiC31 Int vector (expressing the PhiC31 integrase) and one of the pJTI™ platform vectors, containing target sequences (e.g. *attP*) and the Hygromycin B resistance gene to allow for selection of cells containing the R4 target sequences. In addition to these elements, the pJTI platform vector also contains the sequences for resistance against a second selection agent for selecting the retargeting event with the gene of interest (blasticidin, neomycin, or zeocin resistance genes in pJTI/Bsd, pJTI/Neo, or pJTI/Zeo, respectively), but lacks the promoter

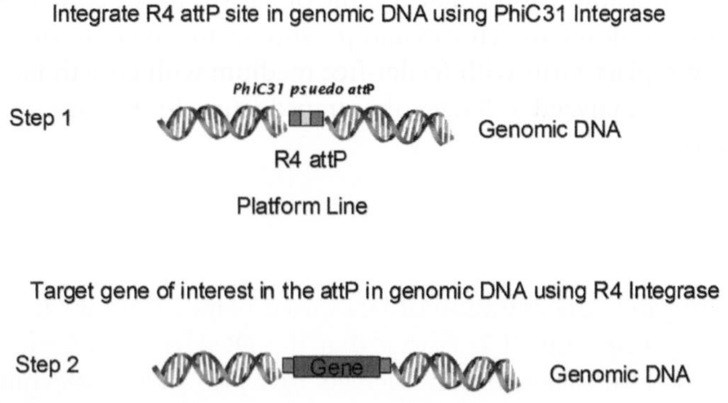

FIGURE 30.2

Overall scheme in engineering cells by site-specific integration by PhiC31 and R4 Integrase. The first step begins with the generation of the R4 platform line containing the R4 *attP* targeting site. The second step results in the retargeting of the gene of interest to the R4 *attP* site.

to express this resistance gene. Transformants containing the desired R4 "*attP* retargeting sequences" are selected using Hygromycin B, and selected for the second selection agent in the retargeting event (Figure 30.3).

Determine Hygromycin B Sensitivity

To successfully generate an R4 platform cell line containing the R4 *attP* retargeting sequence, you need to determine the minimum concentration of Hygromycin B required to kill your untransfected cells. Typically, concentrations ranging from 10 to 400 µg/mL of Hygromycin B are sufficient to kill most untransfected mammalian cell lines. We recommend that you test a range of concentrations to determine the minimum concentration necessary for your cell line of choice.

Culture and Expansion of hPSCs on Feeders

Recommendations for preparing pluripotent cells before transfection:

1. Thaw your hPSC cells onto inactivated MEFs in a 60 mm dish.
2. Expand your cells using collagenase (1 mg/mL) and plating onto inactivated MEFs for 1–2 passages.
3. Passage cells from one 60 mm dish into two 60 mm dishes on Geltrex (feeder free) in MEF-conditioned medium.
4. Passage one more time onto Geltrex (feeder free) in MEF conditioned medium, and expand 2–4 (60 mm) dishes of cells for transfection.
5. When cells become 85% confluent, they are ready to be transfected by electroporation (see Chapter 29).

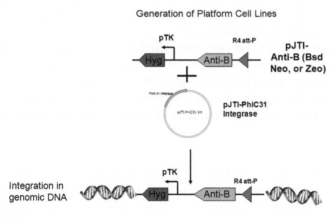

FIGURE 30.3

Generation of the R4 platform line. A plasmid (pJTI/BSD or NEO or ZEO designated as Anti-B) carrying different drug resistant markers is used to generate the re-targeting platform line by introducing an R4 retargeting *attP* site mediated by PhiC31 integrase.

Transfection of Pluripotent Cells with the pJTI and the pJTI PhiC31 Int Vector

Determine the best transfection method for your cell line. Here are some general recommendations that work for the WA09 hESC line with the BTX ECM630 electroporator or the Neon™ microporator.

1. The day before transfection, seed Hygromycin B-resistant MEFs onto 60 mm Geltrex-coated dishes at a density of 3×10^4 cells/cm^2.

2. Prepare DNA in a 1.5 mL microcentrifuge tube, usually 30 μg total DNA per transfection (15 μg pJTI platform vector and 15 μg pJTI™ PhiC31 Int vector). For a transfection control, we recommend using a GFP expression vector starting at 15 μg.

3. Add 5 mL of MEF conditioned medium into a 60 mm dish of inactivated Hygromycin B-resistant MEFs and place back in the incubator.

4. Add 2 mL TrypLE Express to each 60 mm dish of hESCs and incubate for 3–5 minutes at room temperature until the cells dislodge.

5. Gently dislodge cells and pipette up and down several times.

6. Transfer to a 15 mL conical tube, wash with 3 mL of D-PBS (no Mg^{2+}, Ca^{2+}) and centrifuge at 200g for 4 minutes to pellet cells. Repeat 2 times.

7. Resuspend cells in 1 mL of D-PBS (no Mg^{2+}, Ca^{2+}) and take a small aliquot to determine the number of live cells by trypan blue staining.

8. Adjust with D-PBS to obtain 2×10^6 cells/mL when using the BTX electroporator or 1×10^6 cells/mL when using the Neon™ instrument.

9. Add 2×10^6 cells into the 1.5 mL microcentrifuge tube that contains DNA, and transfer the mixture to an electroporation cuvette with a gap of 0.4 cm for the BTX ECM630 electroporator. Alternatively, for the Neon instrument, adjust parameters to 850V, 30 ms, 1 pulse and use 1×10^6 cells/mL with 10 μg of total DNA (5 μg of the pJTI platform vector and 5 μg the pJTI PhiC31 Int vector).

10. Electroporate the cells once using the BTX ECM630 electroporator, with the following conditions: 200V, 10 ms, 2 pulses. For the Neon instrument, adjust parameters to 850V, 30 ms, 1 pulse.

11. Immediately transfer the cells into the 60 mm dish with inactivated MEFs and conditioned medium and incubate in a 37 °C cell culture incubator.

■ Note

The quality and concentration of DNA used plays a central role for the efficiency of transfection. It is crucial that the DNA is free of endotoxins. If using large quantities of DNA, we recommend using commercially prepared plasmid DNA. For smaller quantities, use a commercial kit that delivers pure DNA that is free of endotoxins. Do not precipitate DNA with ethanol to concentrate because it reduces efficiency and viability due to the salt contamination. ■

Selection for Stable Integrants Containing the pJTI Platform

1. Change the medium with fresh MEF conditioned medium the next day.
2. Monitor expression of the GFP control 2 days after transfection under the microscope (Figure 30.4).
3. Start selection when cells reach ~80% confluence with MEF-conditioned medium containing 10 μg/mL Hygromycin (or the appropriate concentration for your cell line).
4. Feed the cells with selective medium every other day until foci can be identified. After foci are identified, change the selective medium every day. Depending on the cell line, colonies will start appearing as early as day 5 of drug selection. Mark the colonies and observe them for an additional period of time (total of 12–21 days under selection).
5. Manually pick single, well-defined colonies and expand using the appropriate selection medium for further analysis.

■ Note

Depending on your purpose, the PhiC31 integrase targets the pJTI plasmid to pseudo *att* sites in the cellular genome. If you wish to identify the

Transfection with a GFP Vector - 24 hours post transfection

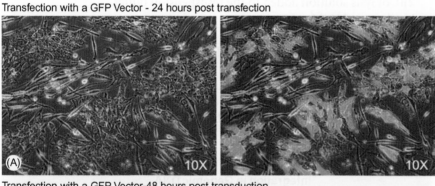

Transfection with a GFP Vector-48 hours post transduction

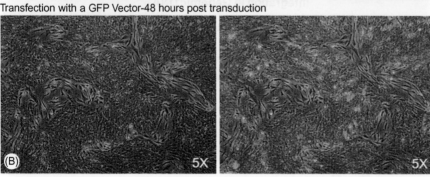

FIGURE 30.4
Transfection with a GFP vector. The pJTI™ R4 vector containing the GFP gene was transfected into hESCs. GFP expression can be observed after 24 hours (A) or 48 hours (B) post transfection.

number of copies of the pJTI plasmid or the specific site in which it integrated in your clones, you will need to expand your clones and perform further analyses as suggested below.　■

Screening R4 Platform Cell Line Clones

We have identified some pseudosites that are considered hotspots for the PhiC31 recombinase (Thyagarajan et al., 2001). Specifically, we have identified a specific pseudosite on chromosome 13q32.3 located in an intronic region of the CYLBL gene (Figure 30.5). We retargeted these clones with a GFP reporter gene and showed that GFP expression is sustained and appropriately regulated over long-term culture, upon random differentiation, and directed induction into neural lineages (Liu, 2009, 2010). If you wish to identify clones that contain the pJTI plasmid integrated in the same pseudosite on chromosome 13q32.3, perform PCR to identify clones that contain a positive PCR band for chromosome 13. See Table 30.1 for primer sequences.

Screening for R4 attP Integration on Chromosome 13

1. Pellet 10,000 to 30,000 cells total.
2. Wash with 500 μL of PBS.
3. Spin down and remove PBS.
4. Resuspend cell pellet in 20 μL of Cells Direct Resuspension Buffer with 2 μL of lysis solution added to the resuspension buffer.
5. Incubate at 75°C for 10 minutes.
6. Spin down 1 minute.
7. Perform a PCR reaction using AccuPrime™ Taq High Fidelity.
8. Add 5 μL of 10× PCR buffer II, 1 μL of Frw primer (10 μM stock, see Screening retargeted clones for sequence), 1 μL of Rev primer (10 μM stock, see Screening retargeted clones for sequence), 1 μL of Taq polymerase, 3 μL of cell lysate, 39 μL of H_2O.

FIGURE 30.5

Screening for pseudosite on chromosome 13q32.3. One pseudosite hotspot identified by Thyagarajan et al. (2008) is on chromosome 13q32.3 located in an intronic region of the CYLBL gene. See Table 30.1 for primers to detect integration at this locus.

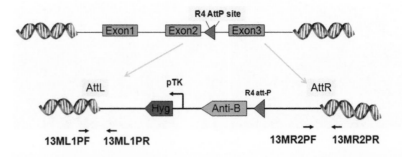

9. PCR cycle conditions: 94°C for 2 minutes, 94°C for 30 seconds, annealing temperature (Table 30.1) for 30 seconds, 72°C for 1 minute, 35–40 cycles, 72°C for 7 minutes.
10. Run reaction on an agarose gel and the expected size of the band is shown in Table 30.1.

■ **Note**

Perform PCR to identify clones that contain a positive PCR band for chromosome 13. After identification, if you wish to confirm that there is only one copy of the pJTI plasmid integrated in your cell's genome, you can use Southern blot analysis to determine the number of integrations in each of your Hygromycin B-resistant clones. ■

Identification of Single Copy Integration by Southern Blot

1. Probe design: we recommend that you use a fragment of the Hygromycin B resistance gene (~1 kb) as the probe to screen your samples. You may amplify the Hygromycin expression cassette from one of the pJTI platform vectors using the appropriate primers. To label the probe, we generally use a standard random priming kit such as the High Prime Kit (Roche). For genomic DNA isolation we recommend using the DNAzol® Reagent to isolate the genomic DNA from the Hygromycin B-resistant clones.
2. Restriction digest: when choosing a restriction enzyme to digest the genomic DNA, we recommend choosing an enzyme that cuts at a single known site outside of the Hygromycin resistance gene in the pJTI platform vector used (such as *Bam*H I or *Hind* III). Hybridization of the Hygromycin probe to the digested DNA should then allow you to detect

Table 30.1 Primer Sequences and Expected PCR Band Sizes for Detection of the Hygromycin B (Hyg) Resistance Gene and One Pseudosite Hotspot Identified by Thyagarajan et al. (2008) on Chromosome 13q32.3 Located in an Intronic Region of the CYLBL Gene

Primer	Sequence	Annealing Temp.	Expected Size
Hyg (Frw)	ATGAAAAAGCCTGAACTCACC	52°C	430 bp
Hyg (Rev)	ATTGACCGATTCCTTGCG		
Chr13 Minus AttL (Frw)	GGGCCAAAGAGAGCAGTAAGCAG	58°C	402 bp
Chr13 Minus AttL (Rev)	AGACAAGCTGTGACCGTCTCCG		
Chr13 Minus AttR (Frw)	AGAACCCGCTGACGCTGCCC	55°C	628 bp
Chr13 Minus AttR (Rev)	GTGGACATGTGTAGCACGCCTGTGC		
Chr13 Plus AttL (Frw)	AGTTAAGCCAGCCCCGACAC	58°C	592 bp
Chr13 Plus AttL (Rev)	TTTTGGCTACCAGTACTAGGCAGG		
Chr13 Plus AttR (Frw)	ACAGAGGGAGCCAACTATTTGCT	52°C	726 bp
Chr13 Plus AttR (Rev)	GGTTATTGTCTCATGAGCGGATACA		

a single band containing the Hygromycin resistance gene from the pJTI™ platform vector if only one integration event has occurred.

3. Southern blot protocol: you may use any Southern blotting protocol. Refer to Chapter 31 for detailed protocols. Protocols are also provided in *Current Protocols in Molecular Biology* (Ausubel et al., 1994) or *Molecular Cloning: A Laboratory Manual* (Sambrook et al., 1989).

■ Note

If you digest genomic DNA from your transfectants with an appropriate restriction enzyme that cuts at a single known site outside the Hygromycin resistance gene, and use a Hygromycin resistance gene fragment as a probe in your Southern analysis, you should be able to easily distinguish between single and multiple integration events. DNA from a single integrant should contain only one hybridizing band corresponding to a single copy of the integrated pJTI platform vector. DNA from multiple integrants should contain more than one hybridizing band. If the pJTI platform vector integrates into multiple chromosomal locations, the bands may be of varying sizes.   ■

Mapping Site of Integration

If you wish to identify integrants at other pseudosites, you can perform plasmid rescue to map the site of integration of the pJTI vector plasmid, Figure 30.6.

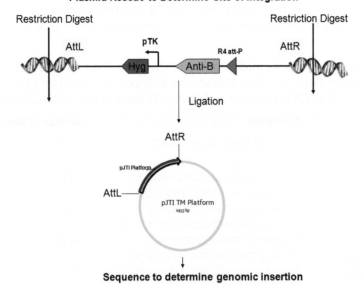

FIGURE 30.6

Plasmid rescue scheme to determine site of integration. To identify other sites of integration, perform plasmid rescue and determine the site of integration by sequencing.

Plasmid Rescue Assay

1. Isolate genomic DNA from individual Hygromycin B-resistant clones grown to confluency using your preferred method.
2. Digest the genomic DNA with a restriction enzyme that does not cut within the pJTI platform vector you have used. Stop the restriction digest by heat inactivation. If the restriction enzyme cannot be heat-inactivated, perform a phenol:chloroform extraction of the genomic DNA and ethanol precipitate.
3. Incubate the restriction fragments with T4 DNA ligase overnight at 16°C under dilute conditions that favor self-ligation.
4. Extract the DNA from the ligation mixture with phenol:chloroform, ethanol precipitate the DNA, and resuspend in water.
5. Electroporate a fraction (25%) of the ligated DNA into DH10B™-T1R electrocompetent *E. coli* using the recommended conditions for the electroporator.
6. Plate electroporated cells on LB-agar plates containing 100 µg/mL ampicillin.
7. Isolate the plasmid DNA from resulting colonies, and sequence with the following primer for the PhiC31 *att*B site: 5′-TCC CGT GCT CAC CGT GAC CAC-3′.
8. Determine the genomic integration site by matching the sequence read to the database at BLAT (www.genome.ucsc.edu/cgi-bin/hgBlat).

Retargeting the Gene of Interest in the R4 Platform Cell Line

Once you have established the R4 platform cell line and confirmed that a single integration event has occurred, you may proceed to retargeting your platform line by cotransfecting the "retargeting expression construct" and the pJTI R4 Integrase vector to generate a stable, isogenic cell line expressing your genetic elements of interest. We recommend the MultiSite Gateway® Technology or GeneArt Seamless Cloning Technology to generate your retargeting constructs.

Constructing the Retargeting Expression Vector

Figure 30.7 shows how the MultiSite Gateway platform can be used to generate your entry clone with one or more elements, followed by generating the final DEST vector to retarget into the R4 platform line.

Targeting the Gene of Interest in the R4 attP Site

The retargeting event is mediated by the R4 integrase expressed from the pJTI R4 Int vector. At this step, the gene of interest is specifically integrated into the platform line genome at the R4 *att*P target site (introduced into the

FIGURE 30.7
Generation of retargeting
expression vector.
Using Multisite Gateway
Technology, clone your
gene of interest in an entry
clone, (A), and subsequently
generate the retargeting
construct by performing
the second cloning reaction
with the pJTI R4 DEST
vector, (B).

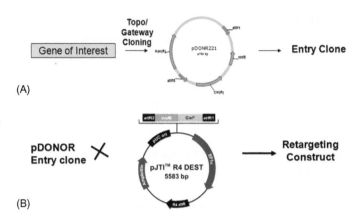

cell line at the first step). This integration event also positions the constitutive human EF1α promoter upstream of the blasticidin, neomycin, or zeocin resistance gene (i.e. "promoterless" selection marker), thus allowing the selection of successfully "retargeted" transformants using the appropriate selection agent (Figure 30.8). For more information on PhiC31 and R4 Integrases, and their uses in targeted integration, refer to Thyagarajan et al. (2001, 2008).

■ Note
After obtaining your retargeting vector, we recommend you use the same procedures for preparing plasmid DNA, transfection methods, maintenance of cells, and determination of antibiotic sensitivity as described above for generating the retargeting line. Depending on which pJTI™ R4 DEST vector you choose (blasticidin, neomycin, or Zeocin resistance genes in pJTI/Bsd, pJTI/Neo, or pJTI/Zeo, respectively), you first need to determine the antibiotic sensitivity of our platform line as described above. ■

Transfection Procedure
1. Passage your platform cell line at least once without Hygromycin B selection prior to transfection.
2. Remember to include negative controls where either the R4 integrase vector or the retargeting expression construct is omitted.
3. It is also important to include a positive control, such as a GFP vector, to gauge transfection efficiency and colony formation.
4. Plate the transformed cells in 60 mm culture dishes containing the appropriate medium and allow the cells to recover without selection until the colonies become well-defined.
5. Follow transfection procedures as described in the section Transfection of pluripotent cells with pJTI platform and the pJTI PhiC31 Int vector.
6. Culture cells in the absence of antibiotics until cells have recovered after transfection (see Figure 30.4).

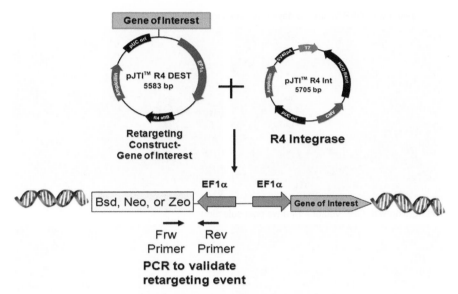

FIGURE 30.8
Target gene of interest in the specific R4 *attP* site. The retargeting event is mediated by co-transfecting the pJTI™ R4 vector with your gene of interest and the R4 integrase. This is followed by screening with the appropriate selection agent to isolate the retargeted clone.

7. Wash the cells and provide fresh medium every day.
8. Each colony recovers at a different rate. Monitor the morphology and size of each colony, and when they reach the confluency shown in Figure 30.4B, start selection with the proper antibiotic.
9. Feed the cells with selective medium every other day until foci can be identified. After foci are identified, change the selective medium every day. Depending on the cell line, colonies will start appearing as early as day 6 of drug selection (Figure 30.9A). Mark the colonies and observe them for an additional period of time (total of 12–21 days under selection).
10. When colonies have fully formed (Figure 30.9B), manually pick single, well-defined colonies and expand using the appropriate medium under selection for further analysis.
11. We recommend that you continue with the Blasticidin, Geneticin®, or Zeocin-based selective pressure even after your retargeted clones have been selected and expanded for downstream experiments. Continuous selective pressure ensures that expression from your gene(s) of interest is maintained.

■ Note

The quality and the concentration of DNA used play a central role in the efficiency of transfection. It is crucial that the DNA is free of endotoxins. If using large quantities of DNA, we recommend using commercially-prepared plasmid DNA. For smaller quantities, use a commercial kit that delivers pure DNA that is free of endotoxins. Do not precipitate DNA with ethanol to concentrate because it reduces efficiency and viability due to salt contamination. ■

Retargeting with GFP Vector - 6 days post selection

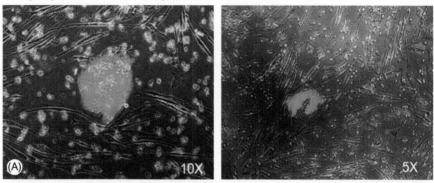

Retargeting with GFP Vector - 14 days post selection

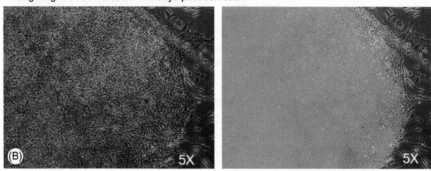

FIGURE 30.9
Retargeting with a GFP vector. The pJTI™ R4 Vector containing the GFP gene was transfected in hESCs. The colony can be observed after being subjected to selection for 6 days (A) or 14 days (B) post transfection.

Screening Retargeted Clones

Upon retargeting your R4 platform line, follow the guidelines below to PCR screen for successful retargeting events using genomic DNA isolated from individual clones. Depending on which pJTI R4 DEST vector you choose, (blasticidin, neomycin, or zeocin resistant genes in pJTI/Bsd, pJTI/Neo, or pJTI/Zeo, respectively) you first need to design primers flanking the antibiotic resistance gene and the EF1α promoter to validate the retargeting event (Figure 30.8).

To detect retargeting between the EF1α promoter and the Zeocin resistant gene, design a forward primer at the 3′ end of the EF1α promoter (see Frw) and Zeocin reverse primer at the 5′ end of the Zeocin gene (Rev) (see primers below).

Frw (GCCTCAGACAGTGGTTCAAAGTTT)
Rev (TGATGAACAGGGTCACGTCGT)
Expected size 500 bp

Screening for a Retargeting Event at the R4 attP Site

1. Pellet 10,000 to 30,000 cells total.
2. Wash with 500 μL PBS.

3. Spin down and remove PBS.
4. Resuspend the cell pellet in 20 μL of Cells Direct Resuspension Buffer with 2 μL of Lysis solution added to the resuspension buffer.
5. Incubate at 75°C for 10 minutes.
6. Spin down 1 minute.
7. PCR reaction using: AccuPrime™ Taq High Fidelity.
8. Add 5 μL of 10× PCR buffer II, 1 μL of Frw primer (10 μM stock, sequence listed above), 1 μL of Rev primer (10 μM stock, sequence listed above), 1 μL of Taq polymerase, 3 μL of cell lysate, 39 μL of H_2O.
9. PCR Cycle condition: 94°C for 2 minutes, 94°C for 30 seconds, annealing temperature 56°C for 30 seconds, 72°C for 1 minute, 35–40 cycles, 72°C for 7 minutes.

■ **Note**

You may use plasmid DNA or the Hygromycin resistance gene as a positive control. A map and a description of the features of each platform vector pJTI/Bsd, pJTI/Neo, and pJTI/Zeo vectors are available online (www.invitrogen.com). ■

■ **Note**

PCR is usually sufficient to confirm the presence of the retargeted sequences in your cell line after transfection. However, you may also perform Southern blot analysis as an additional check to screen for a single copy number. Use the Southern blot protocol of your choice with a radiolabeled probe from the expression vector used to retarget the cells as suggested above. ■

PITFALLS AND ADVICE

1. The quality of the plasmid DNA strongly influences the results of transfection experiments. Use endotoxin-free DNA for all transfections. Make sure that the A260:A280 ratio of the DNA is between 1.8 and 2.0. Do not use phenol:chloroform for extraction, or ethanol precipitation.
2. Make sure cells in culture are at low passage, show minimum differentiation with high viability and are 80–90% confluent before transfection.
3. You must passage your platform cell lines at least once without drug selection prior to transfection.
4. Stem cell platform lines must be passaged at least once as a feeder-free culture in MEF conditioned medium and Geltrex without drug selection prior to transfection.

5. Avoid damaging cells during harvesting. Centrifuge cells at lower speeds (150–200 g). Avoid overexposure to TrypLE, Accutase, or other dissociation reagents. Pipette cells gently.
6. Immediately after electroporation/microporation, transfer cells into pre-warmed medium at 37°C to prevent damage.
7. Low survival rate after transfection may be due to cells being damaged during harvesting, or poor transfection parameters.
8. Poor transfection efficiency may be due to an incorrect ratio of DNA to cell number, either too low DNA or too high cell density. In addition, make sure DNA is free of endotoxin.
9. Start selection when you begin to see colonies with well-defined borders. It is important not to let colonies become too overgrown, otherwise, the untransfected cells will fail to die upon selection.
10. If no colonies are seen after selection is completed, this may be caused by too high of a concentration of antibiotics. Make sure you determine the optimal concentration for your cell type. In addition, make sure the ratio of DNA to cell number is correct.

EQUIPMENT

- Tissue culture hood
- Tissue culture incubator, 37°C, 5% CO_2
- Inverted phase contrast and fluorescent microscopes
- Table-top centrifuge
- Pipetman and pipettors

REAGENTS AND SUPPLIES

Recommended Reagents

Item	Supplier	Catalog #	Note
DMEM/12 + L-Glut + HEPES	Life Technologies	11330-032	
KnockOut Serum Replacement (contains bovine products)	Life Technologies	10828-028	Use at 20%
GlutaMAX™	Life Technologies	35050061	
MEM-Non-essential amino acids	Life Technologies	11140-050	
2-Mercaptoethanol	Life Technologies	21985-023	Best if added day of use

Item	Supplier	Catalog #	Note
PBS (Dulbecco's Phosphate-Buffered Saline without Calcium and Magnesium)	Life Technologies	14190-144	
Recombinant human bFGF	Life Technologies	13256-029	8 ng–12 ng/ml
Pen/Strep (100×)	Life Technologies	15070-063	Optional
DMEM High Glucose + GlutaMAX	Life Technologies	10566-016	
Fetal Bovine Serum	Life Technologies	16141079	
Jump-In TI Platform Kit	Life Technologies	A10897	
Cells Direct Resuspension and Lysis Buffer	Life Technologies	11739-010	
AccuPrime *Taq* DNA Polymerase High Fidelity	Life Technologies	12346-094	
DNAzol® Reagent	Life Technologies	10503-027	
MultiSite Gateway® Pro Plus Kit	Life Technologies	12537-100	
DECAprime II™ Kit	Ambion	1455	
Neon microporator	Life Technologies	MPK5000S	
Hygromycin B resistant inactivated MEFs	Millipore	PMEF-H	
ViraPower™ HiPerform™ Lentiviral Expression kit	Life Technologies	K533000	
Opti-MEM	Life Technologies	31985070	
Collagenase IV	Life Technologies	17104019	
Blasticidin	Life Technologies	A1113902	
Neomycin	Life Technologies	21810031	
Zeocin	Life Technologies	R25001	
Hygromycin B	Life Technologies	10687010	
High Prime Kit	Roche	1585584	
T4 DNA Ligase	Life Technologies	15224017	
DH10B™-T1R electrocompetent *E. coli*	Life Technologies	C640003	

RECIPES

hESC Culture Media

Item	Stock	Final	500 mL
DMEM/F12 + L-Glut + HEPES	100%	80%	400 mL
KnockOut Serum Replacement	100%	20%	100 mL
GlutaMAX	200 mM	1 mM	5 mL

Item	Stock	Final	500 mL
MEM-Non-essential amino acids	10 mM	0.1 mM	5 mL
2-Mercaptoethanol	55 mM	0.1 mM	910 μL
Recombinant human bFGF	25 μg/mL	8–12 ng/mL	240 μL

Fibroblast Feeder Layer Media

Item	Stock	Final	500 mL
DMEM Hi Glucose	100%	90%	400 mL
Fetal Bovine Serum	100%	10%	100 mL
MEM-Non-essential amino acids	10 mM	0.1 mM	5 mL

Preparation of MEF-Conditioned Medium (CM)

1. Plate 9.4×10^6 inactivated MEFs in a T175 flask coated with 0.1% gelatin in MEF medium.
2. The following day, replace the MEF medium with 90 mL ESC medium.
3. Collect MEF-CM from the flasks after 24 hours of conditioning for up to 7 days.
4. Filtered CM can be stored at $-20°C$ until use.
5. At time of use, freshly supplement CM with extra GlutaMAX (2 mM) and bFGF (4 ng/mL).

Preparation of Geltrex-Coated Dishes

1. Thaw a 5 mL bottle of Geltrex (hESC Qualified) at 2–8°C overnight.
2. Dilute 1:1 with cold DMEM/F-12 to prepare 1 mL aliquots in tubes chilled on ice. These can be frozen at $-20°C$ or used immediately.
3. Dilute 1:100 with cold DMEM on ice. Note: an optimal dilution of Geltrex may need to be determined for each line. Try dilutions from 1:30 to 1:100.
4. Cover the whole surface of each culture dish with the diluted Geltrex
5. Incubate in a 37°C, 5% CO_2 incubator for 1 hour. Note: dishes can now be used or stored at 2–8°C for up to a week. Do not allow dishes to dry out.
6. Pipette out diluted Geltrex from the culture container and discard. Cells can be passaged directly into MEF CM on Geltrex coated containers.

■ Note

It is not required to rinse off the Geltrex from the culture container. ■

READING LIST

Liu, Y, Lakshmipathy, T.B., Xue, U, Lieu, H, Fontes, P, MacArthur, A, et al., 2009. Generation of platform human embryonic stem cell lines that allow efficient targeting at a predetermined genomic location. Stem Cells Dev. 18, 1459–1472.

Liu, Y, Lakshmipathy., U., Ozgenc, A, Thyagarajan, B, Lieu, P, Fontes, A, et al., 2010. hESC engineering by integrase-mediated chromosomal targeting. Methods Mol. Biol. 584, 229–268.

Naldini, L., Blomer, U., Gage, F.H., Trono, D., Verma, I.M., 1996. Efficient transfer, integration, and sustained long-term expression of the transgene in adult rat brains injected with a lentiviral vector. Proc. Natl. Acad. Sci. USA 93, 11382–11388.

Pandya, S., Klimatcheva, E., Planelles, V., 2001. Lentivirus and foamy virus vectors: Novel gene therapy tools. Expert Opin. Biol. Ther. 1, 17–40.

Park, F., Kay, M.A., 2001. Modified HIV-1 based lentiviral vectors have an effect on viral transduction efficiency and gene expression in vitro and in vivo. Mol. Ther. 4, 164–173.

Thyagarajan, B., Olivares, E.C., Hollis, R.P., Ginsburg, D.S., Calos, M.P., 2001. Site-specific genomic integration in mammalian cells mediated by phage phiC31 integrase. Mol. Cell Biol. 21, 3926–3934.

Thyagarajan, B., Liu, Y., Shin, S., Lakshmipathy, U., Scheyhing, K., Xue, H., et al., 2008. Creation of engineered human embryonic stem cell lines using phiC31 integrase. Stem Cells 26, 119–126.

Yee, J.K., 1999. Retroviral vectors. In: Friedmann, T. (Ed.), The Development of Human Gene Therapy. Cold Spring Harbor Laboratory Press, Cold Spring Harbour, NY, pp. 21–45.

Zhou, X., Cui, Y., Huang, X., Yu, Z., Thomas, A.M., Ye, Z., et al., 2003. Lentivirus-mediated gene transfer and expression in established human tumor antigen-specific cytotoxic T cells and primary unstimulated T cells. Hum. Gene. Ther. 14, 1089–1105.

Zufferey, R., Dull, T., Mandel, R.J., Bukovsky, A., Quiroz, D., Naldini, L., et al., 1998. Self-inactivating lentivirus vector for safe and efficient in vivo gene delivery. J. Virol. 72, 9873–9880.

READING LIST

CHAPTER 31

Gene Targeting by Homologous Recombination in Human Pluripotent Stem Cells

Sangyoon Han and Ying Liu

EDITOR'S COMMENTARY

Gene targeting by homologous recombination has become a widely used technique for genetically modifying mice. This approach was initiated in the mid-1980s by a handful of pioneering researchers who found that knocking out genes in mouse embryonic stem cells (ESCs) was an effective way to generate mice that had mutations in specific genes. Over the past 20 years, with large investments from funding agencies, variations on this method have resulted in thousands of precisely engineered mouse strains. However, applying the lessons learned from mouse ESC experience to human PSCs has been a challenge. Human PSCs divide more slowly and are more sensitive to transfection techniques and cloning methods that are routinely used for mouse ESCs. However, with adaptations, targeting genes in hPSCs using straightforward homologous recombination methods has been successful in several laboratories. There are alternative methods for gene targeting that have become available in the past few years, and one of these, zinc finger nuclease (ZFN)-mediated homologous recombination, is described in detail in Chapter 32. The nuclease approaches engineer fusions of a restriction enzyme (Fok1) with sequence-specific DNA-binding proteins like ZFs and TALEs (transcription activator-like effectors). The efficiency of targeting protocols using ZFNs and TALE nucleases (TALENs) is generally higher, usually much higher, than conventional gene targeting vectors. One great advantage of conventional gene targeting approaches is that because of the tremendous history of genetic manipulation in the mouse, the construction of targeting vectors is routine at many institutions. There is also a huge toolkit of subtle genetic engineering approaches that have been developed for the mouse, including conditional and inducible systems for controlling the timing and cell type-specificity of the genetic changes. At the moment, the main disadvantage of ZFNs and TALENS appears to be that their design is not easily accessible so they are not widely applied in research laboratories, and commercial sources are too expensive for most grant-funded researchers.

485

J.F. Loring & S.E. Peterson (eds): Human Stem Cell Manual, Second edition.
DOI: http://dx.doi.org/10.1016/B978-0-12-385473-5.00031-X

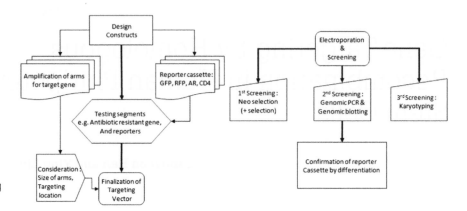

FIGURE 31.1

Workflow for gene targeting in hPSCs.

There is also the unanswered question about whether the nucleases may have significant off-target effects that cause DNA aberrations elsewhere in the genome. The authors of this chapter have both independently targeted genes in hPSCs, and they describe here a method that has worked well for their labs.

OVERVIEW

Here we describe a conventional method used to construct targeting vectors and the workflow for generating reporter lines in hPSCs using those vectors (Figure 31.1). This protocol works well and has been used to generate several reporter lines in multiple hPSC lines. Targeting efficiency is determined by accessibility of the genetic loci in hPSCs, the structure of targeting vectors, and other unidentified factors. This chapter describes the construction of targeting vectors and transfection into hPSCs to make lineage reporters at a relatively high efficiency.

PROCEDURES

Targeting vector design is crucial for the successful creation of any transgenic cell line. Here we outline a method that will decrease false positives and increase efficiency. The strategy described will also greatly reduce laborious work during clonal expansion and screening to identify true positives. Figure 31.1 provides an overview of the process of creating a reporter cell line, and Figure 31.2 illustrates the elements of the targeting construct: 5′ and 3′ arms of homology, positive selection cassette (Neo), reporter (GFP), and negative selection elements (DTA).

Choice of Targeting Vector

To be able to select for cells that have successfully integrated the targeting vector, the vectors are generally constructed to include positive selection

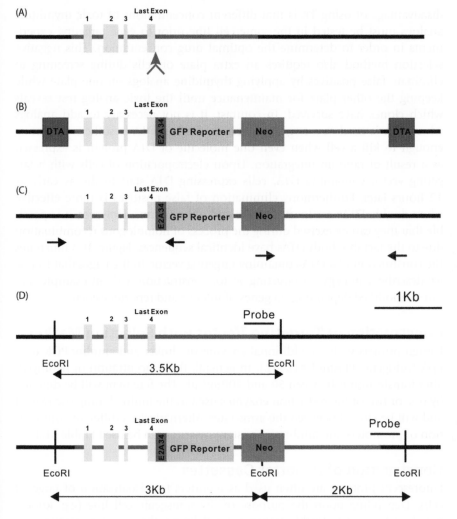

FIGURE 31.2
(A) Schematic of the genomic structure and locus of a hypothetical targeting area. Red line is sequence area used in targeting vector. (B) Design of targeting vector including arms, reporter, and selection markers. (C) The location of primers for genomic PCR. Each set should have one primer outside of the arm and inside of the targeting construct. (D) An example of the probe to be used for Southern blotting and the restriction site for digestion of genomic DNA from the wild type cell line (top). Possible genomic fragment after digestion of a transgenic cell line (bottom).

cassettes, usually genes that confer resistance to drugs such as neomycin, hygromycin, or puromycin. Positive drug selection, however, does not allow selection against random integrations, and strategies using only positive selection often require the screening of many hundreds of colonies to find homologous recombinants.

In order to reduce the likelihood of false positives from random integration, negative selection markers are often added to targeting vector design. Thymidine kinase (TK) and diphtheria toxin A (DTA) are the most common negative selection markers. TK activity is achieved using toxic thymidine analogs (e.g. 5-bromodeoxyuridine, ganciclovir), which are incorporated into DNA only if there is TK activity in the cell via random integration. The disadvantage of using TK is that different concentrations of toxic thymidine analogs must be tested in the same cell line prior to gene targeting experiments in order to determine the optimal drug concentration. This negative selection method also requires an extra plate of cells during screening to eliminate false positives by applying thymidine analogs on one plate while keeping the other plate for maintenance until the toxic analog test reveals which clones have survived. In contrast, it is not necessary to add analogs when DTA is used as a negative selection marker. DTA is sufficiently strong enough to kill a cell when even one molecule of DTA protein is expressed as a result of random integration. Upon electroporation of cells with a targeting vector containing DTA, cells expressing DTA start to die as early as 12 hours later. Furthermore, elimination of false positives is more effective when double DTAs are located at the end of both arms. However, it is possible that they can be excised during the process of homologous recombination due to the fact that both DTAs have identical sequences. Figure 31.3 illustrates the construction of a DTA-containing targeting vector. In the pages that follow we describe a strategy for targeting vector construction with an example that can be modified depending on genes of interest and reporter cassettes.

Construction of Targeting Vector with Selection Marker

Design primers with an additional enzyme site linker and amplify PGK-neo-bpA (Addgene, Plasmid #13442). In general, the concentration of neomycin after transfection is between 50 and 200 μg/mL. The fragment will be digested by one or two of the restriction enzymes used in the multi-cloning site (MCS) and will be cloned between the arms later. Alternatively, a different combination of promoters and antibiotic resistant cassettes can be used (Table 31.1).

Construction of Reporter Cassettes

Fluorescent proteins are often used as reporters for visualization of targeted cells. Depending upon the purpose of the transgenic cell line (e.g. knockout, knock-in, or inducible transgenic cell lines), the reporter cassettes can be located at different sites in the targeting constructs. Here, we specifically discuss knock-in transgenic cell lines in which endogenous expression is not interrupted by the reporter gene. This strategy allows the reporter gene to recapitulate expression of the endogenous gene. A simple method used to achieve this is the introduction of an internal ribosome entry site (IRES) or the self-cleaving 2A peptide fragment at the end of the last exon.

The reporter described here is green fluorescent protein (GFP) with an equine rhinitis A virus 2A domain (ERAV, E2A) linked to the last exon of the gene of interest. The E2A domain can be synthesized with DNA oligos of 13–15 base pairs overlapping with the first 13–15 base pairs of GFP. To achieve this: (1) perform a PCR reaction with synthesized E2A DNA as a

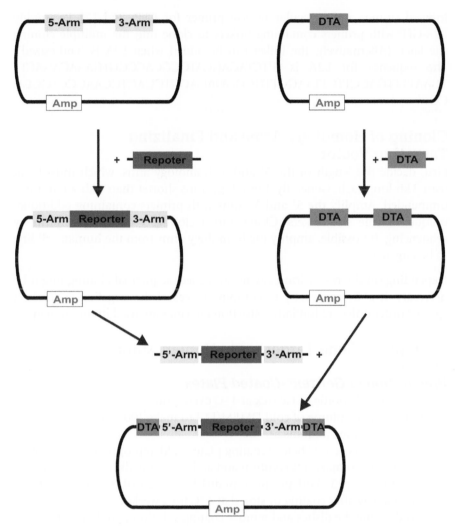

FIGURE 31.3
Cloning strategy. 5′ and 3′ arms are amplified and subcloned with the reporter gene using a positive selection maker such as Neo in the cloning vector. In parallel, negative selection markers are subcloned into the target vector backbone. Later the 5′ arm-reporter-3′ arm fragment is cloned into the targeting vector with a negative selection maker.

Table 31.1 Different Combinations of Promoters with Antibiotic Resistance Genes Described Here Can Be Used for Selection. Each Antibiotic Resistance Gene Has Been Tested in Different Concentrations with hPSCs

Promoters	Antibiotics
EF1a (elongation factor 1a)	Neomycin (G418): 50–200 µg/mL
PGK (phosphoglycerate Kinase)	Hygromycin: 100 µg/mL
SV40 (simian virus 40)	Puromycin: 1 µg/mL
ActB (human beta acin)	

forward primer and a regular reverse primer for GFP; and (2) Re-amplify E2A-GFP with primers containing linkers to clone into the multiple cloning site later. (Alternatively, the linker can be added when E2A is synthesized). The sequence for E2A is 5'TCTAGAGCATGCGCACCGGTGAAACAGACTT TGAATTTTGACCTTCTTAAGCTTGCGGGAGACGTCGAGTCCAACCCTGGGC CC 3'.

Cloning of Homology Arms and Finalizing Targeting Vector

First, decide the length of the 5' and 3' homology arms, which must be at least 3 kb long each. Generally, homology arms shorter than 3 kb are not recommended. Amplify the 5' and 3' arms with primers containing additional unique enzyme site linkers. Clone into a cloning vector and confirm by sequencing. If possible, amplify the homology arms from the human cell line to be targeted.

Depending on the restriction enzyme sites used, sequential cloning might be necessary. It is not necessary to combine pieces of the targeting vector in a specific order unless redundant restriction enzymes are used in the strategy.

Example of Gene Targeting in Human Embryonic Stem Cells
Preparation of Geltrex®-Coated Plates
1. Thaw a whole bottle of Geltrex at 4°C overnight.
2. Add an equal volume of cold DMEM/F12 to make 100× stock solution and store desired aliquots at −20°C.
3. Thaw aliquots at 4°C before coating plates. Add appropriate volume of DMEM/F12 to make 1× solution and add to culture dishes to completely cover plate (e.g. 3–4 mL per 100 mm dish). Coated dishes can be used immediately after coating or stored at 4°C for 2 weeks.
4. Avoid drying the dishes and remove coating solution right before use.

Passage of hPSCs on Geltrex-Coated Plates
■ Note

This is for use with feeder-free, enzymatic passage-adapted hPSCs (see Chapter 3 for feeder-free culture methods)   ■

1. Use medium designed for feeder-free culture (e.g. mTeSR or StemPro).
2. Change medium daily until cells are 80% confluent.
3. Incubate hPSCs with 5 mL of warm dispase at 37°C for 5–10 minutes until the edges of colonies slightly curl up.
4. Aspirate dispase and slowly add warm DMEM/F12 to wash away excess dispase. Repeat once.

5. Add appropriate volume of warm medium (e.g. 5 mL/100 mm dish) and detach colonies by pipetting the medium up and down.
6. Transfer collected colonies into a 50 mL conical tube and pipette vigorously to break apart colonies. In general, pipetting 5–10 times at the fastest speed is enough.
7. Add appropriate amount of warm medium into the tube based on the number of plates. Aliquot cells into dishes.

Electroporation

Different types and brands of electroporation instruments can be used. We outline a procedure for electroporation with the Bio-Rad Gene Pulser® II electroporation system. Methods using other systems are described in Chapter 30.

1. Culture feeder-free hPSCs to 80% confluency, usually cells harvested from two 60 mm dishes ($3\sim6 \times 10^{6)}$ are enough for one electroporation.
2. Add 30 μg linearized DNA to a 1.5 mL microcentrifuge tube.
3. Prepare drug resistant MEF dishes (for positive selection) and add 3 mL of conditioned medium/dish, save in 37°C, 5% CO_2 incubator.
4. Pre-warm 3 mL conditioned medium for each electroporation in a 15 mL conical tube in a 37°C water bath.
5. Add 1 mL Accutase™ to each 60 mm dish and incubate for 3–5 minutes until the cells become dislodged.
6. Collect cells and centrifuge at 200 g for 5 minutes to pellet them.
7. Resuspend cells in 1× PBS as single cells and count.
8. Remove desired number of cells and place in a new 15 mL conical tube.
9. Spin at 200 g for 5 minutes.
10. Remove PBS and resuspend in 800 μL of OptiPro™ SFM or PBS.
11. Add cells into the 1.5 mL microcentrifuge tube that contains DNA, and transfer the mixture to an electroporation cuvette (0.4 cm gap).
12. Electroporate the cells once using the Bio-Rad Gene Pulser II or Xcell system using the following conditions: 250 V, 250 μF.
13. Immediately transfer the cells into the 15 mL conical tube (prepared in Step 4) using a Pasteur pipette and incubate at room temperature for 5 minutes.
14. Transfer the mixture from Step 13 into the 60 mm dish (prepared in Step 3) and incubate in the 37°C, 5% CO_2 incubator.

Picking and expanding clones

1. On the day following electroporation, change medium with 10 mL of fresh, warm culture medium (TeSR™ or equivalent feeder-free medium).

2. Change medium with 10 mL of medium 48 hours later. This will give the cells enough time to settle and any cells with random integration will die.
3. The third day following electroporation, change to medium with G418 at the concentration of 50 μg/mL and increase the concentration by 50 μg/mL every 2 days until the concentration is 200 μg/mL.

■ Note

The gradual increase in G418 concentration allows you to control the rate at which cells are killed so that there is not too much damage to neighboring cells. The concentration of G418 to be used is dependent on the sensitivity of the cell line as well as the power of the promoter driving the drug selection cassette. Cells containing a weak promoter for neo (or a mutated neo) may be unintentionally killed by using too high a concentration of G418. Ideally, a "kill curve" should be established for your cell line. This is done by culturing enzymatically-dissociated cells seeded at low density in several concentrations of G418, ranging from 50 to 400 μg/mL, changing the G418-containing medium each day for 10 days. After 10 days, the lowest concentration at which all of the cells are killed should be obvious. ■

4. Approximately 10–14 days later, there should be colonies big enough to pick manually.
5. Transfer individual colonies into each well of a 24-well plate and continue to culture the dish until more colonies are ready for manual dissection.
6. After 7 days, colonies in the 24-well plate will be ready for splitting into one well of 6-well plate.
7. Add 500 μL of warm dispase into each well of 24-well plate and incubate at 37°C for 5–15 minutes until the edges of the colonies curl up slightly.
8. Remove dispase and slowly add 500 μL of warm medium and aspirate the media. Repeat two times.
9. Add 500 μL of warm medium and lift up all colonies by gentle pipetting. Transfer cells to a 15 mL tube.
10. Add 1.5 mL of warm medium into mixture and gently pipette to break the colonies into small pieces.
11. Transfer into one well of a 6-well plate and incubate in a humidified 37°C, 5% CO_2 incubator until cells are ready to split into one 6-well plate (this will take another week or so).
12. When one well of the 6-well plate is ready to split, save half of the cells for genomic PCR and Southern hybridization to confirm whether the reporter gene is integrated into the right locus.

■ Note

Southern hybridization is the most definitive method to confirm targeted clones. Ideally, the targeting vector should be designed with Southern analysis in mind. It should contain a restriction site that will allow probing of the locus with a probe that recognizes only the targeted sequence. Southern hybridization may also reveal additional targeting construct sequence that has randomly integrated into the genome.   ■

Screening for True Positive Clones
Genomic PCR Screening

Design primers outside of 5' and 3' homology arms with a GC content of 50% and T_m at 62 °C. In order to optimize the genomic PCR conditions, use a set of primers from outside of the 5' and 3' arms in a gradient PCR machine with any high fidelity polymerase. The optimized condition can then be used to screen positive clones. Use a set of primers such as one from the reporter cassette and either primer from outside of the 5' or 3' arm to confirm that the targeting cassette is integrated into the correct locus. Test primers with the same genomic DNA used to amplify the arms using a gradient PCR machine. Internal primers also should have similar GC content and T_m. We have been successful in amplifying DNA when there are true positives using a three-cycle PCR. Steps in three-cycle PCR are: 95 °C for 2 minutes for initial denaturation; followed by 30–35 cycles of 95 °C for 1 minute, the optimal annealing temperature for 1 minute, and 72 °C for 1 minute/kb for extension.

Southern Hybridization

Probes can be prepared using the DIG System (Roche) for non-radioactive probes. Other modified nucleotides are Cy5-dCTP, fluorescein-, rhodamine-, and courmarin-dUTP, and antibodies against these analogs are also commercially available. They can be detected by any chemiluminescent detection system.

Generate the probe with 0.1 ng of template DNA, 0.2 µg of each primer, 5 µL of PCR DIG labeling mix, 5 µL of PCR 10× buffer and 0.5 µL of Enzyme mix and fill to 50 µL with water. The PCR can then be performed for 35 cycles and 1–2 µL of the reaction is then run on a gel to confirm. The probe now can be stored at 4 °C in the short term or −20 °C for long term.

1. Prepare 0.6–0.8% agarose gel in 1× TAE buffer with ethidium bromide (EtBr). Note that TAE buffer resolves larger fragments better than TBE buffer.
2. Run a gel containing the genomic PCR product at 40–60V for 5–6 hours or until the desired fragments are well resolved. Carefully trim the gel to remove the blank space on the edge of the gel that contains no DNA. Take a picture showing a fluorescent ruler.

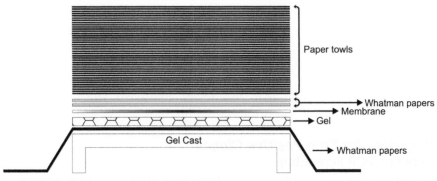

FIGURE 31.4

Southern blot. Blot is set up from bottom to top. A large dish is filled with 0.4N NaOH. Gel is placed on the top of the gel cast up side down. Place two pieces of wick-blotting paper cut to the width of the gel and length such that the wick is in contact with the bottom of the dish. Cut membrane to the exact size of the gel, with a nick in the corner for orientation and place it on the top of the gel. Membrane is now in contact with the bottom of gel. Two pieces of blotting paper cut to the exact size of the gel are placed on the top of the membrane. Bubbles are gently removed by rolling over it with a glass pipette. Stack with similar size paper towels and change it every 2–3 hours.

3. Denature the genomic DNA by submerging the gel in 500 mL Denaturation buffer and rocking it gently for 30 minutes.

4. Neutralize the gel in 500 mL Neutralization buffer by rocking gently for 30 minutes.

5. Set up DNA transfer following the layout illustrated in Figure 31.4.

6. Place gel cast upside down in a large tray.

7. Cut four sheets of Whatman paper (size slightly bigger than the width of gel and long enough to cross gel cast) and soak the papers in transfer buffer.

8. Place two sheets of stacked Whatman papers in one direction and put another two sheets of stacked Whatman paper in the other direction.

9. Put gel with the open wells facing down to ensure that genomic DNA at the bottom of well is close to membrane.

10. Overlay Hybond-N+ membrane (presoaked in 0.4 N NaOH) on top of the gel. Cut or label a corner of membrane to facilitate orientation later.

11. Put two sheets of presoaked Whatman paper on the membrane and stack 5 inches of similar size paper towels. Changing the paper towels every 2–3 hours shortens the transferring time.

12. Put about 500 g of weight on top of paper towels. Make sure the stacks do not fall during transfer.

13. Add plenty of 0.4 N NaOH to the tray and let it transfer for at least 6–8 hours or overnight.

14. Disassemble transfer stack and crosslink genomic DNA using an auto crosslinker. Air-dry the membrane.

15. Incubate the membrane with pre-warmed Pre-hybridization buffer at 42°C for 30 minutes.
16. Denature the probe by heating at 100°C for 5 minutes and chill on ice until hybridization is ready.
17. Change Pre-hybridization buffer with fresh Hybridization buffer with probe and incubate overnight at 42°C.
18. The next day, wash the membrane with Washing buffer I for 10 minutes, 3 times at 68°C.
19. Wash the membrane with Washing buffer II for 10 minutes 1–2 times.
20. Wash the membrane with Washing buffer III for 10 minutes once.
    Note: Washing buffers II and III can be skipped if signals are weak.
21. Wash the membrane with 1 × PBS briefly.
22. Incubate the membrane with blocking solution for 30 minutes on a rocker.
23. Change blocking buffer (2–5% non-fat dry milk in blotting buffer pH 7.4) with antibody solution (anti-DIG-AP antibody at 200 mU/mL in 1–5% non-fat dry milk in Blotting buffer pH 7.4) and incubate for 30 minutes at room temperature with gentle rocking.
24. Wash the membrane at room temperature for 15 minutes with 4–5 changes of blotting buffer.
25. Place the membrane with DNA side up on a transparency film and apply 1 mL of diluted CSPD solution.
26. Incubate the membrane for 15 minutes at 37°C to enhance luminescent reaction and expose to X-ray film or scan image.

REAGENTS AND SUPPLIES

Recommended Reagents

Item	Supplier	Catalog #	Note
One Shot® Top10 Electrocompetent *E. Coli*	Life Technologies	C4040-50	
AccuPrime Pfx Super Mix	Life Technologies	12344-040	
Restriction enzymes	Life Technologies and New England Biolabs	Various	
Zymoclean™ Gel DNA Recovery Kit	Zymo Research	D4007	
Geneticin® G418	Life Technologies	11811	Complete kill curve for accurate dosage
mTeSR1	Stem Cell Technologies	05850	StemPro could potentially be used but has not been tested in this protocol
TeSR	Stem Cell Technologies	05860	StemPro could potentially be used but has not been tested in this protocol

Item	Supplier	Catalog #	Note
0.25% Trypsin-EDTA Solution	Life Technologies	25200-056	
Fetal Bovine Serum – ES cell qualified	Life Technologies	10439	
Dulbecco's Phosphate Buffered Saline without Magnesium and Calcium	Life Technologies	A12856-01	
DMEM/F12 with GlutaMAX	Life Technologies	10565-018	
Dispase	Life Technologies	17105-041	
Geltrex	Life Technologies	12760-013	
50× TAE	Life Technologies	24710030	
DIG System	Roche	1745832	
Anti-digoxigenin (DIG)-AP	Roche	1093274	
CSPD Chemiluminescence System	Roche	1755633	
Recombinant human bFGF	Invitrogen	13256-029	
Collagenase, Type IV, microbial	Invitrogen	17104-019	
Pen/Strep (100×)	Invitrogen	15070-063	Optional

RECIPES

20× SSPE

Component	Amount
NaCl	175.3 g
NaHPO$_4$	26.6 g
EDTA	9.4 g

1. Weigh out and combine the above reagents.
2. Add water to 1.0 L.
3. Stir to dissolve.
4. Adjust pH to 7.4 with NaOH.

Denaturation Buffer

Final Concentration	Component	Amount	Stock Concentration
1.5 M	NaCl	300 mL	5 M
0.5 M	NaOH	40 mL	12.5 N

1. Combine the above reagents.
2. Add water to 1.0 L.

Neutralization Buffer

Final Concentration	Component	Amount	Stock Concentration
1.5 M	NaCl	300 Ml	5 M
0.4 M	Tris-HCl	40 mL	10 M

1. Combine the above reagents.
2. Add water to 1L.

Pre-Hybridization and Hybridization Buffer

Final Concentration	Component	Amount	Stock Concentration
5×	SSPE	250 mL	20×
0.1%	N-Lauroylsarcosine	100 mL	1%
0.02%	SDS	0.2 mL	10%
1×	Blocking buffer	100 mL	10×

Note: blocking buffer is included in DIC kit.

1. Combine the above reagents.
2. Add water to 1L.

Washing Buffer I

Final Concentration	Component	Amount	Stock Concentration
1×	SSPE	50 mL	20×
1%	SDS	100 mL	10%

1. Combine the above reagents.
2. Add water to 1L.

Washing Buffer II

Final Concentration	Component	Amount	Stock Concentration
0.5×	SSPE	25 mL	20×
1%	SDS	100 mL	10%

1. Combine the above reagents.
2. Add water to 1.0 L.

Washing Buffer III

Final Concentration	Component	Amount	Stock Concentration
0.1×	SSPE	5 mL	20×
1%	SDS	100 mL	10%

1. Combine the above reagents.
2. Add water to 1.0 L.

Blotting Buffer

Final Concentration	Component	Amount	Stock Concentration
0.15M	NaCl	30mL	5M
25mM	Tris-HCl	25mL	10M
0.1%	Tween20	1mL	100%

1. Combine the above reagents.
2. Add water to 1.0 L.

READING LIST

Bu, L., Jiang, X., Martin-Puig, S., et al., 2009. Human ISL1 heart progenitors generate diverse multipotent cardiovascular cell lineages. Nature 460, 113–117.

Bu, L., Gao, X., Jiang, X., et al., 2010. Targeted conditional gene knockout in human embryonic stem cells. Cell Res. 20, 379–382.

Davis, R.P., Ng, E.S., Costa, M., et al., 2008. Targeting a GFP reporter gene to the MIXL1 locus of human embryonic stem cells identifies human primitive streak-like cells and enables isolation of primitive hematopoietic precursors. Blood 111, 1876–1884.

Hockemeyer, D., Soldner, F., Beard, C., et al., 2009. Efficient targeting of expressed and silent genes in human ESCs and iPSCs using zinc-finger nucleases. Nat. Biotechnol. 27, 851–857.

Irion, S., Luche, H., Gadue, P., et al., 2007. Identification and targeting of the ROSA26 locus in human embryonic stem cells. Nat. Biotechnol. 25, 1477–1482.

Palmiter, R.D., Behringer, R.R., Quaife, C.J., et al., 1987. Cell lineage ablation in transgenic mice by cell-specific expression of a toxin gene. Cell 50, 435–443.

Pelletier, J., Sonenberg, N., 1988. Internal initiation of translation of eukaryotic mRNA directed by a sequence derived from poliovirus RNA. Nature 334, 320–325.

Ruby, K.M., Zheng, B., 2009. Gene targeting in a HUES line of human embryonic stem cells via electroporation. Stem Cells 27, 1496–1506.

Szymczak, A.L., Workman, C.J., Wang, Y., et al., 2004. Correction of a multi-gene deficiency in vivo using a single "self-cleaving" 2A peptide-based retroviral vector. Nat. Biotechnol. 22, 589–594.

Thomas, R., Capecchi, M.R., 1987. Site-directed mutagenesis by gene tar- geting in mouse embryo-derived stem cells. Cell 51, 503–512.

Urbach, A., Schuldiner, M., Benvenisty, N., 2004. Modeling for Lesch-Nyhan disease by gene targeting in human embryonic stem cells. Stem Cells 22, 635–641.

Xue, H., Wu, S., Papadeas, S.T., et al., 2009. A targeted neuroglial reporter line generated by homologous recombination in human embryonic stem cells. Stem Cells 27, 1836–1846.

Yagi, T., Nada, S., Watanabe, N., et al., 1993. A novel negative selection for homologous recombinants using diptheria toxin A fragment gene. Anal. Biochem. 214, 77–86.

Zou, J., Maeder, M.L., Mali, P., et al., 2009. Gene targeting of a disease-related gene in human induced pluripotent stem and embryonic stem cells. Cell Stem Cell 5, 97–110.

Zwaka, T.P., Thomson, J.A., 2003. Homologous recombination in human embryonic stem cells. Nat. Biotechnol. 21, 319–321.

Genetic Manipulation of Human Pluripotent Stem Cells Using Zinc Finger Nucleases

Tianjian Li, Jason Gustin, Mahesh Dodla, and Ian Lyons

EDITOR'S COMMENTARY

As stem cell research expands and touches other fields of investigation, there is a growing interest in genetically manipulating human pluripotent stem cells (hPSCs), for reasons ranging from introducing or correcting disease-causing mutations to unraveling regulatory pathways. While mouse transgenic and ES cell technologies in development over the past 20 years are being successfully adapted for human cells, there are also relatively new approaches that are increasing the options for modifying the genomes of hPSCs. One of these approaches uses the specificity of DNA-binding proteins to target DNA endonucleases to specific DNA sequences. Zinc finger nuclease (ZFN) technology is the first of these approaches to be commercialized, and other versions of targeted endonuclease tools are gaining interest (such as TALENs (transcription activator-like (TAL) merged to a nuclease (Miller et al., 2001)). The principle of ZFN technology is to take advantage of the specificity of DNA binding by zinc finger proteins. To create a targeting construct, a combination of zinc finger sequences that bind to the region to be targeted are merged with the catalytic domain of a restriction enzyme. The zinc fingers provide the binding specificity, and the nuclease creates a double-stranded break in the DNA. By adding an exogenous template sequence carrying the desired genetic changes (such as mutations that will knock out the gene), the cell's homology-directed repair mechanisms can be exploited to introduce the desired changes into the sequence. In studies done so far, ZFN technology appears to be far more efficient than conventional homologous recombination (HR) based methods for knocking out genes.

Both conventional homologous recombination methods (Chapter 31) and the targeted nuclease methods are described in this book. There has not yet been a direct comparison of the two approaches for introducing the same change into the same gene, so some thought should be invested in deciding what

499

J.F. Loring & S.E. Peterson (eds): Human Stem Cell Manual, Second edition.
DOI: http://dx.doi.org/10.1016/B978-0-12-385473-5.00032-1

method to use. There are caveats to consider for all approaches. Nuclease-based approaches are efficient, but the targeting constructs are difficult to design and produce in the average laboratory, and commercial suppliers are very expensive. Also, efficiency may not be an asset if, for example, the nuclease-based approach introduces an unacceptable level of off-target mutations. Conventional HR methods take advantage of vectors and strategies that are already established over many years for mouse ESCs, and may be a better solution if those resources are readily available and screening large numbers of colonies is a standard procedure in a lab. Conventional HR may also offer more control over the process if the target is sufficiently accessible. The choice of method should depend on the resources and expertise of the researchers.

OVERVIEW

The ability to genetically modify hPSCs increases their usefulness for both clinical and research applications. Several methods have been adopted from mouse embryonic stem cell (mESC) research: introduction of transgenes by viral vectors, lipids, or electroporation, shRNA and siRNA technologies, transient transfections, and homologous recombination (HR) (Giudice and Trounson, 2008; Ma et al., 2003; Zwaka and Thomson, 2003). HR has the advantage in that it can target genetic modifications at specific loci, but it can only do so at low efficiency. In mESC research, the inefficiencies of HR-mediated targeted integration are compensated by the robust culture conditions which can be used to select and grow the targeted cells. HR in hPSCs, however, is more challenging; the rate of HR is orders of magnitude lower than for mESCs, and the culture conditions are not as standardized to select for the modified cells. Multiple groups have achieved successful targeted integration of gene knock-in and knock-out cassettes. Success rates of ~20–30% have been reported, but these estimates are misleading; unlike the convention for mESCs, in which the efficiency is calculated by dividing the number of targeted clones by the number of cells initially treated (usually 10 million), the percentages for hPSCs have been calculated by dividing by the number of colonies that survived the selection process (Giudice and Trounson, 2008; Ma et al., 2003; Zwaka and Thomson, 2003). For hPSC experiments, the number of clones that have the desired genetic modification compared to the starting number of cells transfected is actually in the order of 3 correctly targeted cells per 10 million input cells (0.00003%), which is much lower than the rates reported for mouse cells, which often yield hundreds of correctly targeted clones. In order to increase HR efficiency and decrease the number of clones that need to be screened in order to find a properly corrected clone, several groups (including our own research group) have tested ZFN technology, and found that it has the potential to increase HR efficiencies in hPSCs up to 0.5–3% on a routine basis (Bobis-Wozowicz et al., 2001; Hockemeyer et al., 2009; Soldner et al., 2011; Zou et al., 2009, 2011a, 2011b). This allows

(A) ZFN Candidate design   (B) ZFN Candidate assembly

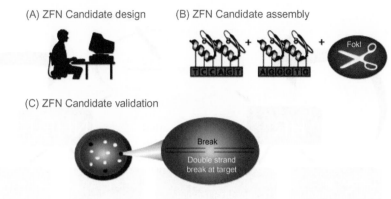

(C) ZFN Candidate validation

FIGURE 32.1
Overview of Zinc Finger
Nuclease Application.
ZFNs are designed using
bioinformatics algorithms
(A), assembled in a modular
fashion (B) and validated in
cell lines (C).

for the identification of targeted clones in a screen of 30–50 stem cell colonies instead of needing to screen thousands of colonies.

ZFNs are engineered restriction enzymes with high specificity for binding to target sequences (Carroll, 2011; Urnov et al., 2010). ZFNs are designed in pairs, each consisting of a zinc-finger DNA binding domain and a split nuclease domain. The DNA binding domain is assembled from tandem repeats of zinc fingers, and the nuclease domain is made of the catalytic domain of the FokI restriction enzyme (Figure 32.1). Each zinc finger recognizes a unique 3 base-pair (bp) motif. ZFNs are engineered to recognize a total of 18–36 bps specified by 6–12 tandem zinc fingers with a 5 or 6 bp gap between the ZFN recognition sites. Given the human genome size (3×10^9 bp), ZFNs allow far better sequence specificity than classical restriction enzymes (e.g. in theory, an 18 bp ZFN recognition site is unique in ~6×10^{10} bp).

ZFNs cut target DNA sequences to create double stranded breaks (DSBs), which activates endogenous DNA repair mechanisms. The two major DSB repair pathways are homology directed repair (HDR) and non-homologous end joining (NHEJ) (Figure 32.2). HDR uses information from either the sister chromatid or an extrachromosomal donor as a template. HDR is conservative in that it faithfully copies information from the template to repair a DSB. NHEJ does not use a template and is therefore error-prone, leading to a repair that may resemble any of the common genetic mutations: insertion, deletion, transition, etc. By manipulating these repair pathways, ZFN technology allows for versatile genome editing capacity depending on the presence or the absence of a donor template, and the choice of the target site. For example, a loss of function mutation can be generated via NHEJ-mediated mutation or introduction of a stop codon by HDR in the coding region. Gain of function mutations can be created by integration of exogenous alleles via HDR or by NHEJ, replicating known activating mutations. ZFNs can be used to alter promoter activities, modulate splice machinery, tag genes, and control microRNA-mediated gene regulation (Figure 32.3) (Cristea et al., 2011; Doyon J.B. et al., 2011; Gutschner, et al., 2011; Moehle et al., 2007).

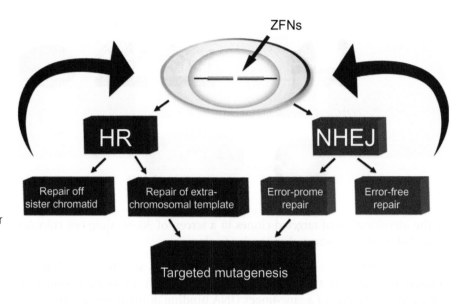

FIGURE 32.2

Mechanisms of ZFN-mediated targeted mutagenesis. Two DNA repair mechanisms, homologous recombination (HR) and non-homologous end-joining (NHEJ), are induced by ZFN-mediated double strand breaks, and can be exploited to engineer genomes

ZFN technology has been successfully applied in many species, including fruit fly, fish, frog, mouse, rat, rabbit, pig, cattle, plants, and various human cells (Beumer et al., 2006; Cui et al., 2011; Doyon et al., 2008; Flisikowska et al., 2011; Hauschild et al., 2011; Shukla et al., 2009; Townsend et al., 2009; Yu et al., 2011). This technology has been successfully applied to hPSCs and has become a practical tool for creating disease models by introducing defined mutations in stem cells and correcting genetic mutations of disease genes (Holt et al., 2010; Lombardo et al., 2007; Soldner et al., 2011). Gene therapy with ZFN technology was successful in a mouse model of Hemophilia B and efficient disease gene correction was achieved in mouse induced pluripotent stem cells (iPSCs) (Li et al., 2011; Sebastiano et al., 2011; Zou et al., 2011b). There are several clinical trials in early stages that use ZFN-modified cells.

As stem cell applications expand, there will be further developments of ZFN targeting specificity and methods of delivery of ZFNs to cells. The specificity of ZFNs is being optimized by a combination of ZFN design and the introduction of a heterodimeric mutation of the FokI domain (Doyon Y. et al., 2011; Miller et al., 2007; Szczepek et al., 2007). ZFNs have been delivered so far to stem cells as plasmids, mRNA, baculovirus, integration defective lentivirus and adeno-associated virus (Handel et al., 2012; Flisikowska et al., 2011; Li et al., 2011; Lombardo et al., 2007; Urnov et al., 2010).

PROCEDURES

One of our first forays into the use of ZFNs in stem cell technology was to make a reporter cell line. We reasoned that a good candidate gene would be

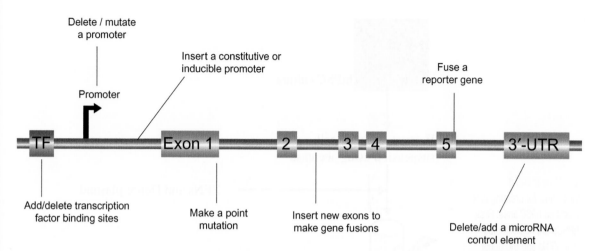

Add/delete transcription factor binding sites

Delete / mutate a promoter

Insert a constitutive or inducible promoter

Fuse a reporter gene

Make a point mutation

Insert new exons to make gene fusions

Delete/add a microRNA control element

FIGURE 32.3
Types of genetic engineering using ZFN-mediated double-stranded breaks. ZFN targeting can be used to add or delete transcription factor binding sites, delete or mutate a promoter, insert a constitutive or inducible promoter, make a point mutation, insert new exons, knock-in a reporter gene, or delete or add a microRNA control element.

one that is expressed in the pluripotent state; hence we first targeted *OCT4*. *OCT4* (*POU5F1*: POU domain class 5 transcription factor 1, also known as *OCT3/4*), is known to play an important role in mouse and human ES cells in maintaining pluripotency (Okamoto et al., 1990; Rosner et al., 1990; Scholer et al., 1990a, 1990b). We targeted OCT4 in human iPS cells with ZFNs to generate a reporter cell line using yellow fluorescence protein (YFP). A ZFN target sequence was identified in the first intron of the gene, and a targeting vector was designed whereby the knock-in disrupts the expression of OCT4 after the first exon and expresses T2A-YFP instead. Using ZFN-mediated HDR, we were able to attain an enriched population of YFP-expressing cells within 2 weeks after introduction of the vectors by Nucleofection. After initial selection for YFP-expressing cells, we used limiting dilution analysis to clone the cells, and subsequently verified clones with Cel-I assay, RFLP, and junction PCR to confirm the presence of the modification. Targeted clones were assayed for characteristics of pluripotency (TRA-1-60 staining, embryoid body assays, and directed differentiation). The workflow for creating ZFN-modified hPSC lines is shown in Figure 32.4 and also see Figure 32.5. Below we describe a method for ZFN targeting that has worked well in our lab.

ZFN Design and Production

Design of specific ZFNs is an important first step for ZFN targeting. There are public web-based tools, such as those provided by the Zinc Finger Consortium (www.zincfingers.org) and the "OPEN" resource (Maeder et al., 2009; Sander et al., 2010), to aid in the design and generation of ZFNs. ZFNs can also be designed and obtained from commercial sources (www.sigmaaldrich.com/life-science/zinc-finger-nuclease-technology/custom-zfn.html).

FIGURE 32.4
Work flow for creating ZFN-modified hPSC lines. First, hPSCs are nucleofected with ZFNs and the donor plasmid. Cells are then expanded in a 6 well plate. From there, some cells are harvested for analysis to confirm the presence of the modification in the population. The assays used are the CEL-I assay, which detects mismatches in double stranded DNA, RFLP to determine whether a restriction site in the donor sequence has been incorporated, and Junction PCR using primers within and outside the donor sequence. The remaining cells are sorted for marker expression (YFP in the example) and single cell cloned. Clones with appropriate modifications are identified using sequencing and the CEL-I, RFLP, and Junction PCR assays. Clones are further analyzed to be sure that they retain pluripotency-associated markers.

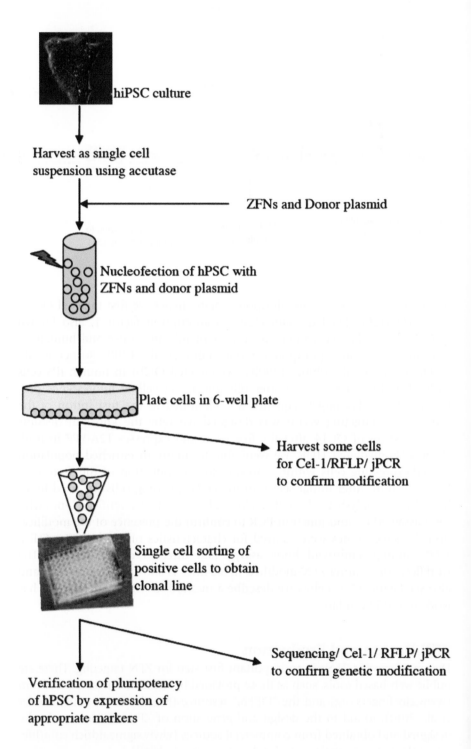

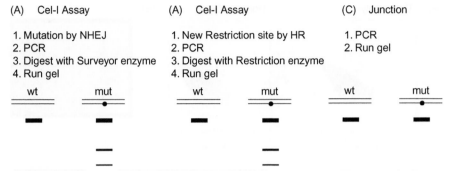

(A)   Cel-I Assay

1. Mutation by NHEJ
2. PCR
3. Digest with Surveyor enzyme
4. Run gel

wt   mut

(A)   Cel-I Assay

1. New Restriction site by HR
2. PCR
3. Digest with Restriction enzyme
4. Run gel

wt   mut

(C)   Junction

1. PCR
2. Run gel

wt   mut

FIGURE 32.5

Overview of assays for genetic modification. (A) The CEL-I assay expected outcome for unmodified (wt) and modified (mut) cells. (B) RFLP assay and expected outcome for unmodified (wt) and modified (mut) cells. (C) Junction PCR and expected outcome. The presence of a band indicates that a genetic modification has occurred.

Culturing and Passaging hPSCs

hPSCs are cultured under feeder-free conditions on extracellular matrix (see Chapter 3 and Recipes). Depending on the size and density of hPSC colonies, cultures are passaged every 4–6 days after seeding using dispase enzyme as described below.

1. Aspirate off the medium and gently rinse the cells with warm DMEM once.
2. Add dispase solution at 1 mg/mL using 4–5 mL for a 100 mm dish or 1 mL for a 6-well plate. Place the dish at 37°C for 7–10 minutes. High-density cultures need longer incubation time.
3. Observe the culture under a microscope. When the edges of colonies start lifting up, aspirate off the dispase solution and gently rinse twice with warm DMEM.
4. Add 4–5 mL of medium and using a cell scraper, gently scrape the colonies off the dish.
5. Use a 10 mL pipette to pipette the cell suspension up and down a few times to break the colonies to smaller aggregates of 4–10 cells. Make sure the cell aggregates do not break up into single cells as it reduces viability.
6. Add the cell suspension to a new 100 mm dish coated with ECM. Each dish can be passaged into 4–6 new dishes depending on cell density. Transfer the dishes back to the 37°C incubator for maintenance.

Preparation of the Cells for Nucleofection

This protocol is intended for one ZFN-mediated targeted integration. For hPSCs, the highest efficiencies are obtained using ZFN mRNA rather than DNA (see Figure 32.6).

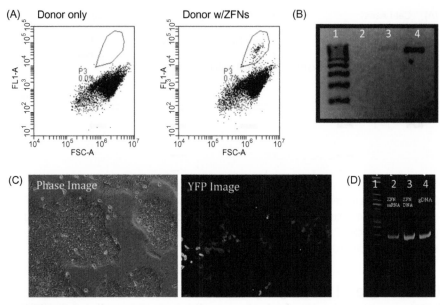

FIGURE 32.6

POU5F1/OCT4-YFP Targeted Integration. (A) hPSCs expressing OCT4-YFP visualized by flow cytometry three days after ZFN-mediated modification. Populations of cells transfected with homologous recombination (HR) construct with or without OCT4 ZFNs were analyzed by flow cytometry. YFP expression is detected only when ZFNs are included (0.7% of cells express YFP). (B) Junction PCR using POU5F1/OCT4- and YFP- specific primers reveals appropriate targeting of the donor construct. Lane 1: Size markers; Lane 2: 5 days after transfection with HR construct alone; Lane 3: 5 days after transfection with HR construct and ZFN; Lane 4: cells transfected with HR construct and ZFN, then enriched by FACS. (C) Cells expressing YFP under control of the endogenous POU5F1/OCT4 promoter remain visible three weeks after targeted integration of the donor construct. (D) Cel-I assay on cells modified with ZFNs. Genetic modification was detected only when ZFNs were introduced as mRNA (Lane2) and not as DNA (Lane3), compared to unmodified cells (Lane 4).

Materials

- 3×10^6 healthy and actively growing cells
- $10 \mu g$ positive control plasmid DNA (CMV-GFP)
- $20 \mu g$ of targeting construct (circular)
- $30 \mu g$ of ZFN mRNA ($1.5 \mu g$/ ZFN)
- Lonza mouse ES Nucleofector kit (VAPH-1001)
- ROCK Inhibitor (Y-27632, 10 mM stock concentration in DMSO, 1000×)
- Hanks Balanced Salt Solution (HBSS)
- Accutase/ROCK Inhibitor (Accutase that contains $10 \mu M$ Y-26732).

1. Harvest cells using Accutase enzyme/ROCK inhibitor (RI) cocktail. Prepare cells by aspirating off growth medium and add 1 mL of Accutase/ RI cocktail, for each well of a 6-well plate. Incubate for 5 minutes at 37°C.

2. Resuspend the cell suspension in HBSS and take a 20 μL aliquot to perform trypan blue exclusion counting using a hemocytometer. Centrifuge the rest of the cells at 200 g for 5 minutes.

3. Count the cells and determine how much volume is required for 3×10^6 cells.

4. Aspirate off the supernatant from the cell pellet and resuspend the cells in 10 mL of HBSS.

5. Transfer the volume of cells necessary for 3×10^6 cells and centrifuge at 200 g for 5 minutes. Meanwhile, add 50 μL mouse ES Nucleofection reagent to three Eppendorf tubes. Add the following to each tube:
   - **Tube 1:** 10 μg positive control plasmid DNA (GFP)
   - **Tube 2:** 10 μg targeting construct (homologous sequence)
   - **Tube 3:** 10 μg targeting construct and 3.0 μg ZFN mRNA pair

■ Note

To adequately assess a targeted integration, three samples should be used for Nucleofections: two controls and one experimental sample. The first control Nucleofection is a positive control GFP expression plasmid. The second control nucleofection is done with a targeting construct only, to assess the random integration background. The third has a targeting construct plus ZFN. This will help to determine the signal to noise ratio of targeted integration over time.   ■

6. After the cells have been pelleted, aspirate off the supernatant and resuspend the cells in 150 μL of nucleofection reagent. Add 50 μL of cell suspension/nucleofection reagent to each of the three tubes containing nucleic acids.

7. Transfer the cell/nucleic acid suspension to a nucleofector cuvette. Perform the nucleofection using the manufacturer's protocol A-023.

8. Immediately, using a transfer pipette, transfer the cell suspension to one well of a 6-well plate containing 4 mL of growth medium and 10 μM ROCK inhibitor.

9. Incubate the cells at 37°C and 5% CO_2.

10. After 48 hours, assay transfection efficiency using fluorescent microscopy. While visualizing cells nucleofected with the positive control, one should observe 75–90% of the cells expressing GFP.

11. After approximately 1 week, transient GFP expression will diminish below background fluorescence and at this point cells can be bulk sorted for YFP expression. This is to prevent positively expressing cells from being out-competed by wild-type cells, especially if the genetic alteration induces a growth disadvantage. Within a week to ten days, single cell sorting can be conducted. Expect a 20% survival rate.

ZFN Activity Verification

Knockout (mutation) or knockin assays at pool or single clone level.

- Knockout assays:
  - CEL-I assay

■ Note

Sequencing can also be used to detect ZFN-induced mutations. ■

- Knock-in assays:
  - RFLP assay
  - Junction PCR

■ Note

Sequencing and selectable markers can also be used to detect knockins. ■

CEL-I Assay
■ Note

The CEL-I assay is a mutation detection assay. CEL I is an endonuclease isolated from celery, that has a high specificity for mismatches, insertions, and deletions in double stranded DNA (Miller et al., 2007). In this assay, a PCR product is amplified from the region that was targeted by the ZFN. The sequences are annealed, and if there is a mismatch, CEL-1 will clip the site. Transgenomics, Inc. has a kit for this assay, which they call the SURVEYOR Mutation Detection Kit for Standard Gel Electrophoresis (Cat. # 706025) (www.transgenomic.com/pd/Surveyor/overview.asp). The following procedure uses this kit. ■

1. PCR amplify the region of interest from a sample of genomic DNA.
2. Take appropriate volume to give 200 ng DNA total (this is an estimate and can be checked by running a small amount of the PCR product against a known standard).
3. Run annealing program (see below) on thermocycler. Keep the samples cool in ice afterwards.
4. Add 1 μL "Surveyor" enzyme and 1 μL "enhancer" (Transgenomic, Inc.)
5. Place at 42°C for 20 minutes.
6. Add loading buffer and run samples on 10% TBE gel, or add 1 μL stop solution if not immediately loading. View gel with ethidium bromide stain (schematic showing procedure and expected outcome is shown in Figure 32.5A).

Annealing Program	
95°C	10 minutes
95°C to 85°C	(−2°C/second)
85°C to 25°C	(−0.1°C/second)
4°C	Hold

RFLP Assays
■ Note
Introduction of a new restriction site allows validation by RFLP.   ■

1. PCR the region of interest using primers outside the arms of homology.
2. Clean up PCR product using a DNA purification column.
3. Digest with a restriction enzyme that cuts only in the donor sequence, for 2 hours.
4. Clean up PCR product using a DNA purification column.
5. Add loading buffer and run samples on 10% TBE gel (Bio-Rad), or add 1 μL stop solution if not immediately loading. View gel with Ethidium Bromide stain (Figure 32.5B).

Junction PCR Assay
1. PCR amplify the region of interest using one primer outside the arm of homology and one primer specific to the donor sequence.
2. Add loading buffer and run samples on agarose gel as described above for CEL-1 assay (Figure 32.5C).

Interpretation of the CEL-I Assay, Junction PCR, and RFLP Assays
The CEL-I assay measures ZFN activity semi-quantitatively. The proportion of cut bands to the total amount of DNA (all bands) gives an estimate of percentage of modified alleles. The sensitivity of this assay is around 1%.

RFLP gives a quantitative measure of targeted integration. The proportion of the mutant band to the total amount of DNA (all bands) is indicative of percentage of targeted integration alleles. The sensitivity of this assay is similar to the CEL-I assay, around 1%.

Junction PCR gives a qualitative measure of targeted integration. It gives a "yes/no" answer, reporting "yes" at frequencies typically above 1/30,000.

Cell Cloning (with or without markers)
1. Dissociate hPSCs with Accutase/RI for 5 minutes. Stop enzymatic digestion with culture medium containing ROCK inhibitor.
2. Centrifuge cells for 5 minutes.
3. Count cells and resuspend at 0.5 cells/100 μL 200g medium with ROCK inhibitor.
4. Transfer 100 μL of cell suspension to each well of a 96-well plate.
5. Repeat for three more 96-well plates.
6. Do not disturb cells for 5–7 days.
7. After 6–7 days, view the plates using a microscope and look for colony formation. Expect only 10–20% of the wells to have cell colonies.

8. Thereafter, feed cells daily by removing only 50 μL media per well and replacing with 50 μL of fresh medium. Be careful not to disturb colonies as they can easily be dislodged by the medium change.

9. By 2 weeks, some colonies may be big enough for enzymatic digestion and transfer to a 24-well plate. Pick the colonies that have maintained typical pluripotent cell morphology and growth characteristics. Harvest the cells using Accutase and transfer to a well of a 6-well plate with ROCK inhibitor. By 3–4 weeks, 10–20 clonal populations should be expected. At this point it is advisable to cryopreserve 2–3 tubes (0.5–1×10^6 cells per tube). Expand the remaining cells in order to harvest genomic DNA for junction PCR and Southern Blot analysis of targeted integration as well as to assess the appropriate phenotype of the marker gene. Once 3–5 clones have met criteria for genotype and phenotype analyses, validate clones for pluripotency and differentiation potential (Figure 32.6).

PITFALLS AND ADVICE

Gene targeting considerations

For targeted integration, the extent of homology can vary. Typically we use 800 bp on each side, however, shorter arms, 200–400 bp, also work and allow for easy detection. At least in some cases, ssDNA oligonucleotides can be used for efficient genome editing.

It is not possible to review all the possible permutations of genetic modifications that can be introduced into hPSCs, but here we consider one of the most frequent objectives: maintenance of expression of a gene while introducing a reporter construct to be co-expressed.

Targeting the 3' end of a gene

It is often useful to target the 3 end of a gene with ZFNs while maintaining expression of the endogenous gene. With ZFNs that cleave within 200 bp of the stop codon, the stop codon can be readily replaced with an IRES or 2A sequence followed by the expression cassette. Usually the expression cassette would code for either a fluorescent protein like GFP to provide visual readout and/or a gene that allows for positive selection by conferring antibiotic resistance to drugs such as puromycin, G418, or blasticidin.

If the tagged gene is expressed in pluripotent cells, then cells with a targeted integration can be readily identified by co-expression of the introduced transgene. However, if the gene is expressed only in differentiated cells, the targeting events will need to be identified by other methods such as PCR. Alternatively, to select for targeting of differentiation-associated genes one can add a second selectable marker to the construct, under the control of a constitutive promoter (e.g. PGK, EF1α), and perhaps flanked by loxP sites to allow its subsequent removal.

Targeting the 5' end of a gene

In general, the 5' ends of genes tend to have higher hit rates for ZFN target design, and many of the approaches used to target 3 ends of genes are applicable. However, it is important to keep in mind that insertion of a selectable marker at the 5' end of a gene may disrupt the expression of the endogenous sequence.

Targeting introns

Constructs can be generated that take advantage of a ZFN that targets an intron. The expression cassette could use a splice donor/splice acceptor strategy to enable transcription of the endogenous gene by alternate splicing, and an IRES or 2A sequence to enable translation of reporter genes.

ZFN mRNA

mRNA is easily degraded. Follow specific precautions for handling mRNA to prevent degradation.

CEL-I Assay Design

The CEL-I assay has a sensitivity of ~1% when detected by Ethidium Bromide staining. One common source of contamination in the CEL-I assay comes from the donor molecule that shares extensive homology to the endogenous locus. Always use a sample transfected with only the donor as a negative control. Try to physically separate the CEL-I assay work area from other lab areas that involve the donor molecules.

FACS

Traditional HDR methods have integration rates of <0.001% of the total nucleofected population, whereas ZFN-mediated HDR leads to integration rates frequently in the range of 1–30%. Regardless of methodology, FACS is a valuable tool to enrich for an integrated population or to positively select a clonal population. We have found that bulk sorting of the cells 7–10 days post-nucleofection is the optimal time to enrich for fluorescent cells. See Chapter 15 for FACS and flow cytometry methods.

Antibiotic Selection

Any of the common eukaryotic antibiotic resistance genes can be used to mark gene expression. Antibiotic resistance genes under the control of pluripotency markers (e.g. *OCT4*) can be used to select for undifferentiated cells. In addition, introducing antibiotic resistance genes into genes associated with specific differentiation pathways will allow selection of specific differentiating cell populations that express the target gene.

Single Cell Cloning

Once cells have been single cell cloned onto a 96-well plate, it is important to note that the cells should not be disturbed for 4–5 days; they do NOT require medium changes. Because there is only one cell per well for several days, the medium is adequate for that amount of time. This reduces the chances of washing away the cloned cells before they have had a chance to adequately attach and multiply several times.

ROCK Inhibitor

For cell sorting and clonal derivation, a single cell suspension of hPSCs must be attained, but this generally leads to a decrease in cell survival. Therefore it is important to use ROCK inhibitor during the enzymatic passaging, FACS sorting, and single cell cloning experiments in order to increase cell survival. Once clones have been selected and expansion begins, ROCK inhibitor is not necessary for everyday medium changes, but during passaging and genetic manipulation, the presence of ROCK inhibitor is highly recommended.

REAGENTS AND SUPPLIES

There are a couple of sources to obtain ZFNs. Protocols are available online for assembling ZFNs yourself, such as those provided by the Zinc Finger Consortium (www.zincfingers.org) and the "OPEN" resource (Maeder et al., 2009; Sander et al., 2010). Alternatively, ZFNs can be designed and purchased from a commercial source (Sigma-Aldrich; www.sigmaaldrich.com/life-science/zinc-finger-nuclease-technology.html).

Item	Supplier	Catalog #
Surveyor enzyme	Transgenomics, Inc.	706025 (25 reactions); 706020 (100 reactions)
Extracellular matrix	Sigma-Aldrich	E1270
mTeSR1 medium	Stem Cell Technologies	05850
DMEM, high glucose	Sigma-Aldrich	D6546
Dispase	Life Technologies	17105-041
Accutase	Sigma-Aldrich	A6964
Phosphate buffered saline	Sigma-Aldrich	P7059
Mouse ES nucleofector kit	Lonza	VPH-5002
Y-27632 dihydrochloride monohydrate (ROCK inhibitor)	Sigma-Aldrich	Y0503
TBE gels	Bio-Rad	345-0053
AccuPrime™ Taq DNA polymerase high fidelity	Life Technologies	12346-094

RECIPES

ECM

1. Thaw the frozen vial of ECM on ice.
2. Aliquot 49.5 mL of DMEM in a 50 mL tube and maintain on ice.
3. Add 500 μL of ECM solution to the 49.5 mL of DMEM and mix.
4. Add 5 mL of the diluted ECM solution to a 10 cm plate and incubate at 37°C for an hour.
5. Remove the ECM solution and add 10 mL of mTeSR1 medium.

Y-27632 Dihydrochloride Monohydrate (ROCK Inhibitor)

1. Add 1.5 mL of DMSO to 1 mg of Y-27632 to obtain a 10 mM solution.
2. Aliquot the solution and store at 4°C. Wrap in aluminum foil to prevent exposure to light.
3. Add 1 μL/mL of medium when needed.

READING LIST

Beumer, K., Bhattacharyya, G., Bibikova, M., Trautman, J.K., Carroll, D., et al. 2006. Efficient gene targeting in Drosophila with zinc-finger nucleases. Genetics 172 (4), 2391–2403.

Bobis Wozowicz, S., Osiak, A., Rahman, S.H., Cathomen, T., et al. 2011. Targeted genome editing in pluripotent stem cells using zinc-finger nucleases. Methods 53 (4), 339–346.

Carroll, D., 2011. Genome engineering with zinc-finger nucleases. Genetics 188 (4), 773–782.

Chen, F., Pruett-Miller, S.M., Huang, Y., Gjoka, M., Duda, K., Taunton, J., et al. 2011. High-frequency genome editing using ssDNA oligonucleotides with zinc-finger nucleases. Nat. Methods 8 (9), 753–755.

Cristea, S., Gregory, P.D., Uknov, F.D., Cost, G.J., et al. 2011. Dissection of splicing regulation at an endogenous locus by zinc-finger nuclease-mediated gene editing. PLoS One 6 (2), e16961.

Cui, X., Ji, D., Fisher, D.A., Wu, Y., Briner, D.M., Weinstein, E.J., et al., 2011. Targeted integration in rat and mouse embryos with zinc-finger nucleases. Nat. Biotechnol. 29 (1), 64–67.

Doyon, J.B., Zeitler, B., Cheng, J., Cheng, A.T., Cherone, J.M., Santiago, Y., et al., 2011. Rapid and efficient clathrin-mediated endocytosis revealed in genome-edited mammalian cells. Nat. Cell Biol. 13 (3), 331–337.

Doyon, Y., McCammon, J.M., Miller, J.C., Faraji, F., Ngo, C., Katibah, G.E., et al., 2008 Heritable targeted gene disruption in zebrafish using designed zinc-finger nucleases. Nat. Biotechnol. 26 (6), 702–708.

Doyon, Y., Vo, T.D., Mendel, M.C., Greenberg, S.G., Wang, J., Xia, D.F., et al., 2000. Enhancing zinc-finger-nuclease activity with improved obligate heterodimeric architectures. Nat. Methods 8 (1), 74–79.

Flisikowska, T., Thorey, I.S., Offner, S., Ros, F., Lifke, V., Zeitler, B., et al., 2000. Efficient immunoglobulin gene disruption and targeted replacement in rabbit using zinc finger nucleases. PLoS One 6 (6), e21045.

Giudice, A., Trounson, A., 2008. Genetic modification of human embryonic stem cells for derivation of target cells. Cell Stem Cell 2 (5), 422–433.

Gutschner, T., Baas, M., Diederichs, S., 2011. Noncoding RNA gene silencing through genomic integration of RNA destabilizing elements using zinc finger nucleases. Genome Res. 21 (11), 1944–1954.

Handel, E.M., et al., 2012. Versatile and efficient genome editing in human cells by combining zinc-finger nucleases with adeno-associated viral vectors. Hum. Gene. Ther. 23(3), 321–329.

Hauschild, J., Petersen, B., Santiago, Y., Queisser, A.L., Carnwath, J.W., Lucas-Hahn, A., et al., 2000. Efficient generation of a biallelic knockout in pigs using zinc-finger nucleases. Proc. Natl. Acad. Sci. USA 108 (29), 12013–12017.

Hockemeyer, D., Soldner, F., Beard, C., Gao, Q., Mitalipova, M., DeKelver, R.C., et al., 2000. Efficient targeting of expressed and silent genes in human ESCs and iPSCs using zinc-finger nucleases. Nat. Biotechnol. 27 (9), 851–857.

Holt, N., Wang, J., Kim, K., Friedman, G., Wang, X., Taupin, V., et al., 2000. Human hematopoietic stem/progenitor cells modified by zinc-finger nucleases targeted to CCR5 control HIV-1 in vivo. Nat. Biotechnol. 28 (8), 839–847.

Li, H., Haurigot, V., Doyon, Y., Li, T., Wong, S.Y., Bhagwat, A.S., et al., 2000. In vivo genome editing restores haemostasis in a mouse model of haemophilia. Nature 475 (7355), 217–221.

Lombardo, A., Genovese, P., Beausejour, C.M., Colleoni, S., Lee, Y.L., Kim, K.A., et al., 2000. Gene editing in human stem cells using zinc finger nucleases and integrase-defective lentiviral vector delivery. Nat. Biotechnol. 25 (11), 1298–1306.

Ma, Y., Ramezani, A., Lewis, R., Hawley, R.G., Thomson, J.A., 2003. High-level sustained transgene expression in human embryonic stem cells using lentiviral vectors. Stem Cells 21 (1), 111–117.

Maeder, M.L., Thibodeau-Beganny, S., Sander, J.D., Voytas, D.F., Joung, J.K., 2009. Oligomerized pool engineering (OPEN): an 'open-source' protocol for making customized zinc-finger arrays. Nat. Protoc. 4 (10), 1471–1501.

Miller, J.C., Holmes, M.C., Wang, J., Guschin, D.Y., Lee, Y.L., Rupniewski, I., et al., 2007. An improved zinc-finger nuclease architecture for highly specific genome editing. Nat. Biotechnol. 25 (7), 778–785.

Miller, J.C., Tan, S., Qiao, G., Barlow, K.A., Wang, J., Xia, D.F., et al., 2011. A TALE nuclease architecture for efficient genome editing. Nat. Biotechnol. 29 (2), 143–148.

Moehle, E.A., Rock, J.M., Lee, Y.L., Jouvenot, Y., DeKelver, R.C., Gregory, P.D., et al., 2000. Targeted gene addition into a specified location in the human genome using designed zinc finger nucleases. Proc. Natl. Acad. Sci. USA 104 (9), 3055–3060.

Okamoto, K., Okazawa, H., Okuda, A., Sakai, M., Muramatsu, M., Hamada, H., 1990. A novel octamer binding transcription factor is differentially expressed in mouse embryonic cells. Cell 60 (3), 461–472.

Orlando, S.J., Santiago, Y., DeKelver, R.C., Freyvert, Y., Boydston, E.A., Moehle, E.A., et al., 2000. Zinc-finger nuclease-driven targeted integration into mammalian genomes using donors with limited chromosomal homology. Nucleic Acids Res. 38 (15), P.e152.

Rosner, M.H., Vigano, M.A., Ozato, K., Timmons, P.M., Poirier, F., Rigby, P.W., et al., 2000. A POU-domain transcription factor in early stem cells and germ cells of the mammalian embryo. Nature 345 (6277), 686–692.

Sander, J.D., Reyon, D., Maeder, M.L., Foley, J.E., Thibodeau-Beganny, S., Li, X., et al., 2000. Predicting success of oligomerized pool engineering (OPEN) for zinc finger target site sequences. BMC Bioinform. 11, 543.

Schöler, H.R., Dressler, G.R., Balling, R., Rohdewohld, H., Gruss, P., 1990a. Oct. 4: A germline-specific transcription factor mapping to the mouse t-complex. EMBO J. 9 (7), 2185–2195.

Schöler, H.R., Ruppert, S., Suzuki, N., Chowdhury, K., Gruss, P., 1990b. New type of POU domain in germ line-specific protein. Nature 344 (6265), 435–439.

Sebastiano, V., Maeder, M.L., Angstman, J.F., Haddad, B., Khayter, C., Yeo, D.T., et al., 2000. In situ genetic correction of the sickle cell anemia mutation in human induced pluripotent stem cells using engineered zinc finger nucleases. Stem Cells 29 (11), 1717–1726.

Shukla, V.K., Doyon, Y., Miller, J.C., DeKelver, R.C., Moehle, E.A., Worden, S.E., et al., 2000. Precise genome modification in the crop species Zea mays using zinc-finger nucleases. Nature 459 (7245), 437–441.

Soldner, F., Laganière, J., Cheng, A.W., Hockemeyer, D., Gao, Q., Alagappan, R., et al., 2000. Generation of isogenic pluripotent stem cells differing exclusively at two early onset Parkinson point mutations. Cell 146 (2), 318–331.

Szczepek, M., Brondani, V., Büchel, J., Serrano, L., Segal, D.J., Cathomen, T., 2000. Structure-based redesign of the dimerization interface reduces the toxicity of zinc-finger nucleases. Nat. Biotechnol. 25 (7), 786–793.

Townsend, J.A., Wright, D.A., Winfrey, R.J., Fu, F., Maeder, M.L., Joung, J.K., et al., 2000. High-frequency modification of plant genes using engineered zinc-finger nucleases. Nature 459 (7245), 442–445.

Urnov, F.D., Rebar, E.J., Holmes, M.C., Zhang, H.S., Gregory, P.D., 2010. Genome editing with engineered zinc finger nucleases. Nat. Rev. Genet. 11 (9), 636–646.

Yu, S., Luo, J., Song, Z., Ding, F., Dai, Y., Li, N., 2011. Highly efficient modification of beta-lactoglobulin (BLG) gene via zinc-finger nucleases in cattle. Cell Res. 21 (11), 1638–1640.

Yusa, K., Rashid, S.T., Strick-Marchand, H., Varela, I., Liu, P.Q., Paschon, D.E., et al., 2000. Targeted gene correction of alpha1-antitrypsin deficiency in induced pluripotent stem cells. Nature 478 (7369), 391–394.

Zou, J., Maeder, M.L., Mali, P., Pruett-Miller, S.M., Thibodeau-Beganny, S., Chou, B.K., et al., 2009. Gene targeting of a disease-related gene in human induced pluripotent stem and embryonic stem cells. Cell Stem Cell 5 (1), 97–110.

Zou, J., Mali, P., Huang, X., Dowey, S.N., Cheng, L., 2011a. Site-specific gene correction of a point mutation in human iPS cells derived from an adult patient with sickle cell disease. Blood 118 (17), 4599–4608.

Zou, J., Sweeney, C.L., Chou, B.K., Choi, U., Pan, J., Wang, H., et al., 2011b. Oxidase-deficient neutrophils from X-linked chronic granulomatous disease iPS cells: functional correction by zinc finger nuclease-mediated safe harbor targeting. Blood 117 (21), 5561–5572.

Zwaka, T.P., Thomson, J.A., 2003. Homologous recombination in human embryonic stem cells. Nat. Biotechnol. 21 (3), 319–321.

PART

6

Stem Cell Transplantation

CHAPTER 33

Intraspinal Transplantation of Human Neural Stem Cells

Lu Chen, Kevin S. Carbajal, Jason Weinger, Ronald Coleman, and Thomas E. Lane

EDITOR'S COMMENTARY

Cell therapy is the first thing most people think of when they consider applications for human pluripotent stem cells. It is critical that the functionality of stem cell derivatives intended for cell therapy be tested in animal models. For neurological disease therapy, the type of transplanted cells and the location of the transplantation are both important issues that must be tested. The first clinical trial using human embryonic stem cell (hESC) derivatives was for spinal cord injury. Geron, Inc., the company that initiated and shepherded the use of hESC-derived oligodendrocyte precursors for spinal cord injury, invested considerable resources in perfecting the technologies and obtaining FDA approval for the trial. However, after four of the proposed eight patients in the Phase I safety trial received transplants, Geron was forced, for financial reasons, to curtail the trial and disband its stem cell program. The company has a commitment to follow up on the Phase I patients for many years, but will not proceed with more patients. In spite of this perceived setback, Geron has done the field a great service by establishing the pathway for obtaining FDA approval for stem cell therapy clinical trials. This chapter describes the first steps required for determining whether cell therapy is a viable approach to treatment of a neurological disease. This chapter outlines how human neural stem/progenitor cells are tested for survival and efficacy in a mouse model. The mouse model in this case is for multiple sclerosis, but the basic concepts could apply to many other neurological diseases.

OVERVIEW

The etiology of the human demyelinating disease multiple sclerosis (MS) is unknown, although genetic factors controlling immune system regulation

519

J.F. Loring & S.E. Peterson (eds): Human Stem Cell Manual, Second edition.
DOI: http://dx.doi.org/10.1016/B978-0-12-385473-5.00033-3

have been suggested to potentially contribute to disease development. In addition, environmental influences, including infectious agents, have also been considered important in the initiation of disease (Ascherio and Munger, 2007a, 2007b; Ebers, Sadovnick, and Risch, 1995; Sospedra and Martin, 2005). The neurotrophic JHM strain of mouse hepatitis virus (JHMV)-infected mouse is a well-established model of viral-induced demyelination. We have shown the intraspinal transplantation of syngeneic mouse neural stem/progenitor cells (NPCs) into JHMV-infected mice is well tolerated and results in extensive remyelination and clinical recovery (Hardison et al., 2006; Totoiu et al., 2004). More recently, we have begun to investigate the therapeutic benefits of human embryonic stem cell (hESC)-derived NPCs using the JHMV model of demyelination. In order to determine whether transplantation of human NPCs is a viable strategy for treating demyelination, it is necessary to better understand the range of environmental conditions that support transplantation-mediated remyelination.

In this chapter, we provide a detailed procedure for intraspinal transplantation of human NPCs including information on cell preparation as well as surgical procedures on mice that enable a successful transplant. We also make note of the critical steps and points one should be wary of, which will help to ensure a successful procedure. In addition, we provide a list of the equipment, supplies, and reagents that have worked well in our laboratory. Note that you must first obtain AALAC approval for doing the procedure in your laboratory.

While establishing this method in your lab, it is important to keep in mind that intraspinal transplants require a team effort. To maximize efficiency, a transplant team manning the four different stations (mouse preparation, laminectomy, injection, and sutures) is desirable. A skilled four-person team can transplant 40 mice in approximately three hours.

PROCEDURES

Preparation of Cells for Transplant

1. Dissociate the cells to be transplanted from tissue culture dishes or flasks using enzymatic dissociation methods; a single cell suspension is critical.

■ Note

The choice of enzymatic dissociation methods varies due to cell types. The gold standard is to achieve a single cell suspension with cell viability ≥90%. ■

2. Wash the cells 3 times with 20 mL cell culture medium in a 50 mL conical tube. The cells are pelleted between the washes by centrifugation at 1500 rpm, 4 °C. Count the cells prior to the final wash.

3. After the final spin down, decant the medium and leave the tube in an upside down position to prevent droplets from reaching the pellet. Using a UV-irradiated, sterile Kimwipe™ to dry the inside walls of the conical tube will help to remove the excess liquid. Do not allow the pellet to dry.

4. Resuspend the cells to a concentration of 100,000 cells/μL by adding in cell culture medium. Transfer the cell suspension to an Eppendorf tube and store on ice.

■ Note

It is essential to remove the residual buffer after the last wash in order to achieve 100,000 cells/μL. One can add in the medium in steps to reach the final desired volume.   ■

5. Check viability of cells if they must be on ice for longer than 2 hours.

Preparation of Mice for Surgery and Transplantation

1. Anesthetize the mouse by intraperitoneal injection of ketamine-xylazine or an equivalent anesthetic. The entire procedure from surgical preparation to suturing (completion) will take 30–40 minutes, so plan accordingly.

■ Note

Mice need to be fully asleep before proceeding to the next steps. It is critical to prevent the mice from moving during the laminectomy and cell injection.   ■

2. Shave the dorsal area of the mouse from the lower back to the neck, and extending 2 cm bilaterally from the midline, with an electric clipper. The hair should be cut as close as possible.

3. To remove the remaining hair, apply a thin layer of hair removal cream (Nair™) with a gauze-tipped applicator.

4. After 1–2 minutes, wipe Nair and the hair off with gauze wetted lightly with soapy water. The prepared area should be clean bare skin without any stray pieces of hair that could get into the wound during the subsequent surgical procedure.

5. Sterilize the prepared area with iodine solution.

Laminectomy

1. Position the mouse dorsal side up with head pointing to the left (if you're right handed). Make a vertical incision (1.3 cm) over the laminectomy site spanning from about thoracic vertebrae T8 to T11.

2. With Graefe forceps held in the left hand, firmly secure the spinal column at thoracic vertebra 8 (T8) and lift the mouse up to exaggerate the spinal curvature. The top of the curvature is T10.
3. Use the scalpel to score the junction between T9 and T10, the space between the two spiny protrusions. Further expose the junction by carefully scraping the muscle layer away to expose the bone.

■ Note

Successful incision between the T9 and T10 junction allows the visualization of spinal cord without further trimming. It reduces bone fragments, which could potentially injure the exposed spinal cord.   ■

4. Use the scissors to further clear muscle away from the lamina around the pedicle with small snips, if the spinal cord can't be seen at this point. Slowly and delicately insert one blade of the scissors into the gap between T9 and T10 and snip the pedicle on both sides. Make sure the curvature of the scissors is always positioned laterally, away from the cord.
5. Lift the lamina to expose the cord and carefully trim it off.

■ Note

Be sure not to leave any free or jagged bone fragments behind.   ■

6. Clean the blood away with sterile cotton swabs.

Injection of Cells

An overview of the spine is shown in Figure 33.1.

1. Assemble the Hamilton syringe and clean it by flushing 200 μL of water though it five times. Repeat the cleaning step with 70% ethanol and HBSS. Insert plunger repeatedly throughout the process.
2. Load the cell suspension into the Hamilton syringe: first, the needle nut and needle need to be loosened from the syringe to prevent backpressure; second, pipette 15 μL of cells and press the pipette tip tightly into the back of the syringe to load it; finally, insert the plunger about 5 mm, tighten the needle nut and depress the plunger until the cell suspension is seen exiting the needle.

■ Note

Make sure there are no bubbles in the syringe and lay the syringe down in the horizontal position to prevent the cells from forming a gradient by gravity. Reload the needle if there are visible cell clumps.   ■

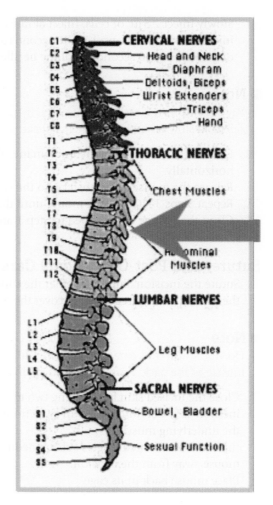

FIGURE 33.1
The spine. *Source:* www
.spinalinjury.net

3. Use a hemostat to grab the spinalis dorsi muscle connecting vertebrae T7 and T8 of the laminectomized mouse.
4. Clamp the hemostat to the left micromanipulator arm in such a position that the mouse's front paws are in the air and its rear paws are lightly touching a platform of sterile paper towels. One should have direct vision of the laminectomy site.
5. Attach the syringe to the right micromanipulator arm at a 70-degree angle and slide the syringe to the lowest position possible before clamping.
6. Stabilize the mouse by pinning its tail against the paper towels.
7. Lower the needle toward the cord until the tip of the needle touches the dorsal midline. Insert the needle 1 mm into the spinal cord. The tip of the needle should be in the gray matter close to the central canal at this position.

8. Slowly inject 2.5 μL of cells. Inject at rate of 1 μL/5 seconds. After injection of the cells, wait for 10 seconds and retract the needle a tenth of a turn every 10 seconds until the needle is out of the cord.

■ Note

Pay attention to possible efflux of cell suspension. Slow injection of the cells and slow retraction of the needle are critical to prevent efflux. ■

9. Detach the syringe from the micromanipulator arm and lay it down horizontally.
10. Release the mouse and transfer it to the suturing table.
11. Repeat Steps 3–10 for each mouse until the syringe is emptied.
12. Clean the syringe as described in Step 1 and reload the needle.

Sutures and Post-Operational Care

1. Suture the incision. Suture together the superficial fascia on both sides of the incision with three knots to cover the exposed spinal cord.

■ Note

Do not suture the cutaneous muscle attached to the skin or the skeletal muscle of the back. Trim the thread as close to the knot as possible. ■

2. Close the incised skin by applying two to three staples depending on the incision size. Carefully pull skin away from the mouse to avoid stapling the underlying muscle.
3. Inject 0.5 mL lactated ringers subcutaneously in the lower back of the mouse away from the incision.
4. Place mouse back in its cage.

■ Note

The mouse should be placed on its side to avoid suffocating. ■

5. Monitor the mouse after anesthesia wears off to ensure bleeding subsides, sutures remain closed, and the mouse return to pre-surgery mobility.

PITFALLS AND ADVICE

Damage to the Spinal Cord

An unsuccessful transplant is often due to damaging the spinal cord during laminectomy. The spinal cord can be easily cut by the scalpel and the scissors used in the procedure. One should ensure that the points of the curved micro-scissors are always facing away from the cord. The sharp bone fragments left behind can also damage the spinal cord. To avoid this, it is critical to accurately locate the junction between thoracic vertebrae T9 and T10 and expose the spinal cord with one incision. Excessive cutting and trimming of the tissue will create small pieces of bones. The laminectomized spine should be examined carefully to ensure that all bone fragments are cleared and that the remaining vertebral structure does not have overtly protruding or jagged edges. Sharp bone edges of the spine will potentially damage the spinal cord during the animal's movement. If an obvious cut is made to the spinal cord, one should make a note and exclude the animal from the study.

Efflux During the Injection

Efflux is another pitfall to avoid during intraspinal transplantation. When the light is shining brightly and directly onto the exposed spinal cord, efflux should be detected easily. The animal should not be used in the study if an efflux occurs. There are a few ways to reduce efflux. Slow injection and slow retraction of the needle are required to avoid efflux. One can wait up to 5 minutes before retracting the needle following injection. Also, smaller gauge needles are preferable. Efflux is more likely to happen with 30-gauge needles than with 33-gauge needles. However, some cells will be lysed when passed through 33-gauge needles. The needle gauge should be chosen by a practical test: run a typical cell suspension through the needle that you would like to use, then compare the viability of the cells before and after their passage through the needle. There should be no loss of viability.

Cell Death Caused by Incubation on Ice

Cells for transplant can undergo lysis if placed on ice for a prolonged period of time (>2 hours) and this will dramatically reduce transplant efficiency. The grouping of mice and the timing for each procedure should be optimized to minimize the time the cells are waiting on ice. For example, we recommend transplanting the mice in groups of four, as each load of the Hamilton syringe contains four injection doses. If everyone in the transplant team is experienced, we suggest that the person injecting the cells starts loading the syringe after the third mouse has been laminectomized and the person preparing the mice should anesthetize the following group after the second mouse of the previous group has been laminectomized. In this manner, a team of 4 people can transplant cells (or control medium) into 40 mice in about 3 hours. One should always check cell viability if the cells have been on ice for more than 2 hours.

EQUIPMENT

- Tissue culture hood: Class II A/B3
- Tissue culture incubator, 37°C, 5% CO_2, in humidified air
- Inverted phase/contrast microscope with 4×, 10× and 20× objectives
- Table-top centrifuge, 1000–1500 rpm, 4°C
- Biosafety hood
- Fiber Optic Illuminator, Fisher Scientific, product # 12-562-36
- Steri 350, Simon Keller AG, product # 06-12287
- Stereotaxic, KOPF Instruments, including: universal holder (model # 1772), electrode holder (model # 1773), animal stereotaxic (model # 902) and left electrode carrier (model# 960)
- 10 µL Hamilton Syringe with removable needle (Hamilton Company, product # 7635-01), Hamilton needles (Hamilton Company, 30G, product # 7803-07, 33G1/2, product # 7803-05)
- Micro-scissors, World Precision Instruments, product # 555500S
- Small Graefe forceps, FST, product # 11053-10
- Hemostat, FST, product # 13010-12
- Olsen-Hegar needle holder, FST, product # 12502-12
- Reflex 7 Wound Clip Applicator, FST, product # 12031-07
- Hair clipper
- Pipettors and pipette aid

REAGENTS AND SUPPLIES

Supplies

- 15 mL and 50 mL sterile conical tubes
- 1.5 mL Eppendorf tubes
- 5 mL, 10 mL, 25 mL sterile disposable pipettes
- Kimwipes
- Gauze tipped applicator
- Gauze
- Sterile cotton swabs
- Scalpels, #10 and #11
- Sutures, Ethicon, product #: 1676G, size 5-0, 3/8″ circle, 19 mm needle, 45 cm braided thread
- 7 mm Reflex wound clips, FST, product #: 12032-07
- 1 mL, 10 mL syringes
- 18G 1/2″ needle

Reagents

Item	Supplier	Catalog #
HBSS	Cellgro	21-022-CM
Ethanol	Gold Shield	
Ketaject	Phoenix Pharmaceuticals	NDC 57319-542-02
Xylazine Hydrochloride	MP Biomedicals	158307
Nair	Church & Dwight Co.	
AddiPak (0.9% NaCl)	Hudson RCI	200-39
Betadine surgical scrub	Purdue Products	Fisher 19-027132
Lactated Ringers	Hospira	NDC 0409-7953-03

Note: *Reagents required for cell dissociation are not listed above, as this will depend on the cell line used.*

RECIPES

Ketamine and Xylazine Anesthetic

Component	Amount	Concentration
Ketaject	0.9 mL per 8 mL solution	80 mg/kg mouse
Xylazine Hydrochloride	0.01 g per 8 mL solution	10 mg/kg mouse

1. Dissolve 0.01 g Xylazine Hydrochloride in 7.1 mL HBSS.
2. Use a syringe to obtain 0.9 mL Ketaject from the sealed bottle and dilute it in the above solution.
3. Filter the solution through 0.22 μm sterile filter.
4. Wrap the tube with foil to avoid the light exposure.

QUALITY CONTROL METHODS

Check Viability of Cells

As we have described in the procedure, it is important to ensure cells are alive before injecting into the spinal cord. We mix 0.4% trypan blue with an aliquot of the cell suspension (1:1, dead cells will be stained blue) and then count the cells with a hemocytometer.

Evaluation of Transplant Efficiency

It is important to evaluate the transplant efficiency, e.g. is it possible to visualize cells within the tissue post-transplant? A variety of methods can be used to accomplish this, including labeling of cells prior to transplant (BrdU), immunohistochemical staining for unique antigen(s), or using cells in which a fluorescent label such as GFP has been genetically incorporated. We have used

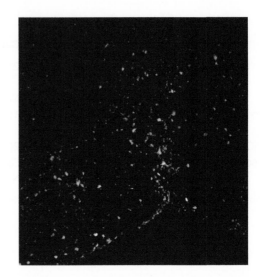

FIGURE 33.2
Transplanted GFP-NPCs survived and migrated within the spinal cords of JHMV-infected mice. Engrafted GFP-NSCs, green; nuclei with DAPI staining, blue.

GFP-positive NPCs (both mouse and human) as a marker to evaluate the transplant efficiency as GFP is, under most circumstances, constitutively expressed and has limited immunoreactivity. In addition, cells can easily be visualized by FITC illumination (Figure 33.2). Importantly, we routinely verify the signal using anti-GFP antibodies as well as transplanting lysed GFP-NPCs to rule out resident cell uptake of GFP.

READING LIST

Ascherio, A., Munger, K.L., 2007a. Environmental risk factors for multiple sclerosis. Part I.: The role of infection. Ann. Neurol. 61 (4), 288–299.

Ascherio, A., Munger, K.L., 2007b. Environmental risk factors for multiple sclerosis. Part II: Noninfectious factors. Ann. Neurol. 61 (6), 504–513.

Ebers, G.C., Sadovnick, A.D., Risch, N.J., 1995. A genetic basis for familial aggregation in multiple sclerosis. Canadian Collaborative Study Group. Nature 377 (6545), 150–151.

Hardison, J.L., Nistor, G., Gonzalez, R., Keirstead, H.S., Lane, T.E., 2006. Transplantation of glial-committed progenitor cells into a viral model of multiple sclerosis induces remyelination in the absence of an attenuated inflammatory response. Exp. Neurol. 197 (2), 420–429.

Sospedra, M., Martin, R., 2005. Immunology of multiple sclerosis. Annu. Rev. Immunol. 23, 683–747.

Totoiu, M.O., Nistor, G.I., Lane, T.E., Keirstead, H.S., 2004. Remyelination, axonal sparing, and locomotor recovery following transplantation of glial-committed progenitor cells into the MHV model of multiple sclerosis. Exp. Neurol. 187 (2), 254–265.

CHAPTER 34

Transplanting Stem Cells into the Brain

Mathew Blurton-Jones, Joy L. Nerhus, and Frank M. LaFerla

EDITOR'S COMMENTARY

Neurological and neurodegenerative disorders have been particularly resistant to traditional pharmaceutical approaches. One of the reasons is that the blood–brain barrier prevents many drugs from reaching the central nervous system, making it difficult to deliver disease-specific therapies. Stem cells are being considered for cell replacement therapy for diseases such as Parkinson's disease, in which a specific neuronal subtype degenerates. But stem cells can play a larger role in treatment of a wider range of neurodegenerative and neurological diseases by acting as delivery systems for growth factors or drugs. These can either be neuron-supporting drugs that help the nervous system survive and regenerate, or drugs that destroy damaging cells in the brain, such as brain tumors. For example, stem cells that carry cancer-destroying drugs are being tested in clinical trials for treatment of aggressive brain tumors like gliomas. This chapter describes methods for delivering stem cells to specific parts of the brain, in this case for testing their potential to help prevent neuronal cell death and improve neuronal function in degenerative diseases such as Alzheimer disease. Cell transplantation methods, like many of the methods in this manual, cannot be learned without hands-on training. This chapter will help prepare a researcher for learning the stereotaxic surgery methods required for accurate delivery of stem cells to a specific region of the brain.

OVERVIEW

A wide variety of approaches have been used to deliver stem cells to the mammalian central nervous system (CNS). By far, the most widely and successfully used strategy involves the direct intracerebral injection of cells via stereotaxic surgery. Although less invasive tactics such as intranasal delivery or venous injection have been explored, there remains little evidence that these approaches provide robust CNS engraftment.

529

J.F. Loring & S.E. Peterson (eds): Human Stem Cell Manual, Second edition.
DOI: http://dx.doi.org/10.1016/B978-0-12-385473-5.00034-5

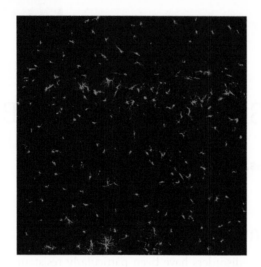

FIGURE 34.1

Human NSCs 6 months after transplantation into an immunodeficient mouse hippocampus.

Procedures for stereotaxic delivery of stem cells to the brain vary considerably from lab to lab. Step-by-step instructions are also extremely difficult to find.

Here we outline protocols for the transplantation of neural progenitor or neural stem cells (NPCs or NSCs) into the adult mouse brain. We provide a detailed account of the equipment and setup required to conduct these experiments. In addition, step-by-step instructions are provided describing the preparation of cells for transplantation and stereotaxic surgery. Transplantation of adherent NSCs is used as an example; however, these same procedures can be readily adapted to transplant virtually any cell type, including both adherent and non-adherent (e.g. neurosphere) populations. Using this approach, our studies indicate that human NSCs can survive and engraft in the adult hippocampus for up to 6 months (Figure 34.1). This chapter also includes a discussion of potential pitfalls and caveats as well as important quality control measures. It is strongly recommended that these aspects be carefully considered before performing your first surgery.

PROCEDURES

It is of utmost importance that cells be prepared and delivered in a manner that maintains their health and viability. Although the procedures outlined below are straightforward, it is critical to begin with a healthy culture of cells and to treat the cells gently throughout the process. If quantification of the cells reveals compromised viability (>10% trypan blue-positive cells), our strong recommendation is to delay surgeries until healthier cells can be obtained. Ideally, viability should be 95% or greater.

Preparation of Adherent Neural Stem Cells for Transplantation

1. Select a healthy 80–90% confluent T75 flask of NSCs for transplantation. A confluent T75 flask will have approximately 6–8 million cells.
2. Rinse the flask with sterile 1× Hank's balanced salt solution (HBSS).
3. Add 3–4 mL of StemPro® Accutase™.
4. Incubate the T75 flask at 37°C for 3–4 minutes, and check for cell dissociation under the microscope. We recommend using the minimum amount of time to dissociate cells.
5. Once the cells have dissociated, neutralize the Accutase with 10 mL of growth medium (no antibiotics).
6. Place the neutralized cell suspension into a 50 mL centrifuge tube.
7. Centrifuge at 1500 rpm (400 g) for 4 minutes.
8. Aspirate supernatant and resuspend cell pellet in 10 mL of antibiotic-free growth medium.
9. Triturate 10 times with a p1000 pipette to break up the pellet, but it is critical to avoid creating bubbles.
10. Centrifuge again at 1500 rpm (400 g) for 4 minutes.
11. Aspirate medium and add 1 mL of vehicle solution (see Recipes).
12. Triturate 10 times with a p1000 pipette to break up the pellet, avoid creating bubbles.
13. Centrifuge for a third time at 1500 rpm (400 g) for 4 minutes.
14. Aspirate buffer and add 50–150 μL of 1× HBSS to cells depending on pellet size. Example: for a larger pellet from a more confluent flask add 150 μL of 1× HBSS.
15. Gently resuspend the cells by trituration six times with a p200 pipette.
16. Estimate the total volume of cells by adjusting a p200 pipette and checking your measurement.
17. Count the cells as described in the section below and adjust the volume by adding the appropriate amount of vehicle to yield 50,000 cells/μL.

Counting and Diluting Cells

1. Add 2 μL of cells (with a p20 pipette) to 198 μL of vehicle to yield a 1:100 dilution.
2. Add 200 μL of trypan blue to the 200 μL of the above cell suspension from Step 1 and add 10 μL to each side of a hemocytometer.
3. For each side of the hemocytometer, count clear cells in all four quadrants and average. Also count dead cells (trypan blue-positive) to estimate cell viability.
4. Average count side A+ average count side B × 1000 = _____cells/μL
5. If the cell density is <50,000 cells/μL, add 10 mL vehicle, re-centrifuge, and resuspend the pellet in a smaller volume and then recount. If cell viability is below 90%, we recommend postponing surgery.

6. Multiply the number of cells obtained in Step 4 by the volume in μL of cell suspension measured in Step 16 above. This is the total number of cells.
7. Calculate how much volume you need to get 50,000 cells/μL and add that much vehicle to the cell suspension (taking into account the volume calculated in Step 16). Example: if you have 6 million cells total then you would need 120 μL total volume, so subtract what volume you already have (measured in Step 16 above) from 120 to determine how much additional vehicle to add.
8. Transfer the cell suspension to a sterile 1.5 mL microcentrifuge tube and place on ice.
9. Prepare three solutions: 1) 95% ethanol (15 mL to clean the needle), 2) vehicle (15 mL to clean the needle), and 3) HBSS (10 mL to keep mouse eyes moist).

Surgical Preparation

1. Clean off the surgical bench and stereotaxic device with 70% ethanol and 10% hypochlorite solution prior to the operation. During the course of the surgery, limit room entry and wear appropriate personal protective equipment.
2. Place a clean cage onto a heating pad for the mouse to have a warm and sanitary environment to recover in following surgery.
3. Prepare a Hamilton syringe. Attach a needle to the syringe. Use either a 33- or 30-gauge needle for cell-based injections. Clean out syringe with 70% ethanol several times followed by vehicle several times. Check for liquid being expelled from syringe to ensure the needle moves smoothly and is not clogged. Be sure to replace the needle with a new one if there is any sign of clogging.
4. Verify the isoflurane absorber (Figure 34.2A) is not full by weighing the canister. For the isoflurane absorber described below, any canister weighing more than 1400 g needs to be replaced with a new isoflurane absorber.
5. Turn on oxygen tank and isoflurane regulator (Figure 34.2B) for anesthesia. Check meter to ensure the level of isoflurane is at or above the midpoint of the two arrows. Add liquid isoflurane if level is below bottom arrow.
6. Set warming plate controller (Figure 34.2C) to 37 °C.
7. Place the mouse into a sealed Tupperware™ container that has a small opening for the isoflurane tubing. Connect the tubing from the isoflurane. To minimize distress, the mouse should be put under anesthesia quickly by turning the isoflurane to the highest setting: 5 L/minute.
8. Once the mouse is anesthetized, turn off the isoflurane and remove the mouse from the Tupperware container, confirm anesthesia by pinching the paw and proceed to the next steps.

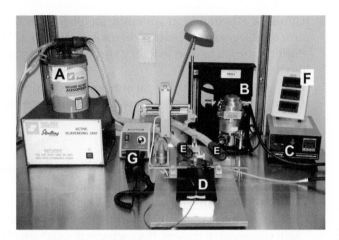

FIGURE 34.2
A typical stereotaxic surgical setup with gas anesthesia. (A) Isoflurane absorber, (B) Isoflurane regulator, (C) warming plate controller, (D) warming plate, (E) gas scavenger tubing, (F) digital display console, (G) Skull drill and controller.

Stereotaxic Surgery

Stereotaxic surgery has been widely used to deliver compounds, place electrodes, and inject viruses into the brain. More recently, stereotaxic surgery has been employed as a practical method for delivering stem cells to the adult mammalian brain. Stereotaxic surgery entails the accurate positioning of a rodent head within a stereotaxic frame and the use of a coordinate system to define specific anatomical targets within 3-dimensional space (reviewed in Athos and Storm, 2001). This approach relies on the availability of detailed brain atlases such as *The Mouse Brain In Stereotaxic Coordinates* (Paxinos and Franklin, 2001). Stereotaxic surgery also relies on the identification of landmarks on the skull, termed Bregma and Lambda, which represent the intersections between the sagittal suture and anterior or posterior sutures (Figure 34.3). Anterior/posterior (AP), medial/lateral (ML) and dorsal/ventral (DV) target coordinates are then defined relative to Bregma. Although it may seem straightforward, positioning a mouse within the stereotaxic frame is one of the more challenging aspects for a novice. We strongly recommend seeking the guidance of an experienced colleague for hands-on training in this critical aspect.

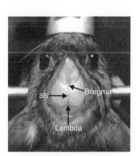

FIGURE 34.3
Correct positioning of a mouse within the stereotax requires the identification of Bregma and Lambda landmarks and accurate placement in "flat-skull position." SS indicates the location of the sagittal suture.

Sterotaxic Set Up

1. Place the anesthetized mouse on a pre-warmed heating plate (Figure 34.2D) and position the teeth over the bite plate. Gently slide the anesthesia mask over its nose.
2. Be sure to position the gas scavenger tubing (Figure 34.2E) near to the anesthesia mask. Set the isoflurane to 2.5 L/minute, and adjust according to breathing rate. If the breathing is fast, turn up the isoflurane (e.g. to 3.0 L/minute). If the breathing is slow, turn down the isoflurane (e.g. to 2.0 L/minute).

3. Lubricate a thermometer with mineral oil and insert into mouse rectum, turn on plate warmer (ensure thermostat is set to 37°C).
4. Position the ear bars so that they lay on the flat surface of the skull just anterior to the ear.
5. Level the top of the skull from anterior to posterior and medial to lateral. Make necessary adjustments by altering the ear bars and level of the collar over the nose.

Surgery

1. Be sure to monitor the mouse's breathing throughout surgery.
2. Trim and shave the top of the head with clippers.
3. Clean off the skin with betadine solution.
4. Pull the skin taut off the scalp behind the ears to make a small hole in the skin with surgical scissors. Place scissors through small hole and with one smooth cut make an incision from the back of the skull to just behind the eyes.
5. Use sterile cotton swabs to spread apart the skin and to clean off the periosteum (clear innermost layer of scalp).
6. Moisten the mouse's eyes with HBSS so they don't dry out.
7. Expose Bregma and Lambda as shown (Figure 34.3).

■ Note

Stereotaxic coordinates are based on a "flat-skull position" in which Bregma and Lambda are located at the same medial/lateral and dorsal/ventral coordinates. Hence, it is critical to check this positioning as follows. ■

8. Move the needle in the AP direction from Bregma to Lambda and back to ensure they are aligned; make adjustments in ear bar placement if the needle does not follow along the sagittal suture (Figure 34.3, SS).
9. Check the dorsal/ventral (DV) plane. Position the needle directly above Bregma and lower the needle until it is just touching Bregma. Zero out all three coordinates by pressing the reset buttons on the digital display console (Figure 34.2F). Lift the needle and move it posterior to Lambda. Lower the needle until it is just touching Lambda. The Z coordinate (dorsal/ventral) on the digital display console should read zero ± 0.10. If not, make adjustments accordingly (e.g. moving the jaw-bar/anesthesia mask up and down) and recheck.
10. Also check the medial/lateral plane by positioning your needle directly above and slightly touching a point on the sagittal suture; zero out Z-coordinates. Pick a lateral point on each side (e.g. ±2.00) lower the needle to the skull and make sure their Z coordinate for both sides is nearly identical (±0.1).

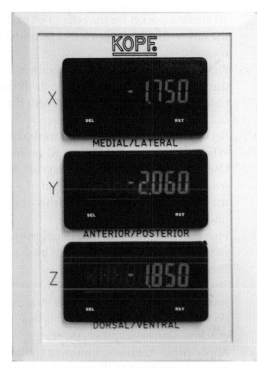

FIGURE 34.4
Coordinates used to target the cell-sparse hippocampal fissure.

11. Position the needle directly above and slightly touching Bregma. Zero out all coordinates and then raise the needle.
12. Adjust the AP and ML coordinates to the designated target. Slowly lower the needle in the DV axis to identify the location on the skull where the bore hole will be drilled. Mark this location with a fine-tipped Sharpie pen.
13. A typical set of coordinates (Figure 34.4) we commonly use to target the cell-sparse hippocampal fissure are: AP: −2.06, ML: ±1.75, DV: −1.85.
14. Drill holes carefully and slowly. Avoid breaching the dura. Double check your drill accuracy by slightly lowering the needle into each hole to ensure that the needle does not touch the edges of the hole. If the needle touches the edges, drill the hole slightly larger to avoid deflecting the needle trajectory. Clean up bone shavings with a sterile cotton swab.
15. Load syringe with the appropriate amount of cells plus an extra 0.5 μL. Eject 0.5 μL from the syringe to make certain air is removed and the needle is not clogged.
16. Lower the needle very slowly into the brain to reach your target DV coordinates.

17. To create a pocket for the cells to sit in, lower the needle 0.1 mm below the target, then slowly (over 30 seconds) bring it back up to the target coordinate.
18. Slowly inject the cells into the brain at a rate of 120 seconds/μL, leave the needle in place and let the cells diffuse for an additional 4 minutes.
19. Using the Dorsal/Ventral adjustment dial very slowly, bring the needle out (1 mm/120 seconds).
20. Repeat Steps 15 and 16 on the contralateral side.
21. When complete, unlock arm and swivel syringe out of the way.
22. Use bone wax to close holes on skull. Brush the bone wax gently over the top with the wooden end of a cotton swab; do not press the bone wax into the holes as this can increase damage to the underlying cortex.
23. Free skin from surface of the skull by pulling up with tweezers (the skin of the scalp tends to stick so it is okay to pull a little hard to release). Press skin together.
24. Stick the tissue of the scalp together with TissueMend. Use cotton swabs to press the incision shut by moving them quickly up and down the incision. TissueMend is heat activated and hardens quickly.
25. Apply Triple Antibiotic ointment (polymyxin B sulfate, bacitracin, and neomycin sulfate) to the wound to help prevent infection.
26. Remove the mouse from the stereotaxic instrument by releasing the ear bars and removing the anesthesia mask. Carefully remove from bite plate by lifting the front teeth from the hole.
27. Inject the mouse subcutaneously with 5 mg/kg of ketoprofen analgesic to minimize discomfort.
28. Place mouse into a pre-warmed clean cage.
29. Monitor mouse for at least 15 minutes to ensure it regains consciousness, normal postural support, and locomotion.
30. Monitor mice daily for any signs of pain or distress and consult with veterinarian if there are any signs of discomfort.

PITFALLS AND CAVEATS

Needle Size/Type

Various needle sizes and types have been utilized for cell transplantation (Blurton-Jones et al., 2009; Cummings et al., 2005; Ebert et al., 2010; Yamasaki et al., 2007). Unfortunately, to date no study has directly compared approaches to identify an optimal needle type or gauge that maximizes cell survival, yet minimizes tissue damage. Rather, researchers have relied on their own unpublished experiences or those of fellow scientists. Without empirical data, we do not want to specifically recommend one approach over another. Although we have had good success using Hamilton microsyringes (Blurton-Jones et al., 2009), colleagues have also achieved excellent engraftment and stem cell survival using beveled glass tips (Cummings et al., 2005). Obviously, the smaller the exterior circumference (higher gauge) of the needle, the less damage to host tissue will occur. However, decreasing inner diameter can also increase shear force on cells and thereby decrease viability.

Anesthesia and Temperature Regulation

Several choices are available for anesthesia, including injectable agents such as avertin and gas-anesthesia such as isoflurane. However, many institutional animal use committees are shifting toward the preferred use of gas anesthesia as it reduces surgical mortality and greatly improves recovery rates. Regardless of which kind of anesthesia is used, mice will quickly become hypothermic without appropriate body temperature regulation. Anesthetic-induced hypothermia should be avoided to reduce perioperative complications and mortality. By using an automated temperature control system coupled to a heating plate and thermometer probe, hypothermia can easily be avoided.

Immunosuppression

Many researchers are interested in transplanting human stem cells into rodent models. For basic biology questions, such an approach may not make the most sense, but for preclinical translational research, xenotransplantation is often necessary. Several strategies can be used to allow transplantation and engraftment of allogeneic or xenogeneic cells into rodents. For researchers studying traumatic brain injury, ischemia, or spinal cord injury, immunodeficient strains can provide an excellent option (Anderson et al., 2011). Using such strains can reduce cost and labor, simplify experimental design, and provide superior long-term engraftment. Commonly used immunodeficient strains including Nod/Scid mice and $Rag2^{-/-}/Il2rg^{-/-}$ double knockouts are available from standard mouse repositories.

Genetically-modified animals, however, are often needed for studies of neurodegenerative disorders. In theory these models can be backcrossed to immunodeficient strains, but in practice this option is both time-consuming and expensive. For disorders that involve robust T- or B-cell responses, such as multiple sclerosis, altering the adaptive immune system could also adversely affect the validity of the model. Hence, other more feasible immunosuppression paradigms are required.

A standard widely-used approach is to deliver calcineurin inhibitors such as cyclosporine-A or FK506 (tacrolimus) via daily intraperitoneal injection. More recently, combinations of FK506 and CD4-neutralizing antibodies have provided improved engraftment (Yan et al., 2006). Unfortunately, we have found that long-term administration of cyclosporine or FK506 causes toxicity in aged mice. Additional promising approaches are being developed, such as combining recombinant CTLA4-Ig, anti-CD40 ligand, and anti-LFA-1 antibodies to target leukocyte co-stimulatory molecules (Pearl et al., 2011). Although these new antibody-based immunosuppressive regimens allow good engraftment with decreased toxicity, they are currently extremely expensive.

EQUIPMENT

- Tissue culture hood: Class II A/B3
- Tissue culture incubator, 37°C, 5% CO_2, in humidified air
- Inverted phase contrast microscope with 4×, 10×, and 20× objectives
- Low-speed centrifuge 300–1000 rpm
- Water bath or Dry Bead Bath, 37°C
- Pipettors, such as Eppendorf p20, p200, p1000
- Pipette aid, automatic pipettor
- Aspirator in the hood
- Refrigerator at 4°C
- Freezers: −20°C, −80°C, and liquid nitrogen cell storage.
- Hemocytometer
- Heating pad
- KOPF Small Animal Stereotaxic Instrument Model 940 including:
    - Digital Display Console (10 micron resolution)
    - Electrode Manipulator with A.P. Slide Assembly Model 960
    - Linear Scale Assembly with Digital Display Model 940-B
    - Electrode Holder with Corner Clamp Model 1770
    - Base Plate Assembly (Base Dimensions 17″ × 10″) Model 900C
    - Mouse Gas Anesthesia Head Holder Model 923-B
    - Bi-Rupture 60 Degree Tip Mouse Ear Bars Model 922
    - Temperature Control System Model TCAT-2LV
    - Heating Plate Model HP-1M
    - Rectal Probe Mouse Model Ret-3-ISO
    - Syringe Holder with Needle Support Foot Model 1772-F-1
- Stoelting Co. Gas Anesthesia Platform plus Mouse mask Catalog #50624
- Stoelting Co. Open Circuit Isoflurane Tabletop System Catalog #50263
- Stoelting Co. Active Gas Scavenger Catalog # 50206
- Stoelting Co. Anesthesia Gas Filters Catalog # 50207
- Stoelting Co. Micromotor Hand piece Drill Kit Catalog #51449
- Stoelting Co. 1.35 mm Drill bit Catalog # 514555
- Hamilton 10 μL Syringe, Model 801
- Hamilton 30-gauge needle 1.0″, Catalog #7803-07
- Clippers, to prepare surgery site

REAGENTS AND SUPPLIES

Supplies

- 15 mL sterile conical tubes
- 50 mL sterile conical tubes
- Sterile pipette tips
- Sterile aspirator pipettes
- T75 flasks
- Mineral oil
- Roboz Moloney Forceps, Catalog # RS-8254
- Roboz Micro Dissecting Scissors 4″ Straight Blunt RS-5980
- Marking pen
- Timer or stopwatch
- Sterile cotton swabs
- Oxygen tank
- Sterile Vacuum Filter Stericup™ Unit 500 mL; pore size 0.22 µm

Recommended Reagents

Item	Supplier	Catalog #	Note
HBSS	Life Technologies	14175-103	Available from other vendors as well
StemPro® Accutase®	Life Technologies	A11105-01	
DMEM/F-12, GlutaMAX™	Life Technologies	10565-018	
N-2 Supplement (100×), liquid	Life Technologies	17502-048	
EGF Recombinant Human	Life Technologies	PHG0311	Available from other vendors as well
Neurobasal Medium (1×), liquid	Life Technologies	21103-049	
GlutaMAX	Life Technologies	35050-061	
B-27® Supplement (50×), liquid	Life Technologies	17504-044	
bFGF – Basic Recombinant Human	Life Technologies	13256-029	
Trypan Blue Solution	Sigma	T8154	
Isoflurane	Western Medical Supply	7263	
Bonewax	Western Medical Supply	1789	
Tissuemend™ II	Western Medical Supply	7372	

RECIPES

■ Note

All solutions should be made in a tissue culture hood.   ■

Vehicle

$1 \times$ HBSS + bFGF (20 ng/mL)

Antibiotic-Free mNSC Growth Medium

1. In a cell culture hood open a 500 mL bottle of DMEM/F12 with GlutaMAX.
2. Add 5 mL of N-2 Supplement.
3. Add EGF (20 ng/mL).
4. Sterile filter.

Antibiotic-Free hNSC Growth Medium

1. In a cell culture hood open a 500 mL bottle of Neurobasal medium.
2. Add 10 mL of B-27.
3. Add bFGF (20 ng/mL).
4. Sterile filter.

QUALITY CONTROL METHODS

Control Groups

As with any experiment, the inclusion of appropriate controls is critical. At the very least a control treatment that closely mimics the surgery and tissue damage caused by stem cell injection should be included. A standard approach is to inject an equivalent volume of vehicle solution into a set of control animals. More rigorous controls include transplantation of an alternative cell type or dead stem cells. For example, fibroblasts have been used as a control for NSC transplantations (Cummings et al., 2005). However, fibroblasts can produce high levels of neurotrophins and thus may mimic some of the neuroprotective effects of stem cell transplantation. An alternative approach is to transplant dead stem cells (Burns et al., 2006). This control can help to determine if there is a role for immune-rejection in any observed effects, although the immune response to dead cells versus live cells is undoubtedly different and thus this control is also subject to varying interpretation. Hence, careful thought should be given as to which controls to include in a given experiment.

Stereotaxic Surgery

Surprisingly, beginning surgeons often find that one of the most difficult aspects of stereotaxic surgery is the accurate placement of mice within the stereotaxic frame. Inexperienced users should seek the guidance of an experienced colleague for hands-on training in this critical aspect. Once one has become comfortable with placing mice within the frame, it is extremely important to perform a pilot study to confirm the accuracy of your target coordinates. For example, perform surgeries on at least six non-experimental mice by injecting either a dye such as Evans Blue or your cells of interest. Mice can be euthanized soon after surgery and the brains analyzed to confirm the correct localization of cells or dye to the targeted region. Every stereotaxic set up is slightly different and such pilot experiments should be performed either when using a new stereotaxic set up or when targeting a new set of coordinates.

Cell labeling

A variety of tactics have been used to label stem cells prior to transplantation to facilitate subsequent identification of the engrafted cells. Unfortunately, some of these approaches lack specificity or can influence cell function or survival. A commonly used approach is to pre-label cells with the thymidine analog bromodeoxyuridine (BrdU). Although this approach can identify engrafted cells, growing evidence suggests that BrdU can also transfer from engrafted cells into large numbers of host neural precursors and glia (Burns et al., 2006). Thus, quantification of stem cell engraftment and differentiation potential via this approach is subject to considerable error. BrdU pre-labeling of NSCs has also recently been shown to affect differentiation potential, cell adhesion, and survival (Lehner et al., 2011). When pre-labeling is required an ideal strategy is to use stem cell populations that stably genetically express a reporter construct such as green fluorescent protein (GFP). For detecting human stem cell populations following transplantation into rodents, several human-specific antibodies are now commercially available. In particular, SC121 (Stem Cells Inc.) provides excellent and specific cytoplasmic labeling of human cells following engraftment (Figure 34.1).

READING LIST

Anderson, A.J., Haus, D.L., Hooshmand, M.J., Perez, H., Sontag, C.J., Cummings, B.J., 2011. Achieving stable human stem cell engraftment and survival in the CNS: is the future of regenerative medicine immunodeficient? Regen. Med. 6, 367–406.

Athos, J., Storm, D.R., 2001. High precision stereotaxic surgery in mice. Curr. Protoc. Neurosci. (Appendix 4:Appendix 4A).

Blurton-Jones, M., Kitazawa, M., Martinez-Coria, H., Castello, N.A., Muller, F.J., Loring, J.F., et al., 2009. Neural stem cells improve cognition via BDNF in a transgenic model of Alzheimer disease. Proc. Natl. Acad. Sci. USA 106, 13594–13599.

Burns, T.C., Ortiz-Gonzalez, X.R., Gutierrez-Perez, M., Keene, C.D., Sharda, R., Demorest, Z.L., et al., 2006. Thymidine analogs are transferred from prelabeled donor to host cells in the central nervous system after transplantation: a word of caution. Stem Cells 24, 1121–1127.

Cummings, B.J., Uchida, N., Tamaki, S.J., Salazar, D.L., Hooshmand, M., Summers, R., et al., 2005. Human neural stem cells differentiate and promote locomotor recovery in spinal cord-injured mice. Proc. Natl. Acad. Sci. USA 102, 14069–14074.

Ebert, A.D., Barber, A.E., Heins, B.M., Svendsen, C.N., 2010. Ex vivo delivery of GDNF maintains motor function and prevents neuronal loss in a transgenic mouse model of Huntington's disease. Exp. Neurol. 224, 155–162.

Lehner, B., Sandner, B., Marschallinger, J., Lehner, C., Furtner, T., Couillard-Despres, S., et al., 2011. The dark side of BrdU in neural stem cell biology: detrimental effects on cell cycle, differentiation and survival. Cell Tissue Res. 345, 313–328.

Paxinos, G., Franklin, K.B.J., 2001. The Mouse Brain in Stereotaxic Coordinates, Second Ed. Academic Press, San Diego, CA.

Pearl, J.I., Lee, A.S., Leveson-Gower, D.B., Sun, N., Ghosh, Z., Lan, F., et al., 2011. Short-term immunosuppression promotes engraftment of embryonic and induced pluripotent stem cells. Cell Stem Cell 8, 309–317.

Yamasaki, T.R., Blurton-Jones, M., Morrissette, D.A., Kitazawa, M., Oddo, S., LaFerla, F.M., 2007. Neural stem cells improve memory in an inducible mouse model of neuronal loss. J. Neurosci. 27, 11925–11933.

Yan, J., Xu, L., Welsh, A.M., Chen, D., Hazel, T., Johe, K., et al., 2006. Combined immunosuppressive agents or CD4 antibodies prolong survival of human neural stem cell grafts and improve disease outcomes in amyotrophic lateral sclerosis transgenic mice. Stem Cells 24, 1976–1985.

Derivation of Human Pluripotent Stem Cells

CHAPTER 35

Development of Human Blastocysts

Natalie Hobson, Ulrich Schmidt, and Steven McArthur

EDITOR'S COMMENTARY

The first human pluripotent stem cells (hPSCs) were embryonic stem (ES) cells derived in Wisconsin from embryos that were donated by patients at Wisconsin and Israeli *in vitro* fertilization (IVF) laboratories. IVF clinics use a variety of assisted reproduction technologies to generate viable embryos. The methods have improved greatly over the years, so that many fewer embryos need to be generated in order to have a good chance at a pregnancy. For example, use of 5-day post-fertilization blastocyst-stage embryos rather than earlier stages allows reproductive biologists to better choose the embryos that are most likely to develop *in utero*. In most countries embryos donated for generation of hESC lines are limited to those that are in excess of the donors' needs. Unlike most of the other chapters, this chapter does not describe a method that is used in a stem cell laboratory. The reason for including this chapter is that I have had the privilege of being invited to observe an embryologist going through the procedures described here to generate human blastocysts. I found it to be so remarkable that I wanted to give other stem cell researchers an idea of what precedes the derivation of an hESC line.

OVERVIEW

Blastocyst-stage embryos consist of two distinct cell types: trophectoderm (TE), which forms extraembryonic tissues, and the inner cell mass (ICM), which develops into both fetal and extraembryonic tissues. The ICM cells exposed to the fluid cavity (blastocoele) are hypoblasts, which develop into extraembryonic membranes. The epiblast cells in the middle of the ICM develop into fetal tissue. Human ES cells arise from an outgrowth of ICM cells, but interestingly are able to make both embryonic and trophoblast *in vitro*.

J.F. Loring & S.E. Peterson (eds): Human Stem Cell Manual, Second edition.
DOI: http://dx.doi.org/10.1016/B978-0-12-385473-5.00035-7

This chapter focuses on developing and culturing embryos in a laboratory up to the point of freezing blastocysts suitable for use in stem cell derivation. This includes a brief overview on the history of *in vitro* fertilization (IVF) as well as the steps of embryo development from oocyte collection and fertilization to blastocyst freezing. It covers the requirements for embryo culture as well as typical vitrification and warming protocols.

LAWS AND ETHICS

Strict laws and guidelines have been developed by most countries to govern the Assisted Reproductive Technologies (ART) sectors of science and medicine. As with all medical procedures, it is essential that there are certain standards and guidelines to ensure the safety of the patients and of those assisting in the procedure. This is even more important when dealing with reproductive technologies, because the creation of embryos adds another level of complexity to the responsibilities of the overseers. Each country or state has individual laws regulating the generation or use of embryos created from human gametes for stem cell derivation. Examples of such guidelines are the "Ethical Guidelines on Assisted Reproductive Technology" (Australia) 2007, and "National Institutes of Health Guidelines on Human Stem Cell Research" (US) 2009.

BACKGROUND

Human Embryo Development *in Vitro*

Many IVF laboratories have dedicated staff embryologists who are responsible for the handling and culture techniques of gametes and embryos. As each embryo has the potential for human life, they must be treated with care and respect and handled accordingly, which is technically challenging due to their microscopic size. At the blastocyst stage, human embryos have a diameter of around 100 μm.

After stimulation, monitoring, and oocyte retrieval the basic steps that are involved in an IVF cycle are insemination, fertilization, embryo culture, and embryo freezing.

Oocyte Insemination and Fertilization

The labeling and identification of gametes and embryos is an extremely important aspect of embryo culture. There must be a visible and accountable chain of custody for all procedures involving gametes, embryos, and patients. All culture dishes and tubes that hold or come into contact with gametes or embryos must have standardized labeling with a minimum of two identifying figures (generally name and a medical number). Most laboratories have implemented a double identification system whereby, whenever possible, two individuals are required to identify each sample.

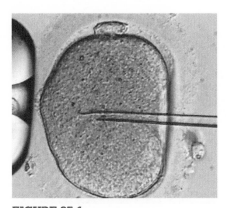

FIGURE 35.1
An oocyte undergoing ICSI injection.

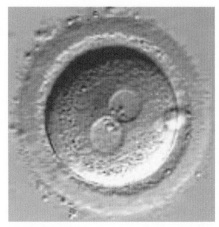

FIGURE 35.2
An embryo containing two pronuclei.

Insemination of the oocytes generally takes place 39–42 hours post-human chorionic gonadotropin (HCG) administration or "trigger" and 3–6 hours post-oocyte retrieval. There are two common methods of insemination; (1) IVF in which a washed and prepared sperm sample is added to the cumulus-enclosed oocyte complexes (COCs) at specific concentrations and left to fertilize, and (2) Intracytoplasmic Sperm Injection (ICSI) (Figure 35.1) whereby individually prepared spermatozoa are selectively chosen by a skilled technician and injected directly into the prepared oocyte. ICSI was developed to be used primarily in situations where there are insufficient sperm to perform standard IVF or the quality of the sperm is questionable. Both techniques achieve similar fertilization outcomes.

While many cellular processes are initiated by the sperm entering the oocyte, this may not always indicate a successful fertilization. At 16–20 hours post-insemination, embryologists are able to detect the presence of two pronuclei which indicates a successful commencement of fertilization (Figure 35.2). The two pronuclei contain the chromosomes from the male and female gametes. Syngamy follows, which involves the combining of the two sets of haploid chromosomes, producing a zygote, which is a single diploid cell.

Embryo Culture

Cleavage begins when the zygote begins its division process initially forming into a two-cell embryo followed by subsequent exponential cell division. By day 2 we expect to see a four-cell embryo (Figure 35.3) and on day 3 an eight-cell embryo (Figure 35.4).

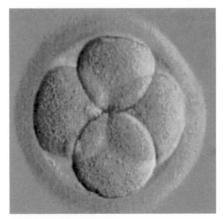

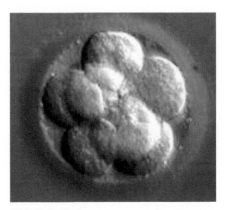

FIGURE 35.4
An eight-cell embryo on day 3.

FIGURE 35.3
A four-cell embryo on day 2.

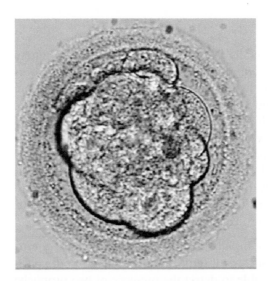

FIGURE 35.5
The morula stage on day 4.

By late day 3 we begin to see compaction of the blastomeres; the outer cells in the embryo form tight junctions in order to seal and isolate the outer cells. At the same time the inner cells form gap junctions or channels that allow small molecules and ions to pass between the cells.

On day 4 the developing human embryo will contain 16–32 cells and the embryo should take on the form of a morula (Figure 35.5). At this point the compacted blastomeres begin to differentiate, with the external cells giving rise to the trophectoderm and the internal cells becoming the inner cell mass. At this stage there may also be signs of cavitation, which is due to the

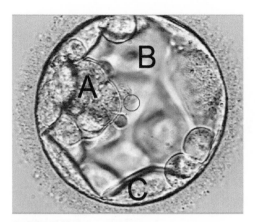

FIGURE 35.6
An early blastocyst on day 5. (A) Inner cell mass;
(B) blastocoele; (C) trophectoderm.

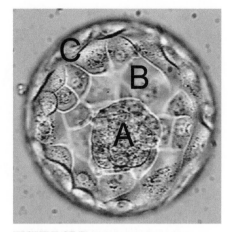

FIGURE 35.7
A fully expanded blastocyst on day 5. (A) Inner cell
mass; (B) blastocoele; (C) trophectoderm.

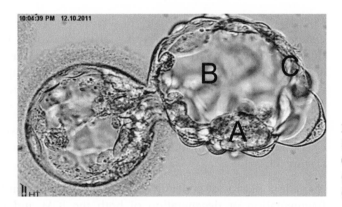

FIGURE 35.8
Hatching blastocyst on day
6. (A) Inner cell mass;
(B) blastocoele;
(C) trophectoderm.

trophectoderm cells beginning to secrete fluid and move towards the edges of the embryo. This is the beginning of the formation of the blastocoele cavity.

On day 5 the blastocoele should be apparent as the embryo develops sequentially from an early blastocyst (Figure 35.6) to a fully expanded blastocyst (Figure 35.7). Cell differentiation will become visible with the clear formation of the inner cell mass (ICM) (A) and the trophectoderm (TE) (C).

The next stage (late day 5 or day 6) is the beginning of the hatching process. Hatching occurs when the embryo frees itself from the surrounding zona pellucida through a series of contractions and enzymatic reactions. Hatching is a natural part of the development of the embryo and is necessary for the eventual implantation into the endometrium (Figures 35.8 and 35.9).

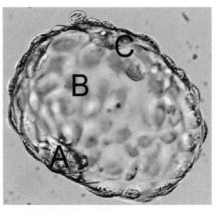

FIGURE 35.9
Fully hatched blastocyst on day 6. (A) Inner cell mass; (B) blastocoele; (C) trophectoderm.

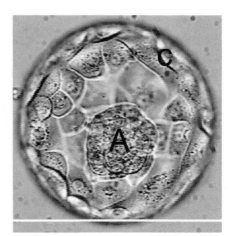

FIGURE 35.10
Grade 1 blastocyst.

Grading of Embryos

Grading of the ICM and TE features of the blastocyst is necessary to determine the overall quality of the embryo, and if being used clinically, its suitability for transfer and freezing. A Grade 1 blastocyst appears as a cavitating blastocyst with a discrete, compact ICM and a good, evenly distributed number of TE cells (Figure 35.10).

A Grade 2 blastocyst shows signs of cavitation with lower cell numbers and/ or a non-discreet ICM and possibly may exhibit some minor degeneration (Figure 35.11).

A Grade 3 blastocyst generally shows signs of significant vacuolization, and fragmentation or degeneration of both the ICM and the TE cells (Figure 35.12). This Grade 3 classification also encompasses any embryo not showing any signs of compaction or cavitation by day 5. Generally most IVF laboratories will freeze Grade 1 and 2 blastocysts, however Grade 3 embryos are generally not considered suitable for freezing.

PROCEDURES

Embryo Culture

The culture system is the most important factor when planning the setup and functionality of an IVF laboratory. The culture system should attempt to mimic the natural environment the embryo would be exposed to in the human body. The type of embryo culture medium and incubation systems

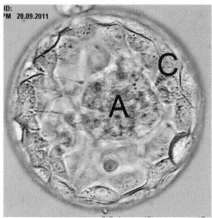

FIGURE 35.11
Grade 2 blastocyst.

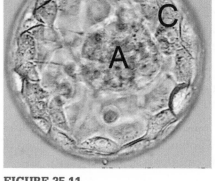

FIGURE 35.12
Grade 3 blastocyst.

used are two of the advances in IVF that have doubled success rates in the last 10 years (Jansen, 2005).

Incubation Systems

Temperature is a critical factor in the successful culture of embryos, with a range between 36.8°C and 37.5°C being the most accepted. Temperatures outside of this range may cause damage to the oocyte or embryo and affect its potential to successfully implant.

pH and osmolarity are also of crucial importance when culturing oocytes and embryos. Most culture medium utilizes a bicarbonate/CO_2 buffer system to keep the pH in the range of 7.2–7.4. The ideal way to control pH levels is to culture in a 5–6% CO_2 environment. Requirements for embryo culture include specific gas mixtures such as a "special mix gas" which consists of 6% CO_2, 5% O_2, and 89% N_2 which is used in commercial bench-top incubating chambers such as the mini incubator or MINC™ for short. It is recommended that all human embryo culture be performed in a reduced oxygen atmosphere (Catt and Henman, 2000).

Embryo Culture Medium

Historically embryo culture medium was a basic salt solution, but pioneers such as Quinn (Quinn et al., 1985) developed this further by incorporating essential components of human fallopian tube fluid. From this research, additions to this basic salt solution included energy sources such as glucose, proteins, and amino acids. There are now a number of companies that specialize in the production of embryo culture medium used to grow embryos.

Further research and refinement (Gardner et al., 2007) has led to the development of sequential embryo culture medium systems that take account of the changing requirements of embryo metabolism over the first 0 to 5 days of development. These systems improved embryo culture by recognizing the importance of small changes in the fallopian tube milieu, including substances that embryos require in a transient manner during early development.

Embryo culture medium containing a phosphate buffer or Hepes buffer is used for procedures requiring handling of gametes outside of the incubator, flushing of follicles, and micromanipulation.

The osmolarity of the culture medium must be in the range of 275–290 m Osm/kg. High humidity in the culture environment ensures that evaporation of the culture medium is kept at a minimum, ensuring a stable osmolarity.

Embryo Culture Procedure

Most laboratories culture in drops of embryo culture medium covered with oil. Droplet sizes of 20 μL and above are usually used for group culture of embryos, while droplet sizes of 10 μL and below are usually used for individual culture of embryos. There are pros and cons for both group and individual culture (Rebollar-Lazaro et al., 2010). Embryos should be cultured under paraffin oil, which prevents evaporation of the medium, preserving a constant osmolarity. The oil also minimizes fluctuations in pH and temperature when embryos are taken out of the incubator for microscopic assessment. Paraffin oil can be toxic to gametes and embryos; therefore it is essential that batches of oil are screened and tested on mouse embryos before use with human embryos. Medium used for the first 3 days of culture typically differs from that used for the next 3 days. We use Embryo Cleavage Medium (Cook Medical) for days 1–3 and Blastocyst Medium for days 4–6.

- Prepare one medium-wash well per embryo on a 4-well dish by pipetting 700 μL of Embryo Cleavage Medium in one well, followed by 300 μL oil. Allow the dish to equilibrate at least 4 hours in the incubator.
- Prepare embryo culture drops of Embryo Cleavage Medium on a 4-well dish by pipetting one or two 10 μL drops into each well, followed by 700 μL Culture Oil on top.
- Place embryo in the medium-wash well and then transfer to the 10 μL drop of Embryo Cleavage Medium for 3 days, observing every day.
- On day 4, wash and culture the embryos in Blastocyst Medium

Vitrification (Freezing) and Warming (Thawing)

Cryogenic storage of embryos in liquid nitrogen or liquid nitrogen vapor (−196°C) provides a "halt" to all biological activity and growth. This allows for indefinite longevity and controlled use of embryos at a suitable time.

Up until 6 years ago, slow freezing was the main method used for cryo-preservation of embyros. Currently, a cryopreservation technique called vit-rification is considered the best method for the freezing of blastocysts. This method employs the use of cryoprotectants to lower the freezing point and protect the cell membranes, and involves freezing the embryo 600 times faster than conventional freezing. Freezing at this rate is so rapid that there is no time for intracellular ice to form. Rather than a physical state change from a liquid to a solid by crystallization, vitrification produces an amorphous state like a "solid liquid" or a glass-like consistency.

Vitrification of Blastocysts
The Cook Blastocyst Vitrification Kit contains:

- A base solution or Cryobase buffer, a 10 mM HEPES buffered medium containing 20 mg/mL HSA (human serum albumin), 0.01 mg/mL gentamycin.
- The permeating cryoprotectant dimethyl sulfoxide (DMSO). This often comes as a separate solution to be added just prior to the vitrification process.
- The non-permeating cryoprotectant ethylene glycol.
- Trehalose, which is a disaccharide that helps cells withstand pro-longed periods of dehydration and also helps prevent disruption of internal cell organelles by effectively splinting them in position.

Preparation
A stepwise set of three solutions is prepared just prior to the vitrification of the embryos to maximize the survival of the embryo. See Recipes: Preparation of Vitrification Medium.

- Solution 1 contains Cryobase buffer only.
- Solution 2 is used for preparation for vitrification and contains Cryobase buffer with 8% ethylene glycol and 8% DMSO.
- Solution 3 is used during cryostorage and contains Cryobase buffer with 0.68 M trehalose, 16% ethylene glycol and 16% DMSO.
- Place the embryo(s) to be cryopreserved into well 1, maximum of three embryos.
- Move the specific embryo to be vitrified into well 2. Have the allocated vitrification device ready.
- Set a timer for 3 minutes but do not start the timer.
- Move the embryo to well 3 for 2 minutes, start the timer.
- With 1 minute remaining on the timer move the embryo to well 4. Complete the next step within 20–30 seconds.

- Using a positive displacement pipette, aspirate $1.5\,\mu L$ of solution, then the embryo and the remaining volume of Vitrification Solution 3 into the pipette. Avoid bubbles.
- Expel the droplet containing the embryo onto the vitrification device and vitrify immediately according to laboratory methods.

Warming of Blastocysts

The Cook Blastocyst Warming Kit contains three medium solutions all based on the formulation of the Cryobase buffer, a 10 mM HEPES buffered medium containing 20 mg/mL HSA, 0.01 mg/mL gentamycin with the addition of Trehalose in various concentrations.

- Solution 1 contains Cryobase buffer with 0.33 M Trehalose.
- Solution 2 contains Cryobase buffer with 0.2 M Trehalose.
- Solution 3 is used for recovery and contains Cryobase buffer only.
- Equilibrate the three Vitrification Solutions to 37°C prior to use.
- Aseptic technique should be used throughout the procedure.
- Equilibrate the three Warming Solutions to 37°C for a minimum of 4 hours before use.
- Prepare a suitable volume of Blastocyst Medium in an incubator with a 6% CO_2 environment at 37°C for a minimum of 4 hours prior to use, for culture of the blastocyst after warming.
- Prepare the Warming Solutions in a 4-well dish by adding $800\,\mu L$ of Warming Solution 1 into wells 1 and 2, $800\,\mu L$ of Warming Solution 2 into well 3, and $800\,\mu L$ of Warming Solution 3 into well 4.
- Identify the vitrification device(s) to be warmed
- Ensure your dish and equipment are all within easy reach
- Immediately extract the embryo from the vitrification device and place into well 1 of your 4-well dish containing your warming solutions and STIR IMMEDIATELY until the bead dissolves.
- Using a pipette, transfer the embryo to well 2. The embryo will shrink to a fully collapsed stage.
- Move to well 3 for 5 minutes. Wash the embryo well.
- Move to well 4 for 5 minutes. Wash the embryo well.
- Place the embryo in the dish containing equilibrated Blastocyst Medium.

EQUIPMENT

- Laminar flow hood for preparation of culture dishes to minimize potential medium contamination (Laftech)
- Stereo dissecting microscope in a gassed warmed chamber for performing day to day tasks of embryo observations

- 4 well plates or other non-toxic petri dishes for culturing embryos (BD Falcon or Nunclon)
- Bench-top incubator or other incubator which maintains 37°C and can have CO_2 or special mix gas pumped into it
- Inverted microscope for performing ICSI and other required manipulations
- Appropriate labels for culture dishes and cryopreservation (Brady)
- Pipettor and tips in appropriate volumes for culture medium (Eppendorf)
- Graduated pipettes and dispenser for oil (BD Falcon)
- Dissecting microscope
- Liquid nitrogen
- Safety PPE wear
- Thermal bucket to hold liquid nitrogen
- Forceps for moving chilled devices
- Vitrification device such as cryotop™
- Mouth pipette and pulled pipettes or similar commercial device
- Storage dewar for long-term storage of vitrified embryos

REAGENTS AND SUPPLIES

Recommended Reagents

Item	Supplier	Catalog #
Cook Medical Sydney IVF Blastocyst Medium	Cook Medical	G20722
Cook Blastocyst Vitrification Kit	Cook Medical	G49621
Cook Blastocyst Warming Kit	Cook Medical	G49626
Cook Medical Sydney IVF Culture Oil	Cook Medical	G32717
Cook Medical Sydney IVF Embryo Cleavage Medium	Cook Medical	G20720

RECIPES

Preparation of Vitrification Medium

This procedure uses the Cook Blastocyst Vitrification kit. Aseptic technique should be used throughout this procedure.

1. DMSO is stored at 2–8°C and should be brought to room temperature prior to use.
2. Snap the neck off the ampule of Vitrification Solution 4 (DMSO) and add 400 µL to 4.6 mL of Vitrification Solution 2 and mix well.
3. Add 1 mL of Vitrification Solution 4 (DMSO) to 5.25 mL of Vitrification Solution 3 and mix well.

4. Prepare the vitrification solutions in a 4 well dish by adding 800 μL of Vitrification Solution 1 into wells 1 and 2, 800 μL of the prepared Vitrification Solution 2 into well 3, and 800 μL of the prepared Vitrification Solution 3 into well 4.

5. Equilibrate the three Vitrification Solutions at 37°C prior to use.

READING LIST

Ethical Guidelines on Assisted Reproductive Technology (Aust.), 2007.

National Institutes of Health Guidelines on Human Stem Cell Research (US), 2009.

Catt, J.W., Henman, M., 2000. Toxic effects of oxygen on human embryo development. Hum. Reproduction 15 (Suppl. 2), 199–206.

Gardner, D.K., Lane, M., 2007. Embryo culture systems. In: Gardner, D. (Ed.), In Vitro Fertilisation: A Practical Approach. Informa Healthcare USA, Inc., pp. 221–282.

Gardner, DK, Lane, M, Stevens, J, Schlenker, T, Schoolcraft, WB., 2000. Blastocyst score affects implantation and pregnancy outcome: towards a single blastocyst transfer. Fertil. Steril. 73, 1155.

Jansen, R.P.S., 2005. Benefits and challenges brought by improved results from *in vitro* fertilisation. Intern. Med. J. 35, 108–117.

Quinn, P, Kerin, JF, Warnes, GM., 1985. Improved pregnancy rate in human in vitro fertilization with the use of a medium based on the composition of human tubal fluid. Fertil. Steril. 44, 493.

Rebollar-Lazaro, I., 2010. The culture of human cleavage stage embryos alone or in groups: effect upon blastocyst utilisation rates and implantation. Reprod. Biol. 10 (3), 227-234.

CHAPTER 36

Derivation of Human Embryonic Stem Cells from Blastocysts

Biljana Dumevska, Julia Schaft, and Ulrich Schmidt

EDITOR'S COMMENTARY

Human embryonic stem cell (hESC) lines were first established in 1998 using very similar culture and conditions that had been developed for the derivation and culture of mouse ESC lines in the 1980s. Since then the methods have improved but not radically changed, but the number of labs deriving hESC lines, and their success rates, have both increased dramatically. Currently, more than 1000 hESC lines are listed on registries such as the International Stem Cell Registry (www.umassmed.edu/iscr; see Chapter 8). Most, but not all, hESC lines are derived from blastocyst-stage embryos, usually from blastocysts that have been cryopreserved and are donated for hESC research because they are no longer needed for reproductive purposes. Many hESC lines carry genetic mutations that were detected by prenatal genetic diagnosis, making them unsuitable for use in reproduction. This chapter describes methods that work very well for derivation of hESCs from blastocyst-stage embryos.

OVERVIEW

In this chapter we focus on protocols for the handling of embryos from the stage when they are thawed to the point they are deemed an embryonic outgrowth and cryopreserved. This chapter will emphasize the methods directly related to the handling of embryos, embryonic outgrowths, and putative hESC lines. This includes methods for embryo handling and culture after thawing, embryo plating, and culture of the initial embryonic outgrowths. The actual *in vitro* fertilization (IVF) including freezing and thawing methods is covered in Chapter 35, as are the ethical issues that surround the use of embryos for the generation of stem cells. Similarly, methods relating to the preparation of mitotically inactivated feeder cells and the expansion, cryopreservation and characterization of putative hESC lines are also described in other chapters (see Chapters 1, 2, 6, and 13–22).

J.F. Loring & S.E. Peterson (eds): Human Stem Cell Manual, Second edition.
DOI: http://dx.doi.org/10.1016/B978-0-12-385473-5.00036-9

The methods we recommend are those that have worked well and are used in our lab. Alternative protocols for zona removal have been added to allow procedures to be modified according to circumstances. We also highly recommend that all protocols should be performed in a biological safety cabinet under sterile conditions using aseptic techniques and preferably disposable consumables.

■ Note

Personal Protective Equipment must be worn when handling harmful substances such as DMSO and liquid nitrogen.   ■

PROCEDURES

Tips for Successful Derivation and Culture

- Avoid the direct use of ethanol around embryos as they are sensitive to harmful compounds. The use of 4% hydrogen peroxide or 1% 7× solution is recommended for cleaning purposes.
- To avoid any fluctuations in pH and O_2 in the culture, gas and temperature, equilibrate all medium for a couple of hours before use or overnight.
- Ideally, thaw the embryo in the same type of medium it was cultured in prior to cryopreservation.
- The criteria for embryo viability for stem cell derivation purposes are different from the clinical criteria for embryo transfer. The presence of only few viable inner cell mass (ICM) cells may be enough for stem cell derivation; hence almost any embryo is worth culturing.
- Use fresh embryos when possible as fresh embryos generally result in better derivation rates.
- Use blastocyst stage embryos for derivation as the derivation rates are much higher due to the presence of the ICM.
- Perform zona removal via cutting immediately after thawing, as the embryo will most likely be collapsed.
- Avoid handling the embryo too much and plate it whole when possible.
- It is important to use fresh feeder layer plates 1–2 days after plating, (see Chapter 2).
- For the initial embryo plating, the 1-well organ culture dish (Figure 36.1) is most suitable as it allows easy access for manipulation.
- Culture all embryo and cell cultures in a controlled humidified atmosphere at 37°C, ideally in an atmosphere of 5–6% CO_2 and 5% O_2, or if not possible, in 5% CO_2 in air. A mini-incubator with separately controlled culture chambers is ideal.
- All cell culture medium should be sterile filtered after preparation and stored in appropriate conditions at 4°C.

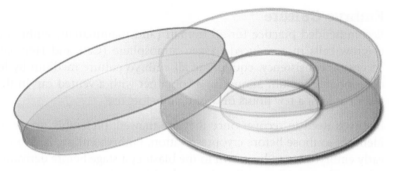

FIGURE 36.1
One-well organ culture dish.

- The start of a stem cell line may be visible anywhere from 3 days after plating up until the first passage; therefore it is worth passaging the outgrowth once, even if there are no promising hESC-like cells visible.
- Observe cultures daily and look for a clump of cells, "bud," or any cells that resemble cobblestones.
- After the first passage it is advisable to maintain the original culture dish for a few days longer and continue observing it. This ensures that any possible loose ICM cells, which may give rise to hESCs, are given the chance to grow.
- Cryopreserve early passage stem cells as early as possible to ensure a back up. Vitrification methods (Reubinoff et al., 2001) are superior to any slow freezing methods, ensuring survival rates of up to 95–100% (see Chapter 35).
- Once vitrified, the hESCs should not be subjected to temperature fluctuations as they can cause the specimens to devitrify and experience chilling injury.
- The use of 20 ng/mL of bFGF in KnockOut™ Serum Replacement (KSR) medium is best for derivation. bFGF can be reduced to the standard concentration of 4 ng/mL once the first colonies are established.

Strategic Planning

The most critical factor in the derivation of hESC lines is the ability to obtain human embryos. The ideal way to achieve this would be through collaboration with an IVF clinic in which couples donate their excess embryos for research purposes. The steps and processes required for this depend on the laws of each state and/or country. It is extremely important to pay special attention to applicable regulatory and ethical guidelines before planning the practical work. Generally, the minimum necessary requirements are written informed consent of all people involved (the donor of all gametes and their respective partners), together with detailed project-specific human ethics (IRB) approval. In some jurisdictions a specific license from regulatory authorities is also required.

Embryo Culture

Recommended practice for all human (and mammalian) embryo culture is to grow cells in a reduced oxygen atmosphere (Catt and Henman, 2000). To ensure consistency, equilibrate all embryo culture medium by leaving an appropriate medium aliquot in a container with a vented cap in the incubator for at least a few hours or overnight before use.

Embryos need to recover once they are thawed, preferably under conditions identical to those before cryopreservation. Embryo culture is also necessary if early embryos are to be grown to the blastocyst stage before derivation. Please refer to Chapter 35 regarding the method for embryo culture.

If using cryopreserved embryos for hESC derivation:

1. Ideally, thaw the embryo in the same type of medium in which it was cultured prior to cryopreservation. The choice of medium for extended culture depends on the clinical practice. Most clinics use a bi-phasic culture approach, in which the medium for days 1 to 3 is different from the medium used for days 3 to 5–6. If the embryo has been cryopreserved at day 3 in Phase I medium (Embryo Cleavage Medium), it is still recommended that it be thawed into the same medium, and only after the initial recovery period moved to Phase II medium (Blastocyst Medium).
2. After the embryo has been thawed using the appropriate thawing/warming method, wash the embryo once in a well containing culture medium and transfer it in a minimum volume into the final embryo culture drop (one embryo per drop). Observe and record the embryo's appearance.
3. Return the dish to the incubator and culture at least 1–2 hours to allow the embryo to recover from freeze–thawing (Figure 36.2).
4. Examine the embryo daily under a dissecting or inverted phase contrast microscope until the desired developmental stage has been reached. Medium changes are not required during culture unless you are using the bi-phasic culture approach (Figure 36.3).

If using fresh embryos for hESC derivation:

1. If the embryo is transferred fresh from the clinical culture to research culture, simply place it into a blastocyst medium drop via one wash in a medium-wash well (see Chapter 35).
2. Examine the embryo daily under the dissecting or inverted microscope until the desired developmental stage has been reached. Medium change is not required during culture.

Removal of Zona Pellucida and Plating of Embryonic Cells

For the highest success rates, it is recommended that embryos are cultured to the blastocyst stage. If the embryo is already hatched then it will require no

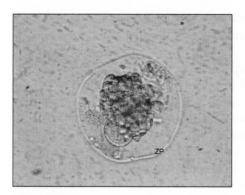

FIGURE 36.2

A collapsed blastocyst 1 hour after thawing. ZP: zona pellucida.

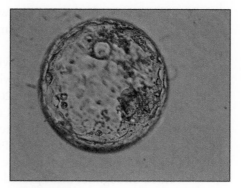

FIGURE 36.3

An expanded blastocyst 1 day after thawing.

further manipulation, or if it is hatching then only minor pipetting of the embryo up and down is required to fully release it from the zona pellucida. Plating is as simple as transferring the embryo over to the feeder layer dish for derivation.

If the zona is still intact at the time when plating is planned, it has to be removed, using enzymatic, acid or mechanical treatment. We recommend the mechanical method over the chemical treatments as it is less time consuming and poses less risk to the embryo. The chemical removal of the zona is explained in the Alternative procedures section.

Embryos can be plated whole (this method is preferable as it poses less threat to the embryo), or the ICM can be isolated. Note the isolation of the ICM is not essential for successful stem cell derivation (Peura et al., 2007). Regular observations and earlier passaging of the outgrowth can minimize the risk of trophectoderm cells inhibiting the growth of ICM-derived stem cells, which is the main concern for whole embryo plating.

The use of feeder cells to support derivation and growth of stem cells is still the preferred option, although feeder-free derivation and culture on extracellular matrices is gaining popularity.

Procedure

1. Assess the embryo microscopically and confirm that it has developed to the blastocyst stage. The embryo grade is less important, as it is possible to derive stem cell lines from poor quality blastocysts. Therefore, it is worth plating every blastocyst.
2. Return the dish to the incubator. Mechanical cutting of zona is least harmful if the blastocyst stage embryo at that time is collapsed, as it

allows more maneuverability with the blade. For example, mechanical zona removal can be done immediately after thawing, when most blastocysts have collapsed.

3. Prepare a feeder dish by removing the feeder medium from feeder cell plate(s) that the embryo will be plated onto and replace with KSR medium supplemented with 20 ng/mL bFGF. Also prepare a 35 mm dish containing 3 mL of warm KSR medium. Return plate(s) to incubator.

4. Pull sterile glass Pasteur pipette(s) in flame to prepare a pipette with a straight-edged large opening. Polish the edges slightly and attach to an embryology-grade mouth hose and keep everything in the laminar flow hood throughout these procedures.

■ Note

You want your pipettes to be slightly bigger than the embryo you are handling so pull a few pipettes with various sized openings to accommodate the varying sizes of the embryos.   ■

5. Prime the pipette with the oil and then the KSR medium before use. Transfer embryo from the embryo culture dish to a 35 mm dish containing warm KSR medium. This acts as a wash step to remove excess oil from entering into the feeder/plating dish.

Zona Pellucida Absent – Hatching or Hatched Embryo

If the embryo is fully hatched, simply transfer the embryo to the feeder/plating dish. If the embryo is hatching then completely release it by gentle agitation with a pipette. This may require gently sucking the embryo up and down using the pipette, taking care not to suck up too much medium as this may cause you to lose the embryo in the pipette or blow out the oil used for priming.

Zona Pellucida Present – Mechanical Removal of Zona Using an Ultra Sharp Splitting Blade

Zona bisection is a very delicate operation and may damage the embryo unless carefully executed. Consequently this technique is recommended only for those with excellent embryo handling skills and steady hands (Figure 36.4).

1. Sterilize an ultra sharp splitting blade with 70% ethanol and allow to dry.

2. Place the 35 mm dish under a stereomicroscope and draw several grooves (very close to each other approximately 10–20 mm apart) onto the bottom of the dish using the sterile blade. Do not cut too heavily, as it will create ridges and furrows that are too deep and encourage the formation of loose plastic spirals. The purpose of the grooves is to secure the embryo, preventing it from slipping and rotating under the blade while attempting to cut it.

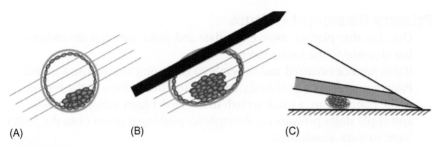

(A)   (B)   (C)

FIGURE 36.4
Graphic representation of the embryo and bisection blade positions for mechanical removal of zona pellucida. (A) The blastocyst placed on top of the grooves drawn on the bottom of the 35 mm dish. (B) The blade is squashing the blastocyst that has been slightly rotated so that the blade can slice the zona diagonally. (C) The position of the tip of the blade for zona slicing as viewed from the side.

3. Rotate the embryo above the created grooves into a suitable position using the blade, not necessarily touching the embryo, but using the flow of medium the blade is creating. A suitable position is where the embryo lies on top of the grooves and the inner cell mass is clearly visible.
4. Press the tip of the blade against the bottom of the dish slightly beyond the embryo, and very gently press the rest of the blade down directly on top of it. The zona pellucida will not be cut cleanly at this point due to its elasticity and resistance, but it will flatten.
5. Once clearly flattened, move the blade carefully sideways, aiming to roll the embryo slightly to its side, positioning the blade so that it is pressing down diagonally on the zona. Keep the tip of the blade gently pressed to the bottom of the dish at all times to give it necessary support.
6. Press the blade down completely, slicing a piece of zona. Always make the cut well away from the inner cell mass, and preferably in the area where the space between the zona and the embryo is the widest.
7. After the zona has been sliced open (if the opening is wide enough), the embryo may be completely released by gentle agitation with a pipette or by applying gentle pressure with the blade. However, often it is necessary to make a few more cuts to the zona in a similar manner to widen the opening.
8. After removal of the last remnants of zona by gentle pipetting, the embryo is ready for plating.
9. Transfer the new feeder dish containing KSR medium to the biosafety cabinet and transfer the embryo whole (if hatched) or the two segments (if accidentally bisected when attempting to remove the zona) to the feeder dish that was prepared earlier.
10. Label the dish and return it to the incubator.

Primary Blastocyst Culture

1. One day after plating observe the plate and make sure that the embryo has attached to the feeder layer.
2. If this has not occurred and the embryo is floating around use a pulled Pasteur pipette with polished edges to direct the embryo to the bottom of the plate. Create a small scratch in the feeder layer using the pipette and apply slight pressure on the embryo, pushing it down onto the feeder layer, to force attachment.
3. Observe cultures daily and carefully change the medium every day or at least every second day.

■ Note

Our experiences suggest changing medium every second day does not have a negative effect on success. However, add fresh bFGF to cultures on the days when medium is not changed, as its half-life in solution is notoriously short.   ■

Passaging of the Embryonic Outgrowth

There is no hard and fast rule on when an embryonic outgrowth needs to be passaged for successful derivation. Observe the cultures daily and note any changes. Cultures may have to be passaged on short notice to obtain the best results.

The following factors contribute to the timing of the first passage:

- Whether hESCs are visible. They can present themselves as a three-dimensional bud-like structure, in which the cells are tightly compacted and appear to be growing up into a ball (Figure 36.5), or they can appear like two-dimensional cobblestones, in which the growth starts off thin and flat (Figure 36.6).
- The age of the feeder cells.
    - The feeder cells' ability to support development will eventually be exhausted. This can be counteracted by starting to use feeder cell-conditioned medium (KSR-medium kept on fresh feeder cells for 24 hours then supplemented with fresh bFGF).
    - Peeling up of the feeders is a problem that is more likely to happen at high feeder densities. The worst-case scenario is having feeders curl up and engulf the embryonic outgrowth, making it practically impossible to recover.

Early passages of embryonic outgrowth are always done mechanically to ensure no loss of cells. The tools for passaging are many and varied and the ultimate choice of tool is completely dependent on the researcher's preferences. When and where to cut is a skill that only experience will teach;

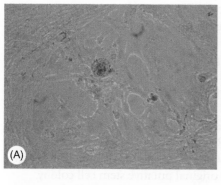
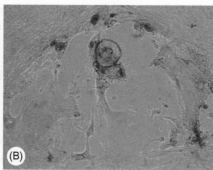

(A)

(B)

FIGURE 36.5
Three dimensional bud
(A) 3 to 5 days and (B) 8
days post-plating.

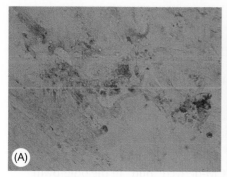
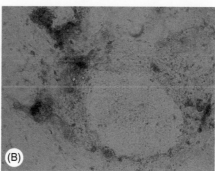

(A)

(B)

FIGURE 36.6
Cobble stone-like cells (A)
6 days and (B) 15 days
post-plating.

generally good outgrowths likely to become stem cells are less problematic than borderline cases, where it is possible to either rescue or destroy a putative stem cell line by the way it is treated.

1. Observe the outgrowths at least every second day to assess the optimal time for passaging. If the embryo was plated whole, watch especially for an epiblast-like outgrowth ("bud") among the trophectodermal cells.

■ Note

ICM and TE-derived cells differ by appearance and are easy to identify; while they first grow as small cells packed tightly together, the latter grow flatter, larger, and more spread out, leading to a "reticular" appearance. Ideally the "bud" is easily distinguished, and if it is not completely surrounded by TE-derived cells, it can be left to grow up to 10 days or even longer. However, if it is tightly surrounded by TE-derived cells, it should be removed from that environment and passaged to a new dish, sometimes even as early as 4 days after plating. ■

2. When the time is right, simply cut the "bud" free of the surrounding cells using the preferred tool with an ultra sharp splitting blade to give the

best control of the cut. Aim at sharp precise cuts, avoiding any "dragging" or ragged edges, as they increase the likelihood of losing precious early putative stem cells (Figure 36.7).

3. Using a flame-pulled, fire-polished glass pipette, gently scrape fragments away from the bottom. This should result in removal of cut colonies without any additional breakage, which is important especially if the "bud" is very small.

4. Transfer the piece(s) either with the same glass pipette or with a p20 pipette to a fresh feeder plate with fresh KSR + bFGF medium. Ideally cut the rest of the outgrowth into few additional pieces and also distribute those pieces evenly in a new dish, away from the original putative stem cell colony.

■ Note

It is also advisable to culture the original dish a few days longer and continue observing it. These measures are to ensure that any possible loose ICM cells, which may give rise to ESC colonies, are given the chance to grow. Since this means additional work, the choice whether to do this depends on the scarcity of the material and how critical it is to try absolutely everything in order to get a cell line from a particular embryo (for example, in the case of a rare disease-specific embryo). ■

5. If it is not possible to clearly see an ICM-derived "bud" in the outgrowth even after culture of 8–10 days, passage the outgrowth anyway, especially if lots of TE-derived outgrowth is present. In this case, cut the whole outgrowth area randomly into pieces and distribute them evenly into a new dish.

6. Continue the medium change routine as before, and observe the development of the Passage 1 outgrowths at least every second day. Eventually if there is no new growth observed in the original dish (for example after a week), it can be discarded. If any promising-looking

FIGURE 36.7
Outgrowth cutting.

colonies emerge (i.e. the 3D bud or 2D cobblestones), passage them either into the Passage 1 dish or into a completely new dish.

7. Perform the second passage in the same manner as the first one. Usually at this stage there are few TE-derived cells still present in the plate. If the outgrowth has grown clearly in size, now is the time to cut it into few pieces.

■ Note

It is difficult to state any absolute rules about early embryonic outgrowth passages, as the decisions about the timing and approach depend strictly on the appearance of the outgrowths. ■

8. Once the outgrowths have clearly transformed to a typical hESC appearance of small, tightly packaged round cells, and when the colony cutting at each passage has resulted in several colonies, the colonies can be treated as any other, established stem cell cultures.

■ Note

We highly recommend that you cryopreserve early passage stem cells as soon as possible to ensure a back up. Vitrification methods (Reubinoff et al., 2001) are superior to any slow freezing methods, as the high survival rates (up to 95–100%) ensure that even one vitrified colony fragment is enough to guarantee the survival of the line in case of loss of ongoing cultures (see Chapter 35). ■

■ Note

It is also a good idea to continue to maintain the master cell banks by manual passaging methods, as they still represent the purest and the least modified early stem cell colonies. See Chapter 1. ■

Medium Change

Changing medium only every second day does not have any negative effects on success. However, add fresh bFGF to cultures on the days when medium is not changed, as its half-life in solution is short.

1. Calculate the amount of medium that is needed and equilibrate accordingly (usually at 37°C and 5% CO_2).
2. Supplement equilibrated medium with bFGF. Higher concentrations of bFGF (20 ng/mL) are advantageous for early outgrowths.
3. Carefully remove cultures, a few at a time, from the incubator and place into the biosafety cabinet (noting not to take cultures that have been passaged on that day).

4. Aspirate and discard medium from culture dishes carefully, not disrupting the feeder cell layer, using either the aspirator or a pipette. Remember to use a different pipette tip for each different cell line.
5. Slowly and carefully add supplemented medium to the culture vessels.
6. Carefully return the culture to the incubator and repeat Steps 3–6 until all cultures have had their medium changed.
7. Record medium change appropriately.

Obtain a Safety Bank of Early Passage hESC

It is important that once you have established and frozen the first initial straw you then expand and obtain a healthy safety stock of your cell line. We suggest vitrifying approximately 20 straws per cell line. This allows you access to early passages in the future and comes in handy if you ever wish to distribute your cell lines or repeat certain experiments. Once the safety bank is obtained, the cell line can be adapted to various ways of passaging (see Chapters 1 and 3) and can be put through vigorous experimentation.

Characterization of Early Outgrowth

Below is how we give our early hESC lines an identity. Generally these tests should be done at the same passage as cells in the safety bank.

Karyotype via Array CGH Analysis

Copy number changes can be detected at a level of 5–10 kilobases of DNA sequence. This method allows us to identify new recurrent chromosome changes such as microdeletions and duplications.

DNA Profiling

DNA profiles are encrypted sets of DNA-based numbers that can be used as a bar code to reflect a cell line's identity. We use this technique to identify our cell lines and make sure there have not been any mix ups.

Immunocytochemistry Analysis of Pluripotency Marker Expression

Different cell types show characteristic gene expression profiles that allow the distinction of stem cells and differentiated cells (see Chapters 16–18). hESCs are generally expected to be positive for specific pluripotency markers (e.g. NANOG, POU5F1/OCT4, SSEA-4) and negative for early differentiation markers (e.g. NESTIN, SSEA-1). This can be assessed by quantitative RT-PCR or preferably by immunofluorescence staining (Chapter 16) as this allows the analysis of individual cells rather than an average of the whole culture. The most elegant method for quantitative analysis of immunofluorescence-stained cells is high-content imaging (e.g. IN Cell imager, GE), which can be performed on small numbers of adherent cells and thus can be performed at a very early stage when only limited cell numbers are available. Other methods include flow cytometry and gene expression analysis (PluriTest) can be used at a later passage when sufficient cell numbers are available (see Chapters 15 and 18).

ALTERNATIVE PROCEDURES

Non-Manual Zona Pellucida Removal
Removal by Pronase

1. Prepare a few drops (20–30 μL) of 4 mg/mL Pronase on a 35 mm Petri dish and cover with oil, allow to warm on a warming stage.
2. Prepare four 35 mm Petri dishes with 3 mL of embryo handling medium each; allow to warm.
3. Transfer the embryo from the culture plate to a Pronase drop in a minimum amount of medium using a suitable transfer pipette.
4. Leave the embryo in Pronase for a few minutes while observing occasionally. Once the zona pellucida appears to be clearly thinned, undulated, and/or expanded, transfer the embryo to the first Petri dish. The zona has not dissolved completely at this stage, but will do so within the next few minutes.
5. Wash the embryo through three dishes allowing the embryo to spend at least 30 seconds in each dish, each time in a minimum volume of medium in order to dilute the Pronase effectively.
6. After the last wash, transfer the embryo to the last Petri dish for bisection and/or plating.

Removal by Acid

1. Prepare a few drops (20–30 μL) of Tyrode's acidic solution in a 35 mm Petri dish and cover with oil, allow to warm on warm stage.
2. Prepare four 35 mm Petri dishes with 3 mL of embryo handling medium each; allow to warm.
3. Transfer the embryo from the culture plate to a Tyrode's acidic solution drop in a minimum amount of medium using a suitable transfer pipette.
4. Observe the embryo constantly, as the effect of acidic solution is very fast, taking usually only 5–10 seconds.
5. Once the zona pellucida appears to be clearly thinned, transfer the embryo immediately to the first Petri dish and wash through the dishes as above.
6. After the last wash, transfer the embryo to the last Petri dish for bisection or plating.

PITFALLS AND ADVICE

- Beware the embryo sticking to the side of the pipette.
  - The embryos are quite sticky, especially once out of their zona, so be sure to use minimal medium volume when transferring it from dish to dish.
  - Priming the pipette in oil and then medium is an important step. This will ensure that the embryo will be picked up in a minimal volume of medium and will prevent it from sticking to the side of the pipette. The oil aids in keeping the embryo in the medium layer and also helps you to know the location of the embryo in the pipette and how much medium can be expelled for the embryo to come out.
  - Be careful not to create bubbles or expel the oil as this will create a visual nightmare and it might make it impossible to locate the embryo. If this does happen, approach the situation calmly and play with the focus. Focus up and down, sifting through the bubbles. The embryo will most likely be floating on top of a bubble.
- Plate all embryos when possible. It only takes one cell to start a hESC line so don't discard embryos solely on poor morphology. Poor quality embryos also produce stem cell lines.
- If the integrity of the trophectodermal layer is compromised by the cutting action when attempting to remove the zona of an expanded blastocyst, the expanded blastocyst will most likely collapse. Don't panic; continue with the removal as it will in all likelihood recover and re-expand just as long as the ICM was not damaged.
- Do not discard platings early on, a stem cell can develop in the later stages of plating, e.g. 10 days after plating.
- Curling of feeder cell layers can cause havoc to a new outgrowth, swallowing it up whole. The older the feeders, the higher the chance of them curling. Observe your platings every day and be careful not to disturb the feeder layer when performing the medium changes. The outgrowth may need to be passaged earlier than recommended just to save the outgrowth in the event of curling feeders.
- Feeder overgrowth can take over a culture. Make sure that the feeders are completely mitotically inactivated.
- Embryos can be lost due to the fact that they still may be floating the day after plating. Make sure the embryo is attached to the feeder layer before performing the first medium change. It is advisable that the first few medium changes be done under a microscope.
- Passaging an outgrowth too small or early can be detrimental to its growth and survival. Allow the outgrowth to form enough cells, e.g. a bud will only get bigger so allow it to grow and transfer the whole bud when possible.

- Be sure that all consents are properly signed and that regulations are followed. If there is a breach in these rules or there are issues concerning consents then a cell line can become unusable and there may be hefty penalties to pay. Have a system in which you have several check and sign off points.
- Keep the embryo environment clear from chemical fumes and smells. Use cleaning and sterilizing agents that don't have an odor and monitor staff and contractors that come in to the lab, making sure to tell them that they are not to wear perfume. An air filter can come in handy especially when unexpected smells filter through the air-conditioning units.

EQUIPMENT

- Stereomicroscope (e.g. Olympus, SZX7).
- Inverted phase/contrast microscope with $4\times$, $10\times$, and $20\times$ objectives
- Tissue culture hood: Class II Biological Safety Cabinet (BSC)
- Tissue culture incubator, 37°C, 5% CO_2, in humidified air
- K-MINC-1000 mini-incubator (Cook IVF, Cat No. K-MINC-1000-US) or Tissue culture incubator, 37°C, 5% CO_2, 5% N2 in humidified air
- Heated stage for the stereomicroscope (LEC instruments, Cat LEC944) and warm stage controller (LEC Instruments, Cat LEC916).
- Pipettors, such as Eppendorf p0.1, p2, p20,p200, p1000
- Low-speed centrifuge 300–1000 rpm
- Water bath, 37°C
- Pipette aid, automatic pipettor for use in measuring and dispensing medium
- Aspirator in the hood
- Refrigerator at 4°C
- Freezers: −20°C, −80°C, and −140°C
- Filter pump
- Retort stand
- Sterile tubing
- Alcohol burner/Bunsen burner

REAGENTS AND SUPPLIES

Supplies

- 5 mL, 10 mL, 25 mL sterile disposable pipettes (Becton Dickinson, Cat# 357543, 357551, 357525)
- Eppendorf Dual Filter tips for the above pipettes (Thermo Fischer. Cat# 0030 077.504, 0030 077.555, 0030 077.571)

- Filter capped culture flasks T25, Primaria T75, T175 (Becton Dickinson, Cat# 353108, 353810, 353112)
- 15 mL and 50 mL sterile centrifuge tubes (Becton Dickinson, Cat# 352096, 352070)
- 35 mm culture dish (Becton Dickinson, Cat# 351008)
- Organ culture dish/IVF dish (BD Falcon Cat# 353037)
- 4-well plates (NUNC, Cat# 179830)
- 0.22 um filters (Millipore, Cat# SVGPB1010)
- Ultra sharp splitting blades (Bioniche Animal Health, Cat# ESE020)

■ Note

These are multi-use blades, which can be cleaned as follows: first, wash the blade by soaking for at least 5 minutes in detergent (e.g. 1% 7× cleaning solution (MP Biomedicals, Cat 76670), then flush with sterile water and finally sterilize with 70% ethanol and allow to air-dry. Just prior to use, flush the blade again with 70% ethanol, allow to air-dry inside the Biological Safety Cabinet, and finally flush in the solution where it is to be used. Alternatively, package the blades individually in autoclave bags and sterilize by irradiation. ■

■ Note

Because of the plastic components these blades cannot be autoclaved or dry heat sterilized! ■

- Embryo transfer pipettes, i.e. glass Pasteur pipettes (Sigma, Cat P 1736), sterilized by autoclaving or dry oven sterilization, flame-pulled and fire-polished just before using. Treat Pasteur pipettes first by acid wash, for example as follows: Soak pipettes overnight in fresh 3% AnalaR HCI (1260 mL HCI in 1 L Milli-Q water), rinse thoroughly in Milli-Q water followed by another overnight soak and rinse in fresh Milli-Q water. Sonicate at 60°C sonicator with Milli-Q water for 10 minutes, rinse once more with fresh running Milli-Q water, and dry in oven. Flame-pull and fire-polish the pipettes just before using to create a suitable size narrow glass pipette.

or

- Disposable sterile embryo transfer pipettes (0.29-0.31 mm, Swemed Int., Cat H-290-310) can also be used.
- Indirect mouth hose (Tek-Event, HPF6237) to be used with flame-pulled Pasteur pipettes.

Recommended Reagents

Item	Supplier	Catalog #	Note
KnockOut-DMEM (lower osmolarity and contains no L-glutamine)	Life Technologies	10829-018	For hESC culture
DMEM-Hepes	Life Technologies	12430-054	For Pronase
KnockOut Serum Replacement (contains bovine products)	Life Technologies	108280-028	Avoid freeze–thawing by aliquotting. Keep thawed aliquots in the fridge for up to 3 weeks only
GlutaMAX™ (200 mM supplied in 0.85% NaCl. Extremely stable in aqueous solution)	Life Technologies	35050-061	
MEM-Non-essential amino acids (100× = 10 mM)	Life Technologies	11140-050	
2-Mercaptoethanol (55 mM in D-PBS)	Life Technologies	21985-023	
Recombinant human bFGF)	Chemicon	GF003	
Pen/Strep (100×)	Life Technologies	15140-122	Avoid freeze–thawing by aliquotting
PBS(−) (Dulbecco's Phosphate-Buffered Saline without Calcium and Magnesium)	Life Technologies	14190-144	
PBS(+) (Dulbecco's Phosphate-Buffered Saline with Calcium and Magnesium)	Life Technologies	14040-133	
Culture Oil	Cook IVF	G32717 or G26708	
Gamete Buffer	Cook IVF	G48258 G48259 or G48260	
Embryo Cleavage Medium	Cook IVF	G20720, G20721 or G19018	
Blastocyst Medium	Cook IVF	G20722 or G20929	
Acidic Tyrode's solution	Sigma	T 1788	Avoid freeze–thawing by aliquotting
Pronase	Sigma	P 8811	

Item	Supplier	Catalog #	Note
Ethanol	Crown Scientific	EA042/2.5	Dilute to 70% v/v with water
7× cleaning solution	MP Biomedicals	76670	Dilute to 1% with water
Hydrogen peroxide 30%	Crown Scientific	HA15425LP	Dilute to 4% with water

RECIPES

Pronase 4 mg/mL (w/v)

- Weigh and dissolve 20 mg Pronase into 5 mL of DMEM-Hepes.
- Filter and aliquot in suitable volumes, e.g. 200 μL each in sterile Eppendorf tubes, and freeze.
- Store at −20°C for up to a year or longer (solution is very stable).
- Avoid thawing and re-freezing! When using, thaw an aliquot and store in the fridge for about a week.

■ Note

The solution does not contain any protein, as Pronase works more effectively without it. ■

KSR Medium

	Ingredient	mL/500 mL	Final Concentration
1	KSR stock aliquot	100	20%
2	MEM-amino acids (100×)	5	1×
3	GlutaMAX (100×)	5	1×
4	β-mercaptoethanol	0.91	0.1 mM
5	Penicillin streptomycin	2.5	50 U (Penicillin) 50 μg/mL (streptomycin)
6	KO-DMEM	386.59	

1. Measure ingredients in numeric order 1–6 (above) into a sterile T175 flask.
2. Fill up to 500 mL with KO-DMEM.
3. Set up the filter pump and retort stand in the laminar flow hood with medium tubing and Millipore GP 0.22 μm filter. Check tubing for deterioration.
4. Discard the first 5 mL of medium, and then filter aliquots of 50 mL into each flask.
5. Label with "KSR Medium," preparation date, and expiration date (10 weeks from preparation)
6. Store in fridge at 4°C.

7. Place a 5 mL aliquot in the incubator for turbidity and mycoplasma testing.
8. 1 week after preparation, check aliquot for turbidity and test for mycoplasma using the MycoAlert kit from Lonza.
9. Release medium for use.
10. Add bFGF on day of use.

QUALITY CONTROL METHODS

Lot-to-Lot Variability of Reagents and Plasticware

Embryos are sensitive to many harmful substances, even to those that are located within the plastics of the consumables you use. It is important that reagents and consumables that are purchased for use with embryos, whether for the culture or for the plating, are tested for embryo toxicity. Most manufacturers will note this on their products but it is also important that you implement your own quality control measures.

Note all lot numbers that are being used. An extra measure could be to test new batches on mouse embryos first before they are released for use on the precious human ones; that way any bad batches are detected well before they are placed into lab rotation.

Established hESC lines are sturdier than embryos. For established hESCs, a minimum of recording lot numbers and testing serum batches and new products (including if it is an already used substance but a different supplier) is enough. Testing new products is as easy as having a side by side study of the old versus the new and it can be as simple as observation for substances such as a different type of FBS.

Cleanliness, Aseptic Technique, and Calibration

- It is important that all equipment is calibrated and cleaned. Being ISO accredited helps as there are protocols to uphold, schedules that need to be maintained, and logbooks to be filled in. Any issue that doesn't pass an internal test or calibration is dealt with externally.
- Test for mycoplasma! Simple mycoplasma testing via an enzymatic detection kit such as the MycoAlert® Mycoplasma Detection Assay (Lonza) and performing a sterility test (placing new medium in quarantine for a week before use) is an easy way to ensure that your cells are free from any contaminants.
- Hands-on training of new staff is imperative, as procedures and standards may differ from lab to lab. A training manual is highly recommended.

Quality Control of Cell Lines

■ Giving your newly derived cell line an identity is essential. Ideally you want to identify your cell line via DNA profiling and karyotyping at the same time/passage as you bank them. This should be done as early as possible to ensure that they are at their purest and least modified stage.

■ It has been proven in our lab that continuous enzymatic passaging can cause chromosomal abnormalities (see Chapters 1 and 3). This has been seen in cultures as young as 10 enzymatic passages. On the other hand the same cell line came back with an identical karyotype after 140+ manual passages. Therefore it is a very good idea to continue and maintain the master cell banks by manual passaging and keep enzymatic passages to a minimal.

READING LIST

Bradley, C.K., Chami, O., Peura, T.T., et al., 2010. Derivation of three new human embryonic stem cell lines. In vitro cellular and developmental biology. Animal 46 (3–4), 294–299.

Bradley, C.K., Scott, H.A., Chami, O., et al., 2011. Derivation of huntington's disease-affected human embryonic stem cell lines. Stem Cells Dev. 20 (3), 495–502.

Loring, J.F., Wesselschimdt Robin, L., Schwartz Philip, H. (Eds.), 2007. Human Stem Cell Manual: A Laboratory Guide. Elsevier, Amsterdam.

Catt, J.W., Henman, M., 2000. Toxic effects of oxygen on human embryo development. Hum. Reprod. 15 (Suppl. 2), 199–206.

Klimanskaya, I., Chung, Y., Becker, S., Lu, S.J., Lanza, R., 2006. Human embryonic stem cell lines derived from single blastomeres. Nature 444, 481–485.

Peura, T.T., Schaft, J., Stojanov, T., 2009. Derivation of human embryonic stem cell lines from vitrified human embryos. In: Turksen, K. (Ed.), Human Embryonic Stem Cell Protocols. Humana Press, Totowa, NJ. pp. 21–54.

Peura, T.T., Schaft, J., Stojanov, T., 2010. Derivation of human embryonic stem cell lines from vitrified human embryos. Methods Mol. Biol. 584, 21–54.

Reubinoff, B.E., Pera, M.F., Vajta, G., Trounson, A.O., 2001. Effective cryopreservation of human embryonic stem cells by the open pulled straw vitrification method. Hum. Reprod. 16, 2187–2194.

Thomson, J.A., Itskovitz-Eldor, J., Shapiro, S.S., Waknitz, M.A., Swiergiel, J.J., Marshall, V.S., et al., 1998. Embryonic stem cell lines derived from human blastocysts. Science 282, 1145–1147.

PART

8

Stem Cells and Society

CHAPTER 37

Stem Cell Patents: What Every Researcher Should Know

Mark L. Rohrbaugh

EDITOR'S COMMENTARY

Most academic researchers know very little about the patent system, and assume that it has no bearing on their lives or work. However, patents on human pluripotent stem cells (hPSCs) have already affected everyone who works in this field, and they promise to have even more impact as practical applications for these cells are developed. In 1998 a patent was issued to James Thomson and assigned to the Wisconsin Alumni Research Foundation (WARF). This patent gave WARF broad rights to all human embryonic stem cells, and it changed this field profoundly. This chapter describes the basic elements of a patent, explains how a patent and a publication differ, and makes it clear how patents affect all human pluripotent stem cell (hPSC) researchers and will continue to have impact on our work in the future.

OVERVIEW

Patenting has become more important to academics in recent years. Patents on biomedical technologies have provided incentives for the development of new products and services throughout the history of the pharmaceutical industry. With the beginning of the biotechnology revolution in the 1970s, however, new challenges had to be sorted out to define the boundary between nature and patentable products derived from nature, such as isolated DNA and recombinant organisms. The Supreme Court seemed to put an end to this controversy in 1980 by ruling that Ananda Chakrabarty's recombinant oil-digesting bacterium was patentable and declaring that the standard for patentability was "anything under the sun made by man" (Diamond v. Chakrabarty, 1980).

More recently, new controversies about the patentability of biological materials have emerged as the science of stem cells has progressed and new legal challenges have been raised in patenting of DNA. In spite of the growing complications, most who watch this process would agree that patents on

J.F. Loring & S.E. Peterson (eds): Human Stem Cell Manual, Second edition.
DOI: http://dx.doi.org/10.1016/B978-0-12-385473-5.00037-0

biological technologies have played an important role in stimulating the growth of the biotechnology industry and hastened the commercial development of new products to treat patients for a wide range of diseases. The controversy plays out most often at the threshold of products of nature; also influential is the manner in which patent rights are used to create incentive or thwart basic research and commercial development. With a focus on human pluripotent stem cells (hPSCs), this chapter will review the core principles of patentability, the patent landscape for stem cells, and how patent rights can be transferred, enforced, and used to promote scientific progress.

WHAT IS A PATENT?

Ever since the founding of the United States, innovation has been recognized as an important driver of economic development and the subsequent improvement of our daily lives. The drafters of the United States Constitution felt so strongly about a national system for the protection of intellectual property that they established it as a power granted to the Congress to enact laws "to promote the Progress of Science and useful Arts, by securing for limited Times to Authors and Inventors the exclusive Right to their respective Writings and Discoveries" (Article I, section 8, clause 8). Patents thus provide incentives to inventors by granting them exclusive rights for a limited period of time to invest in and develop their inventions. The public in return receives instructions on how to make the invention, which otherwise might remain unknown or have to be "reinvented" by each user; access to technologies that inventors develop; and finally free use of the technology after the patent expires. President Lincoln, the only president who was an inventor, famously remarked that "the patent system … added the fuel of interest to the fire of genius in the discovery and production of new and useful things."

Intellectual property rights are distinct from tangible property rights, yet both can be bought and sold. If one buys the latest technological gadget, one owns that particular manifestation of the many inventions used to design and build it (called the "embodiment" of the inventions), but one does not own the patented technology, other than in the gadget that was purchased. Patents are issued by national governments to provide the owner of an invention with the right to exclude others from making, offering for sale, selling, using, or importing the patented invention. In the US, rights in a patent lie with the inventor or the institution to which she has assigned her rights. The proper naming of inventors in the US, unlike the European Union (EU), is important to maintaining a valid patent. A patent can be declared invalid by a court if it can be shown that the list of inventors was made incorrectly with deceptive intent. Unlike conventions for naming authors on a scientific publication, inventorship in the US is based on a legal definition with the order of the names being irrelevant. An inventor is someone who "conceives" of the invention as it is described in at least one claim, which requires understanding not only the idea but also the means of accomplishing it. Those who contribute only to bringing the idea to reality in a laboratory are not inventors, even though they may be authors on a manuscript.

The extent of the rights to exclude others from using an invention ("exclusionary rights") in a patent is defined by the claims; this means that the rights cannot be extracted from a reading of the abstract, but rather are provided by the context of the specification, the understanding of the average person skilled in the art of the technology, and the law as interpreted by the courts. A patent document includes the specification, which must describe or "teach" one skilled in the art to make and use the full scope of the invention without undue experimentation. The patent must set forth the inventors' "best mode" for using it and show that the inventor had possession of the invention at the time of the patent filing (35 USC §112; http://www.uspto.gov/web/offices/pac/mpep/documents/appxl_35_U_S_C_112.htm). For example, discovering one drug that binds to a receptor and claiming all drugs that could possibly bind, but without showing more examples and structure, would not meet this standard. An inventor must claim at least one use for the invention that is "specific, substantial and credible" (Federal Register, Vol. 66, No. 4, pp. 1092–1099), such that a totally speculative claimed use or one that is incidental would not have utility. For example, claiming a cDNA encoding a protein of unknown function to treat people lacking that protein does not show utility. The utility standard, however, does not require scientific proof or regulatory approval.

To be eligible for patent protection, an invention must be comprised of patentable subject matter, have utility, and be novel and non-obvious. Patentable subject matter includes methods of manufacture or use, machines, manufactured materials, compositions of matter (a cell line, for example) or an improvement of these items (35 USC §101; http://www.uspto.gov/web/offices/pac/mpep/documents/appxl_35_U_S_C_101.htm). Claims can be based on materials or living organisms made by human effort, such as isolated and purified microbiological strains, genetically engineered cells, but not laws of nature, materials found in the natural world ("products of nature"), or abstract ideas. Claims for compositions of matter are generally easier to enforce and thus most valuable to companies to protect their products because they cover a product regardless of how it is made. Methods of use for pharmaceuticals are more valuable when products are labeled or sold for the stated use, and there are no competing products available for off-label use. Methods of making a composition are harder to enforce because they are not often apparent when looking at the final product and may not be public knowledge.

The general principle of patentable subject matter has not been controversial in the courts for decades; however, controversy over the boundary line for products of nature has spilled into the courts of late in the case of *ACLU v. Myriad*. A group of patients carrying the BRCA cancer susceptibility gene and physicians treating them have sued Myriad Genetics, which is the sole source for the BRCA diagnostic testing service. They argue that some of Myriad's patent claims are invalid because DNA is a non-patentable product of nature. A New York federal court agreed, but a three-judge panel at the court of appeals reversed that aspect of the decision. The patient groups petitioned the court to rehear the case in light of a recent Supreme Court case, *Mayo V. Promethius*, which ruled that a method of optimizing the dosing of a certain drug was an unpatentable law of nature. While any ruling in this case will be limited to nucleic acids, the rationale for defining a product of nature may well have implications for other products derived from nature such as isolated cells and cell lines.

An invention is not novel, and thus not patentable, if it has been (a) in public use in the US, (b) patented, or (c) described in a printed (paper or electronic) publication anywhere in the world (i) prior to the applicant's invention or (ii) more than 1 year prior to the filing of the invention in the US, termed a "grace period" (35 USC §102). Collectively, these types of public knowledge are referred to as "prior art." Thus, a scientist's published work becomes prior art a year after publication, making it impossible to file a patent on that work. For published articles, inventors should keep in mind that the date of any on-line pre-publication is a relevant date for purposes of establishing prior art. In 2013, under a new law, the standard for relevant prior art will change as described below.

A patentable invention must not be obvious to one of "ordinary skill in the art" based on the prior art at the time of the invention. However, the ease of making the invention does not necessarily make it obvious – obviousness in hindsight alone does not count (35 USC §103). To make a case for obviousness, the patent examiner can use prior art from one or more documents, even the inventor's own articles from more than 1 year before filing, that include various components of an invention and suggestions to combine them together. The non-obvious requirement is complex, particularly for biological inventions, and its interpretation has changed over time under court rulings.

Similar requirements exist in other countries, such as those under the European Patent Office (EPO), where the invention must be susceptible to "industrial application, novelty and inventive step" (European Patent Convention 2006, Art. 52). The novel standard, however, is different outside the US where patent systems are based on the first-to-file, an absolute-novelty system, and there is no grace period for publications prior to the filing date of the invention. In the biotechnology field, where most companies seek an exclusive market share in both the US and Europe, companies and public sector institutions attempt to file applications prior to the date of first disclosure by the inventors in order to retain the option to file the patent in both the US and other high-income countries.

Some countries have used patents laws to enforce moral standards. The EU and India have statutory authority to reject inventions contrary to public order and morality, which has affected patents on pluripotent stem cells. Canada patent law forbidding the patenting of higher life forms has led to the rejection of claims to totipotent but not pluripotent stem cells.

OBTAINING A PATENT ON AN INVENTION

The US Patent and Trademark Office (PTO) has the responsibility to provide evidence that an invention described in a patent application is not patentable. The PTO typically makes arguments for rejecting the patent, and the applicant responds to the argument or amends the claims to overcome the objection. This to and fro process of legal arguments with a patent office is called "prosecution," and if the patent office finds no reason ultimately to disallow a patent, it is issued with a term of 20 years from the earliest date of filing. The PTO staff is chronically overextended and it often takes several years for a patent to be issued. Thus, if it takes 4 years for the patent to be approved, the remaining enforceable term is 16 years.

In the US, the inventors, rather than institutions, are entitled to obtain a patent; that is, the patent is "vested" with the inventors. But in most cases the inventors are employed by a for-profit or academic institution, and have signed a contract that promises to transfer ownership of ("assign") their patents to the institution. When there are multiple inventors, each co-inventor or co-assignee has a right to use and enforce the entire patent. When an invention is made with US Government grant or contract funding (called a "subject invention"), the institution that receives the funding is bound by the terms of the Bayh-Dole Act. This law requires the institution to have a contractual agreement with the inventor to assign the invention to the institution and gives the institution the right to file a patent application for the government-funded invention within certain time constraints. The institution must grant the US Government a royalty-free license to use the invention world-wide for government purposes, such as those activities conducted by the government itself or by its contractors who act on behalf of the government; notably, this right does not extend to grantees who simply receive funds from the government for their own research interests.

If the research resulting in the invention was funded by the US Government, the government retains rights to "march-in" and grant new patent licenses when certain criteria are not met. The most important instances are when (1) the patent owner or its licensee has not or is not likely to take effective steps to achieve practical application of the subject invention or (2) action by the government is necessary to alleviate health or safety needs that are not being reasonably met (35 US §203). An agency funding the subject invention may begin a march-in administrative proceeding when it determines that it has sufficient information to warrant it. If the agency makes a finding against the patent owner and its licensee, the agency may grant a license or force the patent owner to grant a license to others in order to address these deficiencies. The NIH has formally considered the use of march-in rights several times; but in those cases the NIH was able to reach an accommodation with the parties or decided that a march-in proceeding was not warranted under the statutory criteria. For example, the NIH has said that the extraordinary intervention of march-in was not appropriate for addressing drug pricing issues (Norvir, 2004). In spite of having no march-in proceedings, the government's march-in rights are still an effective back-pocket tool to motivate recalcitrant parties to take action to develop an invention or address health and safety needs without the government having to take the formal action of march-in (GAO, 2009).

The challenge for inventors to file patents in parallel in multiple countries has been addressed in the more than 140 countries that have entered into the Patent Cooperation Treaty (PCT) to allow an inventor to file an international "PCT" patent application and designate the countries in which to reserve the ability to file. In practice, for the US, institutions typically file a preliminary application called a "provisional" application that is not examined but serves to establish an official "priority date" if it is converted before the end of 1 year to an ordinary patent application. Alternatively, inventors can file an ordinary application or a PCT application first. When filing a national application first, the applicants have 1 year from the first filing to file a PCT international application. Under the PCT process, the application is published 18 months after the priority date, and the inventor has from 22 to 30 months after the priority date to file in specific countries, depending on the country. The PCT mechanism simplifies the process of filing internationally and allows the patent owner more time to assess the commercial potential of the invention before incurring the high costs of filing in multiple countries.

The right to use a patented invention can be conferred to others by license or assignment (transfer of ownership). Inventors usually assign their rights to their institution, and the institutional patent owners convey rights to use some or all of the claims in a patent under a license agreement. Licenses may be non-exclusive (no restriction on the number of licensees), co-exclusive (two or a set number of parties), or exclusive (one party). A license can grant the right to all or a limited set of "fields of use" of the invention; for example, a patented technology could be licensed exclusively to one group for neurological applications and exclusively to another group for cardiac applications. Exclusive and co-exclusive licenses from public sector institutions typically reserve the right to grant non-exclusive research use licenses to non-profit institutions. The NIH reserves in its exclusive licenses the right to grant non-exclusive internal research use licenses to both for-profit and non-profit institutions. With biomedical inventions that have broader applications, the field of use may be limited to the commercial applications the licensee company plans to pursue. If a company owns a platform technology patent they may choose to exploit it themselves or sublicense some aspects to other companies. In contrast, a public sector research institution is more likely to craft licenses by field of use such as medical application and type of product, e.g. drug, biologic, device, diagnostic, or research-use product or service.

A good example of the complexity of licenses is apparent in the license Wisconsin Alumni Research Foundation (WARF) granted to Geron for rights to the Thomson embryonic stem cell (ESC) patents, which is exclusive for therapeutic and diagnostic products from hESC-derived neural, cardiomyocyte and pancreatic islet cells and non-exclusive for hESC-derived hematopoietic, chondrocyte and osteoblast cells. Geron also has nonexclusive rights to develop research products for hepatocytes, neural cells, hematopoietic cells, osteoblasts, pancreatic islets and myocytes (Rohrbaugh, 2003).

THE AMERICA INVENTS ACT OF 2011

The US patent system is unique in recognizing the inventor who is "first-to-invent." That is, between two inventors claiming the same invention, the one who can demonstrate with evidence that she was first to conceive and reduce the invention to practice is awarded the patent. The reduction to practice date is the earlier of an actual reduction to practice or constructive reduction to practice as of the priority date. Proof of reduction to practice is often a laboratory notebook entry, preferably from a bound notebook that is dated and signed. The process for deciding between two separate (groups of) inventors is an administrative judicial proceeding within the PTO called an "interference." The PTO declares an interference when two or more patent applications with at least one overlapping claim have met the requirements for "allowance" to be issued or when one patent has been issued and there is one or more pending patents in a condition to be allowed. Interference proceedings are often lengthy and costly.

All this will change, however, under the America Invents Act of 2011 for applications filed on or after March 16, 2013 when the US adopts a type of first-to-file system (Fox, 2011). Both the old and new systems will be maintained until all prior applications are abandoned or are issued as patents. The new system is more accurately labeled "first-inventor-to-file" because, unlike other systems requiring absolute novelty on the date of filing, the US will maintain a modified 1-year grace period for only the inventor's publications and public use if the inventor has published the invention. Rather than an interference, a more streamlined "derivation hearing" procedure will be put in place to determine whether the earliest to file inventor, among two or more inventors claiming the same invention, was an actual inventor as opposed to having derived it from someone else. Other changes in the law permit any third party to submit relevant printed publications to the PTO, within certain time limitations, for consideration during the patent prosecution of another party's application. A new post-grant review will allow any third party to request review of a question of validity of a patent within 9 months after the issuance or re-issue of a patent.

PATENT LICENSING POLICIES

If a group that does not hold a license to a patent uses the covered invention, the patent holder may accuse them of "infringement" of the patent. Some countries, but not the US, have laws providing for a general "experimental use" exception that allows non-inventors to use a patented technology or composition of matter without infringing the patent, as long as there is no plan to commercialize the invention. In the US, there is only a very narrow exemption created by the courts to use patented inventions to "satisfy idle curiosity" but not to use inventions for practical applications, including research at a university (*Madey v. Duke, 2002*). There is a specific exemption from infringement under the Hatch-Waxman Act of 1984 for research undertaken to obtain regulatory approval by the Food and Drug Administration (FDA), e.g. for both preclinical development and clinical trials (35 USC §271(e)(1)). Most European countries have statutes that exempt from infringement research intended to further explore the use of a patented invention (Iles, 2005). While non-profit research may constitute patent infringement in some countries, it is rare that a patent owner would threaten, let alone actually sue, a non-profit research institution for infringement. More research might actually enhance the known uses of the product, thereby extending its commercial use, and legal costs of pursuing a law suit would greatly exceed the damages awarded by a court, as well as create bad public relations for the suing party. There are cases, however, particularly with research tools, in which patent owners have threatened to sue in order to restrict the research use of the tool. Without large sums of money to be made from research tools themselves, these patent owners generally want users of the patented tool to take a license that grants back to the patent owner rights to commercially valuable inventions made using the research tool, such as drug candidates tested in an animal model (Hayden 2011a, 2011b). This "reach-through" practice is inconsistent with NIH and most university policies, but some patent holders continue to pursue the possibility of obtaining royalties from commercial products developed using their patented tools.

Public sector institutions, which do not bring technologies much beyond the earliest stages of development, have developed their own policies and practices individually and collectively. Biomedical technologies require additional care in management because of the potential effect on the improvement of public health. The "Nine Points to Consider" (http://essentialmedicine.org/archive/stanford-white-paper) is a proposal that outlines university patenting and licensing practices in order to promote the public interest. The NIH has also taken a leading role in the development of formal policies for its intramural and extramural components to provide incentives for commercialization while maintaining a robust research enterprise that will give rise to more inventive technologies. These policies include the Research Tools Guidelines (Federal Register, Vol. 64, No. 246, Dec 23 1999) and the Best Practices for the Licensing of Genomic Inventions (Federal Register, Vol. 70, No. 68, April 11, 2005), both of which are meant to reduce barriers to material transfers and eliminate license terms that impede research at public research institutions. These policies also encourage the use of exclusive licensing only as needed to provide appropriate incentives for commercialization without placing undue restrictions on public sector research. These goals are particularly important in stem cell research because of the early stage of the field and its many unique biological materials that need to be shared with the research community to ensure that scientific advances and multiple approaches proceed toward clinical testing and commercialization.

MATERIAL TRANSFER

For most researchers, patents only come to their attention if they have an invention to report to their technology transfer staff at their institution. However, patents and proprietary issues may come into play when they require a Material Transfer Agreement (MTA) in order to obtain needed research materials. Whenever ownership and transfer of proprietary biological materials, with or without patent protection, is an issue, researchers may encounter instances in which the terms of the MTA used to transfer the materials may have a negative effect on their ability to conduct research. Terms that restrict research and publication or request reach-through to new inventions should not be used by research institutions providing materials because such terms hinder research, add to administrative burdens, restrict options for optimal licensing, and limit researchers' ability to use multiple materials in the same experiments when providers each demand control over the output. Exclusive options to license new inventions for commercial purposes are often used in agreements with industry that provide proprietary materials for research such as drugs or therapeutics, particularly for new uses of the materials. Fortunately, it is rare that institutions, public or private, try to use their patents to restrict public sector research or to extract fees to conduct research. Usually these disputes are limited to situations in which the public sector is providing clinical services and obtaining reimbursement.

PATENT POOLS

If patents give the owner an exclusive right to control the use of a technology, what happens when more than one owner controls different aspects of the same technology, e.g. a stem cell and the use of that stem cell for a particular therapeutic purpose? In such cases, no one party can use both the material and the method of use without both licenses. With two separate owners, they could block each other's use. The broader composition patent is said to be the dominant patent and the method of a particular use is said to be subordinate since it has a narrower scope and requires access to the composition patent.

Multiple patents owned by various parties can mutually block each other and create a commercialization bottleneck. Industry has sometimes solved this challenge by creating a "patent pool"; the pool is defined by the standard technology that everyone needs, such as MPEG or Radio Frequency Identification (RFID). The patent owners provide cross-licenses to each other and the package of rights to the patents is provided on standard terms to anyone who wishes to enter that market. The only attempt to develop a biomedical patent pool in the US was for SARS, but that effort stalled after infections dropped and industry was no longer interested in developing products related to SARS. There is, however, an international patent pool initiated by Medicines Patent Pool in Geneva to provide licenses to patents needed in the manufacture and sale of anti-retrovirals for the treatment of HIV/AIDS in the least developed countries. Other arrangements have been developed to facilitate the licensing of patents through agreement to common terms. The growing number of patents in the stem cell field make the concept of a patent pool attractive to some potential investors.

STEM CELL PATENTS

The US has issued the largest number of human stem cell patents, whether measured broadly or more narrowly by those limited to "embryonic," "adult," or "somatic" (Bergman and Graff, 2007; Hammersla and Rohrbaugh, 2011). Over 1000 patents have been issued since 1990 with a claim to a human stem cell or a method of using or analyzing them. The number of such patents peaked in 2002 and 2003 and has declined considerably since then. For example, in the US there were 128 patents issued in 2002 with at least one claim containing the words human, stem, and cell, but there was only one in 2010. The number of patents with claims to embryonic stem cells peaked in 2002 at 63 with only 1 issued in 2010. However, the number of patents with somatic or adult stem cell claims, as measured by the words neural, hematopoietic, and stromal or mesenchymal, has increased in the past five years. This trend may be due to scientific progress in moving toward applied technologies using pluripotent stem cells to produce somatic stem cells and the development of therapeutic technologies for particular regenerative medicine applications. What is clearly apparent is that a large number of patents related to stem cell therapeutics must be evaluated by commercial parties engaged in stem cell product development.

The US has the largest number of stem cell patents because it is the largest single market for biomedical products and has treated stem cell patent applications like other biomedical innovative technologies. Some other countries have restricted the ability to obtain such broad claims to stem cells. In considering the patentability of James Thomson's ESC technology owned by WARF, the Enlarged Board of Appeal of the European [Patent] Office in 2008 rejected claims to a method of obtaining primate stem cells because the requisite destruction of a human embryo is contrary to the public order or morality requirement at the EPO (Article 53(a) of the European Patent Convention, case G 0002/06, EPoA). More recently, the Court of Justice of the European Union, which has jurisdiction over the 27 EU members, rejected a patented method for generating neurons from hESCs made by Oliver Brüstle at the University of Bonn, Germany, on the basis that procedures involving hESCs cannot be patented because they involve the destruction of an embryo. This ruling applies not only to methods of creating hESCs but also to previously derived lines. It is not clear yet whether this ruling will reduce commercial interest in hESC therapeutic technologies in Europe or drive more business towards iPSC lines (Callaway, 2011a). It is possible, however, that this ruling opens up more opportunities to explore the underlying technology without the risk of patent infringement while permitting patents and commercial development of therapeutic methods using hESCs (Callaway, 2011b).

HUMAN EMBRYONIC STEM CELL PATENTS

The first hESC patent was issued in1998, based on the derivation of nonhuman primate ES cells by James Thomson and colleagues at the University of Wisconsin; this patent was assigned to the Wisconsin Alumni Research Foundation (WARF; US Patent No. 5,843,780). It was followed by two others that issued in 2001 and 2006 from the same original filing in 1995 (US Patent Nos. 6,200,806 and 7,029,913). These patents have dominated the hESC field because they broadly cover primate (including human) ESC lines created from the inner cell mass of a blastocyst-stage embryo. While certain methods of deriving the lines and maintaining them in culture are claimed, the claims to hESCs themselves as defined by biochemical markers, karyotype, and the ability to divide indefinitely while maintaining a state of pluripotency seem to encompass all useful hESC lines. Much controversy ensued after the granting of these patents due to the breadth of the claims, the lack of other sources of valuable pluripotent stem cells in the early 2000s, and the potential value that these lines hold for regenerative medicine. Also of concern were WARF's restrictive patent and biological material licensing practices to non-profit institutions and the political controversies of what lines would be approved for use with federal funds.

A challenge to the patents was made in 2006 through a re-examination proceeding led by two citizens' groups (Loring, 2007), who argued that that the patents were obvious over the prior art and were impeding scientific progress (Ravicher, 2010). After the PTO responded to the challenge by rejecting all the patent claims (McCook, 2007), WARF voluntarily narrowed its claims to limit the cells to those "derived from a pre-implantation embryo" and the cells to those that would proliferate in an "undifferentiated state." Ironically, this narrowing of the claims to embryo-derived cells assured that the WARF patents would not cover somatic cell-derived induced pluripotent stem cells (iPSCs). The PTO allowed the modified claims in 2008, but upon appeal the PTO found in 2010 that one of the patents, 7,029,913, describing a "replicating *in vitro* cell culture of human embryonic stem cells ..." was obvious in view of the prior art publications (Gallagher, 2010). When the two remaining dominating patents expire in 2015, patents needed for commercialization of hESC technologies will be driven more by patents on differentiated cell types and methods used to obtain and deliver them.

Shortly after US Government policy changed in 2001 to permit the funding of research involving previously isolated hESC lines, the NIH entered into a Memorandum of Understanding (MOU) with WiCell, the subsidiary of WARF managing the hESC rights, to establish terms for the public sector research use of lines developed at the University of Wisconsin and elsewhere. In the case of WARF and WiCell, the original strategy for managing the Thomson ESC patents allowed researchers to use the patent rights with WiCell lines or other parties' lines as long as no cells were used in research where another entity had any rights to inventions that might be made using the cells, even inventions that did not require the use of the ESCs. Companies willing to provide unique research materials, such as biological factors and proprietary compounds, to public sector researchers, often when the companies were not interested in using hESC technologies, were not willing to provide the materials without guarantees of rights in inventions that might result from the testing of their materials in hESCs. Yet, just to obtain the right to use a potential new invention, they would have had to negotiate a license with WiCell and pay fees in the range of tens of thousands of dollars.

WiCell eventually resolved part of this dispute by requiring companies to pay fees and enter into commercial licenses only if they actually use ESCs. Also, WiCell relaxed the restrictions on researchers' ability to distribute derivatives of ESC lines, e.g. those with fluorescent markers, to other researchers (PHS-WiCell MOU, 2008).

This arrangement exemplifies the difference between patent rights and rights in tangible property such as biological materials. The NIH had rights to use the WARF patents under the Bayh-Dole Act by having funded the non-human primate studies in Dr. Thomson's laboratory that gave rise to the primate ESC patents, in which primate was defined as including humans. However, government funding cannot be used to derive hESC lines, and as such the derivation occurred with private funding. The Government had no rights in the tangible property of the hESC lines, and they were not governed by grant policies such as the Research Tools Guidelines. The MOU set forth the terms by which NIH's own researchers could obtain hESC lines from WiCell and from other sources that had a license to the WARF patents, and required WiCell to offer no more onerous terms to NIH funded investigators. NIH then entered into MOUs in 2002 with other providers such as the University of California at San Francisco, BresaGen (Australia), and ESCell International (Singapore), and later agreed upon a model MTA with Mizmedi (Korea).

INDUCED PLURIPOTENT STEM CELL PATENTS

The patent landscape for iPSCs is more complex than for hESCs because iPSCs are not a single, naturally occurring cell type – they can be generated from different cell types using different methods. The first iPSC patent issued in Japan to Shinya Yamanaka and his colleagues of Kyoto University on 12 September 2008 was based on a priority date of 13 December 2005 (Simon et al., 2010). US patents based on the same Japan Patent Office filing were issued on 1 November 2011 (US Patent No.8,048,999) and 15 November 2011 (US Patent No. 8,058,065). On 12 January 2010, the UK Intellectual Property Office issued a patent to Kazuhiro Sakurada and colleagues working for an affiliate of Bayer Schering Pharma, based on a priority date of 15 June 2007 (UK Patent No. GB2450603). The patent has been assigned to iPierian, Inc. of San Francisco. The third patent internationally and the first in the US was issued on March 23, 2010 to Rudolf Jaenisch and Konrad Hochedlinger of the Whitehead Institute of Biomedical Research, but has the earliest priority date of the three, 26 November 2003 (US Patent No. 7,682,828). A second patent was issued to Jaenisch and Hochedlinger on 6 December 2011 (US Patent No.8,071,369).

The various patents to date claim different methodologies for generating iPSCs and compositions of cells. The earlier Jaenisch patent claims primary somatic cells transfected with three "endogenous pluripotency genes" and the second claims: a "composition comprising an isolated primary somatic cell that comprises an exogenously introduced nucleic acid encoding an Oct4 protein operably linked to at least one regulatory sequence". The Yamanaka US patent claims "a nuclear reprogramming factor, comprising an isolated Oct family gene, an isolated Klf family gene, and an isolated Myc family gene," such as Oct3/4, Klf4, and c-Myc, or replacement of the Myc family gene with a cytokine. The Sakurada UK patent claims a method producing iPSCs from "postnatal tissue" by subjecting it to "forced expression of each of the genes Oct3/4, Sox2 and Klf4, but not c-Myc, and culturing in the presence of FGF-2." There are more than 90 pending iPSC patent applications in the US (Resnick and Tse, 2011). There will likely be some overlapping claims that will be subject to interferences, but others also represent various means of inducing pluripotency, including transient viral expression systems (US Patent Application No. 20100311171), transient extra-chromosomal expression (US Patent Application No. 20110061118), recombinant transcription factors (US Patent Application No. 20100233804), and RNA (US Patent Application No. 20110165133).

The outcome of these various pending patents on the issued patents is not yet clear. In addition, there may ultimately be different patent holders in the US (first-to-invent) versus other countries (first-to-file) for the same technology. Unlike the Thomson hESC patents, it is much less likely that there will be a dominant iPSC patent family unless one technology becomes highly preferable over others. This competition for access to alternative iPSC technology for development will hopefully prove beneficial to the public and researchers because researchers and developers will chose not only what works scientifically but also what can be obtained under reasonable terms.

Companies developing iPSC technologies include Fate Therapeutics, iPierian, and Cellular Dynamics International (Resnick and Tse, 2011). Unlike hESC development by companies in the past ten years, the diversity of approaches and the current risk adversity in the biotechnology industry is leading public sector institutions to take some of the lead in bringing iPSC therapies into early stage clinical trials. In doing so, it is critical for clinical and basic researchers to take into consideration the provenance of the cell lines used with respect to any limitations placed upon their ultimate use by the informed consent terms and by intellectual property or material transfer agreements used to obtain cells.

Just as there are search engines to find new publications in scientific journals, there is a search engine for patents. To find out the latest information about issued stem cell patents, use the search functions on the PTO's website: www.uspto.gov/patents/process/search/index.jsp.

READING LIST

America Invents Act, Pub. Law 112-29, effective September 16, 2011 but various provisions come into effect at later dates.

Bayh-Dole Act of 1980, Public Law 96-517, see 35 USC §200-212 en.wikipedia.org/wiki/ Bayh–Dole_Act

Bergman, K., Graff, G.D., 2007. The global stem cell patent landscape: implications for efficient technology transfer and commercial development. Nat. Biotechnol. 25, 419–424.

Callaway, E., 2011. European court bans patents based on embryonic stem cells. Nature <http:// www.nature.com/news/2011/111018/full/news.2011.597.html>.

Gallagher, K., 2010. WARF loses a round in stem cell patent dispute. Milw.-Wis. J. Sentinel <http://www.jsonline.com/business/92682039.html>, (Milwaukee).

Government Accounting Office, Information on the Government's Right to Assert Ownership Control over Federally Funded Inventions, GAO-09-742, July 27, 2009.

Hammersla, A.M., Rohrbaugh, M.L., 2011. Intellectual property claims to stem cell technologies: research clinical testing and product sales, In: Atala A. (Ed.), Progenitor and Stem Cell Technologies and Therapies, pp. 121–146.

Hatch-Waxman Act of 1984, Public Law 98-417. <http://www.fdli.org/pubs/journal/toc/vol54_2. html>.

Hayden, E.C., 2011a. Lawsuit dismissal removes cloud over Alzheimer's research. Nat. News Blog. (12 Aug 2011) <http://blogs.nature.com/news/2011/08/lawsuit_dismissal_removes_clou.html>.

Hayden, E.C., 2011b. Patent dispute threatens US Alzheimer's research. Nature 472, 20.

Iles, K., 2005. A comparative analysis of the impact of experimental use exemptions in patent law on incentives to innovate. Nw. J. Tech. Intell. Prop. 4, 61.

Loring, J.F., 2007. A patent challenge for human embryonic stem cell research. Nat. Rep. Stem Cells. <http://www.nature.com/stemcells/2007/0711/071108/full/stemcells.2007.113.html>.

Madey v. Duke, 307 F.3d 1351 (Fed. Cir. 2002). <http://sippi.aaas.org/ipissues/cases/?res_id=119>.

McCook, A., 2007. Key stem cell patents rejected. The Scientist <http://classic.the-scientist.com/ news/display/53051/>.

Nine Points to Consider in Licensing University Technology. <www.autm.net/source/ NinePoints/ninepoints_endorsement.cfm>.

Norvir March-In Decision, 2004. <www.ott.nih.gov/policy/March-In-Norvir.pdf>.

Patent Convention Treaty, see <www.wipo.int/pct/guide/en/gdvol1/annexes/annexa/ax_a.pdf>.

PHS-WiCell MOU. <www.ott.nih.gov/pdfs/WiCell-rev.pdf>.

Ravicher, D., 2010. Public Patent Foundation. <http://www.pubpat.org/warfstemcell.htm>.

Resnick, D.S., Tse, J.M., 2011. First true induced pluripotent stem cell (iPSC) patent issued by the USPTO. Nixon Peabody <http://nixonpeabody.com/services_pubdetail.asp?ID=4101&SID=83>.

Simon, B.M., Murdoch, C.E., Scott, C.T., 2010. Pluripotent patents make prime time: an analysis of the emerging landscape. Nat. Biotechnol. 28, 557–559.

United States Patent and Trademark Office (USPTO) website to search for issued or pending patents. <www.uspto.gov/patents/process/search/index.jsp>.

World Intellectual Property Organization (WIPO) Patentscope site to search for published PCT applications and patents from several countries. <www.wipo.int/patentscope/search/en/ search.jsf>.

CHAPTER 38

Ethics of Human Pluripotent Stem Cell Research and Development

Mary Devereaux

EDITOR'S COMMENTARY

Human pluripotent stem cell (hPSC) research is not just about the bench science, and researchers who choose to work in this exciting area will at some point find that they are affected, directly or indirectly, by strong public opinions. Most of our research dollars come from the public, in the form of government grants or private foundations, and if we are not paying attention to their concerns, this research will not continue to be supported. Every blunder by a researcher working on hPSCs negatively affects us all, whether it is excessive hype, fraudulent science, or just disregard for the sensitivities of the public. Many people distrust scientists; too often, their concerns are warranted, because we appear to be insensitive to their genuine interest in understanding our research and its impact on their health and well-being. We need to make sure, for the sake of the future of stem cell research, that we understand and adhere to the highest ethical standards. This chapter is written by an ethicist who has been involved from the beginning in the field of hPSC research, and it provides practical guidance to the ethical principles that all of us must understand and follow if stem cell research is to fulfill its promise.

OVERVIEW

In this chapter, we will discuss the relevant laws and review requirements governing stem cell research that will help researchers avoid loss of funding, research misconduct, and other problems. Additionally, we will cover key events in the history of the US debate over federal funding for stem cell research that led to current review requirements. We will also review areas of ethical concern as they affect the rapidly changing field of stem cell research and the resources available to address them. Lastly, we will cover the professional responsibilities for institutional, federal, and state review, careful record keeping, complying with funding restrictions, state and federal law, and the highest standards of scientific integrity.

J.F. Loring & S.E. Peterson (eds): Human Stem Cell Manual, Second edition.
DOI: http://dx.doi.org/10.1016/B978-0-12-385473-5.00038-2

ETHICAL ISSUES FOR HUMAN PLURIPOTENT STEM CELL RESEARCH

Introduction

Scientists working with human embryonic stem cells (hESCs) and/or induced pluripotent stem cells (iPSCs) can best comply with federal and state regulations and anticipate future challenges by appreciating the issues at stake in the public stem cell debate. Scientists working in this field share with large sectors of the US public recognition of the tremendous potential benefits of stem cell research. Already research in this area is advancing our understanding of embryology, including the basic mechanisms of normal (and abnormal) development, and opening new means of disease modeling and pharmaceutical testing. Scientists and laypersons alike hope to see significant advances in the understanding and treatment of a range of diseases such as diabetes, Parkinson's disease, Alzheimer disease, and cancer. Despite this promise, some worry about the means to this end, in particular the use of early stage embryos to derive stem cell lines.

As scientists, we may have little interest in the ethical or political debates generated by beliefs we do not share, preferring to be left alone to do what we do best: science. The practice of science, however, unavoidably takes place in a social context. In the US, the bulk of funding for basic research comes from the federal government, or, as in California, from taxpayer-funded initiatives. With publicly-funded research comes certain legal and regulatory requirements. Like scientists in other fields, stem cell researchers are governed by regulations regarding matters such as biosafety, the welfare of laboratory animals, and the rights of human subjects. Continued taxpayer funding also requires scientists to maintain public trust. Preserving that trust is especially important in the field of stem cell research, given the strength of opposition by some to the methods and materials necessary to advance the field. The US public holds diverse ethical views over embryo research; the use of, or payment for, human oocyte donation; somatic cell nuclear transfer (SCNT); parthenogenesis; and the conditions of informed consent for tissue donation.

Other countries too may apply different ethical – and legal – standards to stem cell research. German law prohibits the creation of new stem cell lines, restricting scientists to lines created before 1 May 2007 (Lenoir, 2000). Britain, Sweden, and many other countries have relatively liberal policies towards embryonic stem cell research, reflecting greater public acceptance than in the US. However, in 2011, the European Court of Justice issued a "binding legal opinion" that denies European patents on inventions that "are even indirectly dependent on human ES cells," a judgment widely opposed by many in the scientific community (Abbott, 2011a, 2011b).

The complexity of the ethical and regulatory landscape surrounding stem cell research thus makes it important that investigators understand and respect the diversity of ethical views held within the international scientific community as well as closer to home. It goes without saying that relevant national and international law remains in flux.

The Embryo Debate

Stem cell research initially drew public attention because early stem cell lines were derived from donated *in vitro* fertilization (IVF) embryos no longer needed or wanted for reproduction. This put the field dead center in a decades-old controversy over the legalization of abortion and the definition of when human life begins. While public support for embryonic stem cell research has grown, disagreement remains over what legal protections, if any, we should assign the early stage embryo, i.e. a blastocyst. Is an undifferentiated cell mass a (potential) person? A being with rights?

Genetically, a human embryo is human. Its cells contain human DNA. Biologically, the embryo is alive, with dividing cells and other metabolic processes. However, the morally relevant question concerns not the DNA or metabolic processes of the cells, but at what stage the organism becomes a person, a being worthy of moral respect and legal protection. For some, those rights begin with conception. This view, rooted in certain religious beliefs, holds that the fertilized human oocyte, at whatever stage of development, is a *person*. As such, a zygote, blastocyst, or embryo should be afforded the same right to (continued) life as a born human child. Not surprisingly, this is a view few in the stem cell community share since it makes embryonic stem cell research ethically forbidden. While most scientists acknowledge that human biological materials should be treated with respect, or at least that the donors of such materials should be, the act of arresting the development of the dividing cells of the human blastocyst *per se* poses no ethical problem.

Not surprisingly, though, those who regard the embryo as a moral person find arguments about the value of the scientific or medical benefits of stem cell research unpersuasive. From their perspective, destroying the 5-day-old blastocyst in the pursuit of such ends, however scientifically worthy, requires sacrificing some (human persons) for the benefit of others. The idea is not that we should *agree with* those who equate the blastocyst or very early embryo with a human infant, but rather *understand* where opponents of embryonic stem cell research are coming from.

The point for those working in stem cell research is simply to understand the debate: that for those who regard moral personhood as beginning with fertilization, no scientific goal will justify embryo destruction. Advocates of stem cell research for their part also advance ethical arguments in justification of embryo use. That the embryos in question come from IVF patients who willingly donate to science what they would otherwise discard or destroy only strengthens the argument that justifies their use in research. For many scientists and large sectors of the lay public, embryonic stem cell research is a social good. It aims at understanding human disease and alleviating suffering.

It is important to note that both sides in the controversy appeal to ethical principles ("do not kill," "relieve human suffering") in justifying their positions. This is not a simple conflict between science and ethics, but between *varying conceptions* of ethics. While most everyone agrees that medical progress is a worthy goal, the issue specific to embryonic stem cell research is over the means to this end.

Funding Restrictions

The US division of opinion over abortion and embryo research led directly to restrictions on public funding of stem cell research. In 1996, Congress moved to restrict the use of federal funds for research that "harms or destroys human embryos." The idea was that such research, while not illegal, would be ineligible for federal funding. This restriction, the so-called Dickey-Wicker Amendment, precluded the use of federal taxpayer dollars in support of research "in which human embryos are created, destroyed, discarded, or knowingly subjected to risk of injury or death greater than allowed for research on fetuses *in utero*." Note that Dickey-Wicker defines a human embryo not merely as the product of sperm and egg, but more broadly as "any organism … *that is derived by fertilization, parthenogenesis, cloning, or any other means from one or more human gametes or human diploid cells*" (Kearl, 2010). In short, the federal funding ban extends to work using a variety of methods, including SCNT and parthenogenesis.

In August 2001, the then President Bush opened federal funding to certain established stem cell lines (those in existence prior to August 9, 2001), while continuing the restriction on federal funding on derivation of or work with new lines. In 2009, President Obama established an NIH Registry to certify stem cell lines eligible for federal funding. Eligible lines must meet strict requirements for informed consent. They must originate in US IVF clinics and be voluntarily donated for research. No payment for donation is permitted and clinics must inform potential donors that they have other options. Interestingly, the legislation establishing the NIH Registry stipulates that "hESC cells are derived from embryos, but are not themselves embryos." As of this writing, the Dickey-Wicker Amendment remains on the books.

Ethical Concerns about Human Donors: Oocytes, Blastocysts, and Somatic Cells

As the field advances, additional areas of ethical concern have emerged. Prior to the discovery of effective reprogramming methods, strong interest in

SCNT drove the demand for human oocytes. Questions arose about the ethical permissibility of putting women at risk "for the common good" of scientific research. And, if so, under what circumstances? And should such donors be paid? In California, for example, state law and the California Institute for Regenerative Medicine (CIRM) regulations prohibit paying women for oocyte donation to research – unlike egg donation for IVF, where payment is legal.

National Academy of Science *Guidelines for Human Embryonic Stem Cell Research* now require consent from both sperm and egg donors before IVF blastocysts may be used for research (NAS, 2005, Recommendation 13, p. 83). Donors of IVF blastocysts should understand that the embryos will be destroyed and that alternatives such as discard, adoption, or freezing exist (NAS, 2005, p.88–89). Donors may not be paid or otherwise influenced by investigators (NAS, 2005, Recommendations 14, 15, & 16, pp. 85–87).

INFORMED CONSENT

Other now standard requirements for informed consent include informing donors that while their donated materials may lead to patents or financial gain for researchers, they themselves will not share in intellectual property rights or financial gain. Donors must also understand and consent to the intended use of their materials in a variety of research applications, including those that may not as yet be known, e.g. translational applications. Other areas covered by informed consent include, but are not limited to, privacy, security, and re-contacting donors for additional medical information. These and other requirements for voluntary and fully informed consent rest upon the principle of autonomy, that is, respect for the freedom of each individual to be self-governing, to make their own decisions regarding health care and participation in research.

What about informed consent requirements for iPSC work? Here the issues seem simpler. At issue are fibroblasts or other somatic cells, often just a small skin biopsy, easily obtained. Yet, while less familiar territory than the ethical issues surrounding embryonic stem cell research, iPSC research raises important ethical and regulatory issues. Aside from concerns about mild pain or infection at the site of biopsy, researchers should be aware that here, too, donors must be fully informed and voluntary participants. Questions arise regarding what donors know – or should know – about the uses to which scientists may put their donated human cells or resulting cell lines. The informed consent process should note that donated tissues or other materials may be used in research involving genetic modification, mingling human and animal cells, or cloning. Research participants should know who will have access to genetic and other personal medical information, how that information will be protected, and that no security system is failsafe.

Another area worth addressing in the original informed consent process involves re-contacting donors. There are two reasons why permission for re-contacting donors might be necessary or desired. The first reason is donor-centered, i.e. would the donor wish to receive information obtained from the study that might be of medical relevance? The second reason is research-centered, i.e. does the donor give permission for re-contact should the research team want further information or testing?

EARLY CLINICAL TRIALS AND STEM CELL TOURISM

As the field advances, the safety and efficacy of stem cell research becomes an increasing concern. Like other emerging biotechnologies, stem cell research requires balancing risks and benefits. Direct consumer marketing for scientifically unproven "stem cell therapies," whether derived from embryos, iPSCs, or other sources, has accelerated in many areas faster than the pace of science-based progress in preclinical research laboratories. Desperately ill patients out of options may pin their hopes on the inflated claims of international (non-FDA approved) stem cell clinics (Devereaux and Loring, 2010). Thus the stem cell community has a professional obligation to guard against encouraging false hope or raising unrealistic expectations about the likely pace of developing, testing, and marketing new therapies. Similar care will need to be exercised in fully informing those enrolling in FDA approved early clinical trials, e.g. Geron's spinal cord injury trial or Advanced Cell Technology's (ACT) Phase 1/2 trial for Stargardt's Macular Dystrophy (SMD). Patients need to know of the risks and low likelihood of individual benefit. Researchers approached about approved (and non-approved) stem cell therapies can direct patients to The International Society for Stem Cell Research (ISSCR) website, an excellent source of expert opinion with well-designed materials on clinical trials and appropriate questions to ask of any "experimental therapy."

Ethical and social issues also arise in regard to cell donation and banking. As the field develops, we need to insure that cell lines adequately represent different ethnic, racial, and disease groups. Fairness also requires that these groups have equitable access to the benefits of taxpayer-funded research. Stem cell researchers and health policy analysts will need to address what to do about so-called "orphan" or rare diseases where the small population of affected patients makes investment in therapies likely less profitable and hence less likely.

REVIEW COMMITTEES

Stem cell research is subject to state and federal regulation. Three review committees oversee stem cell research: the Stem Cell Research Oversight

Committee (SCRO), the Institutional Review Board (IRB), and the Institutional Animal Care and Use Committee (IACUC).

The SCRO Committee

The Oversight Committee (known alternatively as the ESCRO or SCRO) includes experts in the scientific and ethical review of stem cell research. It may also have legal representation, a community representative and/or patient advocate, physicians and/or others with relevant expertise. The ESCRO Committee is charged with the review of proposals to use hESCs and iPSCs in research and teaching, whether federally or non-federally funded. While institutional policies regarding review may vary in details, all proposals to use or derive pluripotent stem cells must be submitted for review. Such proposals might include but may not be limited to: derivation of a stem cell line; purely *in vitro* research utilizing stem cell lines; the introduction of stem cell lines into non-human animals.

In addition to assuring researchers that proposed research is in compliance with applicable laws and funding regulations, and that it follows ethical guidelines and requirements, e.g. for informed consent, the SCRO Committee also recommends policies for compliance and ethical conduct, facilitates education and training of investigators, and advises investigators on required institutional reviews, e.g. the IRB or IACUC. As funding and legal requirements continue to evolve, and are subject to changes in Congress and the Presidency, the SCRO Committee serves the valuable function of keeping researchers undertaking work in this area abreast of new developments in a fast changing regulatory and funding environment.

Institutional Review Board (IRB)

Much of stem cell research involves the use of human materials: embryos, oocytes, sperm, adult stem cells, fibroblasts, iPSCs, and so on. Well-established federal regulations, including the Health and Human Services (HHS) Protection of Human Subjects, 45 C.F.R. 46, Subpart A, establish safeguards for tissue donors and others participating directly or indirectly in research. Researchers should be aware that some forms of stem cell research fall under these regulations. In particular, the regulations apply to materials that can be identified and traced back to the donor. The relevant category is "research involving individually identifiable private information about a living individual." If cells or other materials "can be linked to specific living individuals by the investigators either directly or indirectly through coding systems," IRB review is necessary.

As this is a complex area, researchers should consult federal regulations and their IRB for guidance.

Institutional Animal Care and Use Committee (IACUC)

As in other areas of research, investigators planning to work with animals must submit a proposal to their IACUC. Many funding agencies, including NIH and NSF, require verification of IACUC approval.

SCIENTIFIC RESPONSIBILITIES

Stem cell research holds great potential, both for basic understanding of human development and in the treatment of human disease. Consequently, the field, as noted earlier, has a high public profile. Many of the methods common in the field also draw public attention: work with or deriving cell lines from human embryos, SCNT (or cloning), parthenogenesis, and genetic manipulation of human materials. Stem cell scientists thus work far more in the public eye than those in many other fields. Any misstep is likely to receive wide coverage in the media, affecting careers, institutional reputation, and the credibility of the field. Perceived or real scientific misconduct, conflicts of interest, and so on also provide fodder for those who wish to end public funding and acceptance of stem cell research. In this climate, it is particularly important that the research community follows all requirements for stem cell research, e.g. appropriate review, regardless of funding source. In addition to compliance with regulation, principal investigators are responsible for creating a lab environment in which scientific integrity – honesty, trustworthiness, collaboration, and professional rigor – can flourish. The goal is an open, transparent environment in which all members of the field are encouraged to ask questions, participate in discussion, and hold one another accountable.

READING LIST

Abbott, A., 2011a. German science organizations slam European Court over stem-cell ruling. Nature/News.

Abbott, A., 2011b. Stem cells: the cell division. Nature 480, 310–312.

Devereaux, M., Loring, J.F., 2010. Growth of an industry: how U.S. scientists and clinicians have enabled stem cell tourism. Am. J. Bioeth. 10, 45–46.

Eligibility of Human Embryonic Stem Cells for Research with NIH Funding. <http://stemcells.nih.gov/policy/2009guidelines.htm> (Retrieved 30.01.12.).

Gilbert, S.F., Tyler, A., Zackin, E., 2005. Bioethics and the New Embryology: Springboards for Debate. Sinauer Associates, Inc.

Health and Human Services. Part 45. Protection of Human Subjects. <http://ohsr.od.nih.gov/guidelines/45cfr46.html> (Retrieved 02.04.12.).

International Society for Stem Cell Research (ISSCR). A closer look at stem cell treatments. <http://www.closerlookatstemcells.org/AM/Template.cfm?Section=Home1&Template=/Templates/TemplateHomepage/UnprovenTherapies_1510_20100323T1444.22_LayoutHomePage.cfm> (Retrieved 02.04.12.).

Kearl, M. "Dickey-Wicker Amendment", Embryo Project Encyclopedia (2010) ISSN: 1940-5030. <http://embryo.asu.edu/view/embryo:128106> (Retrieved 02.04.12.).

Lenoir, N., 2000. Europe confronts the embryonic stem cell research challenge. Science 287, 1425–1427.

National Research Council and Institute of Medicine of the National Academies, 2005. Guidelines for Human Embryonic Stem Cell Research. National Academies Press.

NIH Guidelines on Human Stem Cell Research. 1. Scope of Guidelines. <http://stemcells.nih.gov/policy/2009guidelines.htm> (Retrieved 02.04.12.)

Obama, B., 2009. Removing barriers to responsible scientific research involving human stem cells, Executive Order, 13505. <http://www.whitehouse.gov/the_press_office/Removing-Barriers-to-Responsible-Scientific-Research-Involving-Human-Stem-Cells> (Retrieved 02.04.12.).

Index

Note: Page references followed by *t* and *f* denote tables and figures, respectively.

A

Accutase passaging, 30–32, 32*f*
Acetic acid:methanol fixative, 201
Acetylation, 313–314
Acid solution, zona pellucida
    removal using, 569–570
Activin stock solutions, 428, 442
Adobe Photoshop, 262–266
Affinity-based assays, 326, 330–331
Affymetrix, 271, 279, 297, 310–311
AggreWell™ plates, 368
Agilent, 271, 296
Air handling systems, 80, 80*f*, 82
Albumin
    bovine serum, 423, 426, 429–430
    culture media, 61, 64
    secretion assay, 443–444
Alkaline phosphatase (ALP) staining,
    176–177, 177*f*
Allele-specific primer extension
    (ASPE), 328–329
Alpha-1-antitrypsin (AAT) ELISA,
    444–445
Alzheimer's disease, 529
Amaxa nucleofector, 456
America Invents Act of 2011,
    585–586
Amino acids, 55–56
Anesthesia, 527, 537
Aneuploidies, 204
Animal product-free systems, 34–35,
    62, 64

Antibiotics
    avoiding use of, 12, 41, 46–48, 60
    guidelines for use, 47–48, 60,
      99–100
    mycoplasma resistance, 39
    suppliers, 51
    zinc finger nuclease technology,
      510
Antibiotic free NSC growth mediums,
    540
Antibiotic resistance genes, 488, 489*t*
Antibodies
    chromatin immunoprecipitation,
      314–315, 321–322
    dopaminergic neuronal
      differentiation, 394–395
    fetal hepatocytes, 446
    flow cytometry, 226–228, 227*t*
      multiple-color analysis,
        233–241
      suppliers, 246–247
    labeling cells with (*See*
      Immunocytochemistry)
    neural progenitors, 383
Antibody dilution buffer, 270
Anticoagulants, 59
Antimycotics, 100–101
Antioxidants, 60
APEL medium, 423
Apoptosis, 71–72, 408, 525
Ascorbate (vitamin C), 60
Aseptic technique, 47, 48, 575

Assignment of rights, 584–585
Assisted reproduction technologies
    (ART), 545–550, 597–600

B

Background staining, 242, 258–259
Bacterial contamination, 42–43, 43*f*.
    See also Mycoplasma
    recovery from, 100
    testing for, 98, 108
Bands (chromosome), 188–189,
    193. *See also* G-banding
Basic FGF (bFGF) stock solution, 38,
    159, 171, 371, 425–426, 442,
    458–459
Bayer Schering Pharma, 591–592
Bayh-Dole Act, 591
BD Matrigel matrix growth factor
    reduced (GFR) stock solution,
    427
BD Matrigel™ substratum, 30–31,
    35, 36
BeadArray™, 326–327
BeadChip, 205, 214, 327, 335
BEL medium, 430
Bernstein, Alan, 15
Best Practices for the Licensing of
    Genomic Inventions, 587
Best practice standards, 106
Biobanks, 105–125
    Distribution/Working Cell Banks,
      95, 106–108

Biobanks (*Continued*)
  documentation and traceability,
    113–114
  early passage hESC, 11, 568
  future initiatives, 122–123
  laboratory setup, 92–95, 93f, 97,
    102
  Master Cell Bank, 95, 106–108
  optimal culture and preservation
    methods, 108–113
  process outline, 107f
  quality control, 108, 116t–117t
  stem cell providers, 114–121, 115t
Biological hazards, 86, 89
Biological replicates, 275
Biological safety cabinets (BSCs),
  7–8, 49–50, 82–83, 83f
Bioluminescent reporter, 109t
Biopsy, skin. *See* Skin biopsy
Biotin, 56
Bisulfite conversion, 329
Bisulfite sequencing, whole genome,
  326, 330
Blastocoele, 549, 549f
Blastocysts, 545–556
  culture
    equipment, 554–555
    procedures, 550–552, 561–
      562, 564
    reagents and supplies, 555
    recipes, 555–556
    vitrification and thawing,
      552–555, 560
  developmental stages, 547–550,
    549f–550f
  grading of, 550, 550f–551f
  sources of, 545–546, 559
    ethical issues, 597–600
  stem cell derivation from,
    557–576
    equipment, 571
    pitfalls and advice, 570–571
    procedures, 558–570
    quality control, 575–576
    reagents and supplies,
      571–574
    recipes, 574–575
Blastomeres, 548
Blocking buffers, 246, 269

Blotting buffer, 498
BMP4 stock solution, 427, 442
Bone morphogenetic protein (BMP),
  58
Bovine fibronectin coated wells,
  132–133, 133f, 134f, 141
Bovine serum albumin (BSA), 423,
  426, 429–430
Box plot (PluriTest), 301–302, 303f
BPEL medium, 419–420, 423,
  429–430
Bradley, Allan, 15
Brain transplantation, 529–542
  equipment, 538
  pitfalls and caveats, 536–538
  procedures, 531–536
  quality control, 540–541
  reagents and supplies, 539
  recipes, 540
BrdU labeling, 251–252
Bromodeoxyuridine (BrdU), 541
BSA stock solution, 426
BTX ECM630 electroporator, 454–
  456, 470
Buffers, 54
  cell dissociation, 35
  chromatin immunoprecipitation,
    321
  flow cytometry, 246
  Gurr's stock solution, 201
  homologous recombination,
    496–498
  immunocytochemistry, 269–270
Bush, George W., 598

C

Calcineurin inhibitors, 537–538
California Institute for Regenerative
  Medicine (CIRM), 599
Cameras, digital, 262
Cancer-related abnormalities, 39,
  203. *See also* Teratomas
Carbon dioxide gas, 54, 79–80
Carbon dioxide incubators, 83–84,
  88
Cardiomyocyte differentiation,
  413–431
  equipment, 424

pitfalls and advice, 423
procedures, 415–423
  alternative, 423
quality control, 430
reagents and supplies, 424–425
recipes, 425–430
Cavitation, 548–549
CDNA libraries, 271, 290. *See also*
  Gene expression profiling
CEL-I assay, 507–510, 508f
Cell banking. *See* Biobanks
Cell characterization. *See*
  Characterization
Cell count experiments, 22
Cell cross-contamination. *See*
  Cross-contamination
Cell death (apoptosis), 71–72,
  408, 525
Cell differentiation. *See* Pluripotency
Cell Dissociation Buffer, 35
Cell labeling, 541
Cell lines
  available, 105–106, 118t–120t,
    122–123, 143
  derivation from blastocysts (*See*
    Blastocysts)
  disease-specific, 144
  for gene expression profiling,
    273–274
  human fibroblast, 24 (*See also*
    Human dermal fibroblasts)
  identity testing, 97, 109t, 111–112,
    576
  quality control, 576
  somatic cell types, 130, 165
  suitability, 121, 129
  traceability, 113–114
  ZFN-modified, 502, 503f, 504f
Cell sorting, flow cytometric analysis
  using, 233–241
CELLstart CTS, 35
Cell status monitoring, 96–98
Cell surface antigens, 111, 226–228,
  227t
Cellular aggregates
  neural progenitors, 376–377
  oligodendrocyte progenitors,
    403–405, 412
Cellular Dynamics International, 592

Certificate of analysis, 121
Chakrabarty, Ananda, 579
"Challenge experiment" design, 305–306
Chamber slides, 251
Characterization, 108–109, 109t–110t
  embryonic outgrowth, 568–569
  epigenetics (*See* Chromatin immunoprecipitation; DNA methylation assays)
  flow cytometry (*See* Flow cytometry)
  for gene expression profiling, 274–275
  immunocytochemical (*See* Immunocytochemistry)
  karyotyping (*See* Karyotyping)
  molecular diagnostic assay (*See* PluriTest)
  SNP genotyping (*See* Single nucleotide polymorphism genotyping)
  teratoma assays (*See* Teratomas)
Chelex 100 resin, 319
Chemically defined medium (CDM), 441
Cholesterol, 56–57
Chromatin immunoprecipitation (ChIP), 313–323
  antibodies, 314–315, 321–322
  equipment, 320
  example in HiPSCs, 315, 315f
  native, 314
  procedures, 315–319
    alternative, 319–320
  quality control, 322–323
  reagents and supplies, 320–321
  recipes, 321
Chromosomal abnormalities
  cytogenetic analysis, 63–64, 203 (*See also specific method*)
  recovery from, 102
Chromosomes
  bands, 188–189, 193 (*See also* G-banding)
  harvesting, 188–190, 198
    from other species, 196–197
  spreads, 190–193, 192f, 198–199

transgene integration into, 181–183
Cleavage, 547, 548f
Cloning
  homologous recombination, 488–493, 489f
  for karyotype recovery, 102
  lentiviral vector systems, 464–465, 472
  zinc finger nuclease technology, 509–510
Clustering methods, 280f, 280–281
  PluriTest, 302, 304f
CnvPartition, 209–212, 213f, 214–215
Collagenase B, 132–133, 133f, 134f, 141
Collagenase IV, 9, 35, 366
Collagenase IV stock solution, 38
Colony density, 6, 6f
Communal areas, 85
Comparative genome hybridization (CGH), 63, 96, 110t, 112, 203–204, 569
Consensus Guidance for Banking and Supply of hESC Lines for Research, 108–109
Contamination, 41–51
  biobanks, 108
  cross-contamination (*See* Cross-contamination)
  microbial (*See* Microbial contamination)
  prevention of, 47–50, 88, 93
  reagents and supplies, 51
Control groups, 540
Cook Blastocyst Vitrification Kit, 553–555
Copper, 57
Copy number variation (CNV), 63–64, 204–205
  data analysis, 206–215
Core facilities, 115, 115t
  data file formats, 307–308
Corning Synthemax Surface, 35
Cross-contamination, 47, 108
  neural crest cells, 381–382
  skin biopsy culture, 135, 135f
Crosslinking, 314

Cryopreservation, 71–76
  cell banking (*See* Biobanks)
  embryos, 11, 552–554, 560, 568
  equipment, 74
  in feeder-free conditions, 34
  HiPSCs, 151, 153–154
  inventory systems, 87–88
  live culture shipments, 94
  mouse embryonic fibroblasts, 16, 18
  neural stem cells, 390–391, 396
  pitfalls and advice, 74
  procedures, 72–73
  reagents and supplies, 74–75
  recipes, 75
  storage area, 85–86
  thawing (*See* Thawing)
Cryopreservation medium, 26
C-terminus design, 511, 511f
CTK solution, 161
Culture areas
  contamination prevention, 50, 88, 93
  design of, 78–82, 79f
    air handling system, 80, 80f, 82
    incubators, 83–84, 88
    laminar flow workbench, 83, 84f
    for manual manipulation, 82–83, 83f
    sinks, 80, 81f
Culture media, 53–69 *See also specific type*
  commercially available, 61–62
  design of, 53–61
    albumin, 61, 64
    amino acids, 55–56
    antibiotics, 60
    antioxidants, 60
    buffers, 54
    growth factors, 58–59
    heparin, 59
    hormones, 58
    lipids, 56–57
    osmolality, 54–55
    oxygen tension, 59
    salts, 55
    trace elements, 57, 60
    vitamins, 56, 60

Culture media
efficacy assessment, 62–64
embryos, 551, 567–568
feeder cells, 25
feeder-free, 30–34, 60–61
recipes for (*See* Recipes)
xeno-free, 34, 62, 64
Current Good Manufacturing
Practice (cGMP), 35, 64
Cyclosporine, 537
Cytogenetic analysis, 63–64, 203. *See
also specific method*

D
DA1/DA2 mediums, 397
DAPI staining, 198
DAP213 solution, 161–162
Dark field transillumination, 7, 7f
Data analysis
copy number variation, 206–215
DNA methylation arrays, 331–
334, 332f–334f
microarray, 278–282
in PluriTest, 294–298, 309
RNA sequencing, 287–289
Data normalization, microarrays,
278–279, 279f
Demyelination, 519–520
Denaturation buffer, 496
Density
hPSC colony, 6, 6f
MEF plating, 23
oligodendrocyte plating, 409
Density plot (PluriTest), 301, 303f
Dermal fibroblasts. *See* Human
dermal fibroblasts
Developmental drift, 11
Dexamethasone, 58
Dickey-Wicker Amendment, 598
Differential expression
analysis, 289
Differentiation. *See* Pluripotency
Digital images, 262–267
Diphtheria toxin (DTA), 487–488
Dispase dissociation, 9, 10f, 35, 366
Dissecting microscopes, 7, 7f,
82–83, 83f
Dissociation. *See* Passaging

Distribution/Working Cell Banks (D/
WCBs), 95, 106–108
DMEM/F-12 medium, 56, 60
DNA Database of Japan, 282
DNA fingerprinting, 47, 109t
DNA fluorochrome, 109t
DNA methylation assays, 313–314,
325–336
affinity-based, 326, 330–331
equipment, 334
Illumina, 326–329, 327f, 328f
procedures, 329–330
pitfalls and advice, 334
quality control, 335
reagents and supplies, 334
whole genome bisulfite
sequencing, 326, 330
DNA profiling, 111–112. *See also
specific method*
embryonic outgrowth, 568
DNA purification, in chromatin
immunoprecipitation,
318–320
DNA sequencing, 39
DNA shearing, 322–323
DNA staining, mycoplasma, 39, 45
DNA transfection, 451. *See also*
Electroporation
Documentation
biobanks, 113–114, 114f
laboratory, 86–88
Donation, ethical issues, 598–600
Dopaminergic neuronal
differentiation, 385–398
equipment, 393
pitfalls and advice, 393
procedures, 387–392
alternatives, 392–393
reagents and supplies, 393–395
recipes, 395–397
Double stranded breaks (DSBs),
501, 501f

E
Eagle, Harry, 56
Earle's balanced salt solution, 55
EB medium, 395
Ecotropic receptor, 145–149

Ectodermal tissue
embryoid body formation, 367–
368, 368f
teratomas in, 347–351, 347f–351f
EdU labeling, 251–252
Efflux, during intraspinal
transplantation, 525
EGF stock solution, 404
Electricity supply, 81
Electroporation, 451–461
equipment, 457
homologous recombination, 491
lentiviral vector systems, 470,
476–477, 477f
pitfalls and advice, 456–457
procedures, 167–170, 452–454,
454f
alternative, 454–456
quality control, 460–461
reagents and supplies, 457–458
recipes, 458–460
ELISA, alpha-1-antitrypsin, 444–445
Embryoid body differentiation
medium, 372
Embryoid body (EB) formation,
363–373
equipment, 369
neural stem cells, 387–388, 389f
pitfalls and advice, 369
as pluripotency test, 11, 110t
procedures, 364–368
alternative, 368
reagents and supplies, 369–371
recipes, 371–372
spin, 414–423
Embryoid body medium, 372
Embryology. *See* Blastocysts
Embryonal carcinoma cells (ECCs),
357
Embryonic stem cells (ESCs), 144
Embryonic stem (ES) cell-like colony
formation, 176–177, 177f
Embryonic stem cell (ESC) medium,
395, 459
Encyclopedia of DNA Elements
(ENCODE), 272, 289
Endodermal tissue
embryoid body formation, 367–
368, 368f

hepatic, 435–438, 439f
teratomas in, 351–353, 352f–354f
Engelbreth-Holm-Swarm mouse
    sarcoma cell line, 36
Enzymatic activity, during
    immunocytochemistry,
    259–260
Enzymatic dissociation, 9–10, 10f
    Accutase, 30–32, 32f
    collagenase IV, 9, 35, 366
    dispase, 9, 10f, 35, 366
    embryoid body formation,
        365–366
    human dermal fibroblasts, 139
    neural stem cells, 390–391
Epidermal growth factor (EGF),
    399–400
Epidermal growth factor receptors
    (EGFRs), 58–59
Epigenetic changes, 11, 29–30, 98,
    313
    assessment of (See Chromatin
        immunoprecipitation; DNA
        methylation assays)
Episomal plasmid vectors, 166–
    168, 167f, 168f. See also
    Integration-free methods
Equipment. See also specific type of
        equipment
    blastocyst culture, 554–555
    blastocyst stem cell derivation,
        571
    brain transplantation, 538
    cardiomyocyte differentiation,
        424
    chromatin immunoprecipitation,
        320
    cryopreservation, 74
    DNA methylation, 334
    dopaminergic neuronal
        differentiation, 393
    electroporation, 457
    embryoid body formation, 369
    feeder-free culture, 36–37
    feeder layer culture, 12
    fetal hepatocytes, 440
    flow cytometry, 244–245
    gene expression profiling,
        289–290

HiPSC generation
    integration-free method, 170
    pluripotency evaluation, 183
    retroviral vector method, 157
    immunocytochemistry, 267
    intraspinal transplantation, 526
    karyotyping, 199
    lentiviral vector systems, 480
    maintenance of, 88–89, 575
    mouse embryonic fibroblast
        preparation, 24–25
    setup (See Research laboratory;
        specific work area)
    SNP genotyping, 215
    teratoma assays, 341
Ethical issues, 595–603
    assisted reproductive
        technologies, 546, 559,
        596–598
    clinical trials, 600
    donation, 598–599
        informed consent, 599–600
    funding, 596, 598
    medical tourism, 600
    review committees, 600–602
    scientific responsibilities, 602
European Bioinformatics Institute
    (EBI), 282
European Patent Office (EPO), 582,
    589
Exclusionary rights, 581
Expansion
    HiPSCs, 152–153, 169
    incoming cells, 94–95
    mouse embryonic fibroblasts, 19,
        19f
    neural stem cells, 390–391
    oligodendrocyte generation,
        402–403
Extracellular matrices (ECMs),
    36, 435, 504, 513. See also
    Substrata
EZ DNA methylation kit, 329
EZ Passage tool, 10, 10f, 36

F
Fasudil, 427
Fate Therapeutics, 592

Fatty acids, 56–57
Faxitron RX-650, 20, 20f
Fc receptors, 259
FD12 buffers, 246
Feeder-free culture, 29–40
    for biobanks, 108
    equipment, 36–37
    pitfalls and advice, 36
    procedures, 30–34
        alternative, 34–36
    quality control, 38–40
    reagents and supplies, 37, 61–62
    recipes, 38
Feeder layer culture, 3–14
    adapting to feeder-free conditions
        from, 33
    cardiomyocyte generation,
        414–423
    embryoid body formation, 365f,
        365–366
    embryonic cells, 561–562
    equipment, 12
    feeder cells, 24 (See also Mouse
        embryonic fibroblasts)
    for flow cytometry, 228–230
    HiPSCs
        integration-free method, 168
        retroviral vector method,
            150–152
    for lentiviral vector systems,
        469–470
    pitfalls and advice, 11–12
    procedures, 4–8
        alternative, 9–10
    reagents and supplies, 13
    recipes, 13–14
Fertilization, in vitro, 545–547,
    547f
Fetal bovine serum (FBS), 86, 90–91
Fetal hepatocytes, 433–447
    equipment, 440
    morphology, 438, 439f
    pitfalls and advice, 440
    procedures, 435–438
        alternative, 438–439
    quality control, 443–445
    reagents and supplies, 440–441,
        445–446
    recipes, 441–443

Fibroblasts
  human, 24 (*See also* Human
    dermal fibroblasts)
  mouse (*See* Mouse embryonic
    fibroblasts)
Fibroblast growth factor 8 (FGF8),
  386
Fibroblast growth factor (FGF)
  receptors, 58–59
Fibroblast growth factor (FGF) stock
  solutions
  basic (bFGF), 38, 159, 171, 371,
    425–426, 442, 458–459
  FGF2, 38, 403–404, 408
Fibroblast medium, 14, 26, 132–133,
  140–141, 482
Fibronectin, 132–133, 133f, 134f,
  141
File formats, 306–309
First-to-file system (patents),
  585–586
Fixation
  flow cytometry, 241
  immunocytochemistry,
    252–253
Fixatives
  methanol:acetic acid, 201
  PFA, 241, 252, 269–270
FK506, 537
Flow buffers, 246
Flow cytometry (FC), 223–248
  biobanks, 97, 110t, 111
  equipment, 244–245
  fluorescence-activated cell sorting,
    80, 102, 225
  high throughput screening units,
    237–241, 239f, 244
  versus immunocytochemistry, 249
  overview, 223–225, 224f
  pitfalls and advice, 242–244
  procedures, 226–241
    alternative, 241
    multiple-color analysis,
      233–241
  as quality control, 22
  reagents and supplies, 245–247
  work area, 85
Fluorescence-activated cell sorting
  (FACS), 80, 102, 225, 510

Fluorescence in situ hybridization
  (FISH), 96, 203–204
Fluorophores, 236f, 236–237, 255t
Food and Drug Administration
  (FDA), 64, 586
Formaldehyde, 314
Forward light scatter (FSC), 224, 224f
FP medium, 160, 172
Freezer inventory systems, 87–88
Freezing. *See* Cryopreservation
Freezing medium, 75
293FT cells, 145–146, 465–466
293FT medium, 160–161
Functional Annotation of Mouse
  cDNA (FANTOM), 272
Funding, ethical issues, 596, 598
Fungal contamination, 43–44, 44f,
  98, 100–102

G

Gamma irradiation, 19–20
Gateway® vectors, 464–465, 465f
G-banding, 189, 203
  biobanks, 96, 109t, 112
  culture media and, 63
  non-human species, 198
  procedures, 193–195
  resolution, 204
GCTM-2, 228, 230–233
Gelatinized dishes, 21–22
0.1% gelatin solution, 26
Geltrex™ coating, 460, 482, 490–491
Geltrex™ stock solution, 38
Geltrex™ substratum, 30–31, 31f, 32f,
  35, 36
Gene Expression Omnibus (GEO),
  272, 289
Gene expression profiling, 271–291
  critical statistical concepts, 275–
    278, 277f
  culture media and, 63
  embryonic outgrowth, 569
  equipment, 289–290
  experimental design, 273–275,
    276f
  methods, 110t, 204 (*See also*
    specific method*)
  microarray data analysis, 278–282

  in PluriTest, 294–298
  overview, 272–273
  pitfalls and advice, 289–290
Generalized background, 260
Gene targeting. *See* Homologous
  recombination
Genetic drift, 11, 29–30
GenomeStudio®, 206–214, 207f
  PluriTest, 297, 308–310
  quality controls, 216–219
Genome-wide gene expression. *See*
  Gene expression profiling
Genomic tiling arrays, 272
Genotyping, 92, 109t–110t, 111–112,
  187–188, 203. *See also specific
  method*
Germ layers. *See also specific layer*
  embryoid body formation, 367–
    368, 368f
  teratomas in, 347–354,
    347f–355f
Germ line transmission test, 293
Geron, 585
Giemsa stain solution, 201. *See also*
  G-banding
Glass coverslips, 251
Glial restrictive medium (GRM), 404,
  411
Global methylation profile, 98
GlutaMAX™, 55–56
Glutamine, 55–56
Glutaraldehyde:paraformaldeh
  yde (PFA) fixative, 241, 252,
  269–270
Glutathione, 60
Glycosylation patterns, 226
GoldenGate® DNA assay for
  Methylation, 326–327
Good, Norman E., 54
Good laboratory practice
  (GLP), 106
Good manufacturing practice (GMP),
  106
Green fluorescent protein (GFP),
  224, 488–490, 541
Growth factors, 36, 58–59. *See also
  specific factor*
  titration curves, 421, 422f
Gurr's buffer solution, 201

H

Ham's F-12 medium, 56, 57, 60
Hatching, 549, 550f, 562
Hatch-Waxman Act of 1984, 586
Hazardous biological waste, 86, 89
HEPA filters, 80, 80f
Heparin, 59
Hepatic specification, 437–439
Hepatocyte growth factor (HGF)
    stock solution, 442
Hepatocytes. See Fetal hepatocytes
HEPES, 54
Hierarchical clustering, 280f,
    280–281
    PluriTest, 302, 304f
High-efficiency particulate air
    (HEPA) filter, 49
High throughput screening (HTS)
    unit, 237–241, 239f, 244
Histone code, 313
Histone post-translational
    modifications (PTMs),
    313–314
    analysis of (See Chromatin
    immunoprecipitation)
Hochedlinger, Konrad, 591–592
Hoechst staining, 39, 45
Homologous recombination,
    485–498
    efficiency of, 500
    procedures, 486–495
    reagents and supplies,
    495–496
    recipes, 496–498
    workflow, 486, 486f
    zinc finger nucleases (See Zinc
    finger nuclease technology)
Homology directed repair (HDR),
    501, 501f
Hoods (laminar-flow cabinets),
    48–50, 83, 84f
Horizontal-flow clean bench, 49
Hormones, 58
Human Activin A stock solution, 442
Human basic FGF solution, 38,
    159, 171, 371, 425–426, 442,
    458–459
Human dermal fibroblasts (HDFs),
    129–141, 144

ethical issues, 599
HiPSC generation from (See
    Integration-free methods;
    Retroviral vector methods)
isolation
    equipment, 139
    pitfalls and advice, 138–139
    procedures, 131–138
    reagents and supplies,
    139–140
    recipes, 140–141
Human embryonic stem cells
    (hESCs)
    cell lines (See Biobanks; Cell
    lines)
    chromatin immunoprecipitation
    in, 316
    cultured for neuronal
    differentiation, 387, 388f
    derivation from blastocysts (See
    Blastocysts)
    gene targeting in, 490–491
    (See also Homologous
    recombination)
    versus mouse, 3, 485
    patents, 589–591 (See also
    Patents)
Human embryonic stem cell (hESC)
    medium, 75, 423, 429,
    481–482
Human ES cell medium, 160, 172
Human fibroblast lines, 24
Human hepatocyte growth factor
    (HGF) stock solution, 442
Human induced pluripotent stem
    cells (HiPSCs), 143
    available cell lines, 105–106,
    118t–120t, 122–123, 143
    banking (See Biobanks)
    chromatin immunoprecipitation
    in, 315f, 315–316
    disease-specific, 144
    fibroblasts (See Human dermal
    fibroblasts)
    generation methods
    conventional (See Retroviral
    vector methods)
    integration-free (See
    Integration-free methods)

versus human embryonic stem
    cells, 143
    pluripotency evaluation, 175–184
    teratomas derived from, 357,
    357f–358f
Human insulin stock solution, 441
Human leukocyte antigen (HLA), 95,
    97, 122
Human mesenchymal stem cell
    (hMSC) cultures, 57
HumanMethylation450 BeadChip,
    327, 330–331, 335
Human oncostatin stock
    solution, 443
Human pluripotent stem cells
    (hPSCs)
    aggregate generation, 376–377
    banking. See Biobanks
    cardiomyocyte generation from,
    414–423
    cell status monitoring,
    96–98
    characterization of (See
    Characterization)
    cryopreservation of (See
    Cryopreservation)
    culture contamination (See
    Contamination)
    culture media (See Culture media;
    Recipes)
    culture methods (See specific
    method)
    derivation from blastocysts (See
    Blastocysts)
    gene targeting (See Homologous
    recombination)
    hepatocyte generation from,
    435–438
    laboratory setup (See Research
    laboratory)
    versus mouse stem cells, 3
    neural stem cell generation from,
    387–388, 389f
    oligodendrocyte generation from,
    401f, 401–408, 411–412
    standards, 98
    transduction of, 466–467
    transfection methods, 451 (See
    also Electroporation)

Hybridization buffer, 497
Hydrocortisone, 58
Hygiene, 50
Hygromycin B sensitivity, 469, 473*t*
Hypotonic solution, 201

I

.idat file format, 306*f*, 306–309
Identity testing, 97, 109*t*, 111–112, 576
Illumina
    DNA methylation assays, 326–330
        data analysis, 331–334, 332*f*–334*f*
        HT12 chip, 304–305, 305*f*, 310–312
    microarrays, 271, 279, 296–297
    SNP genotyping (*See* Infinium HD super assay)
Image analysis room, 85
ImageJ, 266–267
Immunocytochemistry, 249–270
    biobanks, 96–97, 110*t*
    digital images, 262–267
    embryonic outgrowth, 568–569
    equipment, 267
    neural stem cells, 392
    oligodendrocytes, 400, 400*t*
    pitfalls and advice, 258–262
    procedures, 250–258
    as quality control, 22
    reagents and supplies, 267–268
    recipes, 268–269
Immunofluorescence, visualization of, 256–257
Immunofluorescence kits, 51
Immunoprecipitation
    chromatin (*See* Chromatin immunoprecipitation)
    methylated DNA, 330
    procedures, 316–318
Immunostaining
    embryoid body formation, 367–368, 368*f*
    fetal hepatocytes, 443
    flow cytometry, 231–232, 242–243

multiple-color, 233–235, 234*f*, 235*f*
immunocytochemistry, 252–256, 258–262
pluripotency markers, 11, 22, 63
Immunosuppression, 537–538
Incubation
    antibodies, 253–256
    embryo cultures, 551
Incubators, 83–84, 88
Incyte Genomics, 271
Index system, laboratory, 87
Induced pluripotent stem cells (iPSCs), 144
    ethical issues, 599–600
    human (*See* Human induced pluripotent stem cells)
    mouse, 144–145
    patents, 591–592 (*See also* Patents)
Induced pluripotent stem cells (iPSCs) medium, 395
Infinium DNA methylation assay, 327*f*, 327–330, 328*f*
    data analysis, 331–334, 332*f*–334*f*
Infinium HD super assay
    data analysis, 206–215
    pitfalls and advice, 214
    protocol, 205, 206*f*
    quality controls, 216–219
Informed consent, 599–600
Initial expansion phase, 94–95
Inner cell mass (ICM), 545, 548–550, 549*f*
    isolation of, 561
Insemination, oocyte, 546–547, 547*f*
Institutional Animal Care and Use Committee (IACUC), 16
Institutional animal care and use committee (IACUC), 602
Institutional review board (IRB), 601
Insulin, 58
Insulin-like growth factor receptors (IGFs), 58–59
Insulin receptor (IR), 58–59
Insulin stock solution, 441
Integration-free methods, 165–173
    equipment, 170
    pitfalls and advice, 170

procedures, 166–169
    alternative, 169–170
    reagents and supplies, 171
    recipes, 171–172
Intellectual property rights, 580–581. *See also* Patents
International Society for Stem Cell Research (ISSCR), 600
International Stem Cell Banking Initiative (ISCBI), 106
International Stem Cell Initiative (ISCI), 346
International Stem Cell Registry (ISCR), 105, 557
International System for Human Cytogenetic Nomenclature (ISCN), 189, 195
Intracytoplasmic sperm injection (ICSI), 547, 547*f*
Intraspinal transplantation, 519–528
    equipment, 526
    pitfalls and advice, 524–525
    procedures, 520–524
    quality control, 527–528, 528*f*
    reagents and supplies, 526–527
    recipes, 527
Inventions, patenting of. *See* Patents
Inventory
    freezer, 87–88
    stock and reagent, 87
In vitro fertilization (IVF), 545–547
    ethical issues, 597–600
IPierian, Inc., 591–592
Irradiation, 16, 19–21

J

Jaenisch, Rudolph, 591–592
Japan Patent Office, 591–592
JHM strain of mouse hepatitis virus (JHMV), 520
Junction PCR assay, 508–509, 509*f*

K

Karyotyping, 187–202
    biobanks, 92, 96, 109*t*, 112
    culture media and, 63–64
    embryonic outgrowth, 568

equipment, 199
interpretation of, 195–196, 196f
pitfalls and advice, 198–199
procedures, 189–195
    non-human species, 196–198
for quality control, 11, 39, 96
reagents and supplies, 200
recipes, 200–201
resolution, 196, 197t, 204
spectral, 96, 112
Ketamine, 527
Ki67, immunostaining for, 22
KnockOut™ serum replacement
        (KSR), 4, 13, 86, 91, 246, 383,
        428, 574–575
KOSR medium, 371–372

L

Labeling of cells, 541
Laboratory setup. See Research
        laboratory
Laminar-flow cabinets (hoods),
        48–50, 83, 84f
Laminin, 378
Lectins, 226–228, 227t, 243,
        246–247
Legal issues, 546, 559. See also
        Patents
Lentiviral vector systems, 463–483
    equipment, 480
    mouse ecotropic receptors,
        145–147, 162
    pitfalls and advice, 479–480
    procedures, 464–479
    reagents and supplies, 480–481
    recipes, 481–482
Leukemia inhibitory factor (LIF)
        gene, 145
L-glutamine, 429
Licensing, 584–587
LIF, 15
Lighting levels, 80
Lipids, 56–57
Liver specification, 437–439. See also
        Fetal hepatocytes
Lot numbers, 87
Lot-to-lot variability, 40, 86, 90–92,
        430, 575

Luciferase-based screening, 45
Ludwig, Tenneille, 61

M

Magnetic-activated cell sorting
        (MACS), 225, 240–241
Mann-Whitney-Wilcoxon statistical
        test, 278
March-in rights, 584
Master Cell Bank (MCB), 95, 106–108
Material Transfer Agreement (MTA),
        587
Matrigel™ coating
    cardiomyocytes, 418–420, 419f
    neural progenitors, 379
    oligodendrocyte progenitors, 402,
        405–407, 411
Matrigel™ stock solution, 38
Matrigel™ substratum, 30–31, 35, 36
MCDB media, 57
Mechanical dissociation
    embryonic outgrowth, 565–567,
        566f
    feeder-free culture, 36
    feeder layer culture, 7f, 7–8, 8f
    human dermal fibroblasts,
        138–139
    tools for, 10, 10f, 36
Medical tourism, 600
Medicines Patent Pool, 588
Memorandum of Understanding
        (MOU), 590–591
Mesodermal tissue
    embryoid body formation, 367–
        368, 368f
    teratomas in, 353–354, 354f–355f
Metaphase
    analysis of, 195
    harvesting during, 189–190, 198
Methanol:acetic acid fixative, 201
Methylated DNA
        immunoprecipitation
        (MeDIP), 330
Methylation. See DNA methylation
Microarrays. See also Gene expression
        profiling
    data analysis, 278–282
    in PluriTest, 294–298, 309

methylation (See DNA
        methylation)
Microbial contamination,
        42–45, 112
    bacterial, 42–43, 43f (See also
        Mycoplasma)
    fungal, 43–44, 44f, 98,
        100–102
    human dermal fibroblasts,
        139
    oligodendrocyte progenitors, 408,
        412
    prevention of, 47–50, 88, 93
    recovery from, 99–101
    testing for, 12, 16, 22, 38–39,
        41, 98
        biobanks, 92, 112–113
        incoming cells, 95
        procedures, 45–46
        supplies, 51
    viral, 44
Microporator. See Electroporation
Microscopes
    dissecting, 7, 7f, 82–83, 83f
    for immunocytochemistry, 257
    maintenance of, 89
Microscopy room, 80, 85
Microsoft Silverlight, 296–297
Minimum Information About
        a Microarray Experiment
        (MIAME), 289
Mitochondrial sequences, 98
Mitogen-activated protein kinase
        (MAPK) pathway, 59
Mitomycin C inactivation, 16, 21,
        151–152
Mitomycin C solution, 161
Mitotic inactivation
    methods of, 16, 19–21
    tests for, 22
Mold contamination, 43–44, 44f,
        100–102
Molecular biology area, 78–80, 79f,
        84–85
Molecular diagnostic assay. See
        PluriTest
Monoclonal antibodies (MAbs),
        227–228
Monohydrochloride salt, 427

Morphology, 62–63
    biobanks, 111
    feeder-free culture, 31, 32*f*, 63
    feeder layer culture, 5*f*, 6, 6*f*, 63
    fetal hepatocytes, 438, 439*f*
Morula stage, 548, 548*f*
'Mother stock' creation, 94–95
Mouse antibody pathogen (MAP)
        testing, 16, 22
Mouse ecotropic receptor, 145–149
Mouse embryonic fibroblasts (MEFs),
        15–27. *See also* Feeder layer
        culture
    commercial sources, 23–24
    cryopreservation, 16, 18
    equipment, 24–25
    expansion, 19, 19*f*
    inactivation, 16, 19–21
    isolation, 16–18
    pitfalls and advice, 24
    plating density, 23
    quality controls, 22–23, 23*f*, 86,
        90–92
    reagents and supplies, 25
    recipes, 26
    substratum support, 21–22
Mouse embryonic fibroblast-
        conditioned medium
        (MEF-CM), 410, 459–460, 482
Mouse embryonic fibroblast (MEF)
        medium, 410
Mouse embryonic stem cells, 15
    versus human cells, 3, 485
    pluripotency assays, 293
Mouse induced pluripotent stem
        cells (iPSCs), 144–145
Mouse models
    demyelination, 519–520 (*See also*
        Intraspinal transplantation)
    immunodeficient, 530, 530*f*,
        537–538 (*See also* Brain
        transplantation)
Mouse teratoma assays. *See*
        Teratomas
MRNA expression, 98, 271. *See also*
        Gene expression profiling
MRNA extraction kits, 296
MTeSR1® medium, 34, 61
Multiple antibody staining, 261–262

Multiple-color flow cytometry,
        233–241
Multiple sclerosis (MS), 519
Multiwell plates, 99
Mycoplasma, 12, 44–45
    eradication from labs, 99
    precautions, 46
    testing for, 12, 16, 38–39, 41, 98
        assay kits, 163
        biobanks, 108, 112–113
        HiPSC generation, 162
        incoming cells, 95
        methods, 109*t*
        procedures, 45–46
Myelin, 399

N

National Academy of Sciences
        guidelines, 599
National Center for Biotechnological
        Information (NCBI), 272, 282
National Institutes of Health (NIH)
    patenting policies, 584, 587,
        590–591
    stem cell line registry, 598
Native chromatin
        immunoprecipitation
        (NChIP), 314
Nature Methods, 308–309
Needle size/type, for brain
        transplantation, 536–537
Neomycin resistance, 15
Neon™ Transfection System, 167–
        170, 453–454, 454*f*, 470
Neural crest cells, contamination
        with, 381–382
Neural differentiation medium, 384
Neural induction, 388–390
Neural induction medium (NIM),
        396
Neural islands, isolation of, 379
Neural progenitors, 375–384
    pitfalls and advice, 381–382
    procedures, 376–381
        alternative, 381
        protocols, 376, 376*f*
    reagents and supplies, 382–383
    recipes, 383–384

therapeutic use (*See* Brain
        transplantation)
    yellow sphere formation, 404–
        405, 409, 412
Neural rosettes, manual dissection
        of, 393
Neural stem cells (NSCs), 385,
        387–388
    therapeutic use (*See* Brain
        transplantation; Intraspinal
        transplantation)
Neural stem cell (NSC) mediums,
        396, 540
Neurodegenerative disorders, therapy.
        See Brain transplantation;
        Intraspinal transplantation
Neuroectodermal islands, terminal
        differentiation of, 380*f*,
        380–381
Neurological disorders, therapy.
        See Brain transplantation;
        Intraspinal transplantation
Neuronal cells
    chromatin immunoprecipitation,
        316
    dopaminergic (*See* Dopaminergic
        neuronal differentiation)
Neutralization buffer, 497
Next generation sequencing (NGS),
        273, 282
"Nine Points to Consider,", 587
N2 medium, 384
Non-homologous end joining
        (NHEJ), 501, 501*f*
Novelty Score (PluriTest), 295–296,
        300–301, 302*f*
N-terminus design, 512, 512*f*
Nucleofection, 505–506
Nucleofector® system, 169
Nucleoside analogs, 22
NutriStem™ XF/FF, 62

O

Obama, Barack, 598
OCT4 (POU5F1), 502
Oligodendrocyte progenitor cells
        (OPCs), 399–412
    pitfalls and advice, 408–409

procedures, 400–407
  alternative, 407–408
  quality control, 411–412
  reagents and supplies, 409–410
  recipes, 410–411
Oncostatin stock solution, 443
One-well organ culture dish, 558, 559f
Oocyte insemination, 546–547, 547f
Open pulled straw vitrification, 71, 74
"OPEN" resource, 504, 512
Osmolality, 54–55
OsrHSA, 64
Overstaining, 243, 261
Oxidative stress, 60
Oxygen tension, 59

P

Paraformaldehyde (PFA), 241, 252, 269–270
Parkinson's disease, 529
Passaging
  cardiomyocyte generation, 416–418, 417f, 421–423
  Cell Dissociation Buffer, 35
  embryoid body formation, 365–366
  embryonic outgrowth, 564–567, 565f, 566f
  enzymatic (See Enzymatic dissociation)
  feeder-free culture, 30–32, 32f, 35
  feeder layers, 6–10
  HiPSCs, 152–153
  human dermal fibroblasts, 134f, 134–135, 138–139
  immunocytochemistry, 251
  incoming cells, 94
  lot testing, 90–92
  mechanical (See Mechanical dissociation)
  neural progenitors, 377–378
  neural stem cells, 390–391
  oligodendrocyte generation, 402–403
  zinc finger nuclease technology, 504–505

Patents, 579–593
  America Invents Act of 2011, 585–586
  defined, 580–583
  licensing policies, 586–587
  material transfer, 587
  obtaining, 583–585
  pools, 588
  stem cell, 588–589, 588–590
    human embryonic, 589–591
    induced pluripotent, 591–592
Patent Cooperation Treaty (PCT), 584
Periodic acid Schiff (PAS) assay, 445
Permeabilization, 250, 253
Personal hygiene, 50
Personal protective equipment (PPE), 50, 89
PFA fixative, 241, 252, 269–270
Phenol red, 54
PhiC31, 468, 470–472
Phospholipids, 56–57
Photodocumentation, 89, 262–267
Photolithography, 271
PicoGreen®, 334
PJTI™ platform, 470–474, 471f
Plasmid rescue assay, 474f, 474–475
Plasmocin, 39
Plasmocure, 39
PLAT-E cells, 148–149
PLAT-GP system, 145, 155–156
Plating
  embryonic cells, 560–561
  HiPSCs, 154–155
  incoming cells, 93–94
  mouse embryonic fibroblasts, 23
  multiwell, 99
  oligodendrocytes, 405–407, 409
  skin biopsy, 133, 133f, 137, 138f
Pluripotency, 63
  assessment of, 11, 22, 63, 92, 96–97 (See also Characterization)
  embryonic outgrowth, 568–569
  HiPSCs, 175–184
  methods, 110t, 111
  molecular diagnostic assay (See PluriTest)

teratoma assays (See Teratomas)
cardiomyocytes (See Cardiomyocyte differentiation)
developmental drift, 11
embryoid bodies (See Embryoid body formation)
in feeder-free culture, 31, 32f
on feeder layer, 4–6, 5f, 11, 23, 63
freezing and, 71 (See also Cryopreservation)
growth factors inducing, 58–59
hepatocytes. See Fetal hepatocytes
neural differentiation (See Neural progenitors)
oligodendrocytes (See Oligodendrocyte progenitor cells)
Pluripotency markers, 11, 22, 63, 92, 96–97, 111
  in flow cytometry, 226–228, 227t
Pluripotency Score (PluriTest), 295–296, 298–300, 300f
Pluripotent stem cells (PSCs)
  human (See Human pluripotent stem cells)
  mouse, 3
PluriTest, 11, 293–311
  experimental design and formats, 304–307
  pitfalls and advice, 307–311
  procedures, 296–304
    quality control, 301–304
    results interpretation, 298–301, 299f
    web interface, 296–298, 297f
  workflow, 295, 295f
Political debates, 596–598
Poly-L-ornithine, 378
Polymerase chain reaction (PCR)
  assay kit, 163
  in chromatin immunoprecipitation, 314, 320
  clone screening, 493
  lentiviral vector systems, 473t
  methylation-specific, 98
  mycoplasma testing, 46, 51

Polymerase chain reaction (PCR) (*Continued*)
quantitative (*See* Quantitative PCR)
transgene integration evaluation, 181–183
work area, 85
zinc finger nuclease technology, 508–509, 509*f*
Post-translational modifications (PTMs), 313–314
analysis of (*See* Chromatin immunoprecipitation)
POU5F1 (OCT4), 231–233
Pre-hybridization buffer, 497
Primary antibodies, 250
incubation, 253–254
selection of, 251
species mismatch, 259
Primer design
chromatin immunoprecipitation, 323
homologous recombination, 488, 489*f*
Primer sequences, lentiviral vector systems, 473*t*
Pronase, 569, 574
Protective clothing, 50, 89
Pseudosite hotspots, 472*f*, 472–473, 473*t*
PVA stock solution, 426

Q

Quality control
best practice standards, 106
biobanks, 108, 116*t*–117*t*
blastocyst stem cell derivation, 575–576
brain transplantation, 540–541
cardiomyocyte differentiation, 430
cell characterization (*See* Characterization)
cell status monitoring, 96–98
chromatin immunoprecipitation, 322–323
DNA methylation, 335
electroporation, 460–461

feeder-free culture, 38–40
fetal hepatocytes, 443–445
HiPSC generation, 162*f*, 162–163
intraspinal transplantation, 527–528, 528*f*
microarrays, 278–279, 279*f*
mouse embryonic fibroblasts, 22–23, 23*f*, 86, 90–92
oligodendrocyte progenitor cells, 411–412
PluriTest, 301–304
reagents, 90–92
record keeping, 86–87
SNP genotyping, 216–219
systems for, 89–92, 90*f*
Quantitative PCR (qPRC), 96–97, 110*t*
fetal hepatocytes, 443
primer design, 323
real time, 315
transgene integration evaluation, 181–183
transgene silencing evaluation, 177–181, 178*f*
Quarantine facility, 93

R

RA stock solution, 404
R/Bioconductor, 308–310
Reagents and supplies
blastocyst culture, 555
blastocyst stem cell derivation, 571–574
brain transplantation, 539
cardiomyocyte differentiation, 424–425
chromatin immunoprecipitation, 320–321
contamination testing kits, 51
cryopreservation, 74–75
DNA methylation, 334
dopaminergic neuronal differentiation, 393–395
electroporation, 457–458
embryoid body formation, 369–371
feeder-free culture, 37, 61–62

feeder layer culture, 13
fetal hepatocytes, 440–441, 445–446
flow cytometry, 245–247
HiPSC generation
integration-free method, 171
pluripotency evaluation, 183–184
retroviral vector method, 157–159
homologous recombination, 495–496
human dermal fibroblasts, 139–140
immunocytochemistry, 267–268
intraspinal transplantation, 526–527
inventory and tracking, 86–87
karyotyping, 200
lentiviral vector systems, 480–481
lot-to-lot variability of, 40, 86, 90–92, 430, 575
mouse embryonic fibroblast preparation, 25, 86
neural progenitors, 382–383
oligodendrocyte progenitor cells, 409–410
teratoma assays, 342
testing, 90–92
zinc finger nuclease technology, 512–513
Real-time quantitative polymerase chain reaction (qRTPCR), 315
Recipes. *See also specific solution or medium*
blastocyst culture, 555–556
blastocyst stem cell derivation, 574–575
brain transplantation, 540
cardiomyocyte differentiation, 425–430
chromatin immunoprecipitation, 321
cryopreservation, 26, 75
dopaminergic neuronal differentiation, 395–397
electroporation, 458–460
embryoid body formation, 371–372
feeder-free culture, 38

feeder layer culture, 13–14, 26
fetal hepatocytes, 441–443
flow cytometry solutions, 246
HiPSC generation
    integration-free methods, 171–172
    retroviral vector methods, 159–162
homologous recombination, 496–498
human dermal fibroblast isolation, 140–141
immunocytochemistry, 268–269
intraspinal transplantation, 527
karyotyping, 200–201
lentiviral vector systems, 481–482
neural progenitors, 383–384
oligodendrocyte progenitor cells, 410–411
zinc finger nuclease technology, 513
Recombinant human BMP4 stock solution, 427
Recombinant human/mouse/rat Activin stock solution, 428
Recombinant human SCF stock solution, 427
Recombinant human VEGF stock solution, 427
Record keeping
    biobanks, 113–114, 114*f*
    laboratory, 86–88
Reduced-representation bisulfite sequencing (RRBS), 326, 330–331
Replication, 275
Reporter cassettes, 488–490, 489*f*
Reporter genes, 241
Reprogramming efficiency. *See* Pluripotency
Reprogramming factors, 144–145
    dopaminergic neurons, 386
    episomal vectors, 167*f*, 167–168, 168*f*
    in flow cytometry, 228
    retroviral vectors, 148–149
Research ethics. *See* Ethical issues
Research laboratory, 77–103

cell banking, 92–95, 97, 102 (*See also* Biobanks)
culture contamination (*See* Contamination)
design approach, 78, 78*f*
equipment maintenance, 88–89, 575 (*See also* Equipment)
quality control, 89–92, 90*f* (*See also* Quality control)
    cell status monitoring, 96–98
record keeping, 86–88
safety, 50, 80, 86, 89
storage space, 85–86
work areas, 78–85 (*See also specific area*)
Research Tools Guidelines, 587
Resolution, 196, 197*t*, 204, 212
Restriction enzyme digestion, 326
Retroviral vector methods, 143–163
    equipment, 157
    pitfalls and advice, 156
    procedures, 145–155
        alternative, 155–156
    quality control, 162*f*, 162–163
    reagents and supplies, 157–159
    recipes, 159–162
Reverse transcriptase polymerase chain reaction (RT-PCR), 96–97, 110*t*. *See also* Quantitative PCR
    fetal hepatocytes, 443
    transgene silencing evaluation, 177–181
Review committees, 600–602
RFLP assays, 508
Ringer's solution, 55
RMAExpress, 279
RNA sequencing (RNA-Seq), 273, 282–290. *See also* Gene expression profiling
Robust spline normalization (RSN), 298
ROCK inhibitor (Y27632), 72–73, 427, 510, 513
R4 platform line, 468–469, 469*f*
    clone screening, 472, 472*f*, 473*t*
    retargeting gene of interest, 475–479

S
Safety, lab, 50, 80, 86, 89
Sakurada, Kazuhiro, 591–592
Salts, 55
Sampling, for gene expression profiling, 275–278, 277*f*
SCF stock solution, 427
Scientific responsibilities, 602
Secondary antibodies, 250
    incubation, 255–256
Secondary expansion, 95
Selenium, 57, 60
Self-renewal capacity, 63, 354–355
Serum-containing medium, 430
Serum-free media, 57, 61–62
Short Reads Archive (SRA), 282
Short tandem repeat (STR) genotyping, 47, 95, 97, 111
Side scatter (SSC), 224
Signaling proteins, 385–386
Silverlight (Microsoft), 296–297
Single base extension (SBE), 329
Single copy integration, Southern blot analysis, 473–474, 493–495, 494*f*
Single nucleotide polymorphism (SNP) genotyping, 203–221
    biobanks, 95, 97, 109*t*, 111–112
    cross-contamination testing, 47
    culture media and, 63
    DNA methylation (*See* DNA methylation assays)
    equipment and supplies, 215
    pitfalls and advice, 214
    procedures, 205–213
    quality control, 216–219
    for quality control, 11, 39, 96
    resolution, 212
Sinks, cell culture area, 80, 81*f*
Site-specific integration platform, 468, 468*f*
Skin biopsy
    fibroblast isolation from (*See* Human dermal fibroblasts)
    methods, 131, 131*f*, 132*f*
    plating, 133, 133*f*, 137, 138*f*
    separation of dermal and epidermal layers, 136, 137*f*

Slide preparation
  chromosome spreads, 190–193, 192*f*, 198–199
  for immunocytochemistry, 251
Slow freezing method, 72–73
SMAD signaling, 375–377
SNL (STO-Neo-LIF) cell line, 145, 150–152
SNL76/7 cells, 15
SNL medium, 161
Sodium bicarbonate, 54
Sodium bisulfite conversion, 326
Somatic cell types, 130, 165
Sonication, 314
Sonic hedgehog (SHH), 385–386
Southern blot analysis, 473–474, 493–495, 494*f*
Species mismatch, 259
Spectral karyotyping (SKY), 96, 112
Spin embryoid bodies, 414–423
Spin-inoculation method, 467
SSPE solution, 496
Stage-specific embryonic antigens (SSEA), 227–228, 230–233
Staining
  alkaline phosphatase, 176–177, 177*f*
  antibody (*See* Immunocytochemistry)
  DAPI, 198
  Giemsa (*See* G-banding)
  immunostaining (*See* Immunostaining)
  mycoplasma, 39, 45, 163
  pluripotency markers, 11, 22, 63
Stem cell patents. *See* Patents
Stem cell research oversight (SCRO) committee, 601
Stem cell transplantation
  brain (*See* Brain transplantation)
  spinal (*See* Intraspinal transplantation)
STEMium™, 62
StemPro® hESC SFM, 30–34, 61–62
Stereotaxic surgery, 533*f*, 533–536, 540–541
Sterility, 112–113. *See also* Microbial contamination

STO (SIM immortalized cell line with thioguanine and ouabain-resistance), 15
Stock keeping units (SKUs), 87
Stock solutions *See* Reagents and supplies; Recipes; *specific solution*
Storage space, 85–86
Subchromosomal abnormalities, 39, 63–64, 204
Substrata
  feeder-free culture, 30–31, 31*f*, 32*f*, 35, 36
  mouse embryonic fibroblast support, 21–22
  xeno-free, 35
Suppliers. *See also* Reagents and supplies
  human fibroblasts, 24
  live culture shipments, 94
  lot-to-lot variability, 40, 86, 90–92, 430, 575
  mouse embryonic fibroblasts, 23–24
  stem cell banks, 114–121, 115*t*
  stock keeping units, 87
  surgical instruments, 342
Surfaces. *See* Substrata
Surgical instruments and supplies, 342
  stereotaxic surgery, 533*f*, 533–536, 540–541
Synthemax, 35

T

TALEs (transcription activator-like effectors), 485
Targeted sequencing strategies, 213
Technical replicates, 275
Telomerase, 63, 96–97
Temperature-controlled environment, 80
Temperature regulation, during surgery, 537
Teratocarcinomas, 354–357
Teratomas
  characterization, 345–359
    ectodermal tissues, 347–351, 347*f*–351*f*

    endodermal tissue, 351–353, 352*f*–354*f*
    hiPSC-derived, 357, 357*f*–358*f*
    immature, 354–357, 355*f*–356*f*
    mesodermal tissues, 353–354, 354*f*–355*f*
    procedures, 346
  as pluripotency test, 11, 97, 110*t*, 293, 297
  tumor generation, 337–343
    equipment, 341
    pitfalls and advice, 340
    procedures, 338–340, 341*f*
    reagents and supplies, 342
Tertiary expansion, 95
TeSR™ 2 medium, 61
TG30 (CD9), 228, 230–233
TG343, 240–241
Thawing
  after slow freezing, 73, 73*f*
  embryos, 552–554, 560
  feeder-free culture, 34
  HiPSCs, 154–155
  incoming cells, 93–94
  mouse embryonic fibroblasts, 18
  oligodendrocyte generation, 402
1-thioglycerol stock solution, 426
Thioglycollate medium, 98
Thomson, James, 579, 585, 589–591
Thymidine kinase (TK), 487–488
Tiered cell banking system, 92–95, 93*f*, 102
Tissue culture disposables, 25
Titration
  growth factor, 421, 422*f*
  viral stocks, 466
Tocopherol (vitamin E), 60
TOPO® vectors, 464–465, 465*f*
Traceability, 113–114
Trace elements, 57, 60
Transcription factors. *See* Reprogramming factors
Transfection methods, 451. *See also* Electroporation
Transferrin, 60
Transforming growth factor (TGF), 58–59
Transgene integration

into chromosomes, 181–183
lentiviral vectors (*See* Lentiviral
    vector systems)
Transgene silencing, 177–181
Transportation
    of cultures within laboratory, 50
    live culture shipments, 94
Triiodothyronine (T$_3$), 58
Trophectoderm (TE), 545, 548–550,
    549*f*
Trypsin, 9, 194
Trypsin-like Enzyme (TrypLE), 35
Tumorigenic cells, 39, 203. *See also*
    Teratomas
Tyrode's acid solution, 569–570

U

UEA-1, 228, 230–233, 240–241
UK Intellectual Property Office,
    591–592
Ultra sharp splitting blade, 562–563,
    563*f*
US Patent and Trademark Office
    (PTO), 583, 590, 592

V

Variance stabilizing transformation
    (VST), 298
VEGF stock solution, 427
Ventral foregut specification, 437, 439*f*

Viral contamination, 44
Viral vectors. *See* Lentiviral vector
    systems
ViraPower® HiPerform™ Lentiviral
    Expression kit, 464–466
Vitamins, 56, 60
Vitrification, 71, 74
    versus controlled rate
        freezing, 108
    embryos, 552–555, 560, 568

W

Warming. *See* Thawing
Washing buffers, 497
Water baths, 88
Water supply, 80
Weak staining, 242–243, 261
Whitehead Institute of Biomedical
    Research, 591–592
Whole genome bisulfite sequencing
    (WGBS), 326, 330
Whole genome sequencing, 212–213.
    See also Gene expression
    profiling
WiCell, 590–591
WiCell medium, 4, 13
Wisconsin Alumni Research
    Foundation (WARF), 579, 585,
    589–591
Working cell banks, 95, 106–108

X

Xeno-free media/matrices, 34–35,
    62, 64
X-ray irradiation, 20*f*, 20–21
Xylazine, 527

Y

Yamanaka, Shinya, 591–592
Yeast contamination, 43–44, 44*f*,
    100–102
Yellow fluorescence protein (YFP),
    502
Yellow sphere formation, 404–405,
    409, 412

Z

Zinc, 57
Zinc Finger Consortium, 502, 512
Zinc finger nuclease (ZFN)
    technology, 485, 499–515
    overview, 500–502, 501*f*
    pitfalls and advice, 509–512
    procedures, 502–509, 503*f*
    reagents and supplies, 512–513
    recipes, 513
Zona pellucida, 561, 561*f*
    hatching process, 549, 549*f*, 562
    removal of, 560–563, 563*f*,
        569–570

Printed and bound by CPI Group (UK) Ltd, Croydon, CR0 4YY

03/10/2024

01040311-0003